AMORPHOUS MAGNETISM II

AMORPHOUS MAGNETISM II

Edited by

R. A. Levy
Rensselaer Polytechnic Institute
Troy, New York

and

R. Hasegawa
Allied Chemical Corporation
Morristown, New Jersey

PLENUM PRESS · NEW YORK AND LONDON

Library of Congress Cataloging in Publication Data

International Symposium on Amorphous Magnetism, 2d, Rensselaer Polytechnic Institute, 1976.
Amorphous magnetism II.

Proceedings of the 2d International Symposium on Amorphous Magnetism, held at Rensselaer Polytechnic Institute, Troy, N. Y., Aug. 25-27, 1976.
Includes indexes.
1. Amorphous substances–Magnetic properties–Congresses. I. Levy, Roland Albert, 1944-   II. Hasegawa, Ryusuke, 1940-   III. Title.
QC766.A4I57 1976   530.4'1   77-5377
ISBN-13: 978-1-4613-4180-2   e-ISBN-13: 978-1-4613-4178-9
DOI: 10.1007/978-1-4613-4178-9

Proceedings of the Second International Symposium on Amorphous Magnetism
held at Rensselaer Polytechnic Institute, Troy, New York, August 25–27 1976

© 1977 Plenum Press, New York
Softcover reprint of the hardcover 1st edition 1977
A Division of Plenum Publishing Corporation
227 West 17th Street, New York, N.Y. 10011

All rights reserved

No part of this book may be reproduced, stored in a retrieval system, or transmitted, in any form or by any means, electronic, mechanical, photocopying, microfilming, recording, or otherwise, without written permission from the Publisher

PREFACE

The papers making up this volume represent a summary of the proceedings of the Second International Symposium on Amorphous Magnetism held at Rensselaer Polytechnic Institute on August 25-27, 1976. As a result of the resounding success of the International Symposium on Amorphous Magnetism held at Wayne State University on August 17 and 18, 1972 this symposium was again organized with the purpose in mind of providing a forum for discussion of the most recent theoretical and experimental advances made in the fields of spin glass systems, amorphous magnetic alloys and magnetic oxide glasses.

The symposium was sponsored by the American Physical Society and supported by a grant from GTE Laboratories. Additional support funds were provided by General Electric, Allied Chemical and Ford Motor Company.

The program committee consisted of J. J. Becker (General Electric), P. A. Casabella (RPI), P. J. Cote (Watervliet Arsenal), A. M. de Graaf (Wayne State University), R. Hasegawa, Co-Chairman (Allied Chemical), H. O. Hooper (University of Maine), H. B. Huntington (RPI), R. A. Levy, Chairman (RPI), R. K. MacCrone (RPI), L. N. Mulay (Penn State University), G. L. Salinger (RPI) and J. Wong (General Electric).

The program of the symposium included 7 invited review papers, 53 contributed papers, and 10 additional papers read by title because of time limitations.

The editors wish to extend their deep appreciation to J. J. Becker, A. M. de Graaf, P. Duwez, J. Gustafson, D. L. Huber, U. Larsen, L. N. Mulay, D. I. Paul, J. A. Rayne, N. Rivier and M. Takahashi for efficiently chairing the six sessions of the symposium.

We wish to thank the authors of these papers for their outstanding cooperation in preparing the manuscripts and to the reviewers for their valuable criticism.

Special thanks are due to George M. Low, President of RPI, for delivering the opening remarks, and to John F. Ambrose, Director of Project Planning and P. E. Ritt, Vice President both of GTE Laboratories for their skillful management and support of our efforts.

We wish to express our deep appreciation to P. C. Campbell for skillful administrative assistance, L. P. Winsor for his generous advice on planning activities, J. D. Norman for printing help, P. Cerullo for cocktail and banquet arrangements, F. MacPherson for audio-visual assistance and Mrs. L. Still, (secretary at Allied Chemical) for help with the typing of foreign manuscripts.

The participation of all those that contributed to the success of this symposium is greatly appreciated.

                                                       R. A. Levy
                                                         and

January 7, 1977   R. Hasegawa

CONTENTS

Survey of Theories of Spin Glass.   1
    P. W. Anderson

Some Percolation Scaling Applications to Low Temperature
        Dilute Magnets. 17
    D. Stauffer

μ^+ Studies of Dilute $\underline{Pd}$Fe Alloys. 29
    K. Nagamine, N. Nishida and T. Yamazaki

Spin Waves in Heisenberg Spin Glasses 39
    D. L. Huber and W. Y. Ching

Temperature-Dependent Skew Scattering in an AuMn Spin Glass . 47
    C. M. Hurd and S. P. McAlister

Slowly Relaxing Remanence and Metastability of Spin Glasses . 55
    N. Rivier

Effective Field Theories of Topologically Disordered
        Magnets 63
    T. Kaneyoshi

Spin Glasses and Mictomagnets - Revisited 73
    J. A. Mydosh

Concentration Effects in Spin Glasses 85
    U. Larsen, P. J. Ford, J. S. Schilling
    and J. A. Mydosh

Effect of Pressure on Impurity-Impurity Interactions in
        Spin Glass Alloys 95
    J. S. Schilling, P. J. Ford, U. Larsen and
    J. A. Mydosh

Critical Properties of a Simple Glass Model 105
    A. Aharony and Y. Imry

Magnetic Ordering of $\underline{Au}Cr$: An Ultrasonic Investigation . . . 117
    G. F. Hawkins, T. J. Moran and R. L. Thomas

Two Approaches to the Theory of Spin Glasses: A Comparison
    with Each Other and with Experiment 123
    M. W. Klein

Localized Moment and Spin-Glass-Like Behavior in
    Amorphous YFe_2. 135
    D. W. Forester, W. P. Pala and R. Segnan

Real Space Renormalization Group Calculations for
    Spin Glasses. 145
    A. P. Young

The Remanent Magnetization of Spin Glasses and the
    Dipolar Coupling. 155
    F. Holtzberg, J. L. Tholence and R. Tournier

The Magnetic Behavior of $\underline{Pd}$-Mn Alloys 169
    C. N. Guy and W. Howarth

Some Controversial Aspects of Electronic and Magnetic
    Interactions in the Amorphous Metallic State. . . . 187
    C. C. Tsuei

Magnetic Susceptibility of an Amorphous Non-Transition
    Metal Alloy: $Mg_{70}Zn_{30}$ 197
    B. C. Giessen, A. Calka, R. Raman and D. J. Sellmyer

^{31}P Nuclear Magnetic Resonance Study in the Metallic
    Glass Systems $(Ni_y Pt_{1-y})_{75}P_{25}$ and
    $(Ni_{0.50}Pd_{0.50})_{100-x}P_x$ 207
    H. A. Hines, L. T. Kabacoff, R. Hasegawa and P. Duwez

NMR and Mössbauer Studies of the Amorphous System
    $Fe_{79}P_{21-x}B_x^+$. 221
    K. Raj, A. Amamou, J. Durand, J. I. Budnick
    and R. Hasegawa

Magnetism in Amorphous Zr-Cu(Fe), Zr-Cu(Gd), and Nb-Ni(Fe). . 235
    G. R. Gruzalski, J. W. Weymouth, D. J. Sellmyer and
    B. C. Giessen

From Superconductivity to Ferromagnetism in Amorphous
    Gd-La-Au Alloys 245
    S. J. Poon and J. Durand

The High Temperature Electronic and Magnetic Properties
 of Pd-Alloys in the Glassy, Crystalline and
 Liquid State.................. 257
 H.-J. Güntherodt, H. U. Künzi, M. Liard, M. Müller,
 R. Müller and C. C. Tsuei

Ferromagnetic and Antiferromagnetic Coupling in Amorphous
 $(Ni_{100-c}Mn_c)_{78}P_{14}B_8$ 265
 A. Amamou

Electronic and Magnetic Properties of Amorphous
 Fe-P-B Alloys 275
 J. Durand and M. Yung

Mössbauer Study of a Glassy $Fe_{80}B_{20}$ Ferromagnet 289
 C. L. Chien and R. Hasegawa

The Resistivity of Amorphous Ferromagnets 297
 M. Baibich, R. W. Cochrane, W. B. Muir
 and J. O. Strom-Olsen

Magnetic Regimes in Amorphous Ni-Fe-P-B Alloys......... 305
 J. Durand

Resistivity of Metglas Alloys From 1.5 K to 800 K 319
 J. A. Rayne and R. A. Levy

Electrical Resistivity and Crystallization of Amorphous
 Metglas 2826 and Metglas 2826A............ 327
 W. Teoh, N. Teoh and S. Arajs

Transformation of Some Amorphous FePC Alloys During
 Isothermal Aging.................. 335
 A. S. Schaafsma and F. van der Woude

Perspective on Application of Amorphous Alloys in
 Magnetic Devices.................. 345
 F. E. Luborsky

High Magnetic Permeability Amorphous Alloys of the
 Fe-Ni-Si-B System 369
 T. Masumoto, K. Watanabe, M. Mitera and S. Ohnuma

Magnetostriction of Metallic Glasses............. 379
 R. C. O'Handley

On the Magnetically Induced Anisotropy in Amorphous
 Ferromagnetic Alloys................ 393
 H. Fujimori, H. Morita, Y. Obi and S. Ohta

Application of Domain Wall Pinning Theory to Amorphous
       Ferromagnetic Materials. 403
   D. I. Paul

Magnetic Characterization of Semi-Amorphous Nickel on
       Alumina Dispersions: Correlations with Their
       Methanation and Chemisorption Activities 415
   L. N. Mulay, R. C. Everson, O. P. Mahajan and
   P. L. Walker

Surface Effects in Amorphous Ferromagnets with
       Random Anisotropy. 425
   R. Micnas, A. R. Ferchmin, S. Krompiewski and
   B. Szczepaniak

Spin Wave Excitation and Propagation in Amorphous
       Bubble Films 435
   R. F. Soohoo

Temperature Dependence of the Extraordinary Hall Effect
       in Amorphous Co-Gd-Mo Thin Films 447
   R. J. Kobliska and A. Gangulee

Current Views on the Structure of Amorphous Metals II. 459
   G. S. Cargill III

Structure Simulation of Transition Metal - Metalloid
       Glasses, II. 463
   D. S. Boudreaux

Structural Models for Amorphous Transition Metal
       Binary Alloys. 469
   W. Y. Ching and C. C. Lin

Small-Angle Magnetic Scattering in Amorphous $TbFe_2$ 479
   S. J. Pickart

Magnetism and Structure of Amorphous $Fe_{80}P_{13}C_7$ Alloy 485
   M. Takahashi, M. Koshimura, T. Miyazaki and
   T. Suzuki

The Magnetic and Electronic Properties of Amorphous
       Nickel Phosphorus Alloys 499
   P. J. Cote, G. P. Capsimalis and G. L. Salinger

Structure and Physical Properties of an Amorphous
       $Cu_{57}Zr_{43}$ Alloy 513
   T. Mizoguchi, S. von Molnar and G. S. Cargill III
   T. Kudo, N. Shiotani and H. Sekizawa

CONTENTS

A Proposed Structure Model for Amorphous $Pd_{0.8}Si_{0.2}$ Alloy . . . 521
    T. Fukunga, T. Ichikawa and K. Suzuki

Ground State of an Ising Antiferromagnet with a Dense
    Random Packing Structure 529
    S. Kobe

Magnetic Resonance and Glass Structure 535
    G. E. Peterson

Magnetic Susceptibility and EPR Studies of Reduced
    Titanium Phosphate Glass 549
    C. H. Perry, D. L. Kinser and L. K. Wilson

Characterization of Ferromagnetic Precipitates in
    Glass by Ferromagnetic Resonance 561
    E. J. Friebele, D. L. Griscom and C. E. Patton

Low Temperature Thermal Conductivity of $MnO \cdot Al_2O_3 \cdot SiO_2$
    Glass 571
    A. K. Raychaudhuri and R. O. Pohl

A Calorimetric Investigation of a $CoO \cdot Al_2O_3 \cdot SiO_2$ Glass 577
    L. E. Wenger and P. H. Keesom

Exchange Fields and Crystal Fields in $CoO \cdot Al_2O_3 \cdot SiO_2$
    Glass: A Mössbauer Study 587
    L. H. Bieman, P. F. Kenealy and A. M. de Graaf

Oxidation - Reduction Equilibrium in Glass Forming Melts . . . 597
    A. Paul

Effect of Temperature and Oxygen Partial Pressure on
    Coordination and Valence States of Fe
    Cations in Calcium Silicate Glasses - A
    Mössbauer Study 613
    R. A. Levy

A Mössbauer Study on the Clustering and Crystallization
    Phenomena in $BaO-4B_2O_3$ Glasses Containing
    Dilute Concentrations of Fe_2O_3 627
    K. J. Kim, M. P. Maley and R. K. MacCrone

Magnetic Properties of $(Fe_2O_3-TiO_2)$ in $BaO-B_2O_3-SiO_2$
    Oxide Glasses 643
    L. Trombetta, J. Williams and R. K. MacCrone

Magnetic Properties of an Iron Borosilicate Glass 651
    M. P. O'Horo and J. F. O'Neill

Paramagnetic Impurity Concentrations in Amorphous Polymers. . . 663
    M. Centanni and P. A. Casabella

List of Participants. 673

Subject Index 677

SURVEY OF THEORIES OF SPIN GLASS

P. W. Anderson*

Bell Laboratories

Murray Hill, New Jersey 07974

I hope the organizers of this Conference are as confused about what a "keynote" address should contain as I am. Certainly they have given me no firm instructions, while the politicians have not provided me with the right kind of examples. It would take someone more presumptuous than I to tell you what you should say in the rest of the meeting, and Uri Geller to tell you what you will say. My own response to this charge is, first, to express our pleasure and gratitude at being here to talk about a most fascinating part of physics, and second, to get about doing so as immediately as possible.

Solid state physics has been enormously successful in dealing with many aspects of regular crystalline solids - so successful that many of us feel called upon to branch out and try to find ways of managing systems which don't have simple periodicity properties. The fascination, for me at least, of the general problem of random and amorphous systems is the necessity of developing a whole new conceptual structure not based on conventional band structure and translational symmetry, and requiring newer ideas such as percolation, localization and other forms of the general type of phenomena one might call non-ergodicity, as well as various other new techniques and concepts. The problem of the spin glass is one of the most complex and interesting, and most characteristic, in this new field. Thus it is a very suitable subject for such an opening talk.

_____

* Also at Department of Physics, Princeton University, Princeton, N.J. 08540. Work at Princeton was supported in part by the National Science Foundation Grant #DMR76-00886.

Under the title "Theories of Spin Glass" I can very severely restrict myself, partly because as one of the co-inventors of the term I can use it to mean what I like. My first criterion is that I will discuss only theories which take seriously the possibility that there is something like a genuine thermodynamic phase transition in systems with random exchange interactions, and that this transition is a property of the right kind of randomness and not of some mysterious ad hoc anisotropy field or "clustering" phenomenon. Since no one seems to have expected a phase transition before the experimental results of Canella and Mydosh[1] which appeared, as did the term "spin glass",[2] for practical purposes, at the first of these meetings 3 or 4 years ago in Detroit, this limits me to very recent history. I will also ignore a number of artificial models which have been invented to reduce the spin glass to some trivial result; if there is anything we surely know it is that the spin glass problem is not trivial, and it is easy in each case to see how, in these models, the baby has gone out with the bath water, in that the models do not contain the spin glass physics.

This leaves me with the following outline for my talk:

I. Brief discussion of experiments
    (a) Wide variety of materials, semidilute noble - transition alloys, "ferromagnetic superconductors", glass, CrO_2-VO_2[3] some metallic glasses (see this conference) etc.
    (b) Cusps in X at T_c; frozen in random moments below.
    (c) "Micto" magnetic phenomena starting at T_c.[4]
    (d) No Csp anomaly at T_c, linear at $T \to 0$.[5]
    (e) Computer experiments mimic(b,c,d) with random sign J_{ij}. (Binder,[6] Walker,[7] etc.)
II. Theories type 1: $n \to 0$, "replica" theories: Edwards-Anderson[8], Sherrington-Kirkpatrick,[9] Fischer, etc.
III. Type 2: Thouless, Anderson, Palmer: a mean field theory exact as $Z \to \infty$.
IV. Type 3: attempts at critical behavior: Lubensky and Harris[12] q^3, is "micto" magnetism = marginal dimensionality and/or line of critical points?

Although the characteristic spin glass alloys are the solutions of type Cu-Mn .1-10% or Au-Fe, a great variety of random magnetic materials seems to show the characteristic cusps and mictomagnetic behavior, and I have listed some of these above. I do not place great emphasis on the importance of RKKY oscillatory interactions; what I do think is vital is that the interactions be of random sign, or at least that the different interactions via different paths between two given spins be competing: this competition is the essence of the spin glass phenomenon and is the missing feature of most artificial models. I need hardly show you the characteristic cusp-like susceptibility which you will have seen so often, and which defines experimentally the spin glass phenomenon. One example is shown in Fig. 1. (After Canella, Mydosh and Budnick[1]) At this same T_C at which the cusp appears it is clear from several experimental data such as Mossbauer and μ-meson spectroscopy[3] that an essentially D.C. frozen in moment appears at each site. There is a freezing phenomenon, but all of our experiments as well as computations show that there is no <u>regular</u> structure. Along with the cusp and the frozen moment goes a very high and anomalous degree of field sensitivity of the susceptibility near T_C, as well as the kind of time and history dependence below T_C which we call micto-magnetism.[4]

A few weeks ago I received a letter from Ralph Hudson of the NBS objecting to this term, on the basis that he thought that the only other word in the English language using the same root was "micturation" and the root was Latin for "urine". I think myself that the term is very descriptive: back in the Middle West we used to refer to something as "p---poor" if it was not worth anything more substantial, and that is a good description of this kind of magnetism. The remanance is very small and very sluggish, there are peculiar training phenomena, and the susceptibility is often very history dependent. Unfortunately, I am assured by Collin Hurd and by the OED that Hudson is incorrect and that "micto" is a legitimate Greek root meaning "mixed".

To my mind there are two clear results which must be accounted for by any theory: (1) It is clear from every kind of data that no real ferromagnetism is involved: the frozen-in local moment is many orders of magnitude smaller than in ferromagnetism, and a quite different shape as a function of temperature: it just makes no sense to ascribe the remanance to isolated clusters. (2) There is clearly a definite T_C at which all of the phenomena begin, which is the same T_C as for the susceptibility: again we are confronted with a sharp transition.

My final point about experiments has to do with the other branch of experimental physics, computer experiments. The computations, particularly the exhaustive ones of Binder,[6] very realistically mimic the actual behavior of spin glasses, based on a short-range, random sign J_{ij} (in his case an Ising interaction, for others a more realistic Heisenberg one). These computations, I believe, confirm that the model we are using is correct: what is necessary is to solve it. So on to the theory.

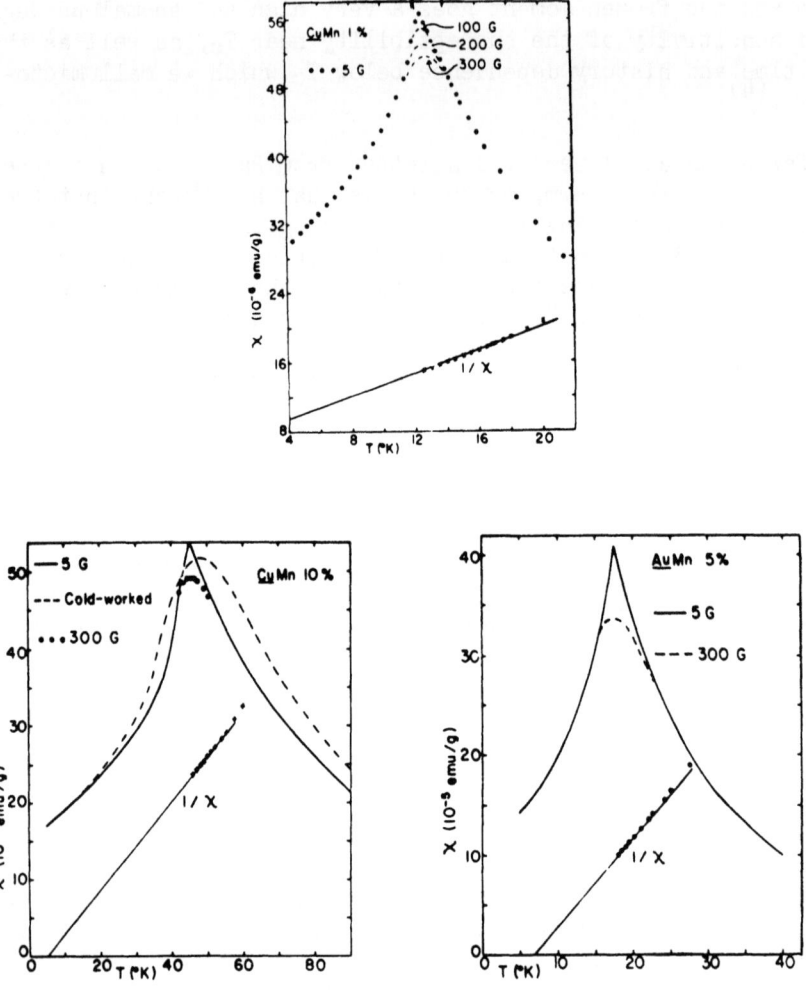

Figure 1 - Typical examples of susceptibility cusps.

The first attempts to understand how such a transition might occur were by Sam Edwards and myself[8] and much the same line has been followed on by Sherrington[9] and associates - Sothern, Kirkpatrick, and by Fischer.[10] Our basic initial contribution was to identify something which might serve as an order parameter for such a transition: namely we decided that since we believed there was no long-range order in space, we had to introduce a long-range order in time:

$$q = \lim_{t' \to \infty} \langle\langle S_i(t=0) S_i(t=t') \rangle\rangle$$

and we assumed that q is zero at high temperatures and that at a sharp transition point T_c it suddenly assumes finite values. (Note that in a finite magnetic field q is always finite so there is no sharp transition!)(See Fig. 2) In all ordinary phase transitions there is time long-range order, i.e. a change to non-ergodic behavior, but there is also an ordinary _spacial_ order parameter. What is fascinating here is the absence of the ordinary kind, and the really exciting question for the theorist is whether that is possible and whether it represents a true phase transition. My own feeling is that it is and does, but that is by no means proven conclusively yet. What is clear to me already is that if the transition occurs, its properties are very markedly different from those at ordinary phase transitions, and my best guess is that those great differences are responsible for the complicated mictomagnetic phenomena which are observed.

The actual formal technique which Edwards brought to the problem is so simple I can show it to you. If you want to calculate the free energy of a system you use

$$e^{-\beta F(T)} = \text{Tr} e^{-\beta H} = Z$$

It is also obviously true that

$$e^{-\beta n F} = (\text{Tr} e^{-\beta H})^n = Z^n$$

for any n.

Now it turns out to be easy to average expressions like $e^{-\beta H}$ over a random distribution of J_{ij}'s; this is true in general but it's particularly true of a Gaussian.

$$P(J_{ij}) = e^{-J_{ij}^2/2\overline{J_{ij}^2}}$$

so if for some integer n we write

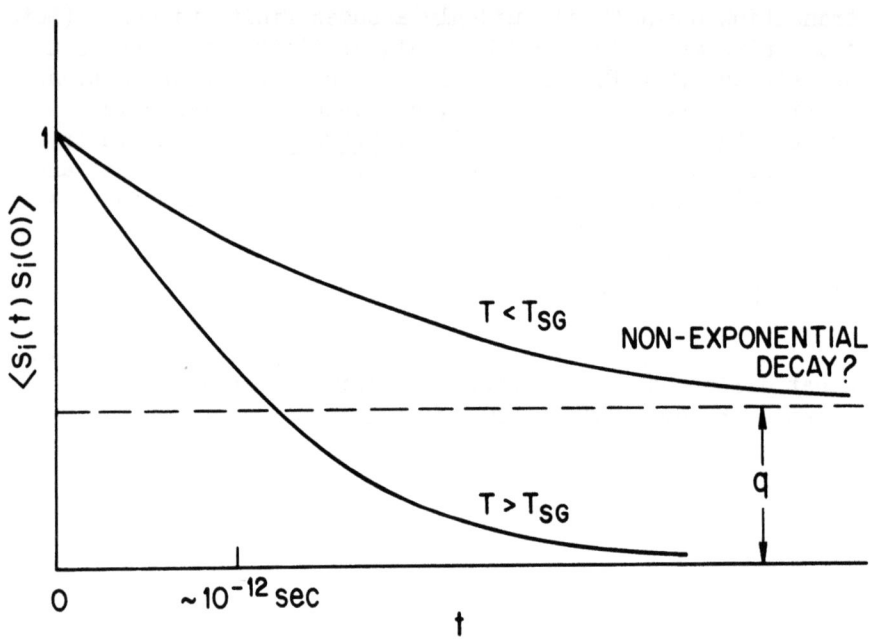

Figure 2 - Schematic diagram of $\langle S_i(t)S_i(0)\rangle$ vs t.

THEORIES OF SPIN GLASS

$$(\text{Tr} e^{-\beta H})^n = \text{Tr}_{S_i(1)} -- \text{Tr}_{(S_i(\alpha))} e^{-\beta \sum_{ij} J_{ij} \sum_\alpha^n S_i(\alpha) S_j(\alpha)}$$

and take the average

$$\langle \quad \rangle_{P(J_{ij})} \sim \text{Tr}_{S_i(\alpha)} \cdots \text{Tr} \; e^{\sum_{ij} \frac{\overline{\beta^2 J_{ij}^2}}{2} (\sum_\alpha^n S_i(\alpha) S_j(\alpha))^2}$$

$$= \text{Tr} \; e^{-\beta \sum_{\alpha\beta} H_{\alpha\beta}}_{S_i^\alpha}$$

$$H_{\alpha\beta} = \sum_{ij} \beta \overline{J_{ij}^2} \; S_i(\alpha) S_j(\alpha) S_j(\beta) S_i(\beta)$$

It looks as though we have just reduced the average free energy problem to a trivial one in that there is no longer any random interaction J_{ij}, but only the mean $\overline{J_{ij}^2}$ which is translationally invariant. But it's not that trivial, since it is not safe to go to the large N, thermodynamic limit of an expression like $e^{-\beta H}$ because H is proportional to N and its fluctuations grow with N: if we did it straightforwardly we'd end up with the free energy of some special favorable configuration of J_{ij}'s, and this is indeed shown by the fact that F isn't independent of n. The only limit in which this average is safe is the limit $n \to 0$, where it becomes $\langle \log Z \rangle$, so the trick is to do the problem for all integer n to get an expression for F(n) and take the limit as $n \to 0$ when finished.

The formal structure which we get, as you see, involves a coupling between the different "replicas" α, and for reasons which are obscure to everyone but Sam Edwards, this replica-replica correlation

$$q_{\alpha\beta} = \langle S_i^\alpha S_i^\beta \rangle$$

seems to behave in exactly the same way as the long-time correlation

$$\langle S_i(0) S_i(\infty) \rangle \; .$$

Most of the work done by this method treats $H_{\alpha\beta}$ strictly within the simplest kind of mean field theory. The most ambitious treatment was that of Sherrington and Kirkpatrick,[9] who realized that mean field theory is traditionally supposed to be exactly correct in the limit as the number of neighbors $Z \to \infty$. Their treatment seems at first sight to have this property. For simplicity they do Ising spins and allow $N = Z \ggg 1$, and they solve exactly for all integer n and extrapolate to $n \to 0$, achieving results which look almost exactly like those of Edwards and Anderson: a cusp both in X and Csp, a linear specific heat as $T \to 0$, etc. The only trouble is a very serious one: it is quite obvious internally that the low temperature results are wrong: the entropy is negative, the free energy is a maximum, not a minimum, and numerical computations have since shown that the ground state energy is too low. Since the <u>only</u> approximate step in the calculation is the $n \to 0$ limit, this has the unfortunate result of showing that that limiting process is not reliable. In fact, Richard Palmer[14] has shown that the crucial step is that they have taken the limit $N \to \infty$ <u>before</u> the limit $n \to 0$, which is not allowable.

In this disastrous situation Thouless, Palmer and I decided that the first essential was to understand this one example thoroughly, and to see if it couldn't be solved by a completely independent method. Rather to our surprise we found that it could, and that the results, while significantly different from those of SK at low temperatures, <u>confirm</u> the transition which SK found, and give the same physical properties at that point. This is no place to go into mathematical detail as to how we did this. The main idea was that we exploited the usual high-temperature diagram expansion in which you expand

$$\ln \mathrm{Tr} e^{-\beta H} = \ln \mathrm{Tr}\left[1 - \beta H + \frac{\beta^2 H^2}{2} + \text{---}\right]$$

which is a power series in 1/T, describing each term as a diagram in which lines are interactions J_{ij} and points are products of spins S_{ij}. In the large Z limit the mean square interaction must fall off as 1/Z to give a finite T_c:

$$\overline{J_{ij}^2} = \frac{\tilde{J}^2}{Z}$$

so that any diagram that has too many interaction lines in it drops out in this limit. Thus we can classify diagrams according to powers of 1/Z. The biggest are the simple chains (Fig. 3) which give the S-K high temperature result:

$$F_{HT} = -N\tilde{J}^2/4T \ .$$

The next biggest, of order N/Z and hence apparently negligible, are

THEORIES OF SPIN GLASS

"CHAINS":

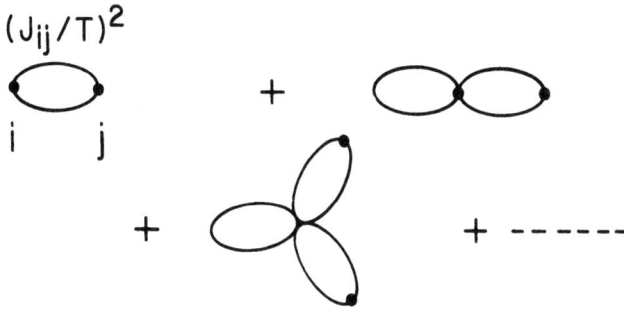

Figure 3 - Chain diagrams summed by the S-K method at high T.

all the simple rings (Fig. 4). These are indeed negligible if they converge, but we find that at $T_c = \tilde{J}$ they <u>diverge</u>, signalling the transition found by S-K. That these diagrams are necessary to find the transition is heartening, since they are the simplest ones which involve competition among different paths.

Below T_c we found that the only way to make our diagram series converge was to introduce a random mean spin m_i at every site satisfying the peculiar mean field equation

$$m_i = \tanh h_i/T$$

$$h_i = \sum_i J_{ij} m_j - \frac{m_i}{T} \sum_j J_{ij}^2 (1-m_j^2)$$

which as you see is not the same as naive mean field theory, which has only the first term, and in fact T_c is exactly 1/2 that given by ordinary mean field theory (but equal to SK, which is n → 0 mean field theory). This equation is extremely complicated to solve - note that it involves the random matrix J_{ij} - but with the help of the computer we believe we did find solutions and some of our results, contrasted with SK's, are shown in Figs. 5 and 6. One

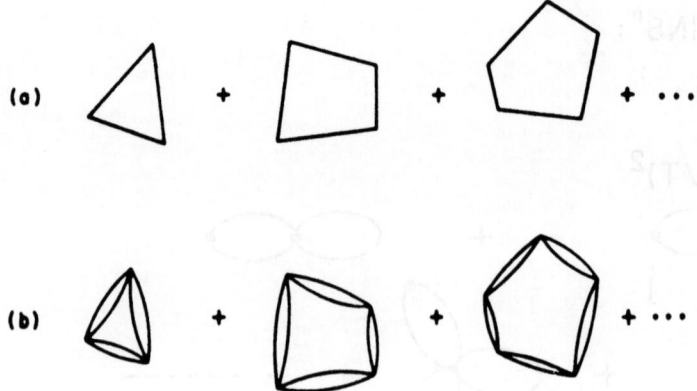

Figure 4 - Single (a) and multiple (b) ring diagrams whose divergence causes the transition.

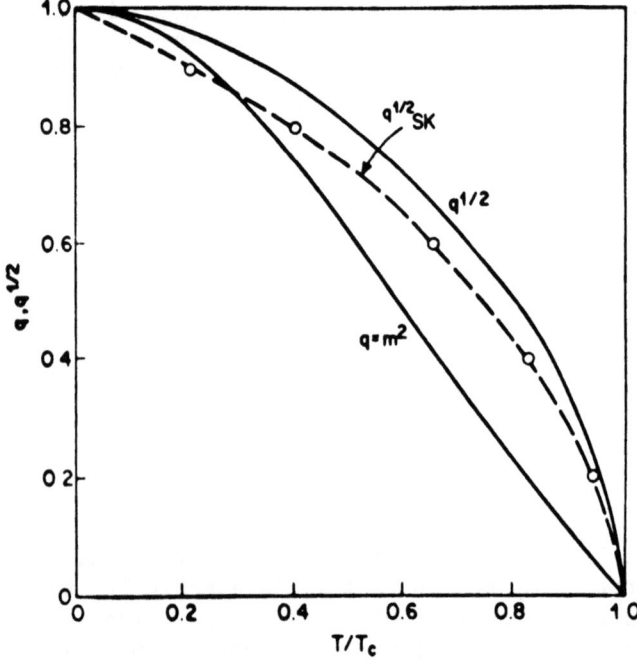

Figure 5 - Experimental results (SK and exact) for $q(T)$.

THEORIES OF SPIN GLASS

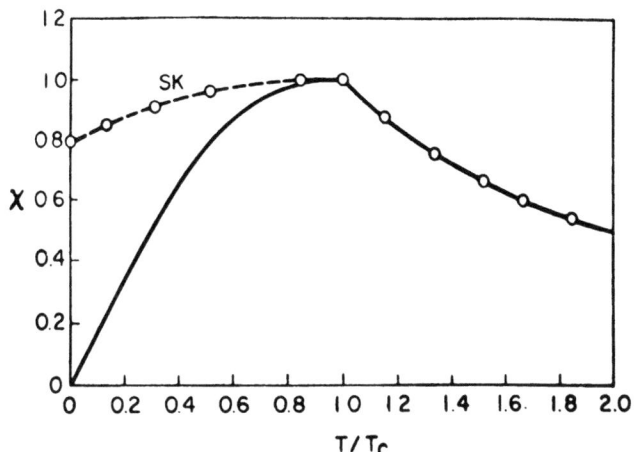

Figure 6 - Same for X(T).

further result is shown in Fig. 7: we found from our numerical computation that as T → 0 the probability distribution $P(h_i)$ of the mean local field becomes linear in h_i: this means the famous linear specific heat would not even result in this Ising model case! Thus the linear specific heat of the EA-SK theories is, as we all suspected, an artifact.

The really spectacular result of our theory, however, is a conjecture which seems to be borne out by our computer solution, and represents what may be a very strange and important property of spin glasses. It is possible for us also to write down an expression for the free energy as a function of the mean magnetization:

$$F_{MF} = - \sum_{ij} J_{ij} m_i m_j$$

$$- \frac{1}{2T} \sum_{ij} (1-m_i^2)(1-m_j^2) J_{ij}^2$$

$$+ T/2 \sum_i [(1+m_i) \ln(\frac{1+m_i}{2}) + (\frac{1-m_i}{2}) \ln(\frac{1-m_i}{2})]$$

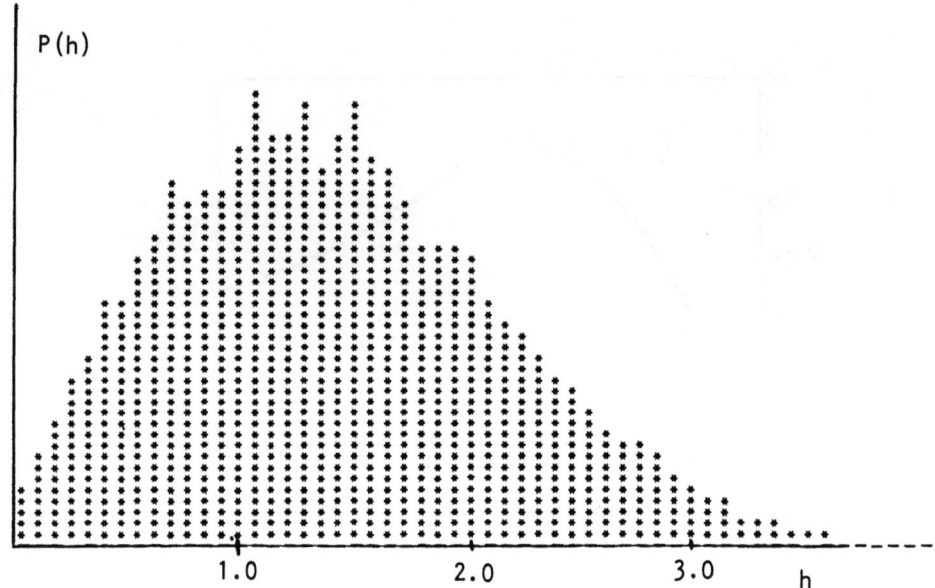

Figure 7 - Histogram of a numerical calculation of P(h).

but this free energy is meaningless unless the ring diagram series converges. We believe in fact that states with m_i^2 less than the largest solution of the mean field equation are forbidden by this ring sum condition: they have anomalously high free energies. What we find, when we look at F in the region near T_c and near $T = 0$ where we can solve the problem, is that it has the peculiar structure shown in Fig. 8: it has a horizontal inflection point at the mean field solution, i.e. $F \sim [(m_i)-(m_i(0))]^3$ <u>at all</u> temperatures. Now obviously one must identify our converging mean field m_i^2 (actually m_i, but that makes no difference to these considerations) with the order parameter q of the other types of theory, and what this says is that <u>at all</u> $T < T_c$

$$F(q) \sim |q-q_0|^3$$

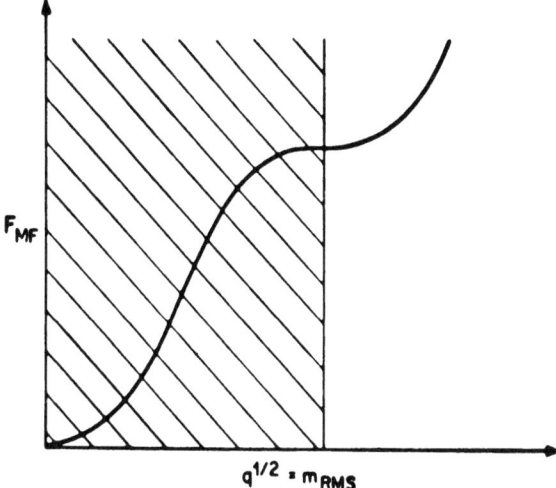

Figure 8 - Schematic of $F(m_i^2)^{1/2}$ below T_c, showing exclusion region.

not, as in usual phase transitions, $F \sim (q-q_0)^2$. If q is to be taken seriously as the order parameter, this means that not just the transition temperature, but all $T < T_c$, has some of the fluctuation properties of a critical point: the critical fluctuations of q persist to all $T < T_c$.

This is the point at which I must go on to the last, and most speculative, part of my talk. So far, we have been discussing results obtained with a very crude model in which, in order to make it exactly soluble, we have assumed infinitely long range interactions - or to put it another way, we are discussing the mean field theory only and not allowing any effects of fluctuations such as nowadays are handled by renormalization group methods. A very important first step towards a real theory was taken by Lubensky and Harris[12] when they started to speculate what kind of renormalization group theory could be based on a mean field theory in which the order parameter is a variable like q. The first thing they found is that such a theory would be very different from the conventional critical point theories because of the nature of the variable q. q, being a correlation function, has the property of being essentially positive. This property can be expressed in terms of the replica version of q by

$$\mathrm{Tr} q^3 = \sum_{\alpha \beta \gamma} S_{i\alpha}^2 S_{i\beta}^2 S_{i\gamma}^2$$

$$= n(n-1)(n-2)$$

One effect of this is that in the mean field $F(q)$ odd power terms such as q^3 can appear, where for an ordinary magnet one may not have M^3, only M^4. This, as Lubensky points out, changes the upper critical dimensionality at which mean field theory is correct from 4 to 6, so that critical fluctuations are much worse. Our TAP mean field theory, incidentally, has exactly this property, not only at T_c, but rather surprisingly a similar behavior at all $T < T_c$ as well. What this means is that even in mean field theory the whole temperature axis below T_c is a line of critical points: a phenomenon which occurs in a few other situations (like the 2-D xy model) but never in mean field theory. Since it is my belief that TAP will in the end prove to serve as a mean field "Ginzberg-Landau" free energy about which fluctuations must be taken into account, fluctuation phenomena will be very large at all $T < T_c$, not just near T_c.

One of the possible consequences of this increase in the strength of fluctuations is an increase in the "lower critical dimensionality" or "marginal dimensionality" below which no ordering is possible. In ordinary systems this is 1 for Ising and 2 for Heisenberg models. I believe I can show that this is 2 for Ising spin glasses and 3 for Heisenberg ones. Thus in physical spin glass systems we are in a very difficult and marginal situation in which the existence of a q is only barely possible if at all, and in which in any case very long range and slow critical fluctuations are occurring at T_c, and even possibly far below it. Under these circumstances I believe it is not at all surprising that the very complicated large-cluster phenomena, which we call Mictomagnetism in spite of Ralph Hudson, occur and dominate the behavior. It may be a long time before they or such things as C_{sp} and field sensitivity, can be understood in detail, but I certainly do not despair of doing so.

In closing let me caution you that a great deal of the last part of my talk is preliminary and speculative. This does not apply to the TAP mean field results, but to all implications I may draw from them as to real systems. But every experimental and theoretical indication leads me to believe that these delightful and unexpectedly elegant mathematical complications may really have a great deal to do with the most inelegantly named phenomena of mictomagnetism.

REFERENCES

1. V. Canella, J. A. Mydosh, and J. I. Budnick, J. App. Phys. **42**, 1689 (1971); V. Canella, In *Amorphous Magnetism* Hooper and

de Graaf, eds., (1973) p. 195; V. Canella and J. A. Mydosh, Phys. B $\underline{6}$, 4220 (1972).

2. P. W. Anderson, *Amorphous Magnetism*, op cit., p. 1.

3. C. Schlenker, B. K. Chakraverty, J. de Physique, to be published (Autrans Conference Proceedings).

4. J. S. Kouvel, J. Phys. Chem. Solids $\underline{24}$, 795 (1963); P. A. Beck, Met. Trans. $\underline{2}$, 2015 (1971).

5. L. E. Wenger and P. H. Keesom, Phys. Rev. B $\underline{11}$, 3497 (1975).

6. K. Binder and K. Schroder, Solid State Communications $\underline{18}$, 1361 (1976); Binder, to be published.

7. L. R. Walker and R. E. Walstedt, International Conference on Magnetism, Amsterdam, 1976 (to be published).

8. S. F. Edwards and P. W. Anderson, J. Phys. F $\underline{5}$, 965 (1975).

9. D. Sherrington and S. Kirkpatrick, Phys. Rev. Lett. $\underline{35}$, 1792 (1975).

10. K. Fischer, Phys. Rev. Lett. $\underline{34}$, 1438 (1975).

11. D. J. Thouless, P. W. Anderson, R. G. Palmer, Phil. Mag., to be published.

12. A. B. Harris, T. C. Lubensky, J.-H. Chen, Phys. Rev. Letter $\underline{36}$, 415 (1976).

13. C. E. Violet and R. J. Borg, Phys. Rev. $\underline{149}$, 540(1966) and other work; A. T. Fiory, this conference.

14. R. G. Palmer, to be submitted.

DISCUSSION

R. Tournier: Do you consider the remanent magnetization an intrinsic property of a true spin glass?

P. W. Anderson: The one thing I know is that these phenomena can almost be infinitely complicated. It is very likely that we have here a question of an enormous range of time scales so that something like the remanence could be basically the same whether you measure it on a time scale of 100 sec or 10^{-5} sec and yet if you measure at infinity it might be zero.

N. Rivier: I would like to make sure I understand correctly what you said when you mentioned in your theory that the slow time dependence of the order parameter was dependent on the fact that the free energy was a saddle point.

P. W. Anderson: Let us put it that that will cause a slow time dependence but that there may be other causes as well.

T. Mizoguchi: Can you comment a little on why the spin glass system appears to be field dependent at the transition temperature?

P. W. Anderson: I think I understand why but I have not proven why the system is so very field dependent. Incidentally, in the Thouless, Anderson and Palmer model we do not get an anomalous field dependence. To get that I think one has to include critical behavior. However, everything one believes says that if you are this close to the lower marginal dimensionality you have what is called an infinite order phase transition in which the correlation length is growing exponentially with $(T - T_c)$ which implies that the field dependence will be logarithmic. The specific heat already tells us that you have something like an infinite order phase transition because that exponent is essentially one over the order of the phase transition so that both the specific heat and the field dependence would be explained if you had something close to an infinite order phase transition.

T. Kaneyoshi: In dealing with the theories of spin glass do you think it is necessary to use a gaussian distribution function?

P. W. Anderson: Not at all, for this infinite range thing it would not matter if you use a gaussian or not. The distribution function would become gaussian because the law of large numbers hold. However, in any case I do not think it is necessary.

SOME PERCOLATION SCALING APPLICATIONS TO LOW TEMPERATURE DILUTE MAGNETS

D. Stauffer

Institute of Theoretical Physics, University

6600 Saarbrücken 11, West Germany

I. INTRODUCTION

This paper deals mainly with theories for localized magnetic moments μ distributed randomly among the sites of a regular lattice. Thus each lattice site is, with a probability p, occupied with a magnetic spin, whereas it is with probability 1-p occupied with a nonmagnetic atom. Only the exchange interaction between the spins is taken into account; all effects due to the nonmagnetic atoms (like lattice vibrations) are neglected. In this model, which is called a "dilute magnet", one now can define "n-clusters" as sets of n spins connected by exchange interactions. This definition makes sense only if the range of interaction forces is finite, as e.g. in a nearest-neighbor model, where an n-cluster is a set of n spins connected by nearest-neighbor bonds.

The number c_n of clusters determines in part the magnetic behavior of the system. The theory which calculates the c_n in this random clustering process is called PERCOLATION theory. A recent elementary introduction to percolation theory was given by de Gennes /1/. (More reviews and many papers on this subject are listed in the author's literature review also given in these Proceedings and thus not cited here again.) Unfortunately, percolation theory so far has not given exact expressions for the cluster numbers in 2 and 3 dimensions for real lattices, except /2/ for n < 20. The phase transition behavior of dilute magnets is connecetd, however, with the unknown behavior of c_n for n → ∞. "Scaling" theories are phenomenological assumptions connected with this n→∞ - limit and with the critical exponents β, γ and δ connected with it. The aim of the present paper is an application of the author's scaling theory of percolation /3/ to the magnetic properties of random magnets.

Sec.II estimates the critical temperature (Curie and Neel point) as a function of concentration p for low temperatures in isotropic magnets; for anisotropic Ising system, a result was suggested already in ref.4 where also the critical behavior around this critical temperature was discussed by a "crossover" theory. Sec.III gives the time-dependent decay of metastable states in anisotropic magnets when the magnetic field is antiparallel to the initial magnetization; simpler relaxations with no change in the sign of the field were already calculated in ref.3. Sec.IV applies percolation theory to spin glasses by a generalization of Smith's Cayley-tree theory /5/ to real lattices.

But before we can start with these new results, we need some review-information on percolation scaling theory and its relation to dilute ferromagnets /3/ since we will apply that information in the following sections. For small concentrations p, only single spins (n=1), pairs (n=2) and very few larger clusters exist; for p close to unity, on the other hand, nearly all spins form one connected "percolating" network (n=∞). Thus there exists a critical concentration, the percolation threshold p_c, such that for $0<p<p_c$ only finite clusters exist whereas for $p_c<p<1$ we have in addition to the finite clusters also one infinite cluster present: the percolating network. If c_n denotes the number of clusters per spin, then we get

$$\sum_n n \cdot c_n = 1 \quad \text{for } p < p_c \tag{1a}$$

since all spins must belong to one of the finite clusters. If an infinite cluster is present, it is not included in the sum $\sum n \cdot c_n$ over all finite cluster sizes n, and thus

$$\sum_n n \cdot c_n < 1 \quad \text{for } p > p_c \tag{1b},$$

with the difference $1-\sum n \cdot c_n$ being the "percolation probability", i.e. the fraction of spins belonging to the percolating network.

Experience with cluster models for other phase transitions suggests for $p \to p_c$ and $n \to \infty$ the assumption /3/

$$c_n = q_0 n^{-\tau} f(\varepsilon n^\sigma) \tag{2a}$$

$$\varepsilon = q_1 (p_c - p) \tag{2b}$$

with positive constants q_0 and q_1. The exponents σ and τ and the analytic function f can be assumed to be "universal", i.e. independent of the specific lattice type (bcc, fcc, sc) but depending on the dimensionality d of the lattice. In general, Eq(1a) then requires, with $x = \varepsilon n^\sigma$

$$\int_0^\infty x^{-1-\beta} \{f(x) - f(0)\} dx = 0 \quad \text{for } p < p_c \tag{2c}$$

where $\beta = (\tau-2)/\sigma$. For $p > p_c$ a reasonable approximation is /3/

$$f(x) \simeq e^x = e^{-|x|} \tag{2d}.$$

For sufficiently large clusters, the function f vanishes exponentially /6,7/. For $p \to p_c$ the critical exponents β, γ and δ are defined by

$$1 - \sum n \cdot c_n \propto (p-p_c)^\beta \quad \text{for } p > p_c \quad (3a)$$

$$\sum n^2 c_n \propto (p_c-p)^{-\gamma} \quad \text{for } p < p_c \quad (3b)$$

$$\sum n \cdot c_n \lambda^n \propto (-\log \lambda)^{1/\delta} \quad \text{for } p = p_c \quad (3c),$$

where λ is a parameter slightly below unity. Eq(2) then gives $\sigma = 1/\beta\delta$, $\tau = 2 + 1/\delta$ together with the scaling law $\beta + \gamma = \beta\delta$. Finally the radius ξ of a cluster with size $n = n_\xi \equiv |\varepsilon|^{-1/\sigma}$ diverges as $|\varepsilon|^{-\nu}$, where ξ is called the coherence length, and $d\nu = \gamma + 2\beta$. Numerically /2/ we have $\beta \approx 0.4$ and $\gamma \approx 1.7$ in three and $\beta \approx 0.14$ and $\gamma \approx 2.43$ in three dimensions.

How are these cluster numbers c_n related to magnetic properties? If we have a spin 1/2 Ising ferromagnet at very low temperatures, then the exchange forces connecting all spins within that cluster align all these spins parallel to each other; thus the spins of the cluster either point all up or all down. The whole cluster therefore acts like a single large spin, as in the theory of superparamagnetism. A single spin (n=1) with magnetic moment μ contributes to the total magnetization M in a magnetic field $H = hkT/\mu$ simply as $\mu \cdot \tanh(h)$. Analogously an n-cluster gives a magnetization $m_n = n\mu \cdot \tanh(nh)$; and the whole magnetization in units of the saturation magnetization is at these low temperatures:

$$M(H, T\to 0) = \sum n \cdot c_n \tanh(nh) \pm (1 - \sum n \cdot c_n) \quad (4).$$

(Thus sums are summations over all finite cluster sizes n; the ± stands for +sign(H) for stable and for -sign(H) for metastable solutions.) The last term in Eq(4), the percolation probability of Eqs(1b, 3a), gives the spontaneous magnetization and comes from the infinite percolating network existing for $p > p_c$ only. Eqs(4) and (2a,b) together give a "scaled" antisymmetric analytic universal cluster-equation-of-state, as it should be /3/. A numerical evaluation with the approximation (2d) shows smooth M-versus-H curves, with the susceptibility decreasing rapidly with increasing H except for a *very* small region near H=0 where the susceptibility-versus-field curve has a horizontal tangent.

This background information on percolation theory should be sufficient for an understanding of the following three sections which can be read separately.

II. VARIATION OF CRITICAL TEMPERATURE FOR $p \to p_c$

This section improves and generalizes a spin-wave argument of Shender and Shklovskii /8/ and gives a Curie temperature vanishing linearly in $p-p_c$.

For uniaxial magnets, the Curie temperature $T_c(p)$ was suggested /4/ to vanish for $p \to p_c^+$ as $-1/\log(p-p_c)$. For isotropic systems like Heisenberg and XY models we now estimate the ferromagnetic $T_c(p)$ from a spin-wave argument /8/. At T=0 for p slightly above p_c, the spontaneous magnetization M_o is given as $M_o \propto (p-p_c)^\beta$ by Eqs (3a,4). At finite temperatures, M_o at the same p is diminished by the amount M_1 due to spin waves of wavevector q and frequency ω_q:

$$M_1 \propto \sum_q \{\exp(\hbar\omega_q/kT) - 1\}^{-1} \approx (kT/\hbar) \sum_q \omega_q^{-1} \qquad (5),$$

where we use the classical approximation of $\hbar\omega_q \ll kT$. The Curie temperature $T_c(p)$ is reached if $M_o = M_1$, when no spontaneous magnetization is left; thus we have as in ref.8:

$$T_c(p) \propto (p-p_c)^\beta / \sum_q \omega_q^{-1} \qquad (6).$$

Dynamical scaling suggests the assumption /9/

$$\omega_q = Dq^2 w(q\xi) \qquad (7),$$

with $w(0) = 1$ ("hydrodynamic regime") and $w(y\to\infty) \propto y^{z-2}$ ("critical regime"). The spin-wave stiffness D varies as /10/

$$D \propto (p-p_c)^{t-\beta}$$

where t is the conductivity exponent for random resistor networks /10/. In the limit $y \to \infty$ all powers of $p-p_c$ should cancel from the frequency. Thus $z - 2 = (t-\beta)/\nu$ (≈ 1.6 in 3 dimensions /10/). Therefore one has $\omega_q \propto q^z \approx q^{3.6}$ for $q\xi \to \infty$, and the main contribution in the q-sum of Eq(6) comes from $q\xi \sim 1$ (for d>2), independent of the upper cutoff. (This argument was missing in ref.8.) Eq(6) thus gives $T_c \propto (p-p_c)^\beta \xi^d \omega_{q=1/\xi}$ or $T_c \propto (p-p_c)^{t-(d-2)\nu}$ as a generalization of the d=3 result of ref.8.

The exponent t was (tentatively) related /11,12/ to the other percolation exponents by

$$t = 1 + (d-2)\nu \qquad d \leq 6 \qquad (8).$$

Assuming this relation to be correct we get from the above result the very simple and thus, we hope, correct prediction

$$\boxed{T_c(p) \propto p-p_c} \qquad p \to p_c \qquad (9)$$

for the ferromagnetic phase transition temperature. This finite slope in the T_c versus p diagram agrees with series expansions /13/. We are not aware of real ferromagnets with short-range forces where this prediction has been tested.

In antiferromagnets, similar arguments give very different exponents. Spin waves now reduce the staggered (sublattice) magnetization (again $\propto (p-p_c)^\beta$ due to the infinite network) and have a frequency $\omega_q = Cqw(q\xi)$ where $w(0) = 1$ and $C \propto (p-p_c)^s$ with $s = 1.1\pm.1$ in three dimensions /10/. Again $\omega_q \propto q^z$ for large $q\xi$, but now:

$z-1 = s/\nu \simeq 1.3$ for $d = 3$, or $\omega_q \propto q^{2.3}$. Thus the q-sum in

$$T_N(p) \propto (p-p_c)^\beta / \sum_q \omega_q^{-1} \tag{10}$$

no longer comes from $q\xi \sim 1$ but from $q\xi \to \infty$: The upper cutoff is relevant now. For $T \to 0$ this cutoff is not a Brillouin zone boundary but given by quantum effects in Eq(5): We need frequencies below ω_{max} where $\hbar\omega_{max} \sim kT$. With $\omega_{max} \propto (q_{max})^z$ we thus have

$$\sum \omega_q^{-1} \sim (q_{max})^d/\omega_{max} \propto (\omega_{max})^{-1+d/z} \propto T^{-1+d/z},$$

and thus at $T = T_N$ from Eq(10): $T_N \propto (p-p_c)^\beta T_N^{1-d/z}$, or

$$T_N(p) \propto (p-p_c)^{\beta z/d} = (p-p_c)^{\beta(1+s/\nu)/d} \tag{11a}$$

$$\boxed{T_N(p) \simeq (p-p_c)^{0.3}} \qquad \text{for } d=3 \tag{11b}.$$

Thus the slope in the T_N versus p diagram should be infinite at p_c, in contrast to some experiments /14/. Further data closer to p_c might clarify the question whether Eq(11b) is correct.

III. QUANTUM NUCLEATION, METASTABILITY AND SPINODALS

The homogeneous nucleation rate due to quantum fluctuations is estimated for the reorientation of the spontaneous magnetization. Neglecting this nucleation rate we find well-defined metastable states without any spinodal lines.

The last term in Eq(4) gives the spontaneous magnetization and is nonzero above p_c only. If this spontaneous magnetization is positive and if the magnetic field is switched from positive (equilibrium) to negative (non-equilibrium) values then first the many finite clusters (starting with the small ones /3/) switch their orientation into the new equilibrium distribution; after some time basically only the infinite network still has the wrong orientation antiparallel to the new magnetic field. This network of connected spins changes its orientation only by forming in its interior small domains oriented parallel to the field ("homogeneous nucleation"); if these domains are larger than some critical size they continue to grow until they extend over the whole place. As in classical nucleation theory at finite temperatures we assume this critical size to be large enough to make bulk concepts like energy $U = u \times$ area and mass $M' = m \times$ area applicable to the domain walls of this critical nucleus. And as in the theory of Lifshitz and Kagan /15/ the phase transition is assumed to be driven by quantum fluctuations instead of thermal fluctuations since the temperature is assumed as low.

This nucleation process can be estimated by the quasiclassical WKB method applied already before /16/ to domain wall motion in finite clusters. Neglecting all (less relevant) preexponential factors, the rate J at which critical-size domains of overturned spins

are formed is then given by
$$J \propto e^{-2S/\hbar} \; ; \; S = \int_0^R \{2M'(r)U(r)\}^{1/2} dr \tag{12}$$
as long as the action S is much larger than ℏ. Here r is the radius of the domain and R is defined as that radius where the energy U vanishes: $U(R)=0$. We have $M' = 4\pi r^2 m$, $U = 4\pi r^2 u - 2N_r\mu|H|$, $N_r = (4\pi/3)r^3\rho$, ρ = number of spins per unit volume. The energy U(r) has a maximum at the "Kelvin" radius

$$r_K = u/|\mu H\rho| \tag{13a}$$

whereas the quantity M'/U is greatest at the "critical" radius

$$r_c = \frac{6}{5} u/|\mu H\rho| \tag{13b}$$

and finally

$$R = \frac{3}{2} u/|\mu H\rho| \tag{13c}.$$

Performing the one-dimensional integration in Eq(12), with $\int_0^1 x^2(1-x)^{1/2} dx = 16/105$, we get for the nucleation rate J:

$$\boxed{\ln(J) \simeq (128\pi\sqrt{2}/105)R^3(mu)^{1/2}/\hbar = (144\pi\sqrt{2}/35)u^{7/2}m^{1/2}|\mu H\rho|^{-3}/\hbar}$$
(14).

Thus ln(J) varies as H^{-3}, not as H^{-2} as it does for thermal nucleation. For a typical preexponential factor /16/ in the nucleation rate, one needs the argument of the exponential in Eq(14) to be around 10^2 in order to have measurable nucleation effects. This condition leads to a critical radius near 6 Angstrom (corresponding to a megagauss field) if /16/

$$m = 10^{-10} \text{ g/cm}^2, \; u = 10 \text{ erg/cm}^2, \; \rho = 10^{22} \text{ cm}^{-3}, \; \mu = 10^{-20} \frac{\text{erg}}{\text{gauss}}$$

We now can <u>define</u> "metastable states" in an unambiguous way by simply neglecting this just calculated nucleation rate for the infinite cluster. Thus in a "metastable" state all finite clusters are in equilibrium with the magnetic field whereas the infinite network remains oriented antiparallel to H. Thus in the metastable equation-of-state, the ± sign in Eq(4) is replaced by -sign(H). This metastable branch is completely symmetric to the stable one, as is evident from Eq(4). Nowhere exists a spinodal line where ∂M/∂H diverges (as it often does in "classical" theories). Instead, above the percolation threshold p_c, we get for one field H two values for M, and for one M two values for H, where at least one value is metastable. Such properties are in general agreement with Binder's conclusions /17/ that: i) the order parameter M alone is not sufficient to characterize a non-equilibrium system like metastable states; ii) one needs for this purpose also fluctuations of the order parameter, i.e. clusters; and iii) spinodal lines do not exist for short-range interactions.

IV. SPIN GLASSES IN THE CLUSTER APPROXIMATION

An application of Smith's Cayley-tree results /5/ to real three-dimensional lattices gives a susceptibility cusp as const + $(T_g-T)^\beta$, $\beta \approx 0.4$, very close to the spin glass phase transition temperature T_g.

In a spin glass like Fe_pAu_{1-p} (see ref.18, or Anderson's paper AA1 here, for a theoretical review), the RKKY interactions are neither ferromagnetic nor antiferromagnetic but oscillate with distance and decay in magnitude as $(distance)^{-3}$. Thus we have no sharply defined clusters in the sense of Sec.I since the range of interaction is infinite. As an approximation, Smith /5/ defines clusters as sets of spins connected by interaction energies larger than the thermal energy kT; similarly to ref.19 he neglects all interaction energies between different clusters which are by definition smaller than kT. Moreover, he takes the interactions within each cluster as very large, thus treating the whole cluster as one superparamagnetic particle rotating rigidly. In this way, also the spin glass is approximated by a cluster model very similar to the percolation cluster description of dilute ferromagnets in Sec.I. This approximation gives /5/ $T_g(p) < p$ for $p \ll 1$.

For ferromagnetic interactions alone, the susceptibility χ is proportional to $(\mu^2/kT)\sum n^2 c_n \propto |p-p_c|^{-\gamma}$ from Eqs(3b,4), also above p_c. More generally, for internally rigid n-clusters with net magnetic moment m_n, one has $T\chi \propto \sum m_n^2 c_n$, which reduces to the first result if all spins are parallel and $m_n \propto n$. In a spin glass a large cluster will have about as many ferromagnetic as antiferromagnetic bonds, and thus will have no net magnetic moment in zero field: $<m_n> = 0$, where the brackets indicate a configurational average over all clusters of the same size n but with different distribution of interactions. E.g. in a uniaxial magnet the distribution of spin orientations ↑ and ↓ will be approximately random such that the number $n_↑$ of up-spins cancels in the average the number $n_↓=n-n_↑$ of down spins: $<m_n> \propto <n_↑ - n_↓> = 0$. But in such a random distribution, the fluctuations are nonzero: $<m_n^2> \propto <(n_↑ - n_↓)^2> \propto n$; thus

$$T\cdot\chi \propto \sum n\cdot c_n \tag{15a}$$

Eqs(1a,3a) thus give for the paramagnetic region, from Eq(17a):

$$T\cdot\chi(p<p_c) \approx const \tag{15b}$$

a simple Curie law as for ideal paramagnets and as required from other theories /20,21/. In the ordered region, i.e. $p>p_c$, we get

$$T\cdot\chi(p>p_c) \propto const - (p-p_c)^\beta + \ldots \tag{15c}$$

from Eq(3a). With Smith's temperature-dependent definition of clusters, it does not matter much whether one increases T (and thus p_c) at fixed p or decreases p at fixed T (and thus at fixed $p_c \propto T$).

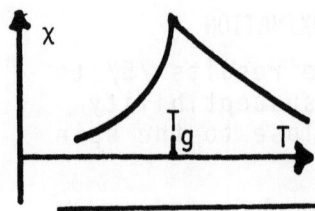

Thus $p-p_c$ can be replaced by T_g-T. If a spin glass is cooled across its transition temperature T_g at fixed composition p, then above T_g it obeys a simple Curie law whereas below T_g the susceptibility has a cusp with infinite slope:

$$\chi \propto \text{const} - (T_g-T)^\beta, \quad \beta \approx 0.4 \tag{16}$$

Theoretically, such a cusp with exponent β is not bad; for χ is related to the Edwards-Anderson order parameter q by /22/

$$\chi = \chi^{ideal}(1-\text{const}\times q) \tag{17}$$

This "order parameter" q vanishes in the theory of ref.23 with an exponent β' not identical but at least similar to the percolation exponent β; e.g. β = β' = 1 for d > 6. (However, Binder and Schröder /21/ questioned the definition of this order parameter and Young /24/ even doubted the existence of a transition temperature T_g.) Experimentally, no such infinite slope in the χ versus T curve was found /25/, in contrast to Eq(16). (Smith /5/ looked at the Bethe lattice (Cayley tree) only where β = 1 and where there is no such infinite slope from Eq(16).)

This discrepancy between theory and experiment might be due (D.A.Smith, private communication) to the long-range RKKY interactions relevant here, which could make the Bethe-lattice theory a better approximation than short-range percolation theory. The temperature-dependent range ξ of the exchange energies J_{RKKY} varies /5/ as $T^{-1/3}$, since the interactions are cut off at

$$kT = J_{RKKY} \propto (\text{distance})^{-3} \tag{18}$$

Analogously to superconducting phase transitions, for ξ_o much larger than the atomic distance, classical Bethe lattice exponents β=γ=1 are expected to be valid then except for a region extremely close to p_c (or T_g) and not accessible experimentally. (Indeed, according to ref.26, the relative width of this region very close to p_c vanishes as $1/\xi^2$ ("Ginzburg criterion") and thus here as $T^{2/3}$.) But this argument of Smith conflicts with computer experiments /21/ for the Edwards-Anderson model, where only nearest-neighbor forces with randomly varying strengths and signs were taken into account and where no evidence for an infinite slope in the susceptibility versus temperature diagrams was found. Ofcourse, also the accuracy of these Monte Carlo simulations /21/ is limited. Thus at present it is not clear whether the phase transition behavior of spin glasses is correctly approximated by a model /5/ of well-defined non-interacting clusters.

V. CONCLUSION

This paper explored some possibilities of applying percolation theory to amorphous magnets (dilute ferromagnets and spin glasses), and its main result are the boxed-in equations in the text:

i) With short range forces, a plausible result was found for the Curie temperature, Eq(9), whereas for the antiferromagnetic Neel temperature of Eq(11b), we suggest further experiments closer to p_c to find out whether the theory is correct.

ii) We calculated a nucleation rate $\propto \exp(-...H^{-3})$, Eq(14), by which the spontaneous magnetization reverses its orientation in a small antiparallel magnetic field H due to quantum fluctuations. The method employed may become relevant for a future understanding of the dynamics of low-temperature magnets and spin glasses; the result itself has not yet been tested experimentally.

iii) For long-range interactions in spin glasses, we generalized Smith's theory /5/ from the Bethe lattice to real lattices and found a questionable infinite slope, Eq(16), in the susceptibility versus temperature curve slightly below T_g.

Quite generally we suggest more experiments /14/ on the rather simple short-range amorphous magnets, like the recent ones of Birgeneau et al/27/. It seems that during the last years the more complicated spin glasses have taken away too much interest from easier-to-understand systems.

We thank Profs. K.Binder, D.A.Smith, P.G. de Gennes, U.Krey and B.I.Shklovskii for their comments on preliminary versions of these calculations.

REFERENCES

/1/ P.G. de Gennes, preprint ("La notion de percolation")
/2/ M.F.Sykes, D.S.Gaunt and M.Glen, J.Phys. A 9, 87, 97, 715, 725, L 43 (1976)
/3/ D.Stauffer, Phys.Rev.Letters 35, 394 (1975), where earlier percolation scaling literature is cited
/4/ D.Stauffer, Z.Physik B 22, 161 (1975)
/5/ D.A.Smith, J.Phys. F 5, 2148 (1975)
/6/ D.Stauffer, Z.Physik B, submitted July 1976
/7/ M.M.Bakri and D.Stauffer, Phys.Rev. B, submitted Dec.1975
/8/ E.F.Shender and B.I.Shklovskii, Phys.Letters 55A, 77 (1975)
/9/ B.I.Halperin and P.C.Hohenberg, Phys.Rev. 177, 952 (1969)
/10/ S.Kirkpatrick, Rev.Mod.Phys. 45, 574 (1973); S.Kirkpatrick and A.P.Harris, AIP Conf.Proc. 24, 99 (1975) and preprint
/11/ A.S.Skal and B.I.Shklovskii, Sov.Phys.Semicond. 8, 1029 (1975)
/12/ P.G. de Gennes, J.Physique 37, L 1 (1976)
/13/ E.Brown, J.W.Essam and C.M.Place, J.Phys. C 8, 321 (1975)
J.S.Reeve and D.D.Betts, J.Phys. C 8, 2642 (1975)

/14/ J.M.Baker, J.A.J. Lourens, R.W.H. Stevenson, Proc.Phys. Soc. 77, 1038 (1961); D.J.Breed, G.Gilijamse, J.W.E.Sterkenburg and A.R.Miedema, J.Appl.Phys. 41, 1267 (1970)
/15/ I.M.Lifshitz and Yu.Kagan, Sov.Phys. JETP 35, 206 (1972); see also the concluding remarks of I.A.Provorotskii, Sov.Phys. Uspekhi 15, 555 (1973)
/16/ D.Stauffer, Sol.State Comm. 18, 533 (1976)
/17/ K.Binder, in: Proceedings of the IUPAP Intern. Conf. Statistical Physics, Budapest August 1975, to be published; K.Binder and D.Stauffer, Adv.Phys. (July 1976), in press.
/18/ G.Heber, Appl.Phys. 10, 101 (1976)
/19/ I.Ya.Korenblit, E.F.Shender and B.I.Shklovskii, Phys.Letters 46 A, 275 (1974)
/20/ C.Domb, J.Phys. A 9, L 17 (1976)
/21/ K.Binder and K.Schröder, Sol.State Comm. 18, 1361 (1976) and Phys.Rev. 14, Sept. 1976, in press; K.Binder and D.Stauffer, Phys.Lett. 57 A, 177 (1976); K.Binder, Magnetism Conference, Amsterdam Sept.1976, to be published in Physica.
/22/ K.H.Fischer, Phys.Rev.Letters 34, 1438 (1975)
/23/ A.B.Harris, T.C.Lubensky and J.H.Chen, Phys.Rev.Letters 36, 415 (1976)
/24/ A.P.Young, J.Phys. C 9, to be published
/25/ C.N.Guy, J.Phys. F 5, L 242 (1975) and private communication
/26/ D.Stauffer, J.Chem.Soc. Faraday Trans. II 72, 1354 (1976)
/27/ R.J.Birgeneau, R.A.Cowley, G.Shirane and H.J.Guggenheim, preprint BNL 21544; see also Breed et al, ref.14.

ADDENDUM

E. F. Shender (J. Phys. C 9, L309 and Zhetf 70, 2251) suggested dynamical scaling similar but not identical to Section II. T. C. Lubensky pointed out an error in the last paragraph of Section II, affecting our ($T_N(p)$ there. For a cluster theory of spin glass different from Section IV see paper by Holtzberg, Tholence and Tournier.

DISCUSSION

R. Tournier: You mentioned that at high temperatures one expects from your theory the Curie law to be obeyed. I wonder what happens to the additional dependence on $1/T^2$ that is commonly observed?

D. Stauffer: The theory predicts that for non-interacting ferromagnets, above T_c the ideal Curie law should be obeyed. The fact that there is an apparent disagreement between experiments and the theory should probably be clearly pointed out in your papers.

PERCOLATION SCALING THEORY - A SURVEY OF RECENT LITERATURE

D.Stauffer

Institute of Theoretical Physics, University

6600 Saarbrücken 11, West Germany

1: REVIEW ARTICLES ON PERCOLATION
De Gennes, preprint (elementary);Kirkpatrick,Rev.Mod.Phys.45,574;
Essam in: Phase Trans. & Crit.Phenomena, Domb & Green, eds, Vol.2.
See also refs. in Bishop,Progr.Theor.Phys.53,50, or paper AA2 here

2: DILUTE MAGNETS AT NONZERO TEMPERATURES
a) Series expansions: Rapaport, J.Phys.C5,1830 & 2813; Rushbrooke et
al,J.Phys.C5,3371. Reeve & Betts,J.Phys.C8,2642; Brown,Essam,Place
J.Phys.C8,321. Cox et al PL 55A,1 & J.Phys.C9,1719.
b) Renormalization group: Lubensky et al PRL 33,1540 & PR B 11,3573;
Krey, PL 51A,189. Khmelnitzki,JETP 41,981; Grinstein & Luther,PR B
13, 1329.
c) Monte Carlo: Ching & Huber, PR B13,2962; Fisch & Harris,AIP Conf.
Proc.('75 MMM Conf.); Landau, Amsterdam Magn.Conf.; Stoll & Schneider, Verhandl.DPG 11,703. Klenin & BLume, PR B14,235.
d) Crossover at $T_c(p)$: Stauffer,Z.Phys.B22,161; Lubensky,preprint;
Stanley,preprint; Young,J.Phys.C9,2103. Shender et al PL 55A,77 &
ZhETF 70,2251; Stauffer, paper AA2 here. Birgeneau et al, preprint

3: PERCOLATION CLUSTER NUMBERS
a) Phenomenology: Essam et al J.Phys.C4, L 228 & 9,365; Domb,J.Phys.
C 7,2677; Harris, PR B 12,203; Grinstein et al. PRL 36,1508.
Stauffer, PRL 35,394; Bakri & Stauffer, preprint. For renorm.group
see Harris et al PRL 35,327 and Priest & Lubensky, PR B13,4159
b) Monte Carlo: Dean & Bird, unpublished 1966 (analyzed by Stauffer,
see 3a). Coey, PR B6,3240; Duff & Canella in: Amorphous Magnetism,
Hooper & de Graaf,eds, p.207; Onizuka,J.Phys.Soc.Jpn.39,527; Quinn
et al, J.Phys.A9, L 9; Leath, PRL 36,921 & preprint. Fremlin,J.Physique 37,813. Müller-Krumbhaar & Stoll, preprint for J.Chem.Phys.
See Kirkpatrick PRL 36,69 for dependence on dimensionality.
c) Series expansions: Sykes et al J.Phys.A9, 87,97,715,725, L 43;
Blease,Essam,Place,preprint; Stauffer,preprint for Z.Physik B

4: OTHER PERCOLATION PROPERTIES
a) Cluster "surfaces": Duff & Canella, AIP Conf.Proc.10,541; Domb et al J.Phys.A8, L 90; Leath (see 3b); Sur et al,preprint for J.Stat. Phys. Phenomenology: Domb (see 3a); Stauffer,J.Phys.C 8, L 172 & (3c)
b) Correlation and size effects: Monte Carlo: Levinshtein et al,ZhETF 69,386; Roussenq et al,J.Physique 37, L 99; Sur et al (see 3c). Phenomenology etc: Reatto & Rastelli,J.Phys.C5,2785; Essam et al, J.Phys.C8,743 & 4219; Stauffer (see 2d); Young & Stinchcombe,J.Phys. C8, L 535; Shur,J.Phys.C9, L 229; Sadovskii,ZhETF 70,1936; Birgeneau et al (see 2d)
c) Conductivity in random resistor networks: Kirkpatrick (see 1 and 3b); Levinshtein et al,ZhETF 69,2203 & 70,2014; Clark et al, Comptes Rendus B281,227; Harris & Kirkpatrick,preprint. Stinchcombe & Watson,preprint; Toulouse,Comptes Rendus B280,629; Skal & Shklovskii,Sov.Phys.Semicond.8,1029 and de Gennes, J.Physique 37, L 1; Stauffer (see 2d).
d) Modified percolation: Müller-Krumbhaar, PL 50A,27; Odagaki et al, J.Phys.Soc.Jpn.39,618; Coniglio,J.Phys.A8,1773 & PR B13,2194. Kawasaki & Tahir-Kheli,Progr.Theor.Phys.55,310. Stephen, PL 56A,149

5: POLYMER GELATION
a) General: Flory,J.Am.Chem.Soc.63,3083,3091,3096 (1941 !); Gordon et al, Pure Appl.Chem.43,1 & Makromolek.Chemie 176,2413. Conference Proceedings in Faraday Disc.Chem.Soc. No.57 (e.g. p.165 there). Biological applications: G.I.Bell,J.Theor.Biol.33,339 & preprint; DeLisi et al,J.Theor.Biol.45,555 & 53,1.
b) Percolation connections: Bell (see 5a),De Gennes,J.Physique 36,1049 and (see 4c), Stauffer,J.Chem.Soc.Faraday Trans.II 72,1354. For Monte Carlo see Falk & Thomas,Can.J.Chem.52,3285; Helfand & Weber, Bull.Am.Phys.Soc. 21,100

6: SPIN GLASSES
a) Monte Carlo: Binder et al, Sol.State Comm.18,1361 & PR B14 (Sept.) & PL 57A,177 & Amsterdam Magn.Conf.; Ching & Huber,preprint. Walker & Walstedt, Bull.Am.Phys.Soc.21,386; Sherrington,AIP Conf.Proc. (1975 MMM Conf.)
b) Other methods: Harris et al, PRL 36,415; Young,J.Phys.C9,preprint. Domb,J.Phys.A9, L 17. Smith,J.Phys.F5,2148 and Stauffer (paper AA2 here).

Abbreviations: PR = Phys.Rev., PRL = Phys.Rev.Lett., PL = Phys. Lett., ZhETF = Zhur.Eksp.Teor.Fiz.(in Russian)
Warning:
This compact-review neglected mean field, effective medium, Bethe, and coherent potential approximations and concentrated only on critical exponents etc. Also, in general only literature from 1974-76 is listed. The review was completed in July 1976.

μ^+ STUDIES OF DILUTE PdFe ALLOYS*

K. Nagamine, N. Nishida and T. Yamazaki

Dept. of Physics, University of Tokyo, Tokyo, Japan and

TRIUMF, Univ. of British Columbia, Vancouver, Canada

ABSTRACT

In order to investigate the ordering mechanism among giant moments around Fe impurities in Pd, μ^+ was used to probe the conduction electron polarization in PdFe alloys above and below the critical concentration of 0.1 at. % (i.e. 0.28 at. % Fe and 0.015 at. % Fe) with reference to pure Pd. Below the ordering temperatures (9.0 K and 0.4 K, respectively), the broadening of the μ^+ field for 0.015 at. % Fe is substantially larger than that for 0.28 at. % Fe, when normalized to the bulk magnetization. The results can be explained in terms of an RKKY spin oscillation in the region outside the giant moment.

1. INTRODUCTION

Metallic Pd with dilute Fe impurities has interesting magnetic properties at low temperatures; the impurity Fe spin strongly polarizes the d-holes on neighbouring Pd sites, forming a large polarized complex ($\sim$10 Å) called the giant moment.[1,2] These giant moments couple to one another to yield long-range ferromagnetism at quite low Fe concentrations. For even lower concentrations, magnetic susceptibility measurements[3] show a marked deviation from the Curie-Weiss law indicating a spin-glass ordering below a critical concentration of 0.1 at. % which corresponds to an average distance between impurity atoms of 15 to 20 Å. However, a recent paper has proposed a different model; namely, a ferromagnetic ordering with a distribution of effective g-values in Pd atoms.[4] In order to gather further insight into this problem, it is quite interesting to study

the difference between the conduction electron polarization above
and below the critical concentration.

Polarized positive muons are used here to probe the conduction
electron polarization in PdFe alloys. When the μ^+ stops at a
random site in a metal with dilute magnetic impurities, it feels
contact fields from the conduction electrons; that is, the contact
field from polarized d-holes or from s-electrons which might be
polarized through s-d hybridization. In addition to this, the μ^+
interacts with dipolar fields from the giant moments. Both of
these have inhomogeneities due to the random distribution of the
field sources. Such magnetic fields and their inhomogeneities can
be measured via the precession frequency and its dephasing time-
constant in the asymmetric positron decay of the μ^+ (μ^+SR method).

As a reference for the PdFe studies, a separate μ^+SR experiment was carried out to measure the μ^+ Knight shift in pure Pd. In
the next section, the result obtained is compared with the relevant
data on hydrogen in Pd. The properties of the μ^+ as a probe to
study magnetic properties of Pd-based alloys is discussed. The
experimental results on PdFe alloys are presented in Section 3 and
their interpretations are given in Section 4.

2. THE μ^+ KNIGHT SHIFT IN Pd

Using known experimental data on the properties of hydrogen in
Pd, we can make reasonable predictions for the properties of the μ^+
in Pd: these diffusion studies as well as neutron experiments[5]
indicate that the μ^+ will preferentially remain at octahedral sites
and might be localized there for its entire lifetime (2.2 µsec)
below room temperature. Recently NMR experiments have been carried
out to measure the proton Knight shift in Pd in the dilute limit,[6]
which could be compared to the μ^+ Knight shift observed at 300 K.

The μ^+ Knight shift experiment on pure Pd was recently carried
out on the M9 channel of TRIUMF. A highly polarized μ^+ (~90%) beam
was stopped in the Pd metallic target (1 mm diam and 50 mm long
wires, loosely banded into a rectangular shape). Stability of the
external field appropriate to the high precision of µSR was
achieved with a Varian 'Fieldial Mark II'. The value of the external field was monitored frequently by replacing the Pd target with
a Cu target and using the known correction for the Knight shift of
μ^+ in Cu (+0.0081%).[7]

The observed local field at the μ^+ site (B_μ) can be decomposed
into a sum of the external field (B_{ext}), the Lorentz field ($4\pi/3$ M)
and the contact field from polarized conduction electrons (H_{int}).
We have neglected the demagnetizing field because of sample
geometry. The dipolar fields from neighbouring atoms are cancelled

because of the cubic symmetry of the μ^+ location which, as mentioned above, is expected to be localized at octahedral sites according to hydrogen data. The observed shift, $(B_\mu/B_{ext} - 1)$, was -0.013 (2) %. The resultant H_{int} is obtained as -0.040 (2) %. The NMR measurements on hydrogen in pure Pd have shown a proton Knight shift of -0.012 (1) % at 343 K for the dilute limit of hydrogen concentration,[6] which corresponds to H_{int} of -0.035 (1) % after the Lorentz field correction. We can say that there is a good agreement between the hydrogen and μ^+ data, which suggests that the local magnetic structure as seen by the interstitial probes is rather independent of the mass of the probe. This result also seems to support our assumptions concerning μ^+ diffusion in Pd.

We have also measured the temperature dependences of the μ^+ Knight shift and the relaxation time in pure Pd down to 0.13 K.[8] Although the statistics were poor, the Knight shift was found to be constant within 0.1% and the relaxation time was always longer than 20 μsec.

A recent polarized neutron experiment on pure Pd[9] revealed a rather large positive spin density at the octahedral site, together with a slightly negative background between octahedral sites and Pd sites. This seems to contradict the above fact that the observed H_{int} is negative. It is also interesting to point out the difference between H_{int} in Pd and that in Ni. After normalizing to the magnetization, the effective hyperfine field X ($= 3H_{int}/4\pi M$) becomes -1.4 (1) for Pd while it is -0.31 for Ni. The Ni results appear consistent with neutron data.[10] These facts indicate that the μ^+ field is mainly determined by the spin-dependent screening mechanism of polarized conduction electrons, which is quite different for Pd and Ni.

3. EXPERIMENTAL RESULTS ON PdFe ALLOYS[8]

The polarized positive muon beam at the Lawrence Berkeley Laboratory 184-inch cyclotron was used for these experiments. The samples were cooled using a ^{3}He - ^{4}He dilution refrigerator in an external field of 1.1 kG applied along the longest axis, which is perpendicular to the μ^+ beam direction. The details of the experimental technique and arrangement were almost the same as in our previous experiments on Ni.[11] The samples used in this experiment were (1) Pd metal with 0.015 at. % Fe impurity (45 mm × 32 mm × 8 mm rectangular shape), and (2) Pd metal with 0.28 at. % Fe impurity (65 mm × 35 mm × 8 mm approximately ellipsoidal shape). The impurity concentrations in these samples have been confirmed by susceptibility measurements down to 1.25 K, in comparison with the existing data.[3] According to the susceptibility data,[3] 0.015 at. % Fe becomes anti-ferromagnetic or spin glass at around 0.4 K, though this interpretation has been questioned,[4] while 0.28 at. % Fe becomes ferromagnetic at 9.0 K.

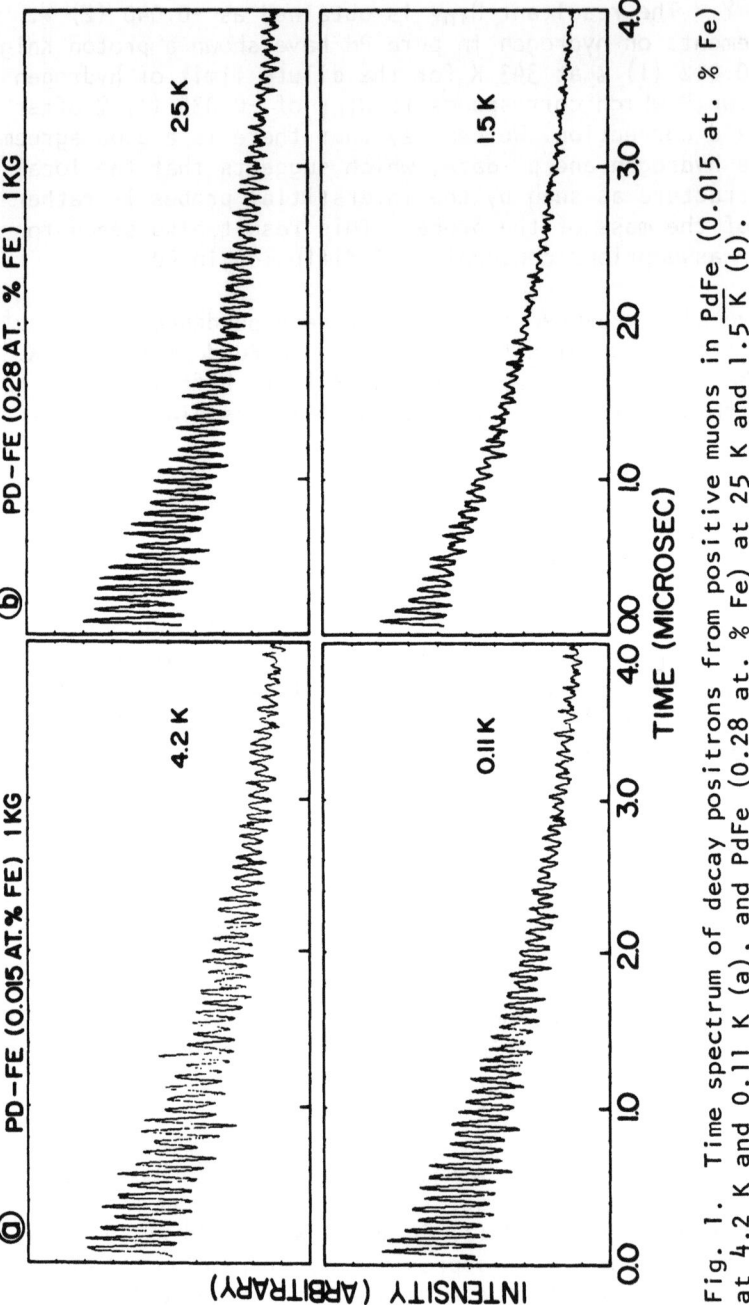

Fig. 1. Time spectrum of decay positrons from positive muons in PdFe (0.015 at. % Fe) at 4.2 K and 0.11 K (a), and PdFe (0.28 at. % Fe) at 25 K and 1.5 K (b).

μ^+ STUDIES OF DILUTE PdFe ALLOYS

The observed time spectra of decay positrons for 0.015 at. % Fe at 4.2 K and 0.11 K and for 0.28 at. % Fe at 25 K and 1.5 K are shown in Fig. 1(a) and 1(b), respectively. One can see a difference in the damping of the precession amplitude as the temperature changes through the transition temperature, in contrast to the temperature-independent relaxation in pure Pd, thus indicating that the observed damping comes from the magnetization induced by Fe impurities. These time spectra yield a local magnetic field ($B_\mu = f(kHz)/13.554$ G) at the interstitial μ^+ and a field inhomogeneity (ΔH). The results of the analysis are summarized in Table I.

Following the same procedure as in pure Pd, the H_{int} was extracted, where the dipolar fields from giant moments inside the Lorentz cavity are taken as averaged to zero because of the random distribution of Fe impurities.[12] In Fig. 2, the temperature dependence of H_{int} and ΔH for these two PdFe alloys is shown. At the lowest temperature, which is well below the ordering temperature, ΔH is almost three times larger than H_{int} for 0.28 at. % Fe while ΔH is 18 times larger than H_{int} for 0.015 at. %. By normalizing H_{int} and ΔH to $4\pi M/3$, we obtain $X = -0.54$ (14) and $\Delta X = 10$ (1) for 0.015 at. % Fe while they are -0.89 (6) and 2.7 (10) for 0.28 at. % Fe. In addition, contrary to the sharp change in ΔH

Table I. Summary of μ^+SR in PdFe alloys

Sample	T (K)	$\dfrac{(B_\mu - B_{ext})}{B_{ext}}$ (%)	$\dfrac{4\pi M/3}{B_{ext}}$ (%)	$\dfrac{H_{int}^*}{B_{ext}}$ (%)	$\dfrac{H_{int}}{4M/3}(=X)$	$\dfrac{\Delta H}{B_{ext}}$ (%)
PdFe 0.015 at. % Fe	77	-0.00 (3)	0.035	-0.03 (3)	-0.8 (8)	0.3 (1)
	4.2	-0.01 (3)	0.058	-0.05 (3)	-0.9 (5)	1.3 (3)
	0.6	+0.03 (3)	0.18	-0.11 (4)	-0.6 (2)	2.2 (3)
	0.11	+0.08 (4)	0.37	-0.20 (5)	-0.54 (14)	3.5 (2)
PdFe 0.28 at. % Fe	77	+0.05 (5)	0.082	-0.02 (5)	-0.2 (6)	0.42 (5)
	25	+0.01 (3)	0.26	-0.2 (3)	-0.8 (12)	1.5 (1)
	4.2	-0.2 (3)	6.9	-5.9 (3)	-0.86 (4)	15 (4)
	1.5	-0.5 (4)	7.0	-6.2 (4)	-0.89 (6)	19 (7)

* We have taken $D = 1.1(4)$ for 0.015 at.% Fe and $D = 0.8(2)$ for 0.28 at. % Fe.

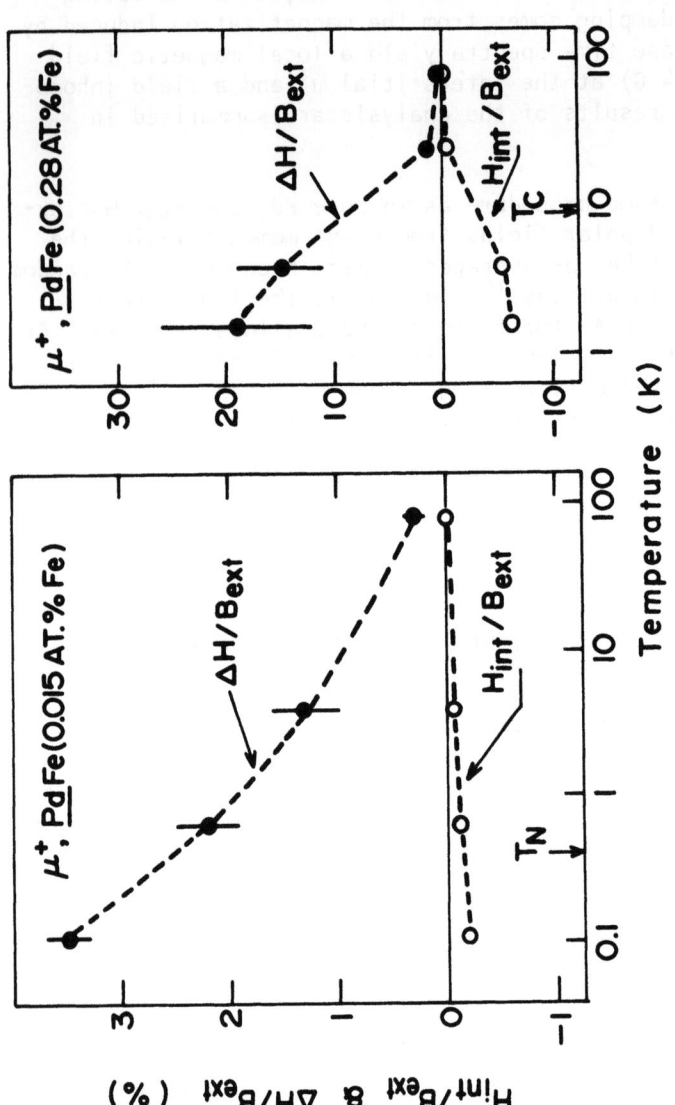

Fig. 2. Temperature dependence of the μ^+ hyperfine field (H_{int}) and the field homogeneity at the μ^+ site (ΔH) for 0.015 at.% Fe and 0.28 at.% Fe, both of which are normalized by the applied field (B_{ext}). The temperature T_N corresponds to an antiferromagnetic or spin-glass transition temperature for 0.015 at.% Fe and T_C corresponds to a ferromagnetic transition temperature for 0.28 at.% Fe, both of which are estimated from the susceptibility data.[3]

and H_{int} at around T_c for 0.28 at. % Fe, there is only a gradual change through the supposed T_N for 0.015 at. % Fe. As indicated by the susceptibility data,[3] this might be due to the applied field of 1 kG, which smeared out the sharp transition. This is similar to the cases of CuMn and AuFe.[13] As clearly shown in the knee of the magnetization curve, the giant moments in 0.015 at. % Fe are aligned almost completely along the 1 kG field at 0.1 K.

4. DISCUSSION

Herein we offer some explanations for our experimental results for PdFe alloys. In the case of 0.015 at. % Fe the average distance between the giant moments is around 40-50 A. If we assume the size of the giant moment in such a low Fe concentration alloy is the same (∼10 A) as for higher concentrations, the distance is much larger than the size of the giant moment. Therefore, most of the μ^+ stay in the region outside the giant moment. On the other hand, as the susceptibility of PdFe alloys increases linearly with the Fe concentration only up to 0.3 at. % Fe, the giant moments are just starting to overlap with each other at 0.28 at. % Fe, so that the contact fields on the μ^+ originate from the polarized d-holes inside the giant moment which is formed by the exchange enhancement effect in the d-band.[14] Superimposed on this and dominant at the region outside the giant moments, we expect a conduction electron polarization from the RKKY exchange interaction with the localized Fe moment without enhancement effect.[14] The conduction electron polarization changes rapidly with position. This spin oscillation is thus responsible for the large inhomogeneity (ΔX) in the field for 0.015 at. % Fe but it does not contribute to a net line shift, resulting in almost the same values of X for these two PdFe alloys. This spin oscillation is related to the mechanism which produces spin-glass ordering of the giant moments in the PdFe alloy with Fe concentrations below 0.1 at. %.

The magnitude of the observed field inhomogeneity can be explained qualitatively using a theory which was intended for the case of great dilution. The main source of the broadening for 0.28 at. % Fe is the dipolar field from the randomly located giant moments. The statistical theory[12] predicts almost close value to ΔH, obtained for 0.28 at. % Fe. But for 0.015 at. % Fe at the lowest temperature, this term will be one order of magnitude smaller than the observed value of 38 (2) G. The broadening due to the RKKY fields from randomly distributed Fe impurities can be estimated from the theory of Walstedt and Walker:[12] The hyperfine coupling constant between conduction electrons and μ^+ which appears in the RKKY amplitude can be calculated from the observed H_{int} for pure Pd at room temperature, corrected for the change of susceptibility. By taking J (exchange coupling strength) = 0.15 eV,[15] n/N (number of d-holes per Pd atom) = 0.36,[16], k_F = 1.25 A^{-1},[14] and

S (Fe spin) = 3.5,[17] we obtain the value which partly accounts for the discrepancy between the dipolar broadening and the observed anomalous broadening in 0.015 at. % Fe.

The static shifts X for PdFe alloys are smaller than those for the pure Pd. If we renormalize X with respect to the induced Pd moments alone (6.5 μ_B out of 10 μ_B), neglecting the contribution to M from the Fe moments at the centres of the giant moments, we find almost the same contact field per average Pd moment in all three cases (within 30%), suggesting that a part of the conduction electron polarization sensed by the μ^+ simply depends on the polarization of Pd atoms no matter whether the latter is formed by an external field or by the Fe impurities.

Based on the specific heat data, Nieuwenhuys[4] proposed a picture of giant moments with a distribution of effective g-values in Pd atoms arising dynamically. In this picture the polarized Pd cloud couples to the local moment on the Fe atom with a characteristic time-constant. The μ^+ may have a possibility to probe the polarization cloud and solute Fe differently.[18] We note that the μ^+ senses a time-period of dynamical behaviour which is somewhat between the range for neutrons and susceptibility measurements. Detailed analysis is required.

5. SUMMARY

Dilute PdFe alloys above and below the critical concentration of 0.1 at. % have been studied by the μ^+SR method. Below the ordering temperature the observed shifts, when normalized to the bulk magnetization, are almost the same. However, the normalized broadening for the spin-glass alloy is substantially larger than for the ferromagnetic alloy. To date, two experimental results have indicated a possible spin-glass ordering below the critical concentration: 1) the magnetic susceptibility deviates from the Curie-Weiss law;[3] 2) below the ordering temperature, the specific heat does not depend very much on the concentration.[17] We think the μ^+ results are further support of spin-glass ordering below the critical concentration.

ACKNOWLEDGEMENTS

We acknowledge the collaboration and contributions of S. Nagamiya and O. Hashimoto at LBL and R.S. Hayano, D.M. Garner, A. Duncan, J.H. Brewer and D.G. Fleming at TRIUMF. We are grateful for the fruitful discussions held with A.S. Arrott, I.A. Campbell and S.I. Kobayashi.

REFERENCES

* Supported by Japan Society for the Promotion of Science, the Toray Science Foundation, the Japanese Ministry of Education, the National Science Foundation, the U.S. Energy Research and Development Administration and the National Research Council of Canada.

1. J. Crangle and W.R. Scott, J. Appl. Phys. **36**, 921 (1965).
2. G.G. Low and T.M. Holden, Proc. Phys. Soc. (London) **89**, 119 (1966).
3. G. Chouteau and R. Tournier, Jour. de Physique **C1**, 1002 (1971).
4. G.J. Nieuwenhuys, Advan. Phys. **24**, 515 (1976).
5. J. Völkl and G. Alefeld, in *Diffusion in Solids*, edited by A.S. Nowick and J.J. Burton (Academic Press, New York, 1975), chap. V.
6. P. Brill and J. Voitländer, Ber. Bunsenges. Physik. Chem. **77**, 1097 (1973).
7. D.P. Hutchinson, J. Menes, G. Shapiro and A.M. Patlach, Phys. Rev. **131**, 1351 (1963).
8. K. Nagamine, N. Nishida, S. Nagamiya, O. Hashimoto and T. Yamazaki, to be published.
9. J.W. Cable, E.O. Wollan, G.P. Felcher, T.O. Brun and S.P. Hornfeldt, Phys. Rev. Lett. **34**, 278 (1975).
10. B.D. Patterson and L.M. Falicov, Solid State Commun. **15**, 1509 (1974).
11. K. Nagamine, S. Nagamiya, O. Hashimoto, N. Nishida, T. Yamazaki and B.D. Patterson, to be published in Hyperfine Interactions.
12. R.E. Walstedt and L.R. Walker, Phys. Rev. **B9**, 4857 (1974).
13. D.E. Murnick, A.T. Fiory and W.J. Kossler, Phys. Rev. Lett. **36**, 100 (1976).
14. T. Moriya, Progr. Theor. Phys. **34**, 329 (1965).
15. S. Doniach and E.P. Wohlfarth, Proc. Roy. Soc. (London) **A296**, 442 (1967).
16. J.J. Vuillemin and M.G. Priestley, Phys. Rev. Lett. **14**, 307 (1965).
17. G. Chouteau, R. Fourneaux and R. Tournier, in Proceedings of 12th Int. Conf. on Low Temp. Phys., Kyoto, Japan (Academic Press of Japan, 1970), 769.
18. We acknowledge Dr. B.H. Verbeek for his suggestion.

SPIN WAVES IN HEISENBERG SPIN GLASSES[*,†]

D. L. Huber[‡]

Sandia Laboratories
Albuquerque, NM 87115

W. Y. Ching

Dept. of Physics
U. of Wis.
Madison, Wis. 53706

ABSTRACT

The existence of spin waves in Heisenberg spin glasses is discussed. It is suggested that propagating modes will contribute to the dynamic susceptibility at low temperatures. At long wavelengths, the frequency is proportional to the wave vector. An evaluation of the spin wave velocity for a classical spin glass with short range interactions is carried out using equation-of-motion methods and compared with the estimate obtained from the second moment of the lineshape function. A general expression for the frequency is derived and is analyzed for the RKKY interaction.

---

[*]Work supported by the U.S. Energy Research and Development Administration.

[†] Supported in part by the National Science Foundation.

[‡]Summer visitor. Permanent address: Dept. of Physics, Univ. of Wisconsin, Madison, WI 53706

I. INTRODUCTION

Recently attention has been drawn to the properties of so-called spin glasses, which are random dilute magnetic alloys with magnetic impurity concentrations on the order of 0.1 to 10 percent. A mean field treatment of a classical Heisenberg spin glass has been discussed by Edwards and Anderson[1] and the corresponding quantum mechanical[2] and Ising[3] calculations have been reported. The phase transition in the Edwards-Anderson model has also been studied using Monte Carlo techniques.[4,5]

The studies mentioned have emphasized the static properties of the spin glass. In this paper, we discuss the possibility of there being spin waves in Heisenberg spin glass systems at low temperatures. Our prediction for the existence of long wavelength, weakly damped spin wave modes is suggested by the similarity to the dynamics of the antiferro-magnetic Heisenberg chain. Despite the absence of long range magnetic order in the linear system, there are well-defined spin waves at low temperatures. At long wavelengths, the modes are insensitive to the details of incipient antiferromagnetic ordering. Their frequency is linear in the wave vector k with a velocity which is weakly temperature dependent.

In the case of Heisenberg spin glasses, we expect similar behavior at small k. Namely, the dynamics of the spin fluctuations should reflect only the conservation of total spin and the average stiffness and susceptibility of the magnetic array. Hence, it is likely that there will be well-defined modes in the dynamic susceptibility $\chi(k,\omega)$.[6] It should be emphasized that these arguments do not hold for arbitrarily small k at finite temperatures, since we expect diffusive behavior in the k → 0 limit.[7] Furthermore, as k increases, the spin wave peaks in $\chi(k,\omega)$ will broaden because of the increase in the bandwidth of the modes contributing to a particular Fourier component of the response.

II. THEORY

An approximate expression for the spin wave velocity can be obtained from the equation of motion for the spin operator $\vec{S}(\vec{k})$ defined by

$$\vec{S}(\vec{k}) = \sum_n e^{i\vec{k} \cdot \vec{r}_n} \vec{S}_n , \qquad (1)$$

where the sum is over all the sites in the system. By using a symmetric decoupling procedure analogous to that developed for one-dimensional magnets[8] we obtain the result

$$\frac{d^2 \vec{S}(\vec{k})}{dt^2} = -V^2 k^2 \vec{S}(\vec{k}) , \qquad (2)$$

in the small k limit. Here, the spin wave velocity, V, is given by

$$\hbar^2 V^2 = \frac{1}{9} \langle \sum_{m,b} J_{jm} J_{jb} \langle \vec{S}_b \cdot \vec{S}_m \rangle r_{mj}^2 \rangle_c$$

$$+ \frac{1}{18} \langle \sum_{m,b} J_{mb} (J_{jm} - J_{jb}) \langle \vec{S}_b \cdot \vec{S}_m \rangle \cdot$$

$$(r_{mj}^2 - r_{bj}^2) \rangle_c , \qquad (3)$$

where $r_{ij}^2 = (\vec{r}_i - \vec{r}_j)^2$ and the J_{ij} are the exchange integrals in the Heisenberg interaction

$$\mathcal{H} = -\frac{1}{2} \sum_{i,j} J_{ij} \vec{S}_i \cdot \vec{S}_j . \qquad (4)$$

The symbol $\langle \cdots \rangle$ denotes a thermal average while $\langle \cdots \rangle_c$ denotes a configurational average. Using Monte Carlo techniques, we have evaluated the various correlation functions appearing in Eq. (3) for a classical spin glass with nearest neighbor interactions. It was found that the dominant contribution (>95 percent) comes from the terms with m = b. Assuming this to hold in general we have

$$V^2 \approx \frac{S(S+1)}{9\hbar^2} \langle \sum_m J_{jm}^2 r_{mj}^2 \rangle_c . \qquad (5)$$

An alternative estimate of the velocity is obtained from $\langle \omega_k^2 \rangle$, the second moment of the normalized lineshape function $\chi''(k,\omega)/\pi\omega\chi(k)$. According to Marshall and Lowde[9] we have the exact expression

$$\langle \omega_k^2 \rangle = \frac{2g^2 \mu_B^2}{3N\chi(k)\hbar^2} \sum_{m,n} J_{mn} (1 - \cos(\vec{k} \cdot \vec{r}_{mn})) \langle \vec{S}_m \cdot \vec{S}_n \rangle , \qquad (6)$$

where g is the electronic g-factor, μ_B is the Bohr magneton, $\chi(k)$ is the static wave vector dependent susceptibility, and N is the number of spins. Assuming the spin wave contribution dominates the dynamic susceptibility and that the modes have a width which is small in comparison with their frequency we have

$$v^2 = \lim_{k \to 0} \langle \omega_k^2 \rangle / k^2 ,$$

$$= \frac{g^2 \mu_B^2}{9N\chi(0)} \sum_{m,n} r_{mn}^2 J_{mn} \langle \vec{S}_m \cdot \vec{S}_n \rangle . \quad (7)$$

In order to compare the two approaches, we make use of Monte Carlo estimates of $\sum r_{mn}^2 J_{mn} \langle \vec{S}_m \cdot \vec{S}_n \rangle$ and $\chi(0)$ for the classical spin glass with nearest neighbor interactions. When the exchange integral has a Gaussian distribution with rms width ΔJ, i.e.,

$$\text{prob}(J_{ij}) \propto \exp[-J_{ij}^2/2(\Delta J)^2] , \quad (8)$$

and the spins are of unit magnitude we obtain the result[4]

$$\chi(0) = (0.45 \pm 0.05) g^2 \mu_B^2 / \Delta J, \quad (9)$$

in the zero temperature limit. For nearest neighbor interactions, the sum in (7) is equal to $-2a^2 U_o(T)$, where a is the lattice constant and $U_o(T)$ is the internal energy. From the Monte Carlo analysis, we find

$$U_o(0) = -(1.84 \pm 0.01) N \Delta J . \quad (10)$$

Using (9) and (10), we obtain the value

$$v^2 = (0.9 \pm 0.1)(a \Delta J/\hbar)^2 . \quad (11)$$

which compares favorably with the classical limit of Eq. (5):

$$v^2 = (2/3)(a \Delta J/\hbar)^2 . \quad (12)$$

That the two methods give approximately the same values for V^2, we regard as evidence in support of our approach for classical spin glass systems with short range interactions. We also expect our analysis to apply at least qualitatively to quantum systems with isotropic long range interactions, as for example Rudermann-Kittel-Kasuya-Yosida (RKKY) exchange where

$$J_{ij}(r_{ij}) = A[\sin(2k_o r_{ij}) - (2k_o r_{ij})\cos(2k_o r_{ij})]/(2k_o r_{ij})^4 \quad . \tag{13}$$

In this case, the symmetric decoupling procedure leads to an expression for the spin wave frequency, $\omega(k)$, of the form

$$\hbar^2 \omega(k)^2 = \frac{2}{3} S(S+1) \langle \sum_m J_{jm}^2 (1-\cos(\vec{k}\cdot\vec{r}_{jm}))\rangle_c . \tag{14}$$

With J_{ij} given by Eq. (13), we obtain the limiting behavior

$$\omega(k)^2 = \frac{2\pi^2 S(S+1) \rho A^2 k^2}{27\hbar^2 (2k_o)^5} \quad , \tag{15}$$

for $k \ll k_o$, ρ being the number of spins per unit volume. In the opposite limit, $k \gg k_o$, $\omega(k)$ approaches a constant value given by

$$\omega(k)^2 = \frac{8\pi^2 S(S+1) \rho A^2}{27\hbar^2 (2k_o)^3} \quad , \tag{16}$$

although in this regime the modes are likely to be overdamped.

III. DISCUSSION

The frequency and damping of the spin waves are both functions of temperature. The latter is difficult to calculate. However, it is plausible that the modes are overdamped before the spin glass transition is reached. On the basis of Eqs. (6)-(7), we expect that the renormalization of the frequency will vary approximately as the square root of the ratio of internal energy to susceptibility. In connection with this, it should be mentioned that, were the susceptibility to follow the Curie law ($\chi(0) \sim 1/T$) down to T = 0, there would be no dynamic response at zero temperature since all the moments of the lineshape function would be zero. The finite susceptibility at T = 0, which is a property of the spin glass phase,[11] gives rise to finite spin wave velocities. Also, we should emphasize the distinction between the spin wave modes discussed here and other types of excitations which are present in spin glass systems.[6] In addition to the long wavelength spin wave-spin diffusion modes, which broadly speaking are the analog of shear waves in real glasses, there are also local excitations between quasi-degenerate configurations of the spin glass system.

ACKNOWLEDGEMENT

The first author would like to thank Dr. P. M. Richards and the other members of the Staff of Sandia Laboratories for their hospitality during the period when this paper was being written.

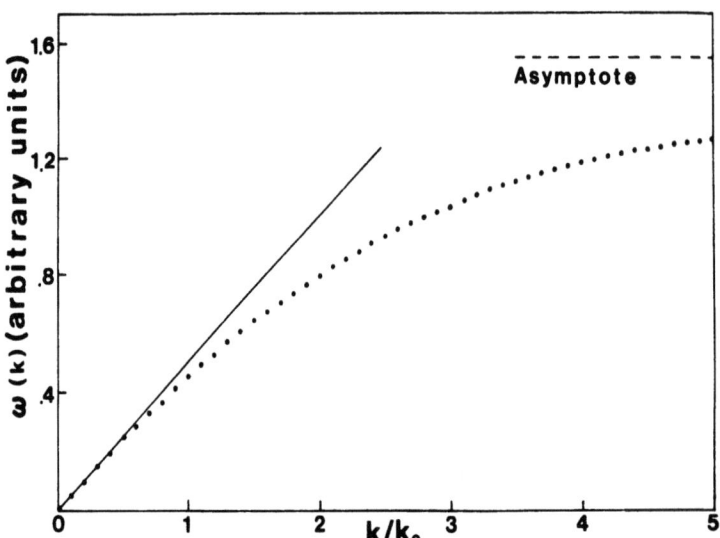

Spin wave dispersion curve, RKKY interaction (Eq.(13)). Dotted line, $\omega(k)$; solid line, linear approximation; broken line, asymptote.

REFERENCES

1. S. F. Edwards and P. W. Anderson, J. Phys. F: Metal Phys. 5, 965 (1975).

2. K. H. Fischer, Phys. Rev. Lett 34, 1438 (1975); Solid State Comm. 18, 1515 (1976).

3. D. Sherrington and S. Kirkpatrick, Phys. Rev. Lett. 35, 1792 (1975).

4. D. L. Huber and W. Y. Ching, AIP Conf. Proc. (to be published).

5. K. Binder and K. Schröder, Solid State Comm. 18, 1361 (1976). K. Binder and D. Stauffer, Phys. Lett A 57, 177 (1976).

6. P. W. Anderson, Amorphous Magnetism, ed. by H. O. Hooper and A. M. de Graaf (New York, Plenum, 1973), p. 1.

7. N. Rivier, Wiss. Z. Techn, Univers. Dresden 23, 1000 (1974).

8. P. M. Richards, Phys. Rev. Lett. 27, 1800 (1971). Strictly speaking Eq. (3) is obtained from a second order decoupling of the equation of motion for the retarded Green's function.

9. W. Marshall and R. D. Lowde, Rept. Progr, Phys. 31, 705 (1968).

10. Our results for $\omega(k)$ for the RKKY interaction differ from those of Ref. 7 in that we obtain a linear dispersion at small k whereas Rivier finds $\omega(k) \sim k^2$. Both theories give $\omega(k) \sim$ const. at large k. However, in Ref. 7, a limiting value for $\omega(k)$ on the order of $\pi \rho AS/2\hbar k_o^3$ is inferred.

11. C. Domb, J. Phys. A: Math. Gen., 9, L17 (1976).

DISCUSSION

M. W. Klein: Have you calculated the contribution to the specific heat from the long wavelength propagating modes associated with the Ruderman-Kittel interaction?

D. L. Huber: No, we have not, our feelings is that to use this picture and calculate the thermodynamic properties may not be applicable. The reason is that you would end up looking at the contribution for those modes whose frequencies are of the order of kT. I suspect these modes are strongly damped by that region and any kind of simple calculation like this may not give the right answer.

TEMPERATURE-DEPENDENT SKEW SCATTERING IN AN AuMn SPIN GLASS

C.M. Hurd and S.P. McAlister

National Research Council of Canada

Ottawa, Ont., K1A 0R9

The Hall effect seen at low temperatures in dilute alloys of a transition metal dissolved in a group-1B noble metal shows[1,2] a large component that clearly does not arise from the classical Lorentz force. This extraordinary component is found generally to increase with decreasing temperature, to be approximately proportional to the solute's concentration, and to show a saturation when the applied field strength B is increased sufficiently.[1] For all but the most dilute alloys,[1] this component far outweighs the classical contribution arising from the Lorentz force. This extraordinary component has been called[3-5] the "skew component" of the Hall effect, and its origin has been attributed[3-6] to the asymmetric (or skew) scattering of the conduction electrons as a result of spin-orbit coupling during their scattering by the magnetic ion.[6] We have studied the temperature dependence of this component in the Hall resistivity ρ_H through the spin glass transition in an Au + 8.1 at.% Mn alloy in applied fields in the range 0.025 - 0.1T. The results show features that are related tentatively to the behaviour of magnetic clusters and 'loose spins'.

As the applied field strength is increased from zero, the skew component, which appears as a nonlinear variation of $\rho_H(B)$ that is superimposed upon the host's linear Lorentz contribution, rapidly becomes the dominant contributor to the total ρ_H (Fig. 1). In the range of fields of interest (<0.1T), it is at least an order of magnitude larger than the Lorentz contribution so it is convenient and adequate in the following qualitative treatment to discuss the behaviour of the total ρ_H, and to avoid the problem of how the minor Lorentz contribution should be subtracted.

The skew component arises from the coupling between the

localised moment's spin and the conduction electron's angular momentum during the electron's temporary residence in the ion's virtual state.[3-8] The result is an elastic scattering event that is asymmetric (or skew) with respect to the plane containing the ion's moment and the electron's incident velocity.[8-10] (This idea was developed from one put forward previously[11,12] to account for an anomalous Hall effect in ferromagnetic metals.) Note that even though the effective Bohr magneton value of Mn in Au ($\sim$5.4) is larger than that of Fe in Au ($\sim$3.7), the skew component per at.% is found[13] to be largest for Fe (Fig. 1). This arises from the stronger resonant coupling between the Fe virtual levels and the conduction electrons; unlike Mn in Au, Fe has a component of its virtual level lying very close in energy to the alloys's Fermi level.[14,15] This stronger resonant coupling is also reflected in the resistivity per at.%, which for Fe in Au is about 3 times larger than that of Mn in Au. In some alloy systems, such as AgMn or CuMn, the resonant coupling is so weak that little or no skew component of ρ_H can be detected.[13]

The underlying purpose in studying the skew component of ρ_H is that since it arises from scattering by the localized moments in the alloy, it should be a useful probe of internal magnetic rearrangement , such as that experienced at the spin glass transition. To test this idea, we have measured the isomagnetic temperature dependence (in fields ranging from 0.025 to 0.100T) of the skew component of ρ_H for the spin glass alloy: Au + 8.10 at.% Mn (Fig. 2). The polycrystalline samples were prepared and characterized as described previously.[16] Prior to measurement each sample was annealed in vacuum at 900°C for 24h before rapid quenching into iced brine, and was maintained at 77 K until measured. A standard cryostat-superconducting magnet combination and dc potentiometric arrangement was used in which, it is important to note, the sample is flipped through 180° about its transverse axis so that without the need to reverse the magnet current the transverse voltage even in B can be determined from the four permutations of applied electric and magnetic field directions. Thus although we use an entirely dc method, flipping the sample means that below the spin glass temperature (T_{SG}) our results — like those of ac susceptibility measurements[17]— relate to the nonequilibrium state.

Figure 2 shows the isomagnetic temperature dependence of ρ_H for each alloy in fields low enough to avoid the complete disruption of the spin glass state.[13] (For graphical convenience the ordinate in each case is $-\rho_H/B$.) Apart from a striking qualitative agreement with the behaviour of the low-field magnetic susceptibility[18] and magnetisation,[19,20] including a clearly defined T_{SG} transition at about 24 K and a flattening of $\rho_H(T)$ in the stronger fields, these results show other notable features. Firstly, a "shoulder" is evident in $\rho_H(T)$ in the temperature range below T_{SG}. A similar feature, which here occurs at a field-dependent temperature that

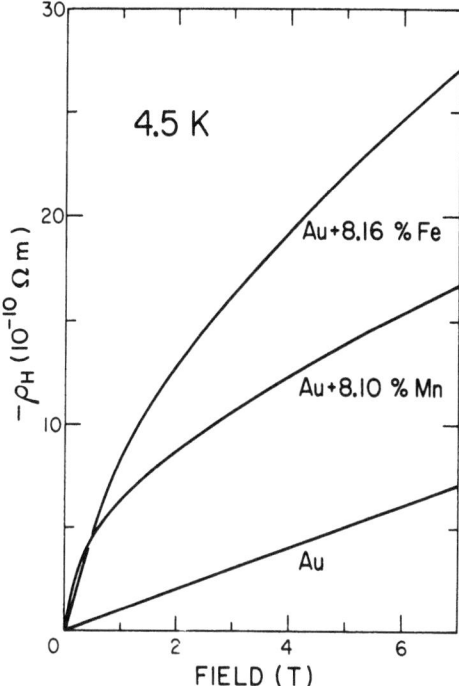

Figure 1: This compares the field dependences of the total Hall resistivity seen in two alloys with that of the pure solvent. The nonlinear field-dependent component in the alloys' results is the anomalous part that arises from the asymmetric scattering.

shifts to lower values as the field increases, has been seen in the magnetic susceptibility[21] of AuFe and in the temperature derivative of the electrical resistivity[22] of a comparable AuMn alloy. In attempting to specify the origin of some of the features seen in Fig. 2, we think the following points are important.

In an ideal substitutional fcc alloy containing 8.1 at.% solute, only 36% of the solute ions have no nearest neighbour of the same kind; 21% form isolated pairs and the rest are in groups of 3 or more. Hence the moments of at least 64% of the solute ions in the alloys of Figs. 1 and 2 are coupled by the d-d contact exchange interaction to form pairs or larger clusters. The classical

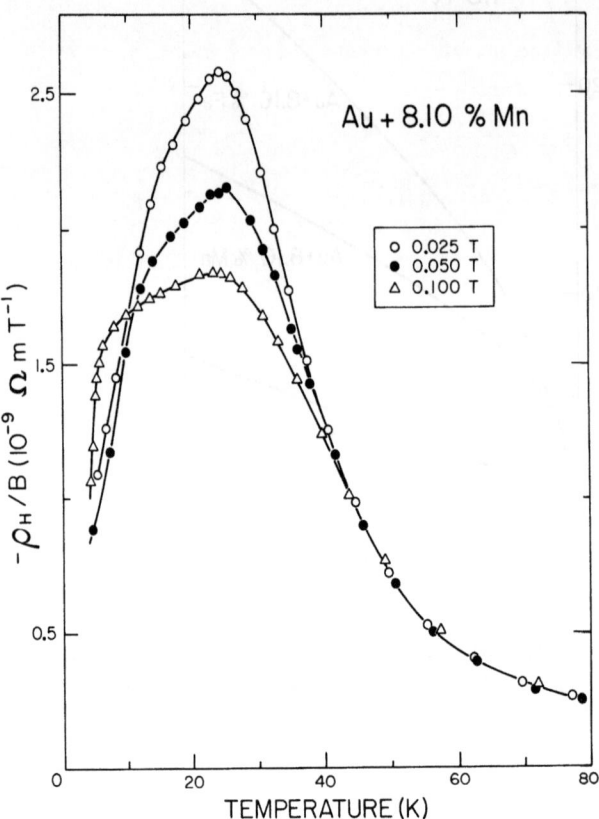

Figure 2: This shows the temperature dependence of the total Hall resistivity (divided by the applied field strength) observed for the AuMn alloy in the fields indicated. The measurements were made with increasing temperature starting from the zero-field-cooled state at ~4.2 K. At the lowest temperatures and field strengths $g\mu_B B/k_B T$ does not exceed ~0.04, so that throughout the experiment the thermal energy $k_B T$ is the major perturber of the spin glass state; the role of the field B is primarily to make the skew scattering effects evident on a macroscopic scale.

picture[23] of an ideal spin glass as a collection of randomly distributed *isolated* moments "frozen" by their RKKY interactions must be inappropriate here, and it is better to consider these alloys in terms of a cluster glass[24] or mictomagnet[20] picture.

A moment localised at a site where the internal field is less than the thermal energy — either through fortuitous cancellation of the RKKY or *d-d* contributions, or because the site is sufficiently isolated from its own kind (the RKKY coupling falls off rapidly as the electrons' average mean free path is reduced[24,25]) — is unlocked from the spin glass matrix at the ambient temperature and is known[24] as a "loose spin"; it is free to align in a weak external field. Thus for any combination of field and temperature that is not completely disruptive, it seems appropriate to regard the state as a mixture of two components:[20,24,25] a cooperatively "frozen" medium of moments and a fraction that are free to reorient their alignments.

Unlike the temperature dependence of the magnetic susceptibility,[17,18] the ordinate in Fig. 2 represents more than just the net alignment of the moments in the field, for it also includes the degree of asymmetry produced per scattering centre. An increase in $|\rho_H|$ could therefore reflect either an increase in the total alignment of the moments or a stronger asymmetric scattering effect per aligned moment. Consequently, if different asymmetric scattering centres dominate different temperature ranges, this will be evident in $\rho_H(T)$, and this may be the origin of the shoulder in $\rho_H(\vec{B},T)$. We suggest that this shoulder separates ranges that are dominated by one type of asymmetric scattering centre — perhaps loose spins in the range below the shoulder and larger clusters in that above. Whatever the cause, the net result is that the skew component of ρ_H becomes less temperature dependent above the shoulder, and this tendency becomes more pronounced with increasing field strength until eventually ρ_H is independent of temperature in the range between T_{SG} and the shoulder. The implication of our suggestion is that the asymmetric scattering cross-section per aligned moment is smaller for a moment that is part of a *d-d* coupled cluster than for one that is isolated. This could represent either the reduction by interference effects of the total scattering cross-section of clusters, or the fact, known from other measurements[19,20], that even with ferromagnetic *d-d* coupling quite a large fraction of the spins forming a cluster are locked antiparallel to those of the main body. Any deviation from strict alignment in a cluster will reduce its effectiveness as an asymmetric scatterer.

The variation with temperature of the magnetisation just above T_{SG} has been attributed to the breaking up of the clusters as the temperature is increased.[20] In Fig. 2 there is supporting evidence for this from the persistence, up to at least 42 K, of a field-dependent component in $\rho_H(T)$. Above this temperature $\rho_H(T)$ is independent of the field, suggesting that the magnetic clusters have all been thermally broken up at this temperature leaving single Curie-like moments to dominate the asymmetric scattering. Further work is in progress over a wider concentration range which we hope will provide more insight into the electron scattering processes in the spin glass state.

We are grateful to G.F. Turner for his careful sample preparation, and to the Analytical Section of the National Research Council of Canada for work done on our behalf.

REFERENCES

1. J.E.A. Alderson and C.M. Hurd, J. Phys. Chem. Solids 32, 2075 (1971).
2. P. Monod and A. Friederich, *Proc. 12th Int. Conf. Low Temp. Phys. Kyoto*, Japan. (1970) p. 755.
3. A. Fert and O. Jaoul, Phys. Rev. Lett. 28 303 (1972).
4. A. Fert and O. Jaoul, Solid State Commun. 11, 759 (1972).
5. A. Fert and A. Friederich, Phys. Rev. B 13, 397 (1976).
6. C.M. Hurd, Contemp. Phys. 16, 517 (1975).
7. Communications by A. Fert, P. Monod, B. Giovannini and C.M. Hurd in *Proceedings of the Saint-Cergue Meeting on the Extraordinary Hall Effect*, edited by G. Cohen, B. Giovannini and D. Sorg (Université de Genève, Genève, 1973).
8. A. Fert, J. Phys. F 3, 2126 (1973).
9. C.M. Hurd and S.P. McAlister, unpublished work.
10. C.M. Hurd, *Electrons in Metals: An Introduction to Modern Topics* (Wiley-Interscience, New York, 1975), p. 128.
11. J. Smit, Physica 24, 39 (1958).
12. J.M. Luttinger, Phys. Rev. 112, 739 (1958).
13. S.P. McAlister and C.M. Hurd, Solid State Commun. (In press), and unpublished work.
14. J. Friedel, Nuovo Cimento. Suppl. 7, 287 (1958).
15. M.R. Steel and P.M. Treherne, J. Phys. F 2, 199 (1972).
16. J.E.A. Alderson, T. Farrell and C.M. Hurd, Phys. Rev. 174, 729 (1968).
17. C.N. Guy, J. Phys. F 5, L 242 (1975).
18. V. Canella and J.A. Mydosh, Phys. Rev. B 6, 4220 (1972).
19. A.P. Murani, J. Phys. F 4, 757 (1974).
20. P.A. Beck, J. Less Common Metals 28, 193 (1972).
21. Unpublished work by V. Canella kindly brought to our attention by J.A. Mydosh.
22. P.J. Ford and J.A. Mydosh, Phys. Rev. B (In press).
23. J.A. Mydosh, Magnetism and Magnetic Materials — 1974, AIP Conf. Proc. No. 24 (1975) p. 131.
24. D.A. Smith, J. Phys. F 5, 2148 (1975).
25. D. Mattis and W.E. Donath, Phys. Rev. 128, 1618 (1962).

DISCUSSION

R. Tournier: Do you observe any hysteresis effects in your samples?

C. M. Hurd: Yes, as I indicate here in the slide representing the case of a Au-Fe sample.

SLOWLY RELAXING REMANENCE AND METASTABILITY OF SPIN GLASSES

N. Rivier

The Blackett Laboratory
Imperial College
London, SW7, UK

Abstract   A microscopic model for the slow relaxation of a spin or cluster of spins in spin glasses is presented. It is shown to be related to Néel's phenomenological theory of magnetic viscosity. The spin dynamics is governed at zero temperature by a Hamiltonian identical to Néel's energy. Néel's anisotropy has a microscopic origin in spin glasses, analogous to the attractive electron-electron interaction in small polarons.

1.   Introduction

Spin glasses exhibit the slowly relaxing magnetization and the various magnetic remanences (isothermal and thermoremanent)[1,2] which are familiar features of rock magnetism and ferromagnetic precipitates. The theory of these phenomena of magnetic viscosity is centered on the work of Néel[3]. This phenomenological theory is based on the existence of an anisotropy energy for the fine magnetic grains which constitute the rock or the lava. The anisotropy, together with the volume of the grain, enter exponentially in the expression for the relaxation rate of the grain, and is therefore of overwhelming importance in the long-time behaviour of the rock off equilibrium.

Although Néel's theory is reasonably successful in explaining the magnetic properties of spin glasses[1] (with the exception of the sharp cusp in the susceptibility), it is difficult to understand the origin of the anisotropy coefficient. Spin glasses are regarded as uniform distributions of magnetic impurities in hosts where crystalline anisotropy is usually negligible. Shape anisotropy is not very likely in spin glasses where the magnetic constituents are the impurity spins themselves. The anisotropy due to mechanical stress should vary markedly from host to host whereas spin glass behaviour does not. Is it possible to obtain an effective anisotropy, and magnetic viscosity from the interaction of the magnetic constituents of the spin glass alone?

This paper is an attempt, 1) to establish a microscopic origin for the anisotropy in spin glasses, 2) to show that a microscopic Hamiltonian is capable of giving rise to magnetic viscosity and slow magnetization creep. The results are encouraging, although the precise link with conventional, equilibrium spin glass theories remains to be worked out.

2. Microscopic origin of the anisotropy

The spin glass is described locally by a magnetic entity S (individual spin, region of short range magnetic order or cluster [4]) interacting with all the other excitation modes a_k^+ of the alloy, which we visualise here as the magnetic diffusion mode encompassing all the magnetic degree of freedom of the spin glass apart from S.* In rock magnetism, S labels one magnetic grain and a_k^+ the phonons (in the case of magnetostrictive interaction) or the weaker magnetic excitations generated by dipolar interactions between grains [5]. The essential feature is the separation of the local structure S from the background of outside excitations. This separation is either microscopic (if S is an individual spin) or the result of some renormalization if S is a magnetic cluster. The outside modes have a density of states $\rho(\epsilon)$. The interaction between S and the outside modes is simply a displacement of the latter, and the Hamiltonian reads ($\hbar = 1$)

$$H = \sum \omega_k (a_k^+ a_k + \tfrac{1}{2}) + \sum g_k (a_k^+ + a_k) S_z - \underline{\Delta} \cdot \underline{S} \qquad (1)$$

The interaction term includes a form factor g_k and is thus dependent on the size of the entity S. In a mean field theory, S is polarised by a random internal field $-\sum g_k \langle a_k^+ + a_k \rangle$ in addition to the external magnetic field $\underline{\Delta}$.

A canonical transformation $\tilde{H}$ = exp (iL) H exp (-iL) with $L = -i S_z \sum \mu_k (a_k^+ - a_k)$ and $\mu_k = g_k/\omega_k$ gets rid of the interaction term and yields the simpler Hamiltonian,

$$\tilde{H} = \sum \omega_k (a_k^+ a_k + \tfrac{1}{2}) - \tfrac{1}{2}\{(\Delta_x - i\Delta_y) S^+ \exp[\sum \mu_k (a_k^+ - a_k)] + h.c.\} - \Delta_z S_z - K S_z^2 \qquad (2)$$

at the cost of an uniaxial anisotropy energy, $-KS_z^2$ with $K = \sum (g_k^2/\omega_k)$. This anisotropy is the analogue of the attractive electron-electron interaction occurring in ordinary polaron theory [6,7]. Coherent shifts of the modes accompany each spin flip in (2) as they do in polaronic motion. Although it may seem absurd to treat the spin of a cluster quantum mechanically, it does carry with it excitations which are quantum mechanical at low temperatures ($\beta \omega_k \gg 1$).

In the absence of a transverse magnetic field $\Delta_\perp = 0$, we have simply

$$\tilde{H} = \sum \omega_k (a_k^+ a_k + \tfrac{1}{2}) - \Delta_z S_z - K S_z^2 \qquad (3)$$

S_z is then a constant of motion. However, if it is taken as a classical entity and made to fluctuate, the fluctuation-dissipation theorem gives a simple method to evaluate the power dissipated by the magnetic cluster into the bath of outside modes, as measured, for example, by the calorimetric experiments of Nieuwenhuys and Mydosh [8]. Let therefore $S_z(t)$ be a classical quantity in the Hamiltonian (1). The power dissipated is given by

$$\dot{Q} = -\langle (\partial/\partial t) H \rangle = -(d/dt) S_z(t) \sum g_k \langle a_k^+ + a_k \rangle(t) \qquad (4)$$

Linear response theory expresses the power dissipated averaged over one cycle of the fluctuating $S_z(t)$, $\langle \dot{Q} \rangle_A$, in terms of the response function

$$\chi_k = \iota \langle [(a_k^+ + a_k)(t), (a_k^+ + a_k)(0)]_- \rangle \theta(t)$$

of the modes, as

$$\langle \dot{Q} \rangle_A (t) = (2\pi)^{-1} \int d\omega \exp(-i\omega t) \tfrac{1}{2} \langle |S_z^2|(\omega) \rangle \omega \sum g_k^2 \, \text{Im} \chi_k (\omega) \qquad (5)$$

where $\langle |S_z^2|(\omega) \rangle$ is the power spectrum of the fluctuations. Since, for an harmonic oscillator of force constant $\varkappa_k$, mass $m_k \simeq m$

$$\text{Im} \chi_k (\omega) = (\pi \omega_k / \varkappa_k) [\delta(\omega - \omega_k) - \delta(\omega + \omega_k)]$$

we obtain

$$\langle \dot{Q} \rangle_A (t) \simeq \langle |S_z^2| \rangle m \int d\epsilon \, \rho(\epsilon) g^2(\epsilon) \cos(\epsilon t) \qquad (6)$$

It is a Fourier transform whose asymptotic behaviour depends on the singularities of the integrand, i.e. of the mode density $\rho(\epsilon)$ [9].
A simple Debye band with cut off at $\epsilon_c = k\theta_D$ yields $\langle \dot{Q} \rangle_A (t) \sim t^{-1}$ times a function oscillating at the cut off frequency, i.e. too fast to be observed directly experimentally. The envelope t^{-1} is observable. If the outside modes are eigenstates of a random matrix, as they are expected to be in spin glasses, their eigenvalue density follows the semi-circular law [10]

$$\rho(\epsilon) = [\tfrac{1}{4}\epsilon_c^2 - (\epsilon - \tfrac{1}{2}\epsilon_c)^2]^{\tfrac{1}{2}} / \tfrac{1}{4}\pi \epsilon_c^2$$

and the envelope of $\langle \dot{Q} \rangle(t)$ goes asymptotically as $t^{-3/2}$. Experimentally, the power dissipated goes as t^{-1} [8], but the present theory is too crude to warrant a detailed discussion of the nature of the excitation modes. The purpose of this discussion was only to show that slow relaxation mechanisms can be obtained from a physically reasonable model by simple arguments. It is not easy to calculate the rate of change of the magnetization by this method.

3. **Dynamics**

The magnetization dynamics is described by the correlation function $K^{\alpha\beta}(t) = \langle S^\alpha(t) S^\beta(0) \rangle$, related to the response function by $\text{Im } \chi^{\alpha\beta}(\omega) = \chi''^{\alpha\beta}(\omega)$ and $\chi''^{\alpha\beta}(t) = \frac{1}{2}[K^{\alpha\beta}(t) - K^{\alpha\beta}(t - i\beta)]$. It is convenient to use an imaginary time representation for K, with analytic continuation $\tau = it$ left to the end of the calculation.

After canonical transformation, the Hamiltonian is given as in (2), and $S^\pm$ becomes $\tilde{S}^\pm = S^\pm \exp[\pm \sum \mu_k (a_k^+ - a_k)]$ while S_z is unaffected. Let $\Delta_y = 0$. Interaction representation is then used with the spin flip terms as the interaction energy, and the correlation function can be written, like the partition function Z, as a series of spin flips in imaginary time ''

$$K^{+-} = Z^{-1} \sum_{n,m=0}^{\infty} (\tfrac{1}{2}\Delta_x)^{n+m} \int_{(\beta,\tau)} d\tau_1 \ldots d\tau_n \int_{(\tau)} d\tau_{n+2} \ldots d\tau_{n+m+1} \, A_N \qquad (7)$$

where (β,τ) labels the integration limits $\beta > \tau_1 > \ldots > \tau_n > \tau$, and the amplitude $A_N(\{\tau_i\})$ of a particular succession (path) of flips is given by

$$A_N(\{\tau_i\}) = \langle S^\pm(\tau_1) \ldots S^+(\tau) S^\pm(\tau_{n+2}) \ldots S^-(0) \rangle_0 \exp\left\{\sum_k \mu_k \sum_{i<j} \alpha_i \alpha_j V_\omega(\tau_i - \tau_j)\right\} \qquad (8)$$

Here, the number of flips is $N = n + m + 2$, with $\tau_{n+1} = \tau$ and $\tau_N = 0$. In (8), the last exponential is the result of the thermal average over outside modes states, which has been performed by using repeatedly the Baker-Hausdorff formula; $\alpha_i = \pm 1$ according to whether τ_i corresponds to S^+ or S^-. The outside modes act therefore as a retarded interaction between spin flips. The interaction potential is

$$V_\omega(\tau) = [\cosh \tfrac{1}{2}\beta\omega - \cosh(\tfrac{1}{2}\beta - \tau)\omega]/\sinh \tfrac{1}{2}\beta\omega \qquad (9)$$

At low temperatures or for modes of high frequencies $\beta\omega \gg 1$, the interaction (9) is a constant for all time intervals $0 \leq \tau \leq \beta$, except for a short dead time $\sim \omega^{-1}$ after each flip. Thus $V(\tau) \simeq 1$ for $\beta\omega \gg 1$, and, given that the number of up and down flips must be equal, $\sum_{i<j} \alpha_i \alpha_j = -\tfrac{1}{2} N$, the contribution of the outside modes at low temperatures is simply a renormalization of the Larmor frequency Δ_x to

$$\Delta_{eff} = \Delta_x \exp\left(-\tfrac{1}{2}\sum_k \mu_k^2\right) \qquad (10)$$

This is Holstein's famous result for the polaron [6]. In view of the subsequent analytic continuation, it is important to retain explicitly the time dependence of the only interval τ not integrated. Thus,

$$K^{+-}(\tau) = K_0^{+-}(\tau) \exp\left[-\sum \mu_k^2 V_{\omega_k}(\tau)\right] \qquad (11)$$

and similarly,

$$K^{22}(\tau) = K_0^{22}(\tau) \qquad (12)$$

where $K_0^{\alpha\beta}(\tau)$ are correlation functions for a spin governed by the simple Hamiltonian

$$H_0 = \Delta_{eff} S_x - \Delta_2 S_z - K S_z^2 \qquad (13)$$

This is indeed the starting point of Néel's theory [3]. The mean field $\sum g_k \langle (a_k^+ + a_k) \rangle$ of (1) has been replaced by a self energy of the cluster, and the two (symmetrical if $\Delta_x = 0$) energy minima of (13) are reminiscent of the order parameter in the mean field theories of spin glasses [4,12].

At high temperatures $\beta\omega \ll 1$, the outside modes are classical, the interaction potential (9) is parabolic, and it is possible in principle to replace the interaction between flips by a Gaussian distribution of magnetic fields.** A complete derivation for $S = \tfrac{1}{2}$ and a treatment of the crossover between low and high temperature regimes can be found in reference 13. The spin is in a double well potential $(-K S_z^2)$ and is subjected to Brownian bombardment. The transverse field is then Δ_x rather than Δ eff.

The transverse correlation function (11) describes the motion of a spin governed by (13), (a simple precession for $S = \tfrac{1}{2}$ where the anisotropy disappears as an operator), modulated by the interaction,

$$\phi(t) = \sum \mu_k^2 V(t) \simeq \int d\epsilon \rho(\epsilon) [g^2(\epsilon)/\epsilon^2][1 - \exp(-i\epsilon t)] = \Gamma(t) + i S(t) \qquad (14)$$

The damping $\Gamma(t)$ increases as t^2, to flatten out with oscillations around the value $\Gamma_0 = \int d\epsilon \rho(\epsilon) [g^2(\epsilon)/\epsilon^2]$ at long times. The frequency shift $S(t)$ decreases to zero as a power of t governed, as in section 2, by the singularities of $\rho g^2/\epsilon$, i.e. either by the behaviour of $\rho(\epsilon)$ at the top of the band ϵ_c

or, if it is a stronger singularity, by the behaviour of $\rho g^2/\epsilon$ at $\epsilon=0$. In the former case, $S(t)$ oscillates with frequency ϵ_c. There are two regimes, separated by $t_c=\epsilon_c^{-1}$. For $t<t_c$, the phase modulation is Gaussian

$$\phi(t) = \tfrac{1}{2}t^2 \int d\epsilon \rho(\epsilon) g^2(\epsilon) + it \int d\epsilon \rho(\epsilon) g^2(\epsilon)/\epsilon \tag{15}$$

while for $t>t_c$, the transients have died out and Γ is constant. It is on this latter regime that slow magnetisation creep is observable.

At arbitrary temperatures, the phase modulation takes the form

$$\phi(t) = \int d\epsilon \, [\rho g^2/\epsilon^2] \{\coth\tfrac{1}{2}\beta\epsilon \, 2\sin^2\tfrac{1}{2}t\epsilon + i \sin t\epsilon\} = \Gamma(t,T) + iS(t)$$

and in the absence of a field, $\Delta = 0$, the transverse response function for $S = \tfrac{1}{2}$ reads simply

$$\chi''_{xx} = \chi''_{yy} = \tfrac{1}{2}\chi''^{+-}(t) = -\tfrac{i}{4} \exp[-\Gamma(t,T)] \sin S(t) \tag{16}$$

The two regimes are clearly visible: at first, the transients dominate through $\Gamma(t)$ as given in (15). Then, for $t > t_c$, the decay is non-exponential, driven by $\sin S(t)$ and dominated by the singularities of $\rho g^2/\epsilon$. Here the outside modes provide the many-body barrier between the different states of magnetization. For $S > \tfrac{1}{2}$, additional effects are associated with the anisotropy, itself a direct consequence of the presence of the outside modes as we have seen above. The absence of direct anisotropy effect and an anomalous entropy of disorder which increases with increasing order parameter [14] for $S = \tfrac{1}{2}$ imply that $S = \tfrac{1}{2}$ spin glasses (if such materials exist) behave differently from the usual ones which all have $S \neq \tfrac{1}{2}$. However, this particular example is instructive as it is an exactly soluble mechanical model giving rise to transients and dissipation.

4. Magnetic viscosity

Given the Hamiltonian (13) a remanent magnetization creeping as the logarithm of the time can be obtained from two points of view which may appear orthogonal at a first glance. The first point of view is that of Richter and Néel [3], and assumes non interacting spins or grains, but a distribution of heights of the energy barrier between the two equilibrium positions in (13), corresponding to a distribution of the volumes of the grains or of the local anisotropy coefficient. Since the barrier height is proportional to the logarithm of the relaxation time prior to which the grain is blocked and after which it is free to rotate superparamagnetically, grains with higher and higher barrier heights are able to relax to equilibrium as the logarithm of the time increases, and the magnetization creeps in the same fashion if the distribution of barrier height is flat in that region.

The second point of view is due to Jaep [5] in his work on magnetic tapes, but has also been proposed by Anderson and Kim to explain the creep of magnetic flux lines in superconductors[15]. If the magnetic grains are interacting, each grain sees a self-consistent field due to all the others so that the barrier height varies, linearly say, with the local magnetization itself. Then, even in the absence of a distribution, the magnetization varies with ln t since,

$$(d/dt)\langle S_z \rangle \propto \exp[-C\langle S_z \rangle/kT] \quad (17)$$

even if the constant of proportionality is itself a slowly varying function of $\langle S_z \rangle$. Both points of view yield $\langle S_z \rangle \sim \ln t$ independently of the model assumed for the spin flip process.

The spin glass has some features of both points of view: a distribution of internal magnetic field, which does depend on the state of local magnetization through the order parameter [4,12]. Although I feel that the latter dominates the dynamics as it dominates the static magnetic properties of spin glasses, the precise correspondence between the parameters of spin glass and magnetic viscosity theories remains to be drawn. The importance of uniaxial anisotropy, essential to produce an energy barrier between the two equilibrium positions, and the Ising character of the Hamiltonian (13) must be emphasized.

I wish to acknowledge many suggestions of value by Dr. C. N. Guy and Professor E. P. Wohlfarth, and am grateful to Drs. Mydosh and Nieuwenhuys for making the results of reference 8 available to me prior to publication.

1. J.L. Tholence, thesis (1973), Grenoble University.

2. C.N. Guy, J.Phys F, to be published; Physica (1977) to appear.

3. L. Néel, Adv. Phys 4, 191 (1955).

   A. Herpin, Théorie du Magnétisme, (PUF 1968).

   E. Kneller, in Magnetism and Metallurgy, A.E. Berkowitz and E. Kneller, eds. (Acad. Press 1969).

4. K.J. Adkins and N. Rivier, J. Physique 35, C4-237 (1974).

5. W.F. Jaep, J.Appl. Phys 42, 2790 (1971).

6. T. Holstein, Ann. Phys 8, 325 (1959).

7. H.B. Shore and L.M. Sander, Phys.Rev. B7, 4537 (1973)
   P.W. Anderson, Phys. Rev. Letters, 34, 953 (1975).

8. G.J. Nieuwenhuys and J.A. Mydosh, Physica (1977) to appear.

9. M.J. Lighthill, Fourier Analysis and Generalised Functions, (Cambridge 1958).

10. M.L. Mehta, Random Matrices (Acad. Press 1967), p 240

    S.F. Edwards and R.C. Jones, J. Phys. A (1976) in press.

11. P.W. Anderson and G. Yuval, in Magnetism, H. Suhl, ed., vol.5, ch. 7 (Acad. Press 1973).

12. S.F. Edwards and P.W. Anderson, J. Phys.F. $\underline{5}$, 965 (1975).

13. N. Rivier and T. Coe, submitted to J. Phys. C.

14. N. Rivier, Physica (1977) to appear.

15. M. Tinkham, Introduction to superconductivity (McGraw-Hill 1975), ch. 5.7.

Notes (added in proof)

* $(a^+ + a)$ should be regarded as a quantization of the local mean field. It is odd under time-reversal symmetry.

** At high temperatures $\beta\epsilon_c \ll 1$, the anisotropy formally disappears from (7 - 9). This is encouraging since remanence and creep effects are absent in spin glasses above the freezing temperature.

EFFECTIVE FIELD THEORIES OF TOPOLOGICALLY DISORDERED MAGNETS

T. Kaneyoshi

Department of Physics, Nagoya University

Nagoya, Japan

In this contribution, effective-field theories of topologically disordered (or amorphous) magnets are studied. An effective field theory of the spin glass in the model of Edwards and Anderson[1] is also presented.

AMORPHOUS MAGNETS

The Hamiltonian for the system is

$$\mathcal{H} = - \sum_i \sum_j J_{ij} S_i \xi_i S_j \xi_j , \qquad (1)$$

where $\sum_i$ denotes a lattice sum (or space integration) throughout the system. The factor ξ_i is a random variable which takes the value 1 or 0 according to whether or not the site i is occupied by a magnetic atom. Here, the sign of J_{ij} is not always positive definite. To find the expectation value of a typical spin S_i, we assume the molecular field approximation. For a spin 1/2 system, the expectation value of a spin S_i is given by

$$\frac{\langle S_i \rangle}{S_i} = \sigma_i = \tanh(\frac{1}{2}\beta \sum_j J_{ij} \sigma_j \xi_j) \qquad (2)$$

with

$$\beta = \frac{1}{k_B T} .$$

Equation (2) is an expression for a particular configuration of magnetic atoms, and hence it is necessary to take an average over all possible configuration. For this purpose, we introduce a new variable defined by

$$m_i = \frac{\langle \xi_i \sigma_i \rangle_r}{\langle \xi_i \rangle_r}, \qquad (3)$$

which means the averaged moment of σ_i. $\langle \cdots \rangle_r$ indicates an average over all possible distributions of magnetic atoms. At this point, following to the method of Kaneyoshi,[2] we transform equation (3) as follows

$$m_i = \frac{1}{\langle \xi_i \rangle_r} \int_{-\infty}^{\infty} d\omega \tanh(\tfrac{1}{2}\beta\omega)\,[-\tfrac{1}{\pi}\mathrm{Im}\langle \xi_i G(\omega+i\varepsilon)\rangle_i] \qquad (4)$$

with

$$G(\omega+i\varepsilon) = \langle (\omega+i\varepsilon - \sum_j J_{ij}\sigma_j\xi_j)^{-1} \rangle_{ir}, \qquad (5)$$

where $\langle \cdots \rangle_{ir}$ indicates a conditional average over all possible distributions of magnetic atoms except the i-th site. $\langle \cdots \rangle_i$ denotes the configuration average of the i-th atom. In order to perform the ensemble average of equation (5), we introduce random functions x_j and X_i defined by

$$x_j = \sigma_j \xi_j \qquad (6)$$

$$X_i = \sum_j J_{ij}\delta x_j \qquad (7)$$

with

$$\delta x_j = x_j - \langle x_j \rangle_{ir} \qquad (8)$$

As has been discussed by Kaneyoshi,[2] we must expand equation (5) with X_i and perform the conditional random average.

APPROXIMATIONS

For clarification, we here introduce two approximations as follows

Approximation I;

$$\begin{aligned}\langle (X_i)^{2\nu}\rangle_{ir} &= [\langle (X_i)^2\rangle_{ir}]^\nu \\ \langle (X_i)^{2\nu+1}\rangle_{ir} &= 0\end{aligned} \qquad (9)$$

EFFECTIVE FIELD THEORIES

This approximation corresponds to those of Handrich[3] and Kaneyoshi[4] studied in amorphous and disordered ferromagnets. In this case equation (4) reduces to

$$m_i = \frac{1}{2\langle\xi_i\rangle_r}[\langle\xi_i\tanh\{\tfrac{1}{2}\beta(W_i+\Delta_i)\}\rangle_i$$
$$+ \langle\xi_i\tanh\{\tfrac{1}{2}\beta(W_i-\Delta_i)\}\rangle_i] \tag{10}$$

with

$$\langle(X_i)^2\rangle_{ir} = \sum_{\ell\ell'}J_{i\ell}J_{i\ell'}\{\langle x_\ell x_{\ell'}\rangle_{ir} - \langle x_\ell\rangle_{ir}\langle x_{\ell'}\rangle_{ir}\}, \tag{11}$$

where W_i and Δ_i are defined by

$$W_i = \sum_j J_{ij}\langle x_j\rangle_{ir} \tag{12}$$

$$\Delta_i = \sqrt{\langle(X_i)^2\rangle_{ir}} \tag{13}$$

For example, if we neglect the positional correlation between magnetic atoms, or if we take the approximations,

$$\langle\cdots\rangle_{ir} \cong \langle\cdots\rangle_r$$
$$\langle x_\ell x_{\ell'}\rangle_r \cong \langle x_\ell\rangle_r\langle x_{\ell'}\rangle_r \quad \text{for} \quad \ell\neq\ell', \tag{14}$$

equation (10) reduces to that obtained by Kaneyoshi[4] in the case of dilute magnets. We can show that the structure fluctuation leads to the reduction both of magnetization and the transition temperature. The method of evaluation for higher-order terms beyond this approximation is given by Kaneyoshi[2] with the help of a diagram technique.

Now, let us take another approximation as follows

Approximation II;

$$\langle(X_i)^{2\nu}\rangle_{ir} = \frac{2\nu!}{(2!)^\nu \nu!}\{\langle(X_i)^2\rangle_{ir}\}^\nu$$
$$\langle(X_i)^{2\nu+1}\rangle_{ir} = 0 \tag{15}$$

According to the diagram language, this approximation just corresponds to all selection of diagrams with crosses of 2-dashed lines (See Kaneyoshi;[2] for example, diagrams (b), (c) and (d) of equation (2.20)). On the contrary, the approximation I corresponds to partial selection of diagrams with crosses of 2-dashed lines (for example, only the diagram (b) of equation (2.20)). In this case, equation (10) is given by

$$m_i = (2\pi)^{-\frac{1}{2}} \int_{-\infty}^{\infty} e^{-\frac{y^2}{2}} \frac{1}{\langle\xi_i\rangle_r} \langle\xi_i \tanh\{\tfrac{1}{2}\beta(W_i + y\Delta_i)\}\rangle_i \, dy \qquad (16)$$

We must pay attention to the fact that equation (16) has a form similar to that of spin glasses derived in the following. In fact, if we take the approximation (14), equation (16) reduces to

$$m = (2\pi)^{-\frac{1}{2}} \int_{-\infty}^{\infty} e^{-\frac{y^2}{2}} \tanh\{\tfrac{1}{2}\beta(mcJ_0 + yJ_1\sqrt{cq - c^2 m^2})\} dy , \qquad (17)$$

where we defined $J_0 = \sum_j J_{ij}$, $J_1^2 = \sum_j (J_{ij})^2$ and $\langle\xi_i\rangle_r = c$. We have also introduced the parameter $q = q_i$ as follows

$$q_i = \frac{\langle\xi_i \sigma_i^2\rangle_r}{\langle\xi_i\rangle_r} \qquad (18)$$

At this point, in order to get the self-consistent equations, we must study the equation of q_i. Within the approximation II, equation (18) is given by

$$q_i = 1 - (2\pi)^{-\frac{1}{2}} \int_{-\infty}^{\infty} e^{-\frac{y^2}{2}} \cdot \frac{1}{\langle\xi_i\rangle_r} \langle\xi_i \,\mathrm{sech}^2\{\tfrac{1}{2}\beta(W_i + y\Delta_i)\}\rangle_i \, dy , \qquad (19)$$

which reduces to, upon using the approximation (14),

$$q = 1 - (2\pi)^{-\frac{1}{2}} \int_{-\infty}^{\infty} e^{-\frac{y^2}{2}} \mathrm{sech}^2\{\tfrac{1}{2}\beta(mcJ_0 + yJ_1\sqrt{cq - c^2 m^2})\} dy \qquad (20)$$

For $c \ll 1$, we can neglect in equations (17) and (20) the term $c^2 m^2$ in comparison with the term cq. Then, equations (17) and (20) do take the forms similar to the simultaneous equations of the spin glass discussed in the following.

SPIN GLASSES

In this part, an effective-field theory of the spin glass is

EFFECTIVE FIELD THEORIES

presented using only the conventional molecular field theory. In the model of Edwards and Anderson (EA), exchange interactions J_{ij} between the spins are random according to a distribution

$$P(\Delta J_{ij}) = [2\pi <(\Delta J_{ij})^2>_r]^{-\frac{1}{2}} \exp[-\frac{(\Delta J_{ij})^2}{2<(\Delta J_{ij})^2>_r}] \quad (21)$$

with

$$\Delta J_{ij} = J_{ij} - <J_{ij}>_r \quad (22)$$

For this model, equation (2) must be rewritten as

$$\sigma_i = \exp[\sum_j \{\sigma_j J_{ji}\}\frac{\partial}{\partial W_i}] \cdot \tanh(\frac{1}{2}\beta W_i) \quad , \quad (23)$$

where W_i and $\{\sigma_j J_{ji}\}$ are defined by

$$W_i = \sum_j <J_{ij}>_r <\sigma_j>_r \quad (24)$$

$$\{\sigma_j J_{ji}\} = \sigma_j J_{ji} - <\sigma_j>_r <J_{ji}>_r \quad (25)$$

At this point, we must pay attention to the important fact that the expectation values of σ_i are also random variables through equation (23). The probability distribution $P(\sigma_i)$ should be obtained from equation (23) in terms of (21). If we denote $<\cdots>_r$ to indicate the ensemble average, we must first perform the random average of exchange interaction and then do the random average of local magnetization;

$$<f(\{\sigma_i\};\{J_{ij}\})>_r = <<f(\{\sigma_i\};\{J_{ij}\})>_J>_\sigma \quad (26)$$

This procedure is essentially equivalent to performing the moment expansion of σ_i, from which we can finally get the $P(\sigma_i)$ as the function depending on the known probability distribution of the J_{ij}. Then, we can easily perform the random average of exchange interaction. We have

$$<\sigma_i>_\sigma = <\exp\{\frac{1}{2}\sum_j \sigma_j^2 <(\Delta J_{ij})^2>_J \frac{\partial^2}{\partial W_i^2} + \sum_j \delta\sigma_j <J_{ji}>_J \frac{\partial}{\partial W_i}\}>_\sigma$$

$$\cdot \tanh(\frac{1}{2}\beta W_i) \quad (27)$$

with

$$\delta\sigma_j = \sigma_j - \langle\sigma_j\rangle_\sigma$$

Now, let us scale $\langle J_{ij}\rangle_J$ and $\langle(\Delta J_{ij})^2\rangle_J$ as

$$\langle J_{ij}\rangle_J = \frac{J_0}{N}, \quad \langle(\Delta J_{ij})^2\rangle_J = \frac{J_1^2}{N} \qquad (28)$$

so that J_0 and J_1 are both intensive. As has been discussed by Kaneyoshi,[5] in the thermodynamic limit $N\to\infty$ equation (27) exactly reduces to

$$m = \langle\sigma_i\rangle_\sigma = (2\pi)^{-\frac{1}{2}}\int_{-\infty}^{\infty} e^{-\frac{y^2}{2}} \tanh\{\tfrac{1}{2}\beta(J_0 m + J_1 q^{\frac{1}{2}} y)\} dy, \qquad (29)$$

where q is defined by

$$q = \langle\sigma_i^2\rangle_\sigma \qquad (30)$$

In order to get the self-consistent equations, we must study the equation of q. As can easily be understood from $\tanh^2 x = 1 - \text{sech}^2 x$, the equation of q is given by

$$q = 1 - (2\pi)^{-\frac{1}{2}}\int_{-\infty}^{\infty} e^{-\frac{y^2}{2}} \text{sech}^2\{\tfrac{1}{2}\beta(mJ_0 + J_1 q^{\frac{1}{2}} y)\} dy \qquad (31)$$

Thus, equations (29) and (31) are nothing but the simultaneous equations obtained by Sherrington and Kirkpatrick,[6] using the Edwards-Anderson replica formalism.

Now, we consider the specific heat in zero applied field. For the conventional molecular field theory, the internal energy U is given by, for $S=1/2$,

$$U = -\frac{1}{4}\sum_{ij}\langle J_{ij}\sigma_i\sigma_j\rangle_r, \qquad (32)$$

which reduces to, upon using the definitions (22)-(25),

$$U = -\frac{1}{4}\sum_{ij}\langle J_{ij}\rangle_J \langle\sigma_i\rangle_\sigma \langle\sigma_j\rangle_\sigma$$

$$- \frac{1}{4}\sum_i [\frac{\partial}{\partial h}\langle\exp(h\sum_j\{\sigma_j J_{ji}\})\rangle_r\Big|_{h=\frac{\partial}{\partial w_i}}] \cdot$$

EFFECTIVE FIELD THEORIES

$$\cdot \tanh(\frac{1}{2}\beta W_i) \qquad (33)$$

If we employ the scaling (28) and take the thermodynamic limit, we can show that equation (33) exactly reduces to

$$\frac{U}{N} = -\frac{1}{4} J_0 m^2 - \frac{1}{8k_B T} J_1^2 q(1-q) \qquad (34)$$

Thus, we can understand that the internal energy obtained by Sherrington and Kirkpatrick is smaller than that obtained by using the conventional molecular field theory.

The zero-field specific heat per spin $C_H = (d/dT)(U/N)$ of spin glasses is obtained from equation (34)

$$\frac{C_H}{k_B} = \lambda^2 q(1-q) + 2\lambda^2 \frac{\partial q}{\partial \lambda}(1-2q) \qquad (35)$$

with

$$\lambda = \frac{1}{2}(\frac{T_g}{T}) \quad ,$$

where the spin glass freezing temperature T_g as determined from $q(T_g)=0$ is given by

$$T_g = \frac{J_1}{2k_B} \qquad (36)$$

For $T>T_g$, the quantity q vanishes and consequently so does the specific heat in zero applied field. When T is infinitesimally less then T_g, we find by using the fact that $q=1-(T/T_g)^2$

$$C_H(T_g^-) = k_B = \Delta C_H \qquad (37)$$

There is a discontinuity in specific heat at the spin glass freezing temperature T_g, which might be characteristic of the conventional molecular field theory. The leading term of the specific heat at low temperatures in the spin glass phase is given by

$$\frac{C_H}{k_B} = \frac{1}{8\pi} + (\frac{2}{\pi})^{\frac{1}{2}} \frac{k_B T}{J_1} (\frac{\pi^2}{12} - \frac{1}{16\pi}) + 0(T^2) \quad , \qquad (38)$$

corresponding to a negative low-temperature entropy which diverges logarithmically as $T \to 0$. This last result may not be unexpected since the internal energy (34) is larger than that of Sherrington and Kirkpatrick.

DISCUSSION

In the previous part, we have discussed the theory of spin glasses. Then, the probability distribution function for the J_{ij} was given by equation (21). At this point, we may have a question; Did the previous results of spin glasses (or equations (29), (30), and (34)) depend on the choice of the distribution function (21)? We can easily show that the results may be independent of the choice of the distribution function under the restriction $P(\Delta J_{ij}) = P(-\Delta J_{ij})$ except the Lorentzian distribution.

For this purpose, instead of equation (21), let us take a distribution function as follows

$$P(\Delta J_{ij}) = \frac{1}{2}\{\delta(\Delta J_{ij}-\Gamma_{ij}) + \delta(\Delta J_{ij}+\Gamma_{ij})\} \tag{39}$$

with

$$\Gamma_{ij}^2 = <(\Delta J_{ij})^2>_J \tag{40}$$

From equation (23), we have in this case

$$<\sigma_i>_\sigma = <\exp\{\sum_j \ln\cosh(\sigma_j \Gamma_{ji}\frac{\partial}{\partial w_i}) + \sum_j \delta\sigma_j <J_{ji}>_J \frac{\partial}{\partial w_i}\}>_\sigma$$

$$\cdot \tanh(\frac{1}{2}\beta w_i) \tag{41}$$

Taking account of the fact that

$$\sum_j <\ln\cosh(\sigma_j \Gamma_{ji}\frac{\partial}{\partial w_i})>_\sigma$$

$$= \frac{1}{2}\sum_j <\sigma_j^2>_\sigma \Gamma_{ji}^2 \frac{\partial^2}{\partial w_i^2} - \frac{1}{12}\sum_j <\sigma_j^4>_\sigma \Gamma_{ji}^4 \frac{\partial^4}{\partial w_i^4} + \cdots$$

and using the method of reference 5, we can easily understand that equation (41) exactly reduces to equation (29), if we employ the scaling (28) and take the thermodynamic limit $N\to\infty$. In this way, after taking the scaling (28) and the thermodynamic limit $N\to\infty$, we can understand that equations (29), (30) and (34) may be independent of the choice of the distribution function under the restriction of $P(\Delta J_{ij})=P(-\Delta J_{ij})$ except the Lorentzian distribution.

REFERENCES

1) S.F. Edwards and P.W. Anderson: J. Phys. F; Metal Phys. $\underline{5}$ (1975) 965.
2) T. Kaneyoshi: J. Phys. C; Solid St. Phys. $\underline{8}$ (1975) 3415.
3) K. Handrich: Phys. Stat. Solidi $\underline{32}$ (1969) K55.
4) T. Kaneyoshi: Prog. Theor. Phys. $\underline{44}$ (1970) 328.
   J. Phys. C; Solid St. Phys. $\underline{6}$ (1973) 3130.
5) T. Kaneyoshi: J. Phys. C; Solid St. Phys. $\underline{9}$ (1976) L289.
6) D. Sherrington and S. Kirkpatrick: Phys. Rev. Lett. $\underline{29}$ (1975) 1792.

SPIN GLASSES AND MICTOMAGNETS - REVISITED[*]

J. A. Mydosh

Kamerlingh Onnes Laboratorium der Rijksuniversiteit

Leiden, The Netherlands

Abstract

A critical survey of the extensive experimental data for spin glass alloys is undertaken with the aim of drawing some basic conclusions and indicating those areas in need of additional investigation. Comparison is made between the spin glasses and the better understood "giant moment" systems in order to gain further insight into the nature of random magnetic alloys. A phenomenological model is presented to connect the various experiments, and some conjectures about the future directions of this research are offered.

Introduction

Several years ago, the concepts suggested by the terms "spin glass" and mictomagnet" first began appearing in the physics and metallurgy literature. Their use was related to the growing interest in problems of an amorphous or random nature. Also, at this time experimental progress in the Kondo alloys was waning, for it was being recognized that at low temperatures impurity-impurity interactions were interfering with the pure Kondo behavior of a single localized moment and its surrounding conduction electrons. Although in a metal the magnetic impurities interact via the conduction electrons, this RKKY interaction is an oscillatory function of the distance between the local moments which, for a random alloy, introduces a mixed coupling (parallel or antiparallel) among the moments. Thus, there can be no "periodic" or "usual" type of long range ferro- or antiferromagnetic order.

At the previous International Symposium on Amorphous Magnetism[1], some of the then controversial experiments and the forerunners of a theoretical description were presented and discussed. In the intervening three years, the proverbial literature explosion has taken

[*] Invited paper.

place with well over 50 experimental and 50 theoretical papers being published in this short time. Since the theoretical area and its latest successes are being reviewed at this Symposium[2], in the present paper we will focus our attention on some of the recent experimental progress and attempt to discuss these observations in terms of simple physical models. An earlier survey of the salient spin glass/mictomagnet properties has already appeared[3].

However, before proceeding, it should be noted that the notion of a random spin freezing at a well-defined temperature offers a powerful and very general description for a wide class of magnetic alloys. The basic ingredients of "good" moments, and a random distribution and interaction of spins are a common characteristic of many systems. For "weak" or fluctuating moments (CuFe,[4] PdCr [5]), these can be "toughened" by a strongly magnetic local environment, i.e., simply by increasing the concentration. The exchange enhanced hosts which become giant moment ferromagnets (Pd with Mn, Fe, Co)[6] when diluted with magnetic impurities, can easily be converted into spin glasses by the addition of H or Ag to reduce the high density of d-states in the Pd host[7,8]. As a matter of fact any mixing of the magnetic interactions, as for example with more concentrated PdMn produces the spin glass state[9]. At present many such alloy systems are now available and all show a common spin glass type of behavior. For it is almost too easy to fabricate another spin glass system, rather than to gain a deeper understanding of the basic physics involved in the archetypal CuMn and AuFe spin glass alloys. Nevertheless, some prudence is perhaps necessary when dealing with such alloys. For while one can always take extreme precautions in the melting or heat treatment of a binary alloy and then assume it to be random, there exists sufficient and hard won experimental evidence that short range or local order may sometimes be present, and this certainly can influence the magnetic ordering. So now that we are achieving a more sophisticated knowledge of these alloy systems, the fundamental metallurgy becomes a very important factor and well-characterized samples are required to fully interpret the variety of experimental measurements.

Recent Experiments

One of the most intriguing spin glass features is the sharp, well-defined freezing temperature T_o which occurs in some of the measurements. Accordingly, we may divide our collection of experiments into two classes: (I) those which show a clear freezing temperature, and (II) those which exhibit a broad change of behavior over a wide temperature range, but for which no distinct value of T_o may be extracted. This is a unique behavior especially when contrasted with the giant moment random alloys of similar concentration[6]. For these latter systems possess an unmistakable and consistent Curie temperature (not, however, unusually sharp) from all of the different measurements, despite the fact that such alloys represent an inhomogeneous type of ferromagnet.

In Table I we summarize the various spin glass experiments (and their references) conforming to these two classes. Let us consider this listing. The cusp-like susceptibility peaks vividly define the freezing temperature.[10] This result has now been shown to be independent of the experimental method (static or low frequency) as long as there is sufficient sensitivity.[11] Here it should be emphasized that the magnitude of the susceptibility around T_o is indeed very small -- $\sim 10^3$ less than in comparable giant moment alloys. Secondly, this magnitude, but not the peak temperature of T_o, is particularly dependent upon the presence of magnetic clusters or giant moments. Three examples of such behavior (the greater the cluster sizes, the larger the susceptibility magnitude) exist: heat treatment and mechanical deformation of CuMn,[12] the annealing of quench condensed CuMn films[13], and the destruction of the residual giant moments in higher concentration PdMn[9]. The very sensitive Mössbauer effect[14] nicely confirms the T_o determinations, while the onset of remanence and irreversibility[15] for $T \leq T_o$ illustrates that a completely different magnetic state is now present -- one that is especially contingent upon the chronology of an external magnetic field[11]. Also strong dependences of the shape or sharpness of χ near T_o result from relatively small fields.[10] The recent observation from muon depolarization experiments[16] of an abrupt appearance of a local internal field gives another determination of T_o. In addition, the smearing of this local field around T_o with increasing external field is in further agreement with the previous collection of measurements. Somewhat unexpected is the sharp behavior found at T_o from the anomalous Hall resistivity[17], since this method requires the application of rather large external fields. Nevertheless, the clear effect here points out the importance in the freezing process of skew-scattering between the local moments or clusters and the conduction electrons.

In contradistinction to the above is this second class of experiments. For example, the magnetic specific heat C_m, a basic thermodynamical variable which lucidly characterizes all of the usual phase transitions, shows nothing of interest at T_o.[18] There is a maximum in C_m, but this occurs at $T \gg T_o$. At Leiden we have been trying to reanalyze a massive amount of early specific heat data. At best we have found some correlation with a knee or maximum in $C_m/T (= dS_m/dT$, where S_m is the magnetic entropy) at the freezing temperature. These comparisons of different sets of data are severely hindered by the use of non-identical samples with various heat treatments for the different measurements. However, in the case of the PdMn spin glass with carefully controlled alloys, good agreement was obtained between the temperature of the peak in the susceptibility and the well-defined maximum in C_m/T [9]. A systematic study of the magnetic resistivity $\Delta\rho$ has revealed no indication of T_o.[19] Guided by critical phenomena theory which shows the critical behavior to be associated with the temperature derivative of a particular transport property, Ford and myself[19] have investigated $d(\Delta\rho)/dT$ for a series of simple spin glass systems. The only correlation, which was detected with a peak or maximum in $d(\Delta\rho)/dT$ at T_o, was for the higher concentration AuFe alloys. Here there are large ferromagnetic clusters and a small electronic mean free path such that significant spin scattering would take place within a well-defined, and static, at T_o, ferromagnetic region. The thermoelectric power (TEP)[10,20] exhibits more complex overall behavior, but it is similar to the resistivity and specific heat with a gradually changing behavior around T_o and visible effects which persist to higher temperatures $T \gg T_o$. The host nuclear magnetic resonance in dilute CuMn $\simeq$ 0.1 at.% shows no abrupt increase in the linewidth at T_o[21] and a slow temperature variation of resonant intensity However, these experiments were carried out on the Cu^{63} host in applied fields greater than 1.8 kOe, and furthermore dynamical effects (relaxation times) are important. Nevertheless, these results concern only a fraction of the host Cu nuclear spins, and extended experiments are now underway to search for the resonance of the remaining nuclei. This brings us now to the ultrasonic investigations which would be expected to show a critical anomaly for a cooperative type of phase transition. For an AuFe single crystal nothing is to be found at T_o.[22] However, some weak indication of the freezing temperature can be seen in AuCr.[23] Thus, our conclusion is that the spin glass freezing is certainly of a more subtle nature that the typically found phase transitions.

There yet remains one important kind of experiment - neutron scattering which cannot at present be adequately classified. A great deal of effort is now being expended upon such measurements and certain preliminary results are available: (1) While it has long been known that no long range magnetic order exists[24], definite correlations or clustering are observed for $T \gg T_o$.[25-27] (2) Neutron polarization analyses[28] have shown a significant difference between the bulk susceptibility and a larger, extrapolated ($\kappa \to 0$) neutron scattering value for $T < T_o$. The interpretation here is in terms of spins frozen in a largely random array. (3) Small angle scattering[29] with separation of the elastic and inelastic components shows a rapid drop in the inverse cluster relaxation time for $T \simeq T_o$. This suggests the beginning of a spin freezing process within the characteristic time of the neutron measurement $\sim 10^{-11}$s. The T_o thus obtained is larger than those resulting from the a.c. susceptibility with a characteristic measurement time of 10^{-1}-10^{-2}s. Therefore, T_o is a dynamical variable depending upon the "time-window" of the given experiment.

Extraction of a Phenomenological Model

Based upon this ever growing collection of experimental data, we can attempt a general, but oversimplified, physical interpretation of the essential phenomena. First, as the temperature is lowered from $T \gg T_o$ many of the randomly positioned and freely rotating spin build themselves into these locally correlated units or "clusters" which can then rotate as a whole. These clusters may simply be ferromagnetically coupled resulting in a giant or superparamagnet moment. Or they may form from a strongly localized overlapping of RKKY interactions in which case the net moment will be proportional to the square root of the number of spins taking part. The remaining spins, those not belonging to any cluster, are independent of each other, but serve to transmit interactions between the clusters allowing for changes in the cluster sizes and their relaxation or response times. Consequently, at these high temperatures we observe a variety of experimental effects. For example, from magnetization measurements[30] one can obtain an indication of the cluster moments and densities. Further, from the specific heat[18] there are large changes in the magnetic entropy as T is reduced towards T_o. It might also be mentioned that in the giant moment systems Pd(Co, Fe and Mn), a similar model is being used[31], however, here the dominant interaction is ferromagnetic with an induced, parallel moment forming on the Pd host.

Now as $T \to T_o$, the various spin components begin to interact with each other over a longer range.[32] The system seeks its ground state configuration for the particular distribution of spins. This means a favorable set of random alignment axes into which the spins or clusters can lock. The existence of such a ground state for randomly coupled spins without any short range order has been theoretically obtained by Edwards and Anderson.[33] The addition of local correlations or cluster formation would only serve to enhance the magnetic phenomena of a random freezing, thereby making them more observable for the experimentalist. We may put these concepts into the language of superparamagnetism by considering this ground state formation to cause a sudden shift in the height of energy barriers for the various cluster. In Fig. 1, we schematically represent this situation. This barrier height is proportional to the product of the cluster moment times the internal field at the cluster site. We know, both from the muon experiments[16] and the calculations of Adkins and Rivier[34], that there is an increase of internal field or shift in the probability distribution P(H,T) to larger fields at $T = T_o$. Therefore, at this "blocking" temperature some of the largest clusters, which also have the greatest barrier heights, now possess insufficient energy to overcome the shift in the barrier, i.e., they are no longer free to rotate or tumble. Due to the intrinsic anisotropy of the clusters (they are certainly far from being spherically symmetric), the axes of the up - down spin orientations become a random directional variable throughout the alloy. Thus, we have this blocking or freezing along a random set of axes resulting in $\Sigma \vec{S}_i = 0$, the spin glass condition. But how do the experiments reflect the freezing process, especially with the known differences between direct magnetic measurements and the indirect

Table I. Collection of Experiments Relating to the Freezing Temperature

Well-defined T_o	Smeared Behavior
Susceptibility[3,10]	Specific Heat[18]
Remanence/Irreversibility[3,15]	Resistivity[19]
Mössbauer Effect[3,14]	Thermo-electric Power[10,20]
μ^+-Precession[16]	Nuclear magnetic resonance[21]
Anomalous Hall Effect[17]	Ultrasonic Velocity[22,23]

? Neutron Scattering?[24-29]

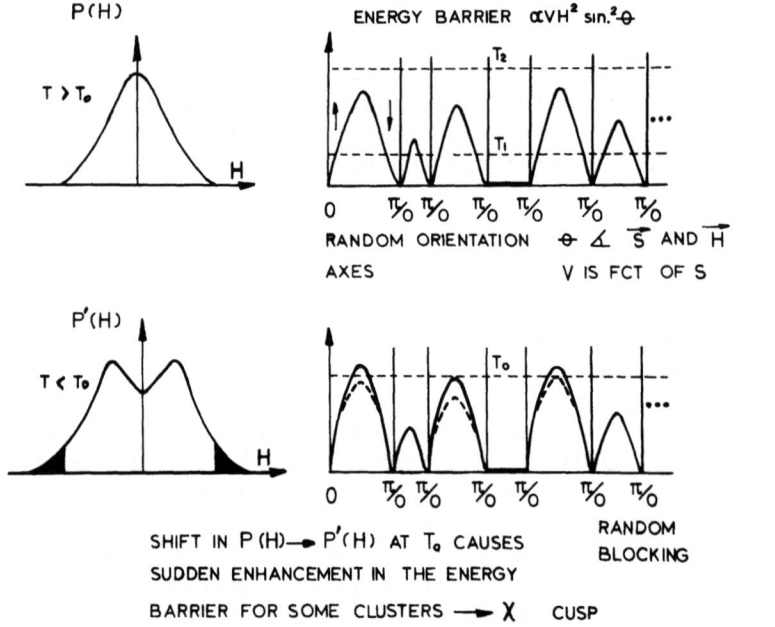

Figure 1. Schematic representation of the freezing process in terms of a distribution of superparamagnetic clusters.

thermal or transport properties? Let us consider one measurement from each class -- the susceptibility, χ, and the specific heat, C_m. From the Curie law valid in spin glasses for $T \gtrsim T_o$, $\chi = N g \mu_B^2 S(S+1)/3k_B T$ where N is the number of clusters and S represents their effective spin. Note here that a few (N small), big (S^2 very large) clusters could be seen in a direct (no background to subtract) magnetic measurement. By contrast for the specific heat, we must first obtain the magnetic contribution usually only a few percent of the total specific heat: $C_m(T) = C_{tot} - \gamma'T - \beta'T^3$ where γ' and β' are the electronic and lattice coefficients of the particular alloy in question (not the pure host). Working in terms of the entropy $S(T) = \int C_m(T) dT/T = NR \ln(2S+1)$, it is clear that a few, big clusters suddenly freezing out at T_o, will not greatly effect the entropy due to the logarithmic dependence of S. So that lacking sufficient experimental sensitivity, an indirect measurement could miss the cooperatife effects at T_o. On the other hand, such experiments do give a coarser overview to a broad temperature range of behavior.

We might finally mention the low temperature $T \ll T_o$ region. Here the specific heat is directly proportional to T and approximately independent of the concentration. This has recently been discussed by Rivier[35] in terms of the probability distribution of internal fields. The relevant distribution is a Lorentzian $P(H_z)$ which is consistent with experiment. In addition, the resistivity (poor man's neutrons) shows the existence of low temperature excitations[19] ($\Delta\rho \propto T^{3/2}$ with a weak inverse concentration dependence). A spin diffusion theory of Rivier and Adkins[36] ascribes this $\Delta\rho(T,c)$ behavior to scattering of the conduction electrons with long-wavelength elementary excitations which are highly damped, noncoherent, localized spin fluctuations. The fact that the susceptibility remains relatively large[3] $\chi(T=0)/\chi(T_o) \simeq 0.6$ further indicates the occurrence of these low temperature excitations. Subsequent neutron scattering measurements are certainly needed to shed more light on such processes and to perhaps disclose the emergence of spin waves.

The Future

Perhaps now that we have surveyed the experimental state of the spin glass problem, we could venture into the futurity of this area. So let us speculate upon a number of topics within which some interesting physics could and is being gained.

1). Although a series of very difficult and clever neutron scattering experiments are underway, more measurements are needed to fully characterize the three distinct temperature regions of a spin glass, i.e., the low temperature excitations (or spin waves), the freezing process at T_o and the build up of the local correlations (clusters) at $T \gg T_o$.

2). There are also two concentration regimes which connect the spin glass issue to other alloy problems. At very low concentrations the Kondo effect will cause a moment weakening fluctuation -- how does this affect the interaction and the freezing of the spins? A nice way to study this competition is through the employment of high pressure[37,38]. For T_o is primarily a function of the concentration, while the Kondo temperature has a strong pressure dependence. At higher concentrations there is the percolation limit -- a transition from a mictomagnetic or "cluster glass" to an inhomogeneous but <u>long range</u> ordered substance. Both ferro and antiferromagnets are possible. The details and the growth (time) dependence of this percolation transition are relatively unexplored, albeit theoretical results[39] are waiting to be tested.

3). The above point at once brings us to the very important spin glass dynamics -- especially the relaxation properties caused by changing an external magnetic field. Long time scale, non-exponential effects have been observed in the magnetization[11,40] and the specific heat[41]. A consistent simulation for many of the experimental results has been obtained via Monte Carlo computer methods.[42] In addition, a number of theoretical efforts, primarily based upon the Edwards and Anderson model,[33] are incorporating a time dependence into the spin glass freezing parameter[43].

4). Over many years, many investigators have claimed or studied the coexistence of superconductivity and magnetism in various alloy systems.[44] Usually, the classification or particulars of the magnetism are omitted. However, with the spin glass concept and a sharp freezing temperature, a framework now exists to describe the magnetic behavior. But why then does this freezing of the moments seem to favor the superconductivity? One possibility is the removal of the strong paramagnetic pair breaking below T_o, while the relatively static frozen clusters do not interact well with the Cooper pairs.

5). At present, most spin glass systems are composed of 3d-magnetic solutes. The rare-earth or 4f - impurities certainly extend the opportunities for which to study magnetic interactions in random alloys. Some noticeable differences are important with these impurities. For example, the 4f wave functions are highly localized, the RKKY interaction[45] is much weaker, and crystal field effects play a significant role. Already interesting spin glass behavior is being observed[46], and the continuation of such studies seems very reasonable. Also here a greater variety of more complicated systems are available to further probe the different properties as with the pseudo binary alloys.[47]

6). Finally, we should devote greater effort to the amorphous or highly disordered lattice spin glasses. For with such fabrication methods as quench-condensation of thin films, splat cooling, and ion implantation, unique and metastable alloys may be produced which have a truely random distribution of the solute atoms. These thin film alloys[13,48] illustrate a number of deviations from the usual spin glass behavior and much could be learned from increased activity with amorphous or micro-crystalline hosts. In addition, the insulating "real" glasses doped with magnetic impurities show strong similarities with the metallic alloys.[49] Further comparison would be most useful and hopefully unite this entire field of amorphous magnetism.

I wish to acknowledge P. J. Ford for a critical reading of this manuscript and S. M. Mydosh-Krysta for sacrificing our vacation.

References
1. Amorphous Magnetism, edited by H. O. Hooper and A. M. de Graaf (Plenum Press, New York, 1973).
2. P. W. Anderson, this conference proceedings. For other recent theoretical reviews see G. Heber, Appl. Phys. 10, 101 (1976), and K. H. Fischer, to be published in Proceedings of the International Conference on Magnetism, Amsterdam, 1976.
3. J. A. Mydosh, A.I.P. Conference Proceedings 24, 131 (1975).
4. V. Cannella and J. A. Mydosh, A.I.P. Conference Proceedings 18, 651 (1974).
5. R. M. Roshko and G. Williams in Low Temperature Physics - LT14, edited by M. Krusius and M. Vuorio (North Holland, Amsterdam, 1975), Vol. 3, p. 274, and J. E. van Dam, thesis, University of Leiden (1973).
6. For a review of these giant moment alloys, see G. J. Nieuwenhuys, Adv. in Phys. 24, 515 (1975).
7. J. A. Mydosh, Phys. Rev. Letters 33, 1562 (1974); A.I.P. Conference Proceedings 29, 239 (1976),and J. P. Burger et al. in Low Temperature Physics-LT14, edited by M. Krusius and M. Vuorio (North Holland, Amsterdam, 1975), Vol. 3, p. 278; J. Physique Lettres (Paris) 37, L-89 (1976).
8. J. I. Budnick et al., A.I.P. Conference Proceedings 18, 307 (1974), and R. A. Levy and J. A. Rayne, Phys. Letters 53A, 329 (1975).
9. H. A. Zweers et al., to be published in Proceedings of the International Conference on Magnetism, Amsterdam, 1976.
10. V. Cannella and J. A. Mydosh, Phys. Rev. B6, 4220 (1972), and Proceedings of the International Conference on Magnetism - Moscow (Publishing House Nauka, Moscow, 1974) Vol. 2, p. 74.
11. C. N. Guy, J. Phys. F 5, L242 (1975), and to be published in Proceedings of the International Conference on Magnetism, Amsterdam, 1976.

12). R. W. Tustison, Solid State Commun. 19, 1075 (1976).
13). J. J. Hauser, Phys. Rev. B5, 110 (1972) and to be published.
14). For a collection of Mössbauer effect references, see B. Window, J. Mag. and Mag. Materials 1, 167 (1975), and Ref. 3.
15). J. L. Tholence and R. Tournier, J. Physique (Paris) 35, C4, 229 (1974), and H. Claus and J. S. Kouvel, Solid State Commun. 17, 1553 (1975).
16). D. E. Murnick, A. T. Fiory and W. J. Kossler, Phys. Rev. Letters 36, 100 (1976), and A. T. Fiory, A.I.P. Conference Proceedings 29, 229(1976), and this conference proceedings.
17). S. P. McAlister and C. M. Hurd, Solid State Commun. 19, 881 (1976), to be published, and this conference proceedings.
18). L. E. Wenger and P. H. Keesom, Phys. Rev. B11, 3497 (1975) and B13, 4053 (1976).
19). P. J. Ford and J. A. Mydosh, Phys. Rev. B14, 2057 (1976) and J. A. Mydosh et al. Phys. Rev. B10, 2845 (1974).
20). P. J. Ford et al., Solid State Commun. 13, 857 (1973).
21). D. E. MacLaughlin and H. Alloul, Phys. Rev. Letters 36, 1158 (1976), and to be published in Proceedings of the International Conference on Magnetism, Amsterdam, 1976.
22). G. F. Hawkins et al., A.I.P. Conference Proceedings 29, 235 (1976).
23). G. F. Hawkins et al., this conference proceedings.
24). A. Arrott, J. Appl. Phys. 36, 1093 (1965) and in Magnetism IIB, edited by G. T. Rado and H. Suhl (Academic Press, New York, 1966) p. 295.
25). H. Sato et al., J. Physique (Paris) 35, C4, 23 (1974), and S. A. Werner et al., A.I.P. Conference Proceedings 10, 679 (1973).
26). A. P. Murani et al., J. Phys. F6, 425 (1976).
27). H. Scheuer et al., to be published in the Proceedings of the International Conference on Magnetism, Amsterdam, 1976.
28). N. Ahmed et al., Proceedings of the International Conference on Magnetism - Moscow (Publishing House Nauka, Moscow, 1974) Vol. 4, p. 91, and Solid State Commun. 15, 415 (1974).
29). A. P. Murani et al., Solid State Commun. 19, 733 (1976). See also A. P. Murani, Phys. Rev. Letters 37, 450 (1976).
30). P. A. Beck, Met. Trans. 2, 2015 (1971); in Magnetism in Alloys edited by P. A. Beck and J. T. Waber (TMS-AIME, New York, 1972) p. 211, and J. Less Common Metals 43, 69 (1975).
31). N. J. Koon, A.I.P. Conference Proceedings 18, 302 (1974), and to be published in Proceedings on the International Conference on Magnetism, Amsterdam, 1976.
32). A percolation (infinite cluster) model for spin glasses has been proposed by D. A. Smith, J. Phys. F4, L226 (1974) and J. Phys. F5, 2148 (1975).
33). S. F. Edwards and P. W. Anderson, J. Phys. F5, 965 (1975).
34). K. Adkins and N. Rivier, J. Physique (Paris) 35, C4, 237 (1974).

35). N. Rivier, Phys. Rev. Letters 37, 232 (1976).
36). N. Rivier and K. Adkins, J. Phys. F5, 1745 (1975).
37). J. S. Schilling et al., to be published in Phys. Rev. B and this conference proceedings.
38). For a theoretical treatment of this problem, see U. Larsen, to be published in Phys. Rev. B and this conference proceedings.
39). D. Stauffer, Phys. Rev. Letters 35, 394 (1975).
40). R. J. Borg and T. A. Kitchens, J. Phys. Chem. Solids 34, 1323 (1973).
41). G. J. Nieuwenhuys and J. A. Mydosh, to be published in Proceedings of the International Conference on Magnetism, Amsterdam, 1976.
42). K. Binder et al., Solid State Commun. 18, 1361 (1976); Phys. Rev. B14, (1976); Phys. Letters 57A, 177 (1976); and to be published in Proceedings of the International Conference on Magnetism, Amsterdam, 1976.
43). N. Rivier, this conference proceedings; S. F. Edwards and P. W. Anderson, to be published; W. Kinzel and K. H. Fischer, to be published, and A. Aharony and B. A. Huberman, to be published.
44). For a review of this area, see Ø. Fischer and M. Peter in *Magnetism V* edited by G. T. Rado and H. Suhl (Academic Press, New York, 1973).
45). The experimental requirements of the RKKY interaction for the 3-d impurities have been investigated by F. W. Smith, Phys. Rev. Letters 36, 1221 (1976) and references therein.
46). B.V.B. Sarkissian and R. B. Coles, Comm. on Phys. 1, 17 (1976).
47). R. J. Trainor and D. C. McCollum, Phys. Rev. B9, 2145 (1974) and B11, 3581 (1975); J. J. Prejean and J. Souletie in Proceedings of the International Conference on Magnetism - Moscow (Publishing House Nauka, Moscow, 1974) Vol. 4, p. 437, and M. H. Bennett and B. R. Coles, to be published in Proceedings of the International Conference on Magnetism, Amsterdam, 1976.
48). D. Korn, Z. Physik 187, 463 (1965).
49). See for example, R. A. Verhelst et al., Phys. Rev. B11, 4427 (1975).

CONCENTRATION EFFECTS IN SPIN GLASSES

Ulf Larsen
H. C. Ørsted Institute, University of Copenhagen
DK-2100 Copenhagen, Denmark

P. J. Ford and J. S. Schilling
Experimentalphysik IV, Universität Bochum
463 Bochum, W. Germany

J. A. Mydosh
Kamerlingh Onnes Laboratorium der Rijks-Universiteit
Leiden, The Netherlands

ABSTRACT

Spin glasses are characterized by a number of directly observable quantities such as: (a) the temperature of the resistance maximum, T_M, and (b) the freezing temperature, T_o, determined by the cusp in the magnetic susceptibility. These composite quantities are shown to be functions of the basic energy scales: (a) the Kondo temperature, T_K, and (b) the root-mean-square RKKY interaction strength, Δ_c. These are determined within the s-d exchange model as functions of the concentration, c, and s-d coupling constant, J. The resulting relations between T_M and T_o and c and J are shown to account qualitatively and quantitatively for the observed concentration and pressure dependences.

INTRODUCTION

During recent years it has been realized that the spin glass condensed state is a widespread phenomenon and a unique state of magnetic matter.[1] In the canonical spin glass alloys, like AuFe, magnetic (Fe) atoms are dissolved in a crystalline, metallic host (Au). The particular disordered condensate and the associated effects that occur in these materials are consequences of the random distribution of magnetic atoms and the long-range, oscillatory nature of their interactions. Because the systems are crystalline solid solutions the disorder only concerns the magnetic degrees of freedom.

Experimentally, there are several directly observable characteristic temperatures in spin glasses. Here we are considering T_M, the temperature of the resistance maximum, and T_o, the freezing temperature identified by a cusp or sharp maximum in the magnetic susceptibility.[1] T_M and T_o depend on a number of experimentally accessible variables, like for example the composition of the alloy, x, the concentration of magnetic atoms, c, or the pressure, P.

In the systems considered here there are two basic energy scales, namely the interaction strength, Δ_c, between different spins on the magnetic atoms, and the Kondo temperature of single spins, T_K. This turns out to be a natural way of separating the physical effects into collective and individual aspects. The systems can be characterized as spin glasses when $\Delta_c > T_K$, and as Kondo alloys when $\Delta_c < T_K$. The transition between these regimes is not a sharp one, and the competitions that take place in this region are of considerable interest, but presently not very well understood.

In general, the observables T_M and T_o do not depend in a simple way on the energy scales. Rather, there exist relations

$$T_M = T_M(\Delta_c, T_K) \; ; \; T_o = T_o(\Delta_c, T_K) , \qquad (1)$$

which have been obtained in the spin glass regime.[2]

The s-d exchange interaction, $-J \vec{\sigma}_i \cdot \vec{S}_i$, between a single spin, $\vec{S}_i$, and the conduction electron spin density, $\vec{\sigma}_i$, at these sites (i) gives rise to both the Kondo effect and to the Ruderman-Kittel-Kasuya-Yosida (RKKY) interaction, $-J(R_{ij})\vec{S}_i \cdot \vec{S}_j$, between two spins a distance R_{ij} apart. The fundamental model parameters are therefore c and J. Also, if scattering reduces the electron mean-free-path, λ, and thereby damps the RKKY interaction at distances longer than λ, it is included in this category. In addition, there are some auxiliary model parameters, such as the electron density of states, $n(E_F)$, at the Fermi energy, E_F, the magnitude of the spins, S, potential scattering phase-shift, δ_V, etc.

As the next step we therefore obtain relations between the scales and the model parameters

$$T_K = T_K(J) \; ; \; \Delta_c = \Delta_c(J,\lambda) \; . \tag{2}$$

The connection between experimental observables and variables is completed by the observation that J depends on the composition of the alloy and very simply on the pressure[3]

$$|J| \simeq |J_0|(1+\varepsilon P) \; . \tag{3}$$

Similarly, λ depends on c, as well as on the concentration of non-magnetic defects, ν, the composition, the annealing-time and -temperature, etc.

$$\lambda = \lambda(c,\nu,x) \; . \tag{4}$$

At present the relationships (3) and (4) are empirical. The advantage of (1) and (2) is that they permit the decoding of the complicated variations of T_M and T_O with c, P, etc. into much simpler variations of the fundamental model parameters, like J. The relations are summarized in Table 1.

Table 1. Summary of Relations.

Observable characteristic temperatures:	$T_M(\Delta_c, T_K) \; ; \; T_O(\Delta_c, T_K) \; ;$ etc.
Basic energy scales:	$\Delta_c(J,\lambda) \; ; \; T_K(J)$
Fundamental model parameters:	$c, J(x,P) \; ; \; \lambda(c,\nu,x)$
Auxiliary model parameters:	$E_F, n(E_F), S, \delta_V,$ etc.
Experimental variables:	$x, c, P, \nu,$ etc.

THEORY

First consider the relations (2). The Kondo temperature is defined by

$$kT_K = E_F \exp[-1/n(E_F)|J|] \quad , \tag{5}$$

and is independent of c.

Because the oscillations in the RKKY interaction tend to average out, the significant quantity determining the scale of the interaction energy is the root-mean-square energy of an average spin (site-i) due to all the other spins (sites-j)

$$\Delta_c = a_S \{\sum_{j \neq i} \Delta(R_{ij})^2\}_{av}^{\frac{1}{2}} \quad , \tag{6}$$

where $\Delta(R)$ is the envelope of $J(R)$ and a_S a spin dependent prefactor of order unity. Taking into account only the dominant ($\ell=2$) partial wave of the s-d interaction one has

$$\Delta(R) = \frac{9\pi J^2 (2\ell+1)^2}{2E_F (2k_F R)^3} e^{-R/\lambda} \quad . \tag{7}$$

Clearly, the closest spins are the more important in (6), and therefore care must be taken with the statistical distribution of the nearest neighbor distances. One ends up with an expression in the form[4]

$$\Delta_c(J,\lambda) = a(S)f(c,\lambda)J^2/E_F \quad , \tag{8}$$

where f is a smoothly increasing function of c and $a(S)=a_S(2\ell+1)^2/4$. Details are given in Ref.4.

Next consider relations (1). In most spin glass theories T_o is proportional to the rms interaction strength, and it therefore follows that

$$T_o = b(S)\Delta_c(J,\lambda) \tag{9}$$
$$= [a(S)b(S)]f(c,\lambda)J^2/E_F \quad .$$

Because such theories disregard the Kondo effect, (9) only holds for $\Delta_c \gg T_K$. For quantum spins in the Edwards-Anderson theory:[5] $[a(S)b(S)] = [(2S+1)^4-1]^{\frac{1}{2}}/12$.

At temperatures above freezing, the interactions produce spin-flip transitions at a rate $\Delta_c/\hbar$, manifesting themselves as "noise" in the spin system.[6] This interrupts the building-up with decreasing temperature of the Kondo resistance and causes the maximum. As in the noise-free model, at temperatures $T \gg T_K$ the parquet approximation is valid.

The relation $T_M = T_M(\Delta_c, T_K)$ has been obtained in the limit $\Delta_c \gg T_K$, where it is shown that[2]

$$T_M \gg T_o \simeq \Delta_c \gg T_K \tag{10}$$

Thus the maximum is situated well above T_K and in addition considerably above the freezing temperature, T_o, in agreement with observation in low-T_K systems.[1,7] Asymptotically, one has

$$T_M \sim \Delta_c \ln(\Delta_c/T_K) \quad ; \quad \text{for } \Delta_c/T_K \to \infty . \tag{11}$$

Details are given in ref.2.

The full relation between T_M and J, c, etc. can be conveniently expressed in terms of dimensionless quantities

$$T_M^* \equiv T_M/E_F = T_M^*(\Delta_c^*, T_K^*) \equiv G(\bar{c}, g),$$

$$\Delta_c^* \equiv \Delta_c/E_F = \bar{c}g^2, \tag{12}$$

$$T_K^* \equiv T_K/E_F = \exp(-1/g),$$

$$g \equiv n(E_F)|J| , \tag{13}$$

$$\bar{c} \equiv a(S)f(c,\lambda)/(n(E_F)E_F)^2, \tag{14}$$

where in particular it should be noticed that the entire dependence on the model parameters reduces to the coupling constant, g, and the <u>effective concentration</u>, $\bar{c}$. The surface defined by $T_M^* = G(\bar{c}, g)$ is shown in Fig.1. The most important feature is the presence of a <u>ridge</u> separating a region where T_M^* increases with g, initially as $T_M^* \simeq \bar{c}g$, from a region where T_M^* decreases with g. Thus eventually (10) and (11) cease to hold when $T_K \to \Delta_c$, and T_M^* passes through a maximum. On the basis of the present calculation T_M^* is conjectured eventually to vanish at the transition from the spin glass into the Kondo regime, at $T_K = \Delta_c$. This line, shown in the (c,g) plane in Fig.1, is given by

$$\bar{c}_o = g^{-2} \exp[-1/g] . \tag{15}$$

The position of the ridge is asymptotically at

$$\bar{c}_r \sim g^{-2} \exp[-1/2g] \quad ; \quad \text{for } g \to 0 , \tag{16}$$

which means that the ridge can be crossed in both the g-direction and the $\bar{c}$-direction.

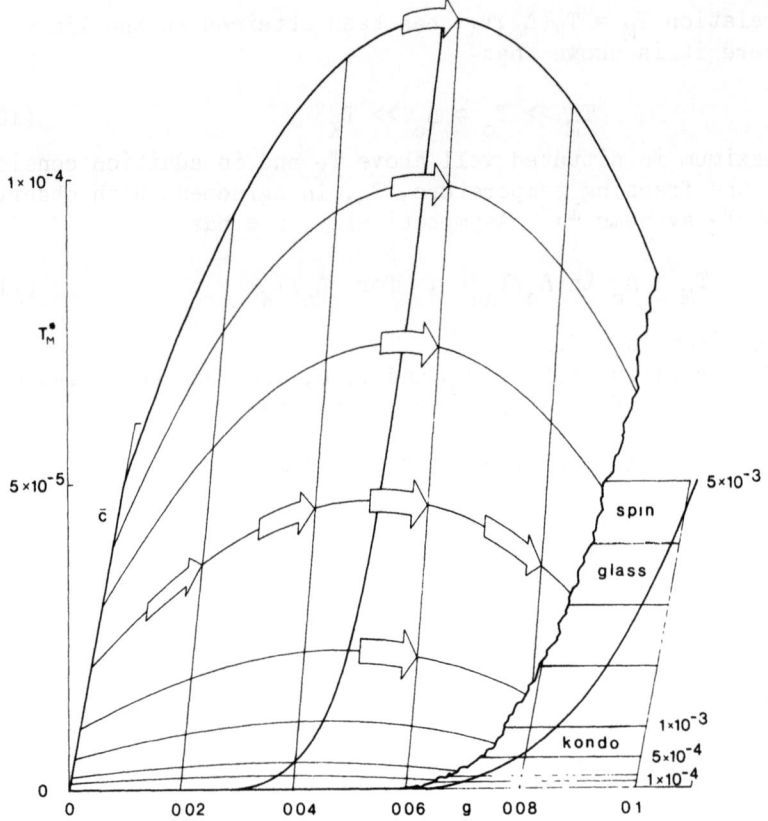

Figure 1 Surface defined by the temperature of resistance maximum, T_M^*, as function of the effective s-d coupling constant, g, and the effective concentration, $\bar{c}$. The full line in the horizontal plane indicates transition between spin glass and Kondo regimes, where the surface of T_M^* is conjectured to reach zero. The full line on surface indicates the ridge, where the slope in the direction of g vanishes. The arrows in the surface illustrate the effect of increasing pressure on T_M^*, for different initial values of g and $\bar{c}$.

This is a formal way of expressing a property of the theory, which in connection with the pressure dependence of J or g, given by (3), has been directly observed in the measured pressure dependence of T_M.[8] In particular, as indicated by the arrows in Fig.1, it follows that for a given effective concentration, dT_M/dP will be positive for systems with small $|J|$, go through zero at intermediate $|J|$ and become negative in systems with larger $|J|$. This is in agreement with observations,[8] shown in Figs. 4 and 5 of the following paper,[3] in the systems AuMn[8], CuMn[8], AuFe[8], LaCe[9] and MoFe.[10]

Also illustrated by a set of arrows in Fig.1 is the possibility to go from negative to positive dT_M/dP by increasing the concentration in a given system (i.e. crossing the ridge with fixed g). In general one can show that $T_M^{-1}\, d\, T_M/dP$ will be universally increasing with the concentration, which is consistent with observations in the systems mentioned above.[8] Another feature, namely the downward curvature of T_M^* as a function of g, leads to the prediction of universal downward curvature in T_M as a function of pressure. This requires measurements over a somewhat larger pressure range to be conclusively verified by the experiments. Further details of the pressure dependence of the resistance in spin glasses are given in the following paper.[3]

CONCENTRATION EFFECTS

In Fig.2 we show a plot of experimental values of T_M and T_o in AuFe taken from a comprehensive survey of the literature. Two important features are apparent. Firstly, contrary to widely held beliefs, neither T_M nor T_o are linear functions of c. Secondly, when the theoretical relation[2] $T_M=T_M(\Delta_c,T_K)$ is used (for $T_K=0.19K$ in AuFe) to calculate from experimental values of T_M the corresponding values of Δ_c, (also shown in Fig.2), it turns out that $\Delta_c=T_o$ over a wide range of concentrations, except at low c where T_o is being depressed by the Kondo effect. This is in agreement with the theoretical relation (9) for $b(S)\simeq1.0$, and also corroborates the relation for T_M in terms of Δ_c and T_K.

Next consider the details of the concentration dependence in Fig.2. In the first place, according to (8)-(9) Δ_c and T_o have a complicated dependence on c given by the function $f(c,\lambda)$. Furthermore, as can be seen in (11), T_M is itself a non-trivial function of Δ_c. These two relations account for the observations in the following way. In (9) the mean-free-path, λ, can be determined[4] from measurements of the resistance,[7,11] and using S=1, $\ell=2$ and $E_F=5.5eV$, the only parameter available to fit the data for T_o is J. Shifting the $f(c,\lambda)$ curve vertically in the log-log plot of Fig.2 to match the freezing temperatures at $c \gtrsim 0.1\%$ gives $|J| = 0.26eV$, whereas $T_K=0.19eV$ determined independently from resistance experiments corresponds to $J = -0.25eV$.[12] For $1\% \lesssim c \lesssim 10\%$ the dominant reduction of mean-free-path is due to scattering against the magnetic atoms themselves (self-damping). This has a significant influence on T_o without which the fit would not be possible.[4] In the low-c range below 0.1% the non-trivial c-dependence is due to the statistical distribution of nearest neighbor distances between spins.[6] It is a striking confirmation of the theoretical relations employed that the values of Δ_c that are derived from entirely different (resistance-)experiments by means of a different theoretical expression, fall precisely on the curve obtained by the fitting of

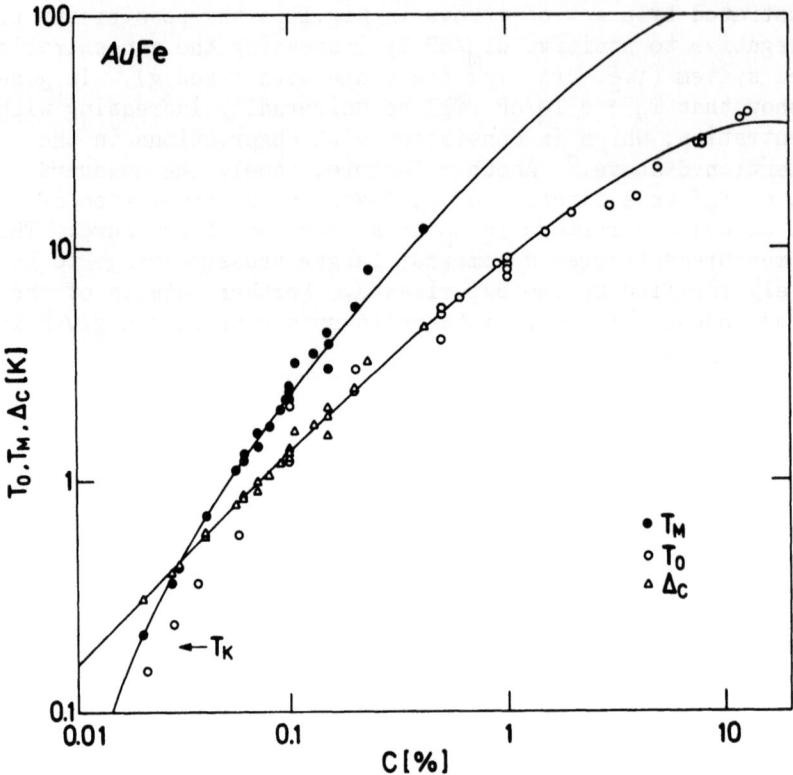

Figure 2 Experimental data for T_M and T_o and derived values of Δ_c as a function of concentration, c, for AuFe. T_K for AuFe at zero pressure has been taken to be 0.19K (J=-0.25eV). The solid lines through the points represent theoretical curves for T_M and T_o.

T_o at the higher concentrations. Conversely, using the theoretical Δ_c from (8) for J=-0.25eV and b(S)=1, gives a theoretical T_M-curve, which is seen in Fig.2 to describe the observed concentration dependence of T_M rather well.

CONCLUSIONS

Theoretical expressions have been obtained that quantitatively account for the measured concentration and pressure dependence of T_M and T_o described in this and the following paper. From our analysis of a substantial amount of data in a series of canonical spin glasses we can draw a number of conclusions.

In the first place, we obtain evidence that the interactions reponsible for the spin glass effects in these systems are indeed the RKKY interaction. Also, these interactions continue to influence in an essential way the physical properties at temperatures above the freezing temperature. Thus our analysis of the resistance measurements clearly indicate the presence of "noise" that causes individual spins to flip at a rate $\Delta_c/\hbar$ and, among other things, explains why $T_M > T_o$. The interaction parameter, Δ_c, in which a number of properties of the alloys are displayed, can now be obtained from resistance measurements, as well as from the freezing temperature, T_o. Thus it is for example possible to determine the magnitude of $J,^3$ which has been difficult previously in systems where it is so small that T_K is outside the observable range of temperatures. Also, from the concentration dependence of Δ_c and T_o we can infer details about the distribution of magnetic atoms and defects. The most important general conclusion that we can draw is perhaps the existence and interplay of the two energy scales Δ_c and T_K. As they both arise from the s-d exchange interaction, they are both indispensable. The properties of the spin glass or Kondo systems are determined by their ratio: Δ_c/T_K. A subject of further study is the transitional region $\Delta_c \approx T_K$ where the competition between spin glass and Kondo effects presumably leads to particularly interesting physical effects.

REFERENCES

1. J.A. Mydosh, AIP Conf. Proc. 24, 131 (1975).
2. U. Larsen, Phys.Rev.B, (1976), to appear.
3. J.S. Schilling, P.J. Ford, U. Larsen, J.A. Mydosh, following paper.
4. U. Larsen, submitted for publication (1976).
5. S.F. Edwards, P.W. Anderson, J.Phys. F5, 965 (1975); K.H. Fischer Phys.Rev.Lett. 34, 1438 (1975); D. Sherrington, B.W. Southern, J.Phys. F5, L49 (1975).
6. I. Riess, A.Ron, Phys.Rev. B8, 3467 (1973).
7. P.J. Ford, J.A. Mydosh, Phys.Rev. B (1976), to appear.
8. J.S. Schilling, P.J. Ford, U. Larsen, J.A. Mydosh, Phys.Rev.B (1976), to appear.
9. F. Zimmer, J.S. Schilling, to be published.
10. P.J. Ford, J.S.Schilling, J.Phys. F6 (1976), to appear.
11. J.A. Mydosh, P.J. Ford, M.P. Kawatra, T.E. Whall, Phys.Rev. B10, 2845 (1974).
12. J.W. Loram, T.E. Whall, P.J. Ford, Phys.Rev. B2, 857 (1970).

EFFECT OF PRESSURE ON IMPURITY-IMPURITY INTERACTIONS

IN SPIN GLASS ALLOYS[*]

J. S. Schilling and P. J. Ford
Experimentalphysik IV, Universität Bochum
463 Bochum, W. Germany

Ulf Larsen
H. C. Ørsted Institute, University of Copenhagen
DK-2100 Copenhagen, Denmark

J. A. Mydosh
Kamerlingh Onnes Laboratorium der Rijks-Universiteit
Leiden, The Netherlands

ABSTRACT

The electrical resistivity of dilute (c<1at%) spin glass alloys such as Au:Fe, Au:Mn, Cu:Mn, and Mo:Fe has been measured from 1.2-40K at pressures between 0 and 100 kbar. In these alloys the cooperative locking-in of the impurity spins at a temperature T_o leads to a resistivity maximum at $T_M = T_M(\Delta_c, T_K)$ which is a function of both the impurity-impurity interaction strength Δ_c and the Kondo temperature T_K. Whereas T_M always increases with the impurity concentration, both the sign and magnitude of the pressure dependence of T_M are found to depend in a complicated way on the particular system studied. From $T_M(P)$ and the known positive pressure dependence of T_K it is possible using a theory of Larsen to derive the pressure dependence of the average interaction strength $\Delta_c(P)$. The analysis lends support to the view that in these systems the long range RKKY-oscillations represent the dominant interaction mechanism.

PRESSURE AS A VARIABLE IN DILUTE MAGNETIC ALLOYS

We have recently found that the high pressure technique is an important complement to studies of the concentration dependence of magnetic alloys. Varying the concentration of the magnetic component frequently leads to a drastic change in its magnetic state as can be seen for example in Au:Fe where one observes changes from paramagnetic to spin glass and finally to ferromagnetic behavior. However, once concentration studies have established the general "lay of the land," pressure measurements can then supply detailed information about the nature of the impurity-conduction electron and the impurity-impurity interactions in the alloy. By using pressure we are able to vary the basic system parameters in a reversible manner. This, together with the fact that a <u>single</u> sample with a unique configuration of magnetic impurities is used, is an important advantage of the technique allowing a meaningful comparison between experiment and theory.

As an example of the application of the high pressure technique, let us briefly consider the effect of pressure on the interaction between conduction electrons and magnetic impurities in very dilute Kondo alloys. The spin-scattering resistivity ρ_{kondo} in such alloys is a function of a number of parameters, among them, J the effective exchange parameter, S the impurity spin, δ_V the potential scattering at the impurity site, and n the density of states. Theory in general predicts a universal resistivity law $\rho=\rho(T/T_K)$, where $T_K=T_F\exp(1/nJ)$. The temperature dependence of ρ_{kondo} is shown in the lower part of Fig. 1. The inflection point of this curve marks the Kondo temperature T_K but is in practice difficult to determine accurately either because it lies below the accessible temperature range or is masked by the rising phonon scattering. On the other hand, <u>changes</u> in T_K on a particular alloy, such as for instance are caused by application of high pressures, can be determined quite accurately. In a study of the resistivity of Cu-110 ppm Fe under pressure,[1] ρ_{kondo} is found to shift bodily on a logarithmic temperature scale to higher temperatures, confirming the law $\rho=\rho(T/T_K)$. This is shown in Fig. 2. The change in T_K with pressure is given by $d\ln T_K/dP \simeq +1.1\%/kbar$, and is due to an increase in $|J|$ with pressure, while δ_V, S, and n stay essentially constant.[1] Similar studies have been carried out on a wide number of dilute alloys (e.g. Au:Mn,[2] Au:Fe,[2] Ag:Mn,[3] Cu:Mn,[3] Y:Ce,[4] and La:Ce,[5]) and in all cases T_K and $|J|$ are found to increase with pressure. This information will be of importance in subsequent studies of the pressure dependence of interactions between impurities in spin glasses.

EFFECT OF PRESSURE ON IMPURITY-IMPURITY INTERACTIONS

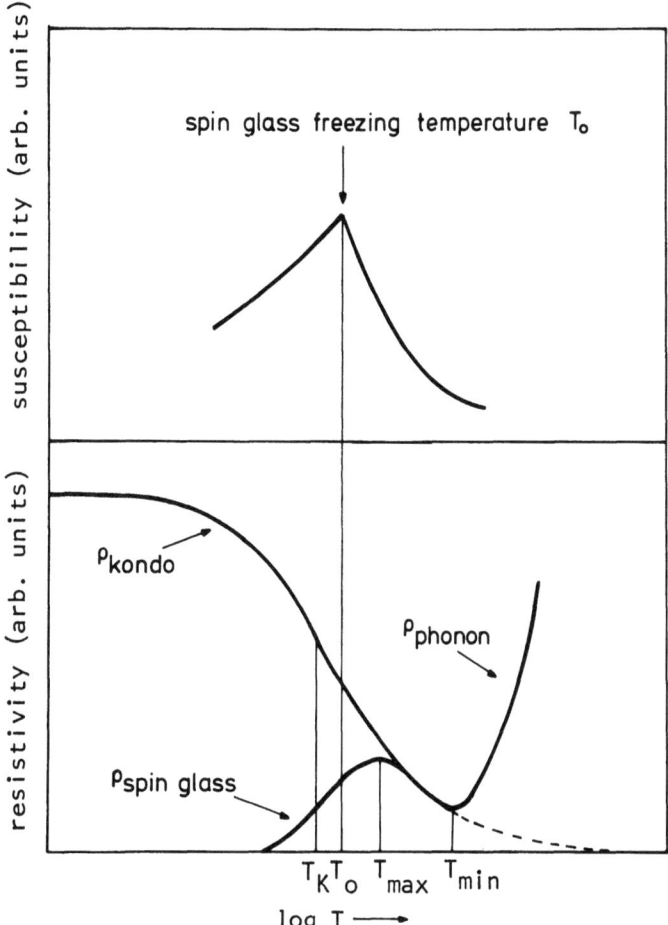

Figure 1 a.c. magnetic susceptibility of a spin glass, and Kondo and spin glass resistivities versus log T.

RESISTIVITY MAXIMA IN SPIN GLASSES: PRESSURE DEPENDENCE

If the concentration of such very dilute magnetic alloys is increased, interactions between impurities gain an importance and lead to magnetic ordering phenomena. In spin glasses the randomly distributed impurity spins behave paramagnetically at high temperatures. Below the so-called spin glass freezing temperature T_o, the spins become locked in place with random spin directions which is signalled by a sharp peak in the a.c. susceptibility,[6] shown in

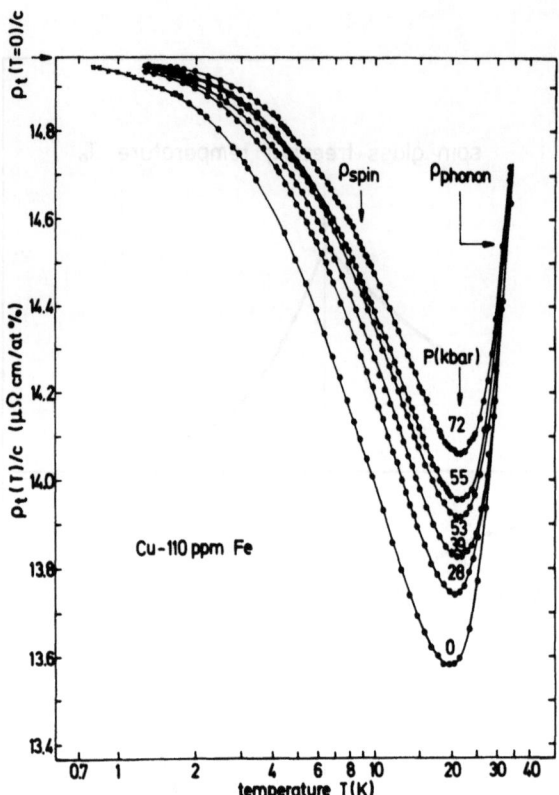

Figure 2 Total measured resistivity versus log T for a dilute Cu:Fe alloy, see Ref. 1.

Fig. 1. Before this happens, at temperatures above T_O, the presence of the interactions leads to a reduction in the spin-flip scattering channels and therefore to a bending over of the rising Kondo resistivity with decreasing temperature. Consequently, the resulting resistivity maximum at T_M does not occur at T_O but is always located at a somewhat higher temperature. Although both T_O and T_M invariably increase with the impurity concentration, the exact relationship between these two temperatures is complicated.

We have carried out a series of electrical resistivity measurements on the spin glass alloys Au:Fe,[7,8] Au:Mn,[8] Cu:Mn,[8] and Mo:Fe[9] between 1 and 40K and in the pressure range 0-100 kbar. The results for Mo:Fe and Au:Mn are shown in Fig. 3. Pressure is seen to shift T_M strongly to lower temperatures in Mo:Fe whereas it moves to slightly higher temperatures in Au:Mn. The results for all the data are summarized in Fig. 4 where T_M is plotted as a function of relative volume, V/V_O. The observed T_M shifts, which are reversible in pressure, are seen to differ markedly both in magnitude and

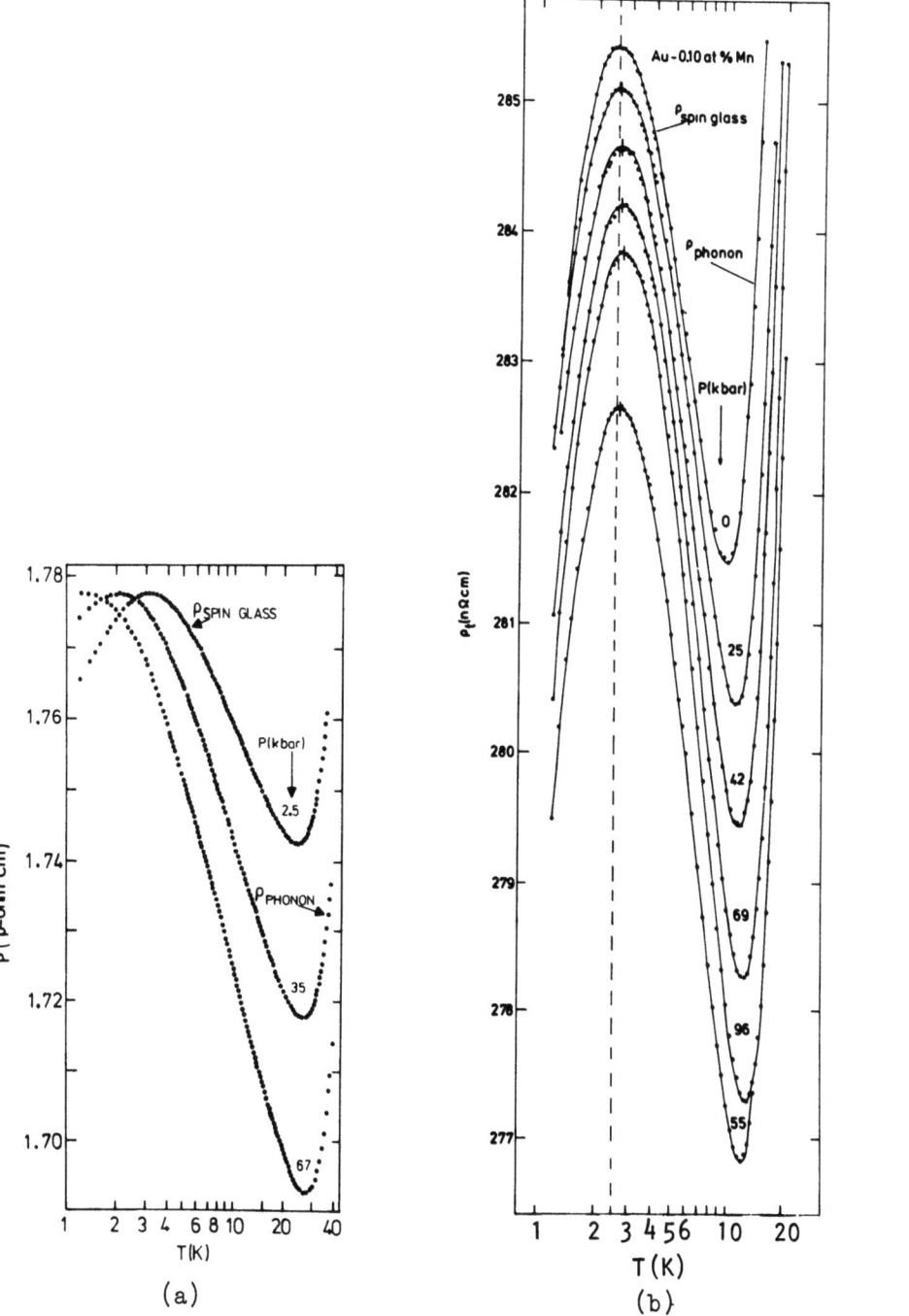

Figure 3 Total measured resistivity versus log T for (a) Mo-1 at% Fe and (b) Au-0.10 at% Mn spin glass alloys. The data are shifted vertically for clarity, see Refs. 8 and 9.

direction and to depend on concentration. Also included are some
very recent data on La:Ce.[10] Noteworthy are the large negative values
of the slope dT_M/dP for Mo:Fe and La:Ce. A positive slope for
Ag:Mn and a negative slope for Cu:Cr have been reported.[3] Other
pressure effects which are common to data on all systems, are an
increase in the magnitude of the resistivity at T_M and a broadening
of the maximum with increasing pressure.

The widespread assumption that $T_M \propto J^2$ would imply that T_M should
always increase with pressure, contrary to the present experiments.
A solution of this discrepancy is indicated if one notices that all
alloys with negative values of the slope dT_M/dP (Au:Fe, La:Ce, Mo:Fe,
Cu:Cr) have relatively high Kondo temperatures, whereas positive
slope alloys (Au:Mn and Ag:Mn) have extremely low T_K's, with Cu:Mn
an intermediate case. An accurate treatment of the Kondo effect
is clearly necessary to allow extraction of quantitative information
from the resistivity maximum.

COMPARISON WITH THEORY

In a recent theory by Larsen[11] T_M is shown to be a function of
both the root-mean-square RKKY interaction, Δ_c, and T_K, i.e. $T_M = T_M(\Delta_c, T_K)$. The resulting dependence of T_M on $n|J|$ for a fixed concentration is shown qualitatively in Fig.5, where the arrows represent
a pressure increase of about 100 kbar. The Kondo temperature of a
particular system defines on which part of the T_M "hill" it rides
(see also Fig. 1 of preceding paper), determining the expected pressure
dependence of T_M. For example for Au:Mn the pressure arrow points
uphill, for Cu-0.15 at% Mn horizontal, and for Au:Fe, La:Ce, and
Mo:Fe downhill, in agreement with the results shown in Fig.4. The
correlation observed above between the magnitude of T_K and the
pressure dependence of T_M is now apparent: the higher T_K, the more
negative is dT_M/dP. The downward concavity of the T_M curve in
Fig.5 is reflected in the curved theoretical fits to the data in
Fig.4. With increasing concentration the entire T_M curve pinned at
$T_M=0$ is stretched to the right and upwards (see previous paper) and
in general $T_M^{-1} dT_M/dP$ becomes more positive with increasing concentration. This can be clearly seen in Fig. 4 for Cu:Mn. The
theory is also able to account for the observed increase in the
resistivity at T_M and the broadening of the resistivity maximum.

Using the expression $T_M=T_M(\Delta_c, T_K)$, from the measured pressure
dependence of $T_M(P)$ and $T_K(P)$ one can determine $\Delta_c(P)$. The proportionality of the RKKY interaction to J^2 implies that Δ_c should
increase with pressure in all systems, and this is indeed observed.
Due to the lack of accurate $T_K(P)$ data for the other systems at the
time of writing, only Au:Fe and Au:Mn will be considered in the
following. From $T_K(P)$ one obtains $J(P)$ and, using the known compressibility to give the volume dependence one defines the linear
relations $[\Delta_c(v)]^{\frac{1}{2}} = [\Delta_c(0)]^{\frac{1}{2}}(1+\alpha v)$ and $|J(v)| = |J(0)|(1+\varepsilon v)$,

EFFECT OF PRESSURE ON IMPURITY-IMPURITY INTERACTIONS

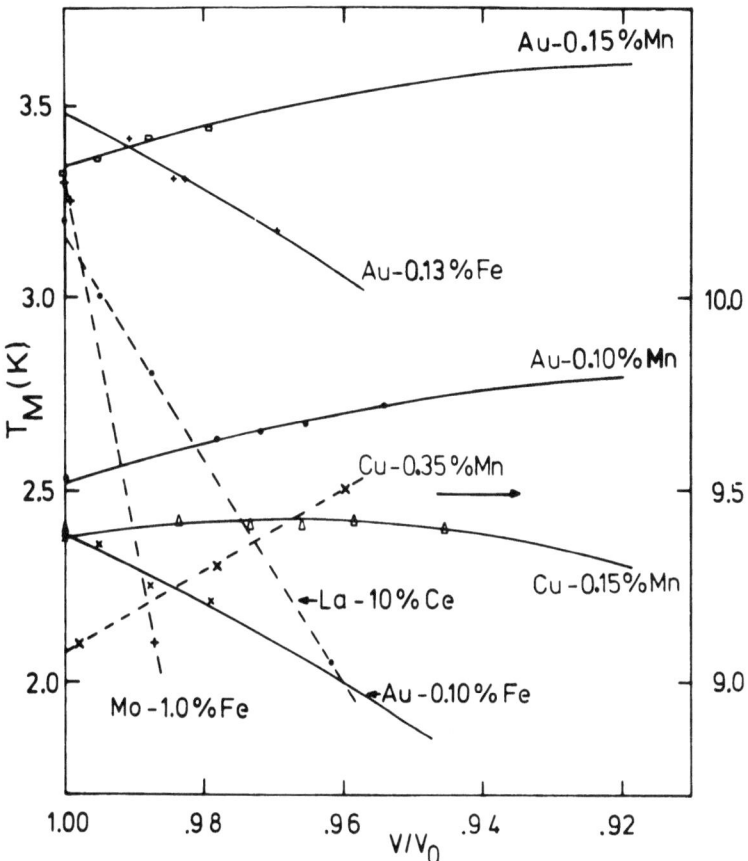

Figure 4   Temperature of resistivity maximum versus relative volume V/V_0 for all systems studied. Cu-0.35%Mn uses temperature scale to the right. All concentrations are in at%. The solid lines are fits using a theory of Larsen[11] (see preceding paper). The dashed ones are preliminary straight-line fits.

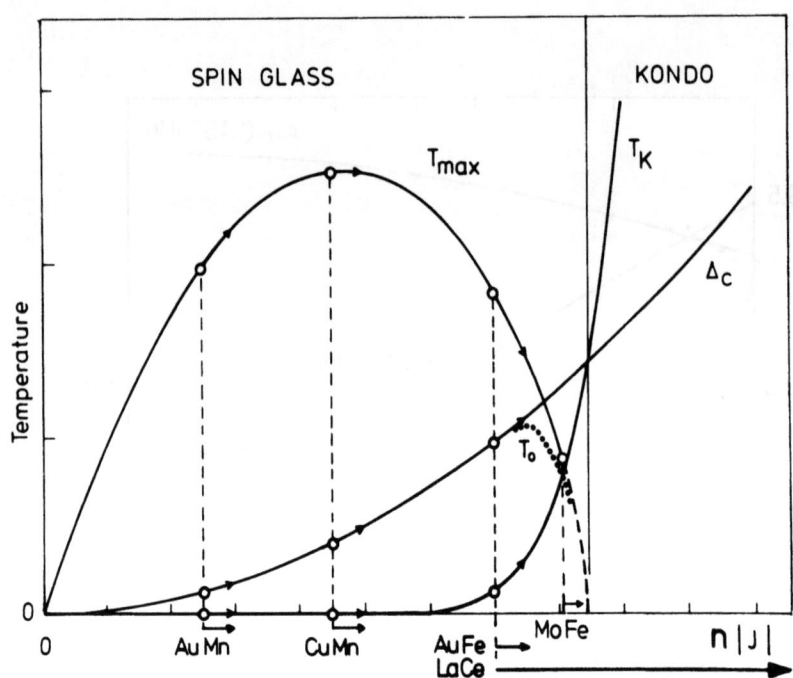

Figure 5 Expected functional dependence of T_M, Δ_c, T_K, and T_o on $n|J|$ at fixed impurity concentration in both the spin glass $\Delta_c > T_K$ and Kondo $\Delta_c << T_K$ regimes.

where $v = (V_o - V)/V_o$. When fitted to the data the parameters $\Delta_c(0)$, α, and ε depend to some extent on the choice of $J(0)$, i.e., on the initial value of $T_K(0)$ chosen. A set of values for Kondo temperatures quoted in the literature [12-15] and the corresponding values of the parameters are shown in Table 1. It is seen that $\alpha = \varepsilon$ can be well satisfied with reasonable values of the Kondo temperature which implies $\Delta_c(P) \propto [J(P)]^2$. Also, the values of $\Delta_c(0)$ are of the right magnitude to be rms-RKKY interaction strengths.[11] This result provides evidence that the RKKY interaction is indeed responsible for the observed spin glass effects.

Table 1.

Alloy	$T_K[K]$	$-J(0)[eV]$	ε	α	$\Delta_c(0)[K]$
AuFe-.13%	0.24 a)	0.25	1.5	2.1	2.1
	0.12 b)	0.24	1.4	1.4	1.8
AuMn-0.10%	7×10^{-5} c)	0.15	2.6	3.5	0.44
	8×10^{-12} d)	0.09	2.5	2.5	0.18

a) Ref.12, b) Ref.13, c) Ref.14, d) Ref.15.

In conclusion we believe that the rather striking variation in both the direction and the magnitude of the shift with pressure of T_M gives a useful insight into the nature of impurity-impurity interactions in spin glasses. Using the pressure technique, whereby J can be varied in a quasi-continuous fashion, therefore provides a useful complement to the earlier concentration studies. We are presently using the technique to study other types of spin glasses and the results of these investigations will be reported at a later date.

REFERENCES

1. J.S. Schilling and W.B. Holzapfel, Phys.Rev. B8, 1216 (1973).
2. J. Crone and J.S. Schilling, Solid State Commun. 17, 791 (1975).
3. J. Crone, private communication.
4. M.B. Maple and J. Wittig, Solid State Commun. 9, 1611 (1975).
5. K.S. Kim and M.B. Maple, Phys.Rev. B2, 4696 (1970).
6. V. Cannella and J.A. Mydosh, Phys.Rev. B6, 4220 (1972).
7. J.S. Schilling, J. Crone, P.J. Ford, S. Methfessel and J.A. Mydosh, J.Phys. F 4, 2116 (1974).
8. J.S. Schilling, P.J. Ford, U. Larsen and J.A. Mydosh, Phys. Rev. B (to appear)(1976).
9. P.J. Ford and J.S. Schilling, J.Phys.F Letters (to appear)(1976).
10. F. Zimmer and J.S. Schilling, to be published.
11. U. Larsen, Phys.Rev. B (to appear)(1976), see also preceding paper of this conference.
12. J.W. Loram, T.E. Whall and P.J. Ford, Phys.Rev. B2, 857 (1970).
13. J.W. Loram, A.D.C. Grassie and G. Swallow, Phys.Rev. B2, 2761 (1970).
14. K. Matho and M.T. Béal-Monod, Phys.Rev. B5, 1899 (1972).
15. J.W. Loram, T.E. Whall and P.J. Ford, Phys.Rev. B3, 953 (1971).

*This work supported in part by the Deutsche Forschungsgemeinschaft.

CRITICAL PROPERTIES OF A SIMPLE SPIN GLASS MODEL

A. Aharony[*]

Tel Aviv University, Ramat Aviv, Israel

Y. Imry

Tel Aviv University, Ramat Aviv, Israel
and Brookhaven National Laboratory
Upton, New York 11973[**]

1. INTRODUCTION

The spin glass state and the phase transitions into it have recently drawn much theoretical attention (1-9). In this phase the magnetic moments are supposed to be frozen but pointing in random directions, due to a competition between e.g. quenched ferromagnetic and antiferromagnetic interactions. In many systems of experimental interest (10-13) the magnetic ions are embedded in a metallic matrix and interacting with a RKKY coupling which becomes random in magnitude and sign due to the distribution of the inter-ion spacing. However, in many theoretical models simpler short range random exchange interactions are assumed (2-9). At the present stage the theories appear not to be conclusive due to the use of mean field theory which neglects critical fluctuations and difficulties with the $n = 0$ (2,14) trick at low temperatures (7,8). The more recent renormalization group calculations (6) have used an expansion around $d = 6$ (d is the number of space dimensions) which would not lead to accurate results at $d = 3$. In view of those difficulties it seems worthwhile to have an exactly soluble model for the spin glass transition which although not realistic enough to correctly represent the systems under consideration may nevertheless give one some new insights on the problem. Such a model was recently suggested by Mattis (9), where by assuming a particular distribution of the exchange interactions, the spins can be redefined, leading to a uniform exchange model. Thus, for a vanishing magnetic field all the properties of the Mattis model can be calculated in terms of those of the ordered model. A uniform magnetic field on the original model will become a random magnetic field (15,16) on the transformed spins. This leads to a number of predictions and conjectures (17) based on what is known on the random field model (16,18,19). The fact that the uniform magnetic field when applied on the spin glass has the same effect as a random field on an ordered uniform system is an exact result for the Mattis model. However, this type of result may be qualitatively valid also for more realistic spin glass systems. If indeed in the spin glass phase all spins are frozen and point in well-defined but random directions, an "up" magnetic field will appear to be of a random direction as viewed from the randomly varying "ordering" frame of reference of each spin. In fact, quite generally, the magnetic field tends to break up the spin glass ordering and turn it into a ferromagnetic type ordering. Our work highlights the importance of the magnetic field as a variable in the study of spin glasses. This has also motivated us to construct a scaling theory for the effects of the magnetic field, whose range of validity is much broader than the original Mattis model.

In section 2 we describe the Mattis model as following from a particular quenched random solid solution picture, and discuss its zero-field properties. The random field model is reviewed in section 3. The application to the spin glass problem is made and the more general scaling theory presented in section 4. Section 5 contains concluding remarks, generalizations and a discussion of the limitations of the model.

2. THE MODEL AND ITS ZERO FIELD PROPERTIES

Consider (17) a quenched magnetic alloy of the composition $A_p B_{1-p}$. Its Hamiltonian in a magnetic field B is

$$\mathcal{H} = -\sum_{<i,j>} J_{ij} \vec{S}_i \cdot \vec{S}_j - B \sum_i S_i^z , \quad (1)$$

where $\vec{S}_i$ is an n component classical spin vector with a unit length (including the Ising case, n=1) and J_{ij}, the nearest neighbors exchange, is given by J_{AA}, J_{BB} or J_{AB}, depending on whether the pair of ions on sites i and j are AA, BB or AB, respectively. It is easily seen, by defining a variable ε_j to be 1 (-1) if the site j is occupied by an A(B) ion, that in the case where $J_{AA} = J_{BB} = -J_{AB} = J$, which we shall henceforth assume, (1) becomes

$$\mathcal{H} = -J \sum_{<i,j>} \varepsilon_i \varepsilon_j \vec{S}_i \cdot \vec{S}_j - B \sum_i S_i^z . \quad (2)$$

This is the model suggested by Mattis (9). In the classical case one may redefine the spins (9)

$$\vec{\tau}_i = \varepsilon_i \vec{S}_i , \quad (3)$$

($\vec{\tau}_i$ is again an n-dimensional unit vector) and obtain:

$$\mathcal{H} = -J \sum_{<i,j>} \vec{\tau}_i \cdot \vec{\tau}_j - B \sum_i \varepsilon_i \tau_i^z \quad . \tag{4}$$

This is a uniform exchange model, but the magnetic field, while remaining in the z direction, is random in sign. For the case which we shall consider, where $p = 1/2$ and $\bar{\varepsilon} = 0$, one has $\overline{B\varepsilon_j} = 0$ (a bar over the quantity denotes an average over the disorder configurations). Clearly, all the zero field properties of this model are expressible in terms of those for the uniform exchange (ordered) model. In particular, all the thermodynamics, including the value of T_c and the specific heat singularity of the model (2) are exactly the same as those of the ordered model (4). The thermal average of a particular spin, in a given configuration of the ε's, is:

$$M_i = <S_i^z> = \varepsilon_i <\tau_i^z> = \varepsilon_i M_n(T) \propto \varepsilon_i |t|^{\beta_n} \quad , \tag{5}$$

for $T<T_c$, where M_n is the magnetization of the pure n component system and $t \equiv (T-T_c)/T_c$. z is assumed to be the direction along which the τ-system is ordered (20) and the subscript n refers to the critical exponents of the pure n-component system. Thus, clearly (for $\bar{\varepsilon} = 0$):

$$\bar{M}_i = \bar{\varepsilon} M_n(T) = 0 \quad . \tag{6}$$

The average magnetization is zero, while each spin, in a given configuration of the ε's, points along $\varepsilon_i \hat{z}$, which is the reason why the uniform field B appears as a random field to the spins as viewed around their random local average directions.

One may now define the usual spin-glass order parameter (2-6)

$$Q \equiv \overline{|<\vec{S}_i>|^2} = \overline{|<\vec{\tau}_i>|^2} \propto |t|^{2\beta_n} \quad . \tag{7}$$

This is not unique. One may also consider other order parameters, for example

$$\tilde{Q} \equiv \overline{|<S_i>|} \propto |t|^{\beta_n} \quad . \tag{8}$$

We shall have more to say about this arbitrariness in choosing the order parameter later.

The zero field spin-spin correlation function is given by

$$\Gamma^{\alpha\beta}(r_{ij}) \equiv \overline{<S_i^\alpha S_j^\beta> - <S_i^\alpha><S_j^\beta>} = \overline{\varepsilon_i \varepsilon_j} \, \Gamma_n^{\alpha\beta}(r_{ij}) \quad , \tag{9}$$

where $\Gamma_n(r_{ij})$ is the correlation function of the pure model. We can assume no long range correlations in the atomic positions: $\overline{\epsilon_i \epsilon_j} \to \overline{\epsilon}^2$ as $r_{ij} \to \infty$. Thus $\Gamma^{\alpha\beta}(r_{ij})$ will tend to a finite limit at long distances only if $\overline{\epsilon} \ne 0$ ($p \ne 1/2$) - which will then imply ferromagnetic long range ordering. In the case of interest to us, $\overline{\epsilon} = 0$ and we can assume that $\overline{\epsilon_i \epsilon_j}$ either vanishes for $i \ne j$ or decays to zero as a function of r_{ij} on some finite microscopic scale, of the order of a lattice constant. The correlation (9) is thus short ranged, its singular behavior is given by $t^{1-\alpha_n}$ for $T > T_c$ and by $|t|^{2\beta_n}$ for $T<T_c$. The latter behavior is due to the subtracted $\overline{<S_i>\,<S_j>}$ part in the definition (9).

The zero field magnetic susceptibility is given by

$$\chi = \frac{\partial M}{\partial B} = \frac{1}{TN} \sum_{i,j} \sum_\alpha \Gamma^{\alpha\alpha}(r_{ij}) = \frac{1}{T} \sum_\alpha \{\Gamma_n^{\alpha\alpha}(0) + \frac{1}{N} \sum_{i \ne j} \overline{\epsilon_i \epsilon_j} \Gamma_n^{\alpha\alpha}(r_{ij})\}. \quad (10)$$

For $p \ne 1/2$ this will diverge like $|t|^{-\gamma_n}$ as $T \to T_c$. However, for the pure-spin-glass ordering case, $p = 1/2$, χ has an assymetric cusp at T_c, behaving as $\frac{1}{T} + O(t^{1-\alpha_n})$ for $T>T_c$ and like $- O(|t|^{2\beta_n}) + O(|t|^{1-\alpha_n})$ for $T<T_c$. The $|t|^{1-\alpha_n}$ corrections follow when there are short range correlations in the ϵ's.

Below T_c, χ is a sum of $(1-M_n^2)/T$ and a term proportional to E_n/T due to possible short range correlations in the ϵ's. Thus, as $T \to 0$, χ (eq. 10) vanishes exponentially for the Ising model and as a power of T for $n \ge 2$. It should be emphasized that those predictions are specific for the Mattis model and may well change when the exchange distribution allows for weakly coupled spins and clusters.

The singular behavior of the resistivity, ρ, is a quantity which is relatively easy to measure. If there are short range correlations in the ϵ's, one finds here a cusp-like behavior of ρ characterized both above and below T_c by the exponent $1-\alpha_n$ (21).

It is interesting to note that although the model considered here should not be expected to describe real systems, the cusps in χ and ρ appear to agree qualitatively with experiments (10-12) and with Monte Carlo simulations (22). The behavior of the specific heat does not agree with the experiments of ref. 13a, where no anomalies in the specific heat were found. However, some anomalies were found for different materials in ref. 13b. What is perhaps more serious is that the low temperature behavior of the specific heat in this model is identical to that of the pure case. The special $O(T)$ [or $O(T^2)(8)$] term in the specific heat at low temperatures (1,7) does not appear, probably due to the lack of "loosely bound clusters" in this model.

As noted by Mattis (9), the nonlinear magnetic susceptibility diverges more strongly than the linear one as $T \to T_c$. By symmetry $M''(B=0) = 0$, while $M'''(B=0)$ is proportional to the fourth cumulant average of M

$$M'''(0) = -\frac{6}{T^3 N} \sum_{i,j} \Gamma_n^{zz}(r_{ij})\{\Gamma_n^{zz}(r_{ij}) - 2M_n(T)^2\} \propto |t|^{-\gamma_{3,B}} . \quad (11)$$

Assuming scaling for the pure correlation function Γ_n^{zz}, one finds that

$$\gamma_{3,B} = 2\gamma(n) - 2 + \alpha(n) . \quad (12)$$

$\gamma_{3,B}$ is thus positive and M''' diverges for $d<4$. It is interesting to note (17), that the divergence is related to the correlation function of the order parameter Q, $\Gamma_Q^{\alpha\beta}(r_{ij}) = \{\Gamma_n^{\alpha\beta}(r_{ij})\}^2$, so that $\gamma_{3,B} = \gamma_Q$. Note that since $\nu_Q = \nu_n$, $\beta_Q = 2\beta_n$, $\eta_Q = d-2+2\eta_n$, the critical exponents associated with the order parameter Q satisfy all the usual scaling relations.

On the other hand, one can also easily evaluate the correlation function of the order parameter $\tilde{Q}$ (eq. 8). It is easily seen that $\Gamma_{\tilde{Q}}^{\alpha\beta}(r_{ij}) = \Gamma_n^{\alpha\beta}(r_{ij})$, thus all exponents of $\tilde{Q}$ are identical to those of the pure system - $\nu_{\tilde{Q}} = \nu_n$, $\eta_{\tilde{Q}} = \eta_n$, $\gamma_{\tilde{Q}} = \gamma_n$, $\beta_{\tilde{Q}} = \beta_n$. These exponents again satisfy all the usual scaling relations. It is interesting to note that $\gamma_{\tilde{Q}} > \gamma_Q$; in fact, above $d=4$ $\gamma_{\tilde{Q}}=1$ while χ_Q does not diverge. Thus the instability of $\tilde{Q}$ is more pronounced than that of Q, which suggests that for this particular model $\tilde{Q}$ may be a more appropriate choice of the "main" order parameter than Q. While this question clearly needs further study, we remark that the order parameter $\tilde{Q}$ (which is equal to $\langle\tau\rangle$) appears in fact to be analogous to the one suggested by Kosterlitz, Thouless and Jones (8b) in another exactly solvable model. There the exchange matrix is diagonalized by a linear transformation on the spins and the most unstable mode, corresponding to the maximum eigenvalue, is suggested as the natural order parameter. In our case this mode is trivially associated with the variable τ.

3. PROPERTIES OF THE RANDOM FIELD MODEL

Eq. (4) with $B \neq 0$ contains a quenched random magnetic field. The general problem where the field conjugate to the order parameter is random and quenched was considered for the ideal Bose gas ($n\to\infty$) in ref. 15, and for general n in ref. 16. Such models appear to be physically realizable at present for spin glasses, as discussed here, and for displacive and charge-density-wave transitions; their critical properties are sufficiently different from the usual ones to warrant a careful study. Two main general results were obtained in ref. 16. Heuristic arguments were given to show that in systems with a continuous symmetry ($n \geq 2$) an arbitrarily weak random field should eliminate long range ordering for $d \leq 4$. This can be seen as due both to the diverging fluctuations perpendicular to the assumed direction of ordering and to the instability of the ordered state towards domain formation following the statistical fluctuations of the random field. The latter argument indicates that for $n = 1$ the random field will break the long range order for $d \leq 2$. Also, the dimensionality at which the critical behavior becomes mean-field-like was found to be $d=6$ (independent of n) and expansions for the critical exponents were found to order $\varepsilon = 6-d$. These expansions are identical to those of the pure (no random field) model around $d = 4$. This latter identity has been proven to $O(\varepsilon^3)$ by Grinstein (18), who also found that the hyperscaling law $\nu d = 2-\alpha$ is replaced by $\nu(d-2) = 2-\alpha$ and conjectured that this reduction of d by two might hold more generally. This conjecture was proven to be valid to all orders in perturbation theory in ref. 19, where heuristic arguments were also given to interpret it physically. It appears that the lowering of the effective dimensionality by two can be obtained approximately by assuming (19) that the dominant cause for disorder close enough to T_c is not thermal fluctuations but the disordering energy due to statistical fluctuations in the random field. A separate heuristic argument was given (19) for $n=1$ using Thompson's (23) picture. For $n=1$ the reduction of d by two is not valid for lower dimensionalities (the critical dimension below which long range order is lost decreases from $d=2$ with the random field to $d=1$ with no random field). This particular qualitative argument yields $\nu = (d+2)/4(d-2)$ for $2<d<6$, i.e. $\nu=5/4$ at $d=3$ and $\nu\to\infty$ when $d\to 2$. Although one should not trust this prediction numerically, it is suggested that the critical behavior with the random field at $d=3$ and $n=1$ will be extremely interesting and very different from the usual one. Experiments on Ising like spin glasses in a magnetic field might probe this interesting new critical behavior, while the usual Heisenberg case should verify the ideas about the elimination of the long range order in a random field.

4. APPLICATION TO THE SPIN GLASS IN A FIELD

In the first part of this section we shall apply the results of section 3 to the Mattis model in a field. Although the results are strictly valid only for this particular case, our above arguments indicate that they may have a wider range of validity. In the second part of this section we present a more general scaling theory for the effect of a magnetic field on a spin glass.

The fact that a random field disrupts the long range ordering in continuous symmetry systems suggests that the sharp phase transition will also be eliminated. This latter statement is not obviously valid however, due to the fact that a sharp transition to another spin-glass-like phase cannot be ruled out, although one would not expect a sharp transition with a nonzero field present which is conjugate to the ordering mode. Assuming that the random field (in the τ system) does indeed smear the transition - one would predict the cusps in ρ and χ at T_c to be smeared out in the presence of the field. Such smearing does indeed occur experimentally, for systems that approximate Heisenberg spin-glasses (12). A real test for our theory would be if the measurements of ρ(T) and χ(T) with and without the field woulb be made for an Ising-like spin glass. Our prediction in this case would be that the cusps in ρ and χ <u>should not</u> be smeared in a d = 3 system, because for Ising-like systems the random field disrupts ordering only below d = 2 ! Another important prediction would be the unusual values of the critical exponents for an Ising-like spin glass in a field at d = 3.

A smearing of the cusp in χ at a finite field has also been predicted by the previous theories (1-8) of the spin glass transition. We remark however that a $|t|^{1-\alpha}$ singularity is an extremely delicate critical effect which can be hardly expected to be accurately obtained from a mean field theory. Thus, further work is needed in order to ascertain the validity of the cusp in χ as obtained in mean field theory, as well as its smearing in a magnetic field. It should also be noted that in the mean field theory, similar results would be obtained for Heisenberg and Ising systems. The smearing is thus predicted in these theories to occur also in the Ising case. As mentioned above, experiments on the latter systems would be invaluable for clarifying this question.

Since the magnetic field is an important variable for spin glasses, it is natural to write down a scaling theory for the singular part of the free energy as a function of t and B. This has the form (choosing B^2 as the variable, due to symmetry reasons)

$$F(t,B^2) \propto |t|^{2-\alpha} f(B^2 |t|^{-\phi}) \tag{13}$$

where α is the specific heat index of the system without a field and φ is the cross-over exponent. For the particular case of the Mattis model $\alpha = \alpha_n$ and (16,17) $\phi = \gamma_n$. However, eq. 13 should be generally valid even for realistic systems, albeit with different values of α and φ. From a scaling relation like (13) one can easily obtain the critical singularities in the higher order non-linear susceptibilities (24). The kth magnetic field derivative of M diverges as $|t|^{\gamma_{k,B}}$ where

$$\gamma_{k,B} = \alpha - 2 + (k + 1)\phi/2 \tag{14}$$

Eq. (12) for the divergence of M''' in the Mattis model is a special case of this with $k = 3$, $\alpha = \alpha_n$, $\phi = \gamma_n$. In the general case measurements of χ and M''' (which should be much more strongly divergent!) should yield the correct exponents α and ϕ. Similar scaling assumptions can be made for the correlation functions.

5. CONCLUDING REMARKS

It should be emphasized that the particular model embodied in Eq. (2) and obtained for the AB mixture when $J_{AA} = J_{BB} = -J_{AB}$ and $p = 1/2$, is indeed very specialized. If the above restrictions are relaxed, one can easily show that many of the additional terms are relevant in the renormalization group sense. Thus, the critical behavior of real spin glasses at $B = 0$ can be expected to be different from that of the pure system. However, the difference between our results and those of Ref. 6 may not be totally due to the difference in the models considered. For the particular case of the Mattis model one can see that the usual truncation of the expansion of the effective $n \to 0$ Hamiltonian in the cumulants of the exchange distribution does not yield the correct results at any finite order. The possible necessity of summing the cumulants to all orders could also affect the result of a renormalization group calculation (25), although it might well be an artifact of the Mattis model.

Having a solvable model allowed us to obtain some insights on the importance of the magnetic field and the identification of the "main" order parameter. A more general scaling theory for the former effect was given. Experiments on real spin glasses with magnetic fields should be invaluable to clarify the phase diagram, the nonlinear properties in the field and the question of the smearing of the cusps. Finally, the critical behavior of an Ising-like spin glass in a finite field is suggested as an outstanding problem both experimentally and theoretically.

The model considered here can also be used in studies of the time dependence of the magnetic properties and of the transformation between quenched and annealed behaviors (26).

We conclude by mentioning that random single site anisotropies which are of interest can be handled by similar models. A particular model might be given by

$$\mathcal{H} = - J \sum_{\langle ij \rangle} (\hat{x}_i \cdot \vec{S}_i)(\hat{x}_j \cdot \vec{S}_j) - B \sum_i S_i^z, \tag{15}$$

where x_i is a randomly oriented unit vector. Denoting $\tau_i = x_i \cdot \vec{S}_i$ yields an Ising like model with random field which should be similar to the model of Ref. 27 (28). Also, random anisotropy fields might play a role similar to that of the random magnetic field in breaking long range order for $n \geq 2$.

ACKNOWLEDGMENTS

We thank Professor D.C. Mattis for a preprint of Ref. 9 and Professors M. Blume, M.E. Fisher, P.G. de Gennes, T.C. Lubensky, A.P. Young and especially S. Ma for helpful discussions.

REFERENCES

*Supported by a grant from the United States - Israel Binational Science Foundation (BSF), Jerusalem, Israel.

**Supported by USERDA.

1. C. Held and M.W. Klein, Phys. Rev. Lett. 35, 1783 (1975) contains many previous references; see also M.W. Klein, unpublished.

2. S.F. Edwards and P.W. Anderson, J. Phys. F 5, 965 (1975).

3. D. Sherrington and B.W. Southern, J. Phys. F 5, L49 (1975).

4. K.H. Fischer, Phys. Rev. Lett. 34, 1438 (1975) and Solid State Comm. 18, 1515 (1976).

5. D. Sherrington and S. Kirkpatrick, Phys. Rev. Lett. 35, 1792 (1975).

6. A.B. Harris, T.C. Lubensky and J.H. Chen, Phys. Rev. Lett. 36, 415 (1976).

7. For a recent review, see D. Sherrington, AIP Conf. Proc. (in press).

8a. D.J. Thouless, P.W. Anderson, E. Lieb and R.G. Palmer, unpublished.

8b. J.M. Kosterlitz, D.J. Thouless and R.C. Jones, Phys. Rev. Lett. 36, 1217 (1976).

9. D.C. Mattis, Phys. Lett. 56A, 421 (1976).

10. Many experiments are reviewed in J.A. Mydosh, AIP Conf. Proc. 24, 131 (1974); T.A. Fiory, AIP Conf. Proc. (in press).

11. C.N. Guy, J. Phys. F$\underline{5}$, L242 (1975).

12. D.E. Murnick, A.T. Fiory and W.J. Kossler, Phys. Rev. Lett. $\underline{36}$, 100 (1976).

13a. L.E. Wenger and P.H. Keesom, Phys. Rev. B$\underline{11}$, 3497 (1975).

13b. B.V.B. Sarkissian and B.R. Coles, Comm. on Physics $\underline{1}$, 17 (1976) and references therein.

14. V.J. Emery, Phys. Rev. B$\underline{11}$, 239 (1975).

15. P. Lacour-Gayet and G. Toulouse, J. Phys. (Paris) $\underline{35}$, 425 (1974).

16. Y. Imry and S. Ma, Phys. Rev. Lett. $\underline{35}$, 1399 (1975). In this paper the spin redefinition is also mentioned as possibly leading to a random magnetic field.

17. A. Aharony and Y. Imry, Solid State Comm. (in press).

18. G. Grinstein, unpublished.

19. A. Aharony, Y. Imry and S. Ma, unpublished.

20. It may be argued that $\langle \tau_i^z \rangle = 0$, on the grounds that the field conjugate to the order parameter $\langle \tau \rangle$ is strictly zero, while an infinitesimal symmetry breaking field is needed, which is conventionally sent to zero <u>after</u> the thermodynamic limit is taken. While this remark is correct in principle, its validity may be restricted only to infinite time averages. Although the <u>infinite</u> time average of $\vec{\tau}_i$ is zero, the average $\vec{\tau}$ should point, within any reasonable macroscopic time, in a well defined direction which we choose as the z-axis.

21. M.E. Fisher and J.S. Langer, Phys. Rev. Lett. $\underline{20}$, 665 (1968). Note that in the Born approximation, the leading singularity in the resistivity below T_c is $|t|^{1-\alpha_n}$ and not $|t|^{2\beta_n}$ due to the fact that ρ is determined by the average $\overline{\langle SS \rangle}$ correlations without the subtraction in Eq. 9. We thank M.E. Fisher and S. Alexander for discussions on this point. See also, e.g., S. Alexander, J.S. Helman and I. Balberg, Phys. Rev. B$\underline{13}$, 304 (1976).

22. K. Binder and D. Stauffer, Phys. Lett. $\underline{57A}$, 177 (1976).

23. C. Thompson, J. Phys. A$\underline{9}$, L25 (1976).

24. Y. Imry, Phys. Rev. B$\underline{13}$, 3018 (1976).

25. See, in this connection, A. Aharony, Y. Imry and S. Ma, Phys. Rev. B13, 466 (1976).

26. The annealed case of the present model was considered by E. Domany, Phys. Lett. 49A, 339 (1974). See also, A. Aharony and B.A. Huberman, J. Phys. C9, L465 (1976).

27. R. Harris, M. Plischke and M.J. Zuckermann, Phys. Rev. Lett. 31, 160 (1973).

28. Chen and Lubensky (private communication) found an Ising spin glass phase for this system, using the methods of Ref. 6.

MAGNETIC ORDERING OF AuCr: AN ULTRASONIC INVESTIGATION

G.F. Hawkins, T.J. Moran[*] and R.L. Thomas

Department of Physics, Wayne State University

Detroit, Michigan 48202

INTRODUCTION

The magnetic properties of the noble metal-3d transition metal spin glass alloys show a number of unusual characteristics, as summarized in the experimental review paper by Mydosh.[1] In particular, the sharp peaks in the magnetic susceptibility which scale systematically with transition metal concentration are suggestive of the onset of some kind of magnetic ordering. Recently a careful search has been made for evidence of such ordering in the specific heat, both for AuFe[2] and CuMn[3] spin glass alloys. No sharp changes were observed in the vicinity of the susceptibility peak temperature, T_o, although the previously reported[4,5] linear dependence of the low temperature magnetic contribution to the specific heat, C_m, was confirmed, and in the case of CuMn,[3] rounded maxima were observed at temperatures above T_o. Results of measurements of the changes in magnetic entropy in the same experiments[2,3] indicate that fewer than one third of the transition metal ions participate in magnetic ordering between T_o and 0 K. Specific heat measurements of AuCr have also been carried out recently.[6] Broad maxima were observed for a range of Cr concentration 0.02 at.% < C < 0.1 at.%, at temperatures which scale linearly with C. This region of concentration does not quite overlap that of the existing susceptibility data[1,7] (0.2 at.% < C < 10 at.%).

A number of molecular field theories[8-12] have been proposed to explain the sharp peak in the low field susceptibility and the linear term in $C_m(T)$ at low temperatures. A common feature of these theories, however, is the prediction of a cusp or discontinuity at T_o, in apparent disagreement with experimental evidence to date. An earlier proposal[13] that a spin density wave state could account for the susceptibility peak also predicts a cusp in the specific heat.

Ultrasonic experiments have provided useful information in the study of magnetic phase transitions. It can be shown from general thermodynamic arguments that the temperature dependence of the sound propagation is simply related to that of the magnetic specific heat, and sharp maxima in the ultrasonic attenuation and minima in the ultrasonic velocity have been observed for such transitions.[14] Relative velocity changes can be measured with a resolution of a part in 10^7, so that depending upon the strength of the magnetoelastic coupling, the ultrasonic experiments provide a sensitive test for specific heat anomalies of the sort expected theoretically.

In an earlier paper,[15] the authors investigated the temperature dependence of the ultrasonic velocity in a single crystal of $Au_{0.92}Fe_{0.08}$ which was cut from the same boule as that used for the specific heat measurements.[2] No evidence was found for a velocity minimum near T_o. The present paper extends this ultrasonic investigation of spin glass alloys to the case of AuCr.

EXPERIMENTAL RESULTS

Measurements were made of the longitudinal wave ultrasonic velocity as a function of temperature at 13 MHz for three polycrystalline alloys of AuCr for which low field susceptibility measurements have previously been reported.[1,7] We have repeated the susceptibility measurements in order to check the quantitative agreement of the ultrasonic data with T_o. The samples were cold worked in order to reduce the polycrystalline grain size before making ultrasonic measurements. Despite this fact, grain boundary scattering limited the measurements to a single transmitted ultrasonic pulse. Measurements were made using a phase comparison technique.[14] Commercially calibrated Pt and Ge thermometers were used.

Figure 1 shows the temperature dependence of the 13 MHz longitudinal wave sound velocity for $C = 10.8$ at.%. The maximum deviation from the background is seen to occur at $T_m = 95$ K, in close agreement with the position of the susceptibility peak $T_o = 96$ K. This deviation is plotted as a function of reduced temperature, $\varepsilon = (T - T_m)/T_m$, in Fig. 2. Also shown in Fig. 2 are the corresponding minima for $C = 6.1$ at.% and $C = 3.7$ at.%. For each of the three samples studied the position of the velocity minimum was found to correspond in temperature to that of the susceptibility

MAGNETIC ORDERING OF AuCr

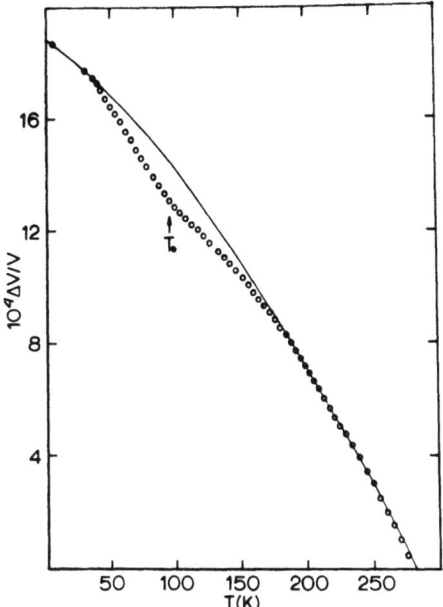

FIG. 1. Temperature dependence of the longitudinal wave sound velocity at 13 MHz in Au 10.8% Cr. The solid curve is an estimated background. The susceptibility peak temperature, T_o, is indicated.

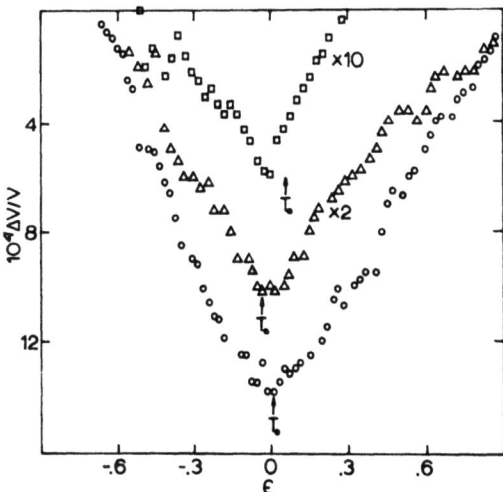

FIG. 2. Deviations of the sound velocity from background as a function of reduced temperature.

- ☐ $C = 3.7$ at.%, $T_m = 39$ K, $T_o = 41.5$ K
- △ $C = 6.1$ at.%, $T_m = 58$ K, $T_o = 56.2$ K
- ○ $C = 10.8$ at.%, $T_m = 95$ K, $T_o = 96$ K

peak to within the experimental uncertainty. It may also be seen in Fig. 2 that the magnitude of the velocity minimum increases rapidly and its width broadens somewhat with increasing chromium concentration. If one compares these minima with those observed for typical cooperative antiferromagnetic or ferromagnetic phase transitions (for which the uncorrected velocity curves usually display a characteristic inverted lamda minimum[14]) they can be regarded as being quite rounded. On the other hand, if one compares the present data with minima observed in this laboratory[16] for insulating magnetic aluminosilicate glasses (which also display sharp susceptibility peaks), they are comparatively narrow. Deviations from the background velocity temperature dependence in the aluminosilicate glasses, for example, extend to $|\varepsilon| \sim 5$, whereas the present data for AuCr exhibit deviations only for $\varepsilon \lesssim 1$.

These data represent, indirectly, the first experimental evidence for the existence of peaks in the specific heat at temperatures close to T_o for a spin glass. Calorimetric measurements for AuCr alloys for which sharp susceptibility peaks have also been observed would be of interest to confirm this result. The shapes of the minima in the sound velocity reported here do not suggest that the specific heat will show a cusp or discontinuity. A definitive resolution of this important point, however, awaits further experimental study.

ACKNOWLEDGMENTS

The authors are grateful to Dr. V. Cannella for the loan of the samples and to Dr. Cannella and Professor A.M. de Graaf for useful discussions.

REFERENCES

*Present Address: AFML/LLP, Wright-Patterson AFB, Dayton, Ohio 45433.

1. J.A. Mydosh, AIP Conf. Proc. 24, 131 (1975).

2. L.E. Wenger and P.H. Keesom, Phys. Rev. B11, 3497 (1975).

3. L.E. Wenger and P.H. Keesom, AIP Conf. Proc. 29, 233 (1976).

4. J.E. Zimmerman and F.E. Hoare, J. Phys. Chem. Solids 17, 52 (1960); L.T. Crane and J.E. Zimmerman, J. Phys. Chem. Solids 21, 310 (1961).

5. F.J. du Chatnier and J. de Nobel, Physica 28 181 (1962).

6. T. Kemény, G.J. Nieuwenhuys and H. Algra, Low Temperature Physics - LT 14, Vol. V, ed. M. Krusius and M. Vuorio (to be published).
7. V. Cannella and J.A. Mydosh, AIP Conf. Proc. $\underline{18}$, 651 (1974).
8. K. Adkins and N. Rivier, J. Phys. (Paris) $\underline{35}$, C4, 237 (1974).
9. S.F. Edwards and P.W. Anderson, J. Phys. F$\underline{5}$, 965 (1975).
10. K.H. Fischer, Phys. Rev. Letters $\underline{34}$, 1438 (1975).
11. D. Sherrington and B.W. Southern, J. Phys. F$\underline{5}$, L49 (1975).
12. B.W. Southern, J. Phys. C$\underline{8}$, L213 (1975).
13. A.W. Overhauser, J. Phys. Chem. Solids $\underline{13}$, 71 (1960).
14. B. Lüthi, T.J. Moran and R.J. Pollina, J. Phys. Chem. Solids, $\underline{31}$, 1741 (1970).
15. G.F. Hawkins, T.J. Moran and R.L. Thomas, AIP Conf. Proc. $\underline{29}$, 235 (1976).
16. T.J. Moran, N.K. Batra, R.A. Verhelst and A.M. de Graaf, Phys. Rev. B$\underline{11}$, 4436 (1975); T.J. Moran, N.K. Batra, F. Bucholtz and R.L. Thomas, 1974 Ultrasonic Symposium Proceedings, IEEE Cat. #74 CHO 896-1SU (1974).

TWO APPROACHES TO THE THEORY OF SPIN-GLASSES: A COMPARISON WITH EACH OTHER AND WITH EXPERIMENT

Michael W. Klein[+]

Department of Physics, Bar-Ilan University

Ramat-Gan, Israel

We use the selfconsistent Mean-Random-Field (MRF) approximation to derive the properties of spin-glasses in the random molecular field approximation. For the Gaussian distributed exchange potentials discussed by Sherrington and Kirkpatrick (SK) we obtain identical magnetic properties and a somewhat similar specific-heat to that obtained by SK who used the $n \to 0$ expansion of the free energy derived by Edwards and Anderson. For the RKKY-potential treated in the Ising model, the MRF-method gives the very low temperature specific heat and magnetic susceptibility in excellent agreement with experiment. It similarly gives a spin-glass transition, where the transition temperature $T_c \propto c^\alpha$, where c is the impurity concentration and $\alpha \approx 0.66$ for the Ising model and preliminary results give that $\alpha \approx 0.61$ for the Heisenberg model. Similarly we find that there are small but important concentration dependent correction to the "concentration independent" low temperature specific heat. A possible reason for the agreement of the Ising like model and the disagreement of the Heisenberg like prediction with experiment is given.

1. INTRODUCTION

The method of the Mean-Random-Field approximation is a selfconsistent method developed by the author[1] to obtain the properties of a random magnetic system in the molecular field approximation. The method is different from the usual Weiss-molecular-field approximation used in a ferromagnet in the following respects. Whereas in the Weiss-molecular field approximation the molecular field has constant value (or possibly several values in the case of a system with more than one sublattice) throughout the system below T_c and is zero

[+] Part of this work supported by USAFOR Grant #73-2430B

above T_c, where T_c is the transition temperature, in the MRF-approximation the field varies from site to site and has an infinite set of values below the transition temperature. In the original derivation of the MRF-method[1] for the RKKY[2]-interaction, the distance of closest approach between the impurities was allowed to go to zero. This has exaggerated the importance of large internal fields and gave a transition for all temperatures, thus representing a physically unrealistic system at high temperatures. Recently Riess and Klein[3] have shown that if there is a cutoff in the interaction potential (i.e. the maximum allowed value of the RKKY-potential is limited by its near-neighbor value) the MRF-approximation gives a spin-glass-transition with a well defined spin-glass order parameter m. The order parameter m arises naturally in the theory, and the width of the probability distribution is proportional to m. Thus they[3] find that above the transition temperature the probability distribution of the internal fields is a δ-function centered about the field H = 0. Below the transition temperature the width of the the probability distribution increases monotonically with decreasing temperature. Such a point of view is consistent with the recent μ-meson depolarization experiments of Fiory et. al.[4] performed on several spin-glasses. Whereas there is some ambiguity in the Riess-Klein[3] derivation with regards to the concentration dependence of the transition temperature, an appropriate modification of the MRF-method removes this ambiguity and gives that $T_c \propto c^\alpha$, where $\alpha \approx 0.66$ for the Ising distribution of fields. Preliminary results indicate that $\alpha \approx 0.61$ for the Heisenberg distribution of fields.

The probability distribution $P(\vec{H})$ of the field $\vec{H}$ in a Heisenberg like distribution was derived by Adkins and Rivier[5] for T = 0 using a somewhat different method, and the probability distribution of the magnitude of the field was derived by Held and Klein[6] using the MRF-approximation. Whereas the Ising like distribution gives that the probability distribution P(H) for zero field, P(H=0), is finite, the Heisenberg distribution gives $P(H \approx 0) \propto H^2$. Thus these two distribution predict markedly different low temperature thermal properties. This difference exhibits itself in the low temperature specific heat, the low temperature magnetic susceptibility and the low temperature resistivity. One of the dilemmas, which is at present not yet understood, is that whereas the Ising-like distribution explains the experimentally observed data on the low temperature specific heat[7,8], the low temperature magnetic susceptibility[9] and the low temperature resistivity[10], the presumably more correct Heisenberg model distribution function explains none of these experimentally measured quantities[11,12].

An alternative approach to the theory of spin-glasses was recently proposed by Edwards and Anderson (EA)[13] and Sherrington and Kirkpatrick (SK)[14]. They used a set of Gaussian distributed random potentials to obtain the free energy of a spin glas system.

Using the MRF-method to obtain the thermodynamic properties of a set of Gaussian distributed exchange potentials of SK[14], we obtain the identical magnetic properties and a somewhat similar specific heat to that of SK[14].

A summary of this paper is as follows. We first compare the MRF method to the results obtained by the $n \to 0$ expansion of the free-energy of Sherrington Kirkpatrick. Next we discuss in detail the nature of the spin-glass transition in the MRF-approximation. Finally we give the results for the low temperature specific heat and magnetic susceptibility and the low T resistivity in the Heisenberg and Ising MRF-approximation. Our model also predicts small but important concentration dependent corrections to the "concentration independence" of the Ising-model low T specific heat.

2. MOLECULAR FIELD THEORY FOR THE GAUSSIAN DISTRIBUTED EXCHANGE POTENTIAL

In this section we show that for the Gaussian distributed exchange potentials used by SK[14] the random molecular field model gives identical magnetic properties to that determined by SK.

Consider the Hamiltonian $\mathcal{H}$ of the form

$$\mathcal{H} = -\tfrac{1}{2} \sum_{i \neq j} J_{ij} S_i S_j \qquad (2.1)$$

where J_{ij} is a random variable. SK[14] use a probability distribution of J_{ij} of the form

$$P(J_{ij}) = (2\pi J^2)^{-1/2} \exp\{-(J_{ij} - J_o)^2/(2J^2)\} \qquad (2.2)$$

for each J_{ij}.

Using the molecular field approximation[1] we have for the internal field at site i, H_i,

$$H_i = \sum_j J_{ij} \langle S_j \rangle \qquad (2.3)$$

where $\langle S_j \rangle$ is the thermal average of the spin at site j in the field H_j as is discussed in Ref. (1) (Henceforth denoted by I). Using the statistical model of Margenau[15] in the form developed in Eq. (2.7) of I we obtain

$$P(H_i) = \int P(R) d^3R \, \delta(H_i - \sum J_{ij} \langle S_j \rangle), \qquad (2.4)$$

where for the case considered by SK[14]

$$P(R) d^3R = \prod_{i \neq j} P(J_{ij}) dJ_{ij} \qquad (2.5)$$

where $P(J_{ij})$ is given by Eq. (2.2).

Substituting Eq. (2.5) in Eq. (2.4) and letting $\delta(x) = (2\pi)^{-1} \int \exp(i\rho x)d\rho$ gives

$$P(H_i) = \frac{1}{2\pi} \int \exp[i\rho H_i] d\rho \prod_{j=1}^{N} \int P(J_{ij}) dJ_{ij} \exp[-i\rho J_{ij}\langle S_j\rangle]$$

$$= (2\pi)^{-1} \int \exp\{i\rho[H_i - J_o \sum_j \langle S_j\rangle] - (J^2\rho^2/2) \sum_j \langle S_j\rangle^2\} \quad (2.6)$$

We define the average magnetization M by the relation,

$$M = \langle\langle S_i\rangle\rangle_c \equiv \int P(H_i) \tanh\beta H_i \, dH_i \quad (2.7)$$

where $\langle\langle S_i\rangle\rangle_c$ indicates an average of $\langle S\rangle$ over the coordinates of the system. We also define the spin-glass order parameter m

$$m = \langle\langle S_i\rangle^2\rangle_c \equiv \int P(H_i) \tanh^2\beta H_i \, dH_i \quad (2.8)$$

where as a selfconsistency condition we require that $zM = \sum \langle S_i\rangle$ and $zm = \sum_i \langle S_i\rangle^2$, where z is the number of effective neighbors. This result in the self consistent solution that $P(H_i)$ is independent of i.

We remark that our M corresponds to SK's magnetization and our m corresponds to SK's order parameter q. Using the definitions Eqs. (2.7) and (2.8) in Eq. (2.6) and integrating gives

$$P(H) = (2\pi m z J^2)^{-\frac{1}{2}} \exp\{-[H-J_o zM]^2/(2J^2 mz)\} \quad (2.9)$$

where z is the effective number of neighbors.

Using Eq. (2.9) in Eq. (2.7), and changing variables of integrating gives

$$M = (2\pi)^{-\frac{1}{2}} \int \exp(-x^2/2) dx \tanh\beta\{J\sqrt{zm}\, x + J_o zM\}. \quad (2.10)$$

Using Eq. (2.9) in Eq. (2.8) and letting $\tanh^2 x = 1 - \text{sech}^2 x$, we obtain

$$m = 1 - (2\pi)^{-\frac{1}{2}} \int \exp(-x^2/2) dx \, \text{sech}^2\beta\{J\sqrt{zm}\, x + J_o zM\} \quad (2.11)$$

The results for the magnetization M and the spin-glass order parameter m arising from the MRF-approximation are thus identical with that obtained by SK[14] from the evaluation of the free-energy using the n-expansion.

The low T specific heat C is in the MRF-model

$$C_v(T) = N \int P(H) C_v(H) dH. \quad (2.12)$$

The internal energy U is $U = -(H/2)\tanh\beta H$ and $C_v(H) = [H^2/k_BT^2)]\cdot\text{sech}^2\beta H$. Using this result in Eq. (2.12) and changing variables of integration gives

$$C_v(T) = \frac{N}{2k_BT^2} \int_{-\infty}^{\infty} \frac{e^{-x^2/2}}{\sqrt{2\pi}} \{\sqrt{mz}\, x + J_o zM\}^2 \text{sech}^2\beta\{\sqrt{mz}\, x + J_o M\} dx \tag{2.13}$$

For the spin glass transition with $M = 0$ we obtain, for the leading term in the specific heat at very low temperatures,

$$C_v = Nk_B\left(\frac{k_BT}{z^{1/2}J}\right)\cdot\left(\frac{2}{\pi}\right)^{1/2}\left[\left(\frac{\pi^2}{12}\right) - A\right] \tag{2.14}$$

where $A = 0$ for our case and $A = (2\pi)^{-1}$ in SK's derivation. The agreement of the specific heat with that of SK[14] is limited to low temperatures, for at high temperatures SK[14] obtain that $C_v \propto T^{-2}$ and the molecular field approximation gives $C_v = 0$ for $T > T_c$, where T_c is the transition temperature to be discussed later on. However the magnetic properties from the two methods are identical.

3. SPIN GLASS TRANSITION WITH THE RKKY INTERACTION

In the original derivation of the MRF-approximation[1] the probability distribution, P(H), of the field H was derived by allowing the distance of closest approach between the impurities to be zero. The derivation exaggerated the importance of large fields and gave a Lorentzian P(H), and thus an infinite transition temperature. For a realistic physical system the distance of closest approach between the impurities is limited to a near neighbor distance r_{nn}. It was shown by Riess and Klein[3] that when one introduces a cutoff in the interaction at r_{nn} one obtains a concentration dependent spin glass transition. We give next a derivation of the spin-glass transition somewhat modified from that presented by Riess and Klein[3]. Let

$$m = \int_{-\infty}^{\infty} P(H) |S\, B_S(H)|\, dH \tag{3.1}$$

where the vertical brackets indicate absolute values. $B_S(H)$ is the Brillouin function experiencing a field H. Then the selfconsistent probability distribution P(H) is[3]

$$P(H) = \pi^{-1} \int_0^{\infty} \cos \rho H \exp[-(4\pi/3)c\, n_o a\rho f\{n_o a\rho(d/r_c)^3\}]\, d\rho \tag{3.2}$$

where n_o is the number of sites per unit cell, d is the lattice constant, a is the strength of the interaction at a distance of one lattice constant, r_c is the cutoff distance, i.e. the distance of closest approach between the impurities, c is the fractional impurity concentration, and

$$f(x) = \int_0^x [(1 - \cos t)/t^2] \, dt \qquad (3.3)$$

Equation (3.2) gives the probability distribution for the Ising system. In Eq. (3.2) we have shown explicitly the dependence of P(H) on the cutoff distance r_c. With the probability distribution given in the form Eq. (3.2) we remove the ambiguity in the concentration dependence of the transition temperature obtained by Riess and Klein[3]. We note that the integral, Eq. (3.2), depends upon the parameter m, where m itself is given as an integral of the probability distribution. Thus Eq. (3.2) represents an integral equation for P(H).

For the case when the cutoff distance $r_c \to 0$, we obtain that $f(x) \to \pi/2$, and we recover the Lorentzian probability distribution of I.

When $r_c = d$, Eq. (3.2) takes the particularly simple form,

$$P(H) = (\pi a n_o m)^{-1} \int_0^\infty \cos x [\frac{H}{a n_o m}] \exp[-(4\pi/3)cxf(x)] dx \qquad (3.4)$$

Let the magnitude of the k-th moment of the probability distribution be η_k, then

$$\eta_k = \int_{-\infty}^\infty P(H) |H^k| \, dH \qquad (3.5)$$

$$= (2/\pi)(a n_o m)^k \int_0^\infty \int_0^\infty \cos(xy) y^k \exp[-(4\pi/3)cxf(x)] dx \, dy$$

$$\equiv \eta_k(0) m^k$$

where $\eta_k(0)$ is independent of the temperature and is a function of c, a and S. The value of m defined in Eq. (3.1) becomes small for high temperatures. We will show later on that $m = 0$ for $T \geq T_c$, where T_c is the spin glass order parameter. Thus it turns out that whereas m enters our selfconsistent solution naturally, it has all the properties of an order parameter. The physical meaning of m is directly obtained from Eq. (3.4), where we find that only $H/(a n_o m)$ enters the integral for P(H). Thus the width of the probability distribution-function is proportional to m. This shows that when $m \to 0$, the width of the probability distribution also approaches zero, and P(H) approaches a delta-function at $H = 0$. Thus all the spins experience zero fields for $T > T_c$.

Since for high temperatures m is small, the probability distribution is strongly concentrated near $H = 0$. This allows the expansion of Eq. (3.1) is a power series of (βH) to give

TWO APPROACHES TO THE THEORY OF SPIN GLASSES

$$m = \int P(H) \, dH \left[\frac{S+1}{3} \beta H - \frac{(S+1)(2S^2 + 2S + 1)}{90} (\beta H)^3 + .. \right]$$

$$= \frac{S+1}{3} \beta \eta_1(0) m - \frac{(S+1)(2S^2 + 2S + 1)}{90} \beta^3 \eta_3(0) m^3 + O(m^5) \qquad (3.6)$$

Equation (3.6) follows directly from Eq. (3.5). Solving Eq. (3.6) we find that $m = 0$ for $T > T_c$, where

$$T_c = \frac{S+1}{3} \eta_1(0) \qquad (3.7)$$

$\eta_1(0)$ itself is proportional to S as can be seen from Eq. (3.1), since $m(T = 0)$ is proportional to S. We have calculated the concentration dependence of $\eta_1(0)$ with a computer and found that

$$\eta_1(0) \propto c^\alpha \qquad (3.8)$$

where $\alpha \approx 0.66$ for the Ising case for $1\% < c < 5\%$ and $\alpha \approx 0.5$ for $c > 6\%$.

Equations (3.7) and (3.8) hold for the Heisenberg distribution as well. Preliminary estimates show that α for this case is approximately 0.61 for $1\% < c < 5\%$.

We can solve for all $m(T)$ as a function of T selfconsistently using Eq. (3.1). To obtain m for T near T_c it is sufficient to use Eq. (3.6), neglecting powers of m higher than m^3 in the moment expansion of $B_S(x)$.

We remark that the value of m is not usually measured in an experiment, and can be obtained indirectly only, by examining the temperature at which the probability distribution of the fields changes from a delta function to a function having finite width. Such a measurement was presented by Fiory et. al.[4] who measured the μ-meson depolarization in spin glasses and found a sharp increase of the depolarization at some fixed temperature associated with T_c.

We next obtain the magnetic susceptibility for the spin-glass system.

The magnetization M for $N = N_o c$ spins is given by

$$M = N_o c \int P(H) [SB_S(H+B)] \, dH \qquad (3.9)$$

where B is the externally applied field. The dependence of P(H) upon B has to be considered also, however, since P(H) depends only upon B^2 its derivative will not contribute to the magnetic susceptibility.

In the limit $B \to 0$, we obtain from Eq. (3.9) that M is identically zero for all T, since P(H) is symmetric in H. The magnetic susceptibility, χ, is obtained by differentiating M with respect to B, we have

$$\chi = N_o c \int P(H) \lim_{B \to 0} \frac{d}{dB} \{SB_S(H+B)\} dH \qquad (3.10)$$

Let $P_{eff}^2 = S(S+1)(g\mu_B)^2$. Then for temperatures near T_c we expand Eq. (3.10) in a power series of H and obtain

$$\chi = \frac{N_o c P_{eff}^2}{3k_B T} \int P(H) dH [1 - \frac{2S^2+2S+1}{10(k_B T)^2} \beta^2 H^2]$$

$$= N_o c P_{eff}^2 / (3k_B T) \{1 - \frac{2S^2+2S+1}{10(k_B T)^2} \eta_2(0) m^2\} \qquad (3.11)$$

where we have used Eq. (3.5) to obtain the second of Eq. (3.11).

For $T \geq T_c$, $m = 0$ and we obtain

$$\chi = N_o c P_{eff}^2 / (3k_B T) \qquad : \quad T \geq T_c \qquad (3.12)$$

we thus obtain a T^{-1} dependence of the paramagnetics susceptibility rather than the $(T - T_c)^{-1}$ dependence for ferromagnets.

Solving Eq. (3.6) for m^2 and substituting into Eq. (3.11) gives

$$\chi = N_o c P_{eff}^2 / (3k_B T) [1 - \{3\eta_2(0)/\eta_3(0)\}(T_c - T)] \qquad (3.13)$$

Examining Eq. (3.13) as $T \to T_c$ shows that χ is continuous at $T = T_c$, however χ has a discontinuous derivative at $T = T_c$, and

$$\frac{d\chi}{dT}] = \frac{N_o c P_{eff}^2}{3k_B T} [\frac{3\eta_2(0)\eta_1(0)}{\eta_3(0)} -1]; \quad T \lesssim T_c \qquad (3.14)$$

Equation (3.14) shows that the slope of the susceptibility at T just below T_c is only a function of the first three moments of the distribution function evaluated at $T = 0$.

Using Eq. (3.8) in Eq. (3.12) gives that

$$\chi(T_c) \propto c^{1-\alpha} \qquad (3.15)$$

where $\alpha \approx 0.66$ for the Ising model, and preliminary results indicate that $\alpha \approx 0.61$ for the Heisenberg-like distribution.

We remark that it is expected that the molecular field description will not give the correct critical indices and the appropriate behaviour of the system in the critical region very close to T_c, however the description of where the phase transition occurs as well as the behaviour of the system above the transition temperature is expected to be given reasonably well by the molecular field approximation. The Riess and Klein[3] derivation is the first one that gives the description of the system for a realistic Ruderman-Kittel[2] potential. Further detailed calculations of the behaviour of the system near the transition region is given by the author elsewhere[11,12].

4. THE LOW TEMPERATURE PROPERTIES

Finally we give the very low temperature properties of the RKKY spin-glass system in the MRF-approximation. All the results are obtained by expressing the thermodynamic property of a single spin in an effective field H and integrating over the probability distribution of all fields. The results were obtained by Fisher and Klein[11] and Klein[12].

Examining the expression for P(H) given in Eq. (3.4) we find that for very low impurity concentrations and small internal fields ($an_0m \gg H$) the probability distribution is approximately Lorentzian, whereas for large internal fields ($an_0m \ll H$) P(H) drops off like $\exp[-H^2/\sigma^2]$. Since the very low temperature properties are determined by small values of H we use a Lorentzian to calculate these properties. A complete calculation shows that there are small, but important, concentration dependent corrections to the very low temperature thermodynamic properties due to the cutoff in the potential at a near neighbor distance.

We first neglect these corrections and later discuss how they modify the low T specific heat.

The low temperature low field magnetization M in the Ising model is[11] approximately ($k_B T \ll \gamma c$, $\mu_B B \ll k_B T$)

$$M = \frac{N_0 B(g\mu_B)^2}{\gamma}\left\{\frac{2}{\pi} + \frac{4}{\pi}\ell n(2S+1)\left(\frac{k_B T}{\gamma c S}\right) + K(S)\left(\frac{k_B T}{\gamma c S}\right)^2 + O(T^3)\right\} \quad (4.1)$$

where $\gamma = 2\pi^2 an_0/3$, and $K(S) = 16[\ell n(2S+1)]^2/\pi^3 - 4\pi/[3(2S+1)]$. Equation (4.1) predicts that as $T \to 0$, M is independent of the impurity concentration and increases linearly with T for small T. K(S), the curvature of the susceptibility is positive for $S > 2$, is approximately zero for $S = 3/2$ and is negative for $S \leq 1$. The behaviour of M was found to be in agreement with the first two terms of Eq. (4.1) by Franz and Sellmyer [8]. The spin dependent curvature remains to be compared with experiment.

The low T high field susceptibility in the Ising model ($k_B T \ll \gamma c$, $\mu_B B \gg \gamma c$) is given

$$M = N_0 c\, g\mu_B S\left\{1 - \frac{2\gamma Sc}{\pi g\mu_B B}\left[1 + \frac{2\pi^2}{3(2S+1)}\right]\left(\frac{k_B T}{g\mu_B B}\right)^2\right] + O(c^3)\right\} \quad (4.2)$$

This result is in qualitative agreement with the experiments of

Hou and Coles[15]. The reader should be cautioned that there is still a controversy on the low T high B (B >> γc) behaviour as was pointed out recently by Smith[16].

The low temperature resistivity of certain spin glasses measured by Ramos[10] is found to be linear in T and independent of the impurity concentration c, again in agreement with the Ising model MRF-prediction obtained by Harrison and Klein[17].

We next return to the correction to the low temperature specific heat C_H and magnetic susceptibility χ due to the cutoff at r_{nn}. This cutoff reduces the probability for large field and hence increases the probability for fields near H = 0. For the case with no cutoff we have that $P(H = 0) = (\pi\Delta)^{-1}$, where $\Delta = \gamma mc$. When we consider the correction due to the cutoff we obtain[18]

$$P(H = 0) = \exp\{(4\pi/3)\sqrt{2}\ c\}\ \{\pi\Delta\}^{-1} \qquad (4.3)$$

where Eq. (4.3) was obtained for an fcc lattice[18] with $r_c = r_{nn}$. Since the very low T specific heat is from the MRF-model proportional to $cP(H = 0)$, the model predicts that the very low T specific heat is linear in T and proportional to $\exp[4\pi\sqrt{2}\ c/3]$. Previous calculation gave that C_H is independent of c. The small concentration dependent correction, if conformed experimentally, gives an important correction which exhibits an additional aspect of the probability distribution.

5. ISING VS. HEISENBERG MODEL

Before we conclude we would like to comment further on the question of the Ising vs. Heisenberg model. We found that the experimental low T properties of the system are reasonably well explained by an Ising model but disagree completely with the Heisenberg MRF-predictions[11]. We are thus faced with the dilemma that the presumably incorrect Ising model agrees with experiment, whereas the presumably correct Heisenberg model does not. We have no clearcut explanation for this difficulty. However, we wish to make some possible suggestions for the reason for this difficulty.

It was shown in Sect. 3 that a spin glass transition occurs at some temperature T_c, where the average magnetization is zero, however the square of the local magnetization is non zero. This transition exists in the Heisenberg as well as the Ising model MRF-approximation. Now in the Heisenberg model, the direction of the vector spin changes from site to site, however each spin, regardless what its direction is locally, contributes to building up the nonvanishing order parameter below T_c. At T near zero the spins are frozen-in in their random directions. The magnetization is a function of $\vec{r}$ and is a vector quantity $\vec{M}(\vec{r})$ depending upon $\vec{r}$, such that $\int \vec{M}(\vec{r}) d\vec{r} = 0$. Now consider a spin at $\vec{r} = \vec{r}_i$. $\vec{M}(\vec{r}_i)$ has a certain well defined direction, because the field $\vec{M}(r_i)$ has a well defined direction, the direction being dictated by the orientations and positions of $N_o c - 1$ other spins. The Heisenberg distribution was obtained by assuming that each spin may be oriented in an arbitrary direction, thus resulting in a degeneracy of the field as a function of the angle, and thus the probability for the field H is proportional to $H^2 dH$. When the system orders, some particular component $\vec{H}(\vec{K})$ of the vector field $\vec{H}$ predominates the system, presumably resulting in some complicated spiral-spin-like structure of the spins. Since the system is now ordered, the local spin at a particular site $\vec{r}_i$, $M_i(\vec{r}_i)$, will not have an arbitrary direction of orientation [i.e. will not have all possibly allowable orientations with equal probability] but will be frozen-in in the direction determined by the order parameter locally. Thus when calculating the probability distribution, the volume element will no longer be $H^2 dH$, but rather closer to an Ising like distribution. Clearly these arguments are at best hand-waving and no proof for a realistic system.

I wish to acknowledge Professor Joel L. Lebowitz for many helpful discussions.

REFERENCES

1. M.W. Klein, Phys. Rev. 173, 552 (1968): Ibid Phys. Rev. 188, 933 (1969)
2. M.A. Ruderman and C. Kittel, Phys. Rev. 96, 99 (1954): T. Kasuya, Prog. Theor. Phys. (Kyoto) 16, 45 (1956): K. Yosida, Phys. Rev. 106, 893 (1957)
3. I. Riess and M.W. Klein, to be published
4. A.T. Fiory, D.E. Murnick, M. Leventhal, and W.J. Kossler, Phys. Rev. Lett. 33, 969 (1974): D.E. Murnick, A.T. Fiory, and W.J. Kossler, Phys. Rev. Lett. 36, 100 (1976)
5. K. Adkins and N. Rivier, J. de Physique 35, C4-237 (1974); Amorphous Magnetism, Edited by H.D. Hooper and A.M. de Graat, Plenum Press, New York, 1973, page 215
6. C. Held and M.W. Klein, Phys. Rev. Lett. 35, 1783 (1975)
7. J.E. Zimmerman and F.E. Hoare, J. Phys. Chem. Solids 17, 52 (1960)
8. J.M. Franz and D.J. Sellmyer, Phys. Rev. B8, 2083 (1973)
9. B. Dreyfus, J. Souletie, J.L. Tholence and R. Tournier, J. Appl. Phys. 39, 846 (1968)
10. E.D. Ramos, Journ. of Low Temp. Phys. 20, 547 (1975)
11. B. Fischer and M.W. Klein, to be published
12. M.W. Klein, to be published
13. S.F. Edwards and P.W. Anderson, J. Phys. F5, 965 (1975)
14. D. Sherrington and S. Kirkpatrick, Phys. Rev. Lett. 35, 1792 (1975)
15. H. Margenau, Phys. Rev. 48, 755 (1935)
16. F.W. Smith, Phys. Rev. Lett. 36, 1221 (1976)

17. R.J. Harrison and M.W. Klein, Phys. Rev. 154, 540 (1967), ibid. B1, 940 (1970), Erratum: Phys. Rev. 167, 875 (1968)
18. M.W. Klein, to be published

LOCALIZED MOMENT AND SPIN-GLASS-LIKE BEHAVIOR IN AMORPHOUS YFe$_2$

D. W. Forester

Naval Research Laboratory, Washington, D.C. 20375

W. P. Pala and R. Segnan[†]

American University, Washington, D.C. 20016

INTRODUCTION

Mössbauer studies reported in this paper were made on a bulk amorphous sample of YFe$_2$(a-YFe$_2$) prepared by rapid sputtering techniques at Battelle Northwest Laboratories[1]. Rare earth-Fe$_2$ alloys prepared by that group have been shown to be structurally amorphous using neutron diffraction[2] and x-ray diffraction[3] techniques. Previous bulk magnetization[4] and neutron diffraction measurements[5] made on this same a-YFe$_2$ sample have revealed anomalous magnetic behavior. The magnetization data taken in fields above 1000 Oe showed no evidence of a magnetic phase transition down to 3.5K, the lowest temperature used. Small angle neutron scattering[5] showed an intense magnetic component at low temperatures with a broad peak at T $\cong$ 40K for the smallest used momentum transfer vector, q=0.04Å^{-1}. Below T $\cong$ 50K a small magnetic hysteresis and remanent magnetization developed which increased at lower temperatures. The hysteresis, remanent magnetization and scattering peak were attributed to possible anisotropy effects. High field magnetization measurements at 1.5K and 60 kOe produced a moment of 1.1μ_B per iron atom, considerably less than the saturated value of about 1.7μ_B. This result was interpreted as indicating itinerant-moment rather than localized moment behavior for the iron magnetization. Spin-spin correlation lengths inferred from the neutron work varied from 8Å to 6Å between T$\sim$60 and 175K.

We present here Mössbauer results which show a magnetic transition at T $\cong$ 55K below which the magnetic system displays spin-glass-like behavior. The anomalous magnetic behavior observed earlier can be associated with this "spin-glass" phase.

EXPERIMENTAL

The Mössbauer absorber was prepared in powdered form by grinding in an inert argon atmosphere. The powder was immediately combined with an organic binder and pressed into a pellet. Spectra were taken in the constant acceleration mode and in the zero velocity, single-channel mode where thermal scans were made to search for possible magnetic ordering transitions.

The temperature dependence of the Mössbauer spectra is shown in Figs. 1 and 2. Above $T \cong 58K$ the spectra exhibit a quadrupole doublet which is characteristic of all RE-Fe$_2$ amorphous alloys studied to date[6-7]. As the temperature is lowered below 58K a magnetic hyperfine splitting develops. The average magnetic hyperfine field $\bar{H}_{eff}(T)$ grows smoothly with decreasing T and follows closely an S=1 Weiss molecular field dependence as shown in Fig. 3. The measured value $\bar{H}_{eff}(5K)=233 \pm 6$ kOe is comparable with that in a-GdFe$_2$[8] ($\bar{H}_{eff}(5K)=248$ kOe) which has the largest hyperfine field of any heavy rare earth-Fe$_2$ alloy.

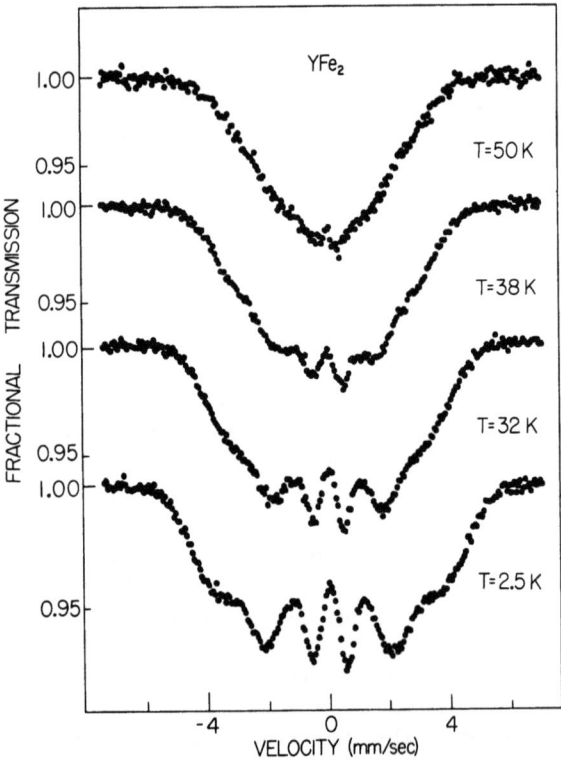

Fig. 1. ^{57}Fe Mössbauer spectra in amorphous YFe$_2$.

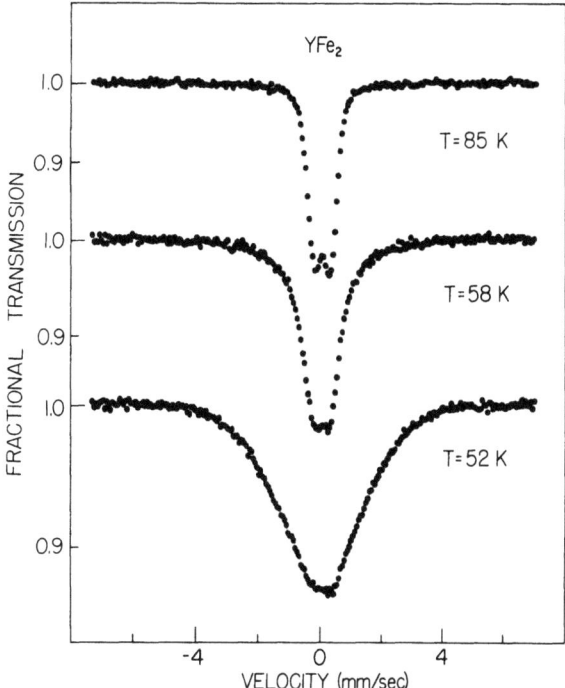

Fig. 2. ^{57}Fe Mössbauer spectra in amorphous YFe$_2$.

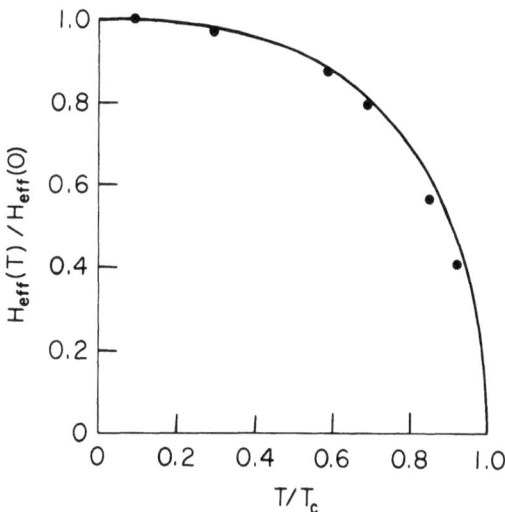

Fig. 3. Reduced average hyperfine field versus reduced temperature of amorphous YFe$_2$ with T_c=55K and $\bar{H}_{eff}(0)$=233 kOe. The temperature dependence closely follows a Weiss Molecular Field model for S=1 (solid curve).

As noted above, the transition near T=55K was not detected in earlier studies. To determine the properties of the low temperature magnetic phase we performed several experiments. First of all, the transition temperature is by no means sharply defined as seen in Fig. 2. Data taken with zero velocity-single channel temperature scans is given in Fig. 4. A thermal hysteresis is noted when the transition region is spanned quickly enough. These data show that the transition is sluggish. To eliminate hysteresis, scans must be taken at a rate of 5 minutes per degree or slower. The hysteresis is not reflected in the spectra of Figs. 1 and 2 since each of these were taken over a period greater than 10 hours. Preliminary low field (~10 Oe) susceptibility measurements[9] on this sample show a cusp near 55K which is completely washed out at 1000 Oe.

To further define the low temperature state, spectra were taken in external magnetic fields up to 70 kOe. For purposes of comparison, similar spectra were taken on a-GdFe$_2$. Gd was chosen since it is an S state ion and therefore its anisotropy contribution should be minimal. This alloy also has the highest Curie temperature[10], T_c=500K, of any of the heavy rare earth-Fe$_2$ materials. Although not a rare earth, Y has been found to be a good rare earth substitute[11]. Like Gd, Y has no orbital anisotropy but unlike Gd it carries no magnetic moment. Spectra taken on a-GdFe$_2$ at T=20K are shown in Fig. 5. As expected, a field of

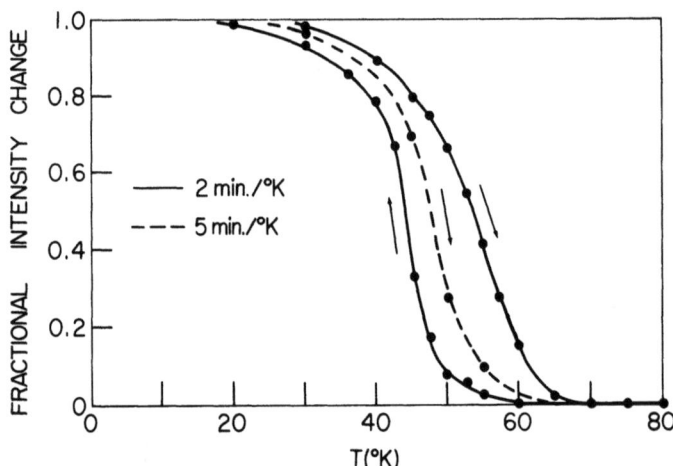

Fig. 4. Fractional intensity change versus temperature at zero relative velocity between source and absorber. The solid curves were taken at a sweeep rate of 2 min/°K and illustrate thermal hysteresis near the transition temperature. The dashed line was taken slowly enough to eliminate the hysteresis effect.

7 kOe, sufficient to overcome the sample demagnetizing field, completely aligns the moments in this low-anisotropy ferrimagnet. (We know the Fe moments are aligned since the $\Delta m_I = 0$ transition are eliminated in this geometry with the gamma-ray propagation direction taken parallel to the applied magnetic field.) When the same experiment is performed with a-YFe$_2$ at T=10K as shown in Fig. 6, completely different and anomalous behavior is observed. At 10 kOe the spectra are virtually unchanged while at 20 kOe (not shown) they broaden only slightly. This behavior is not characteristic of amorphous ferromagnetism but is similar to that recently defined[12] as "speromagnetism". At higher temperatures $\overline{H}_{eff}$ is fractionally reduced as is the intensity of the $\Delta m_I = 0$ lines indicating a partial alignment of the moments with the external field. The decrease in $\Delta m_I = 0$ intensity, although not apparent because of broad overlapping lines, is confirmed by the two theoretical fits shown as solid lines through the data in Fig. 6. This partial alignment allows the earlier reported[4] growth of low-temperature bulk magnetization with applied field to be described in terms of local moments rather than an itinerant moment model. From the theoretical intensities in Fig. 6 we deduce what crudely can be called an average angle, $\overline{\theta}$, which the magnetic moments make with the applied field. At 60 kOe we obtain $\overline{\theta} \simeq 50°$. Comparing this with the previous bulk magnetization result and using $1.1\mu_B = \mu_{Sat}\cos\overline{\theta}$ we calculate $\mu_{Sat} \simeq 1.7\mu_B$ in agreement with the expected saturated Fe moment.

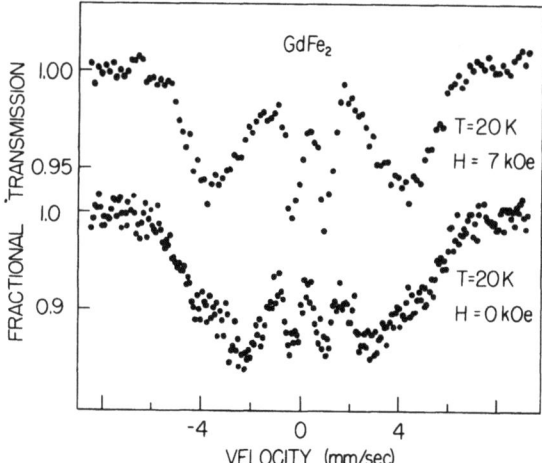

Fig. 5. ^{57}Fe Mössbauer spectra in amorphous GdFe$_2$ at T=20K. The top spectrum was taken in an external field of 7kOe along the gamma-ray propagation direction. Elimination of the two $\Delta m_I = 0$ transitions near ± 2 mm/sec shows that the Fe moments are completely aligned with the external field.

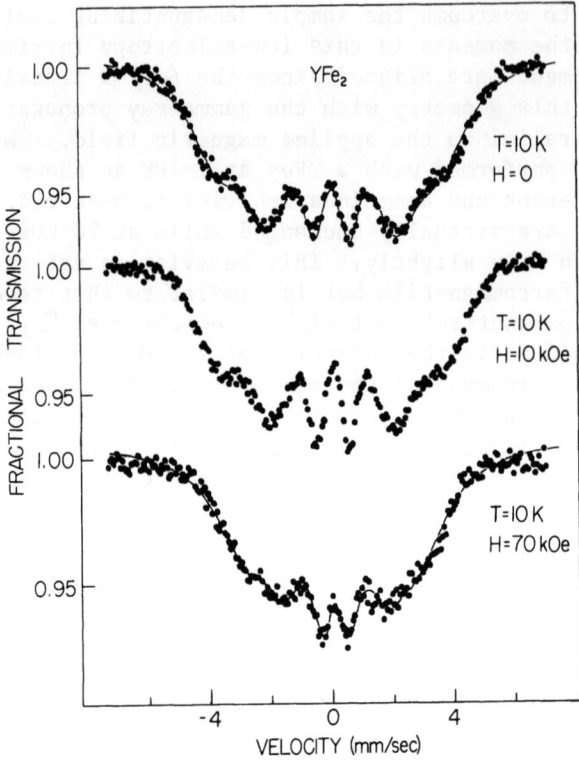

Fig. 6. ^{57}Fe Mössbauer spectra in amorphous YFe$_2$ at T=10K. The magnetic field was applied parallel to the gamma-ray propagation direction. At H=10 kOe there is very little difference in the spectrum from that at H=0. At H=70 kOe the spectrum is collapsed slightly and the Δm_I =0 intensity is reduced. The solid curves through the data are computer least squares fits described in the text.

Spectra were also taken at a number of temperatures and magnetic fields above the 55K transition. A typical spectrum taken at T=163K and H=70 is shown in Fig. 7. It is made up of a magnetic and a "non-magnetic" component. Although the magnetic component collapses as the temperature is raised, the magnetic and non-magnetic components appear to maintain the same relative areas. A plot of the magnetic component's splitting versus H/T indicates superparamagnetic clusters with a distribution of sizes up to roughly 6Å in diameter, containing about 50 Fe spins. This is in agreement with the spin-spin correlation lengths observed in the neutron scattering studies[5]. For the range of H/T values available to us the "non-magnetic" component could be either small (S≤2) superparamagnetic clusters with ferromagnetic intra-cluster exchange or they could be distributed size clusters with antiferromagnetic intra-cluster exchange.

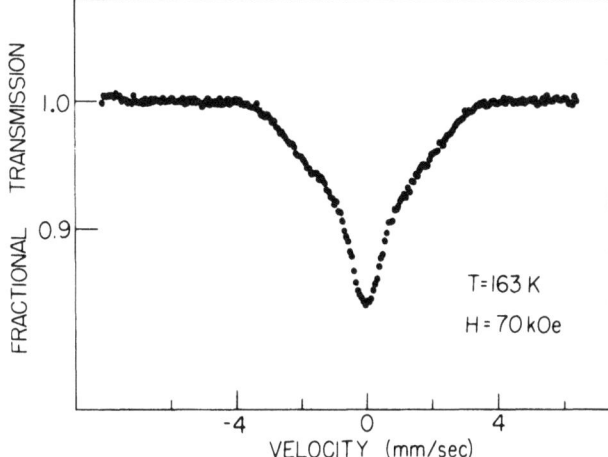

Fig. 7. ^{57}Fe Mössbauer spectrum in amorphous YFe$_2$ at T=163K and H=70 kOe. This spectrum and others at different H/T values for T>55K show a broad "magnetic" hyperfine spectrum superimposed on an unsplit "non-magnetic" component. The magnetic component has an H/T dependence corresponding to superparamagnetic clusters with a maximum size containing approximately 50 Fe spins.

DISCUSSION AND CONCLUSIONS

Unlike the heavy rare earth-Fe$_2$ alloys which are amorphous ferrimagnets, a-YFe$_2$ displays characteristic spin-glass behavior. First of all, the Mössbauer results show that below T=55K, the "freezing" temperature, the moments are frozen in random directions in 3-dimensional space. The picture which seems most appropriate here is a glassy state dominated by a distribution of magnetic exchange with competing ferromagnetic and antiferromagnetic interactions. This picture is supported by the very large magnetic fields required to even partially align the Fe moments. Magneto-crystalline anisotropy is apparently ruled out since it is difficult to justify a larger anisotropy in a-YFe$_2$ than in a-GdFe$_2$ which was aligned in very low fields. Secondly, the magnetic transition at 55K reflects a "viscous" freezing and exhibits irreversibility and time dependent effects frequently observed in spin-glass systems. The presence of a cusp in the low field susceptibility which washes out at higher fields is further indication of such behavior. Put in this context, the earlier observations of magnetic hysteresis and remanence also imply a spin-glass state. The earlier magnetization studies failed to show a magnetic transition because the minimum field used, 1000 Oe, is enough to suppress it.

Although it cannot be proven from the data, it is interesting to speculate on the magnetic and "non-magnetic" parts of the spectra above 55K as illustrated in Fig. 7. Perhaps these are residual short-range ordered ferromagnetic and antiferromagnetic clusters which remain above the "freezing" temperature.

The Mössbauer results remove the need for an itinerant moment explanation of the earlier low temperature, high field magnetization results. The Fe magnetic moment has been found to scale roughly as the hyperfine field in these alloys[13]. Since $\bar{H}_{eff}(5K)$ in a-YFe$_2$ is comparable to that in a-GdFe$_2$, their moments are expected to be roughly the same, $1.7\mu_B$. The Mössbauer data indicate that Fe has this moment in zero external field. The enhanced moment at high fields, then, is due to a physical rotation of the moments arising from a torque produced by the field which competes with the collective ferromagnetic and antiferromagnetic exchange in the glassy state.

REFERENCES

1. We wish to thank Dr. A. Clark for providing us with this sample.

2. J. J. Rhyne, S. J. Pickart, and H. A. Alperin, AIP Conf. Proc. 18, 563 (1974).

3. G. S. Cargill, AIP Conf. Proc. 18, 631 (1974).

4. J. J. Rhyne, J. H. Schelleng and N. C. Koon, Phys. Rev. B10, 4672 (1974).

5. S. J. Pickart, J. J. Rhyne, and H. A. Alperin, Phys. Rev. Lett. 33, 424 (1974).

6. D. Sarkar, R. Segnan, E. K. Cornell, E. Callen, R. Harris, M. Plischke and M. J. Zuckerman, Phys. Rev. Lett. 10, 542 (1974).

7. D. W. Forester, R. Abbundi, R. Segnan, and D. Sweger, AIP Conf. Proc. 24, 115 (1975).

8. D. W. Forester, Int. Conf. on Structure and Excitations of Amorphous Solids (to be published in AIP Conf. Proc. 1976).

9. D. W. Forester, W. P. Pala, R. Segnan, N. Koon, J. J. Rhyne and J. H. Schelleng (paper in preparation).

10. J. J. Rhyne, AIP Conf. Proc. 29, 182 (1976).

11. W. C. Koehler in *Magnetic Properties of Rare-Earth Metals*, R. J. Elliott, editor, Plenum Press, New York, London (1972).

12. J.M.D. Coey and P.W. Readman, Nature 246, 476 (1973).

13. N. Heiman and K. Lee, Phys. Lett. A55, 297 (1975).

† Work supported in part by Office of Naval Research Contract No. N00014-75-C-0736.

DISCUSSION

N. D. Heiman: I would like to add that we have also looked at $ZrFe_2$ and from the results we have gotten so far it looks as though $ZrFe_2$ which is electronically different from YFe_2 has exactly the same behavior. But on the other hand $LaFe_2$ which is electronically the same than YFe_2 is a good ferromagnet. The only difference between the two that is striking is the size effect. La is a very big ion while Y and Zr are smaller. So it seems that the size effect of the rare earths determines whether the alloy will be a spin glass or a ferromagnet.

D. W. Forrester: In these rare earths-Fe_2 alloys, if you have a rare earth incorporated, it seems to enhance the Fe-Fe interaction. The Fe-Fe interaction from a number of experiments seems to be the dominant factor with the rare earth-Fe_2 being next in line and the rare earth-rare earth lowest. However, if you replace that rare earth with a non-rare earth the Fe-Fe interaction becomes very small as you can see by the transition going way down, and what looks to be a distribution of exchange between ferromagnetic and antiferromagnetic that is quite small. It is unusual and unexplained.

R. W. Cochrane: Have you tried any different composition than that of YFe_2?

D. W. Forrester: No, I have not, but Neil Heiman has.

D. D. Heiman: We looked at a series of compositions up to YFe_5 and they all look more or less the same. We find our results to be in agreement with Don Forrester's as far as the Mössbauer effect is concerned. We find the internal field has gone up as we increase the Fe content implying the Fe moment is going up but we still see the spin glass effect behavior.

REAL SPACE RENORMALIZATION GROUP CALCULATIONS FOR SPIN GLASSES

A.P. YOUNG

Institut Laue-Langevin

156X Centre de Tri, 38042 Grenoble Cédex, France

ABSTRACT

The Edwards and Anderson spin-glass model is investigated by a real space rescaling transformation for Ising spins in two and three dimensions. A transition to a spin glass phase is found in three dimensions but no transition is predicted for d = 2.

1. INTRODUCTION

The possibility of phase transitions in magnetic systems with competing ferromagnetic and antiferromagnetic interactions at random has recently received considerable attention. There is evidence [1,2] that a sharp phase transition does occur in so called spin glass systems where randomly positioned dilute magnetic impurities in a non magnetic host metal are coupled by the RKKY interaction and the competition between ferromagnetism and antiferromagnetism arises because of the oscillatory dependence of the RKKY coupling with distance. On the other hand much recent theoretical effort has gone into studying a model, first proposed by Edwards and Anderson [3], (hereafter known as EA) where the spins lie on a regular lattice and the interactions between nearest neighbour pairs are assumed to be independent random variables which can take positive or negative values. While the EA short range model and real spin glass systems differ in many ways both have the important feature of competing ferromagnetic and antiferromagnetic interactions at random and are expected to display the same general behaviour.

The Hamiltonian for the EA model with Ising spins can be

written as

$$H = -\frac{1}{2} \sum_{i,j} J_{ij} S_i S_j \qquad (1)$$

where $S_i = \pm 1$, the interaction J_{ij} is between nearest neighbours only and each J_{ij} is an independent random variable with a probability distribution $P(J)$ assumed symmetric. The 'replica' technique used by EA and several subsequent authors [4,5] predicts a transition to a state where the spins are locked in fixed orientations but there are as many spins pointing up as down so the net magnetization is zero. For Ising spins the transition temperature T_c^{mf} is predicted to be

$$T_c^{mf} = \langle J^2 \rangle^{1/2} z^{1/2} \qquad (2)$$

where z is the lattice coordination number. The replica technique has problems at low temperatures [4] but has recently been shown [6] to predict the transition temperature correctly in the limit of infinite range interactions. For finite range interactions the situation is less clear and in particular it has not proved possible to develop an expansion in $(1/z)$ away from the mean field solution of reference 6. Computer simulations on a square [7] and a simple cubic [8] lattice indicate that a transition does occur but at a temperature T_c where $T_c < T_c^{mf}$ and for a Gaussian distribution

$$\frac{T_c}{T_c^{mf}} = \begin{cases} 0.50 & \text{square} \\ 0.61 & \text{simple cubic} \end{cases} \qquad (3)$$

However the relaxation to equilibrium was extremely slow and followed a power law or logarithmic variation rather than an exponential decay. Consequently there is a possibility that the simulations see a rapid increase in the spin relaxation time rather than a singularity in the true equilibrium properties. There are also complicated dynamical and irreversible effects observed experimentally [9] and it is still not definitely established whether or not the experiments indicate a genuine phase transition.

The present paper attempts to answer the question of whether a phase transition occurs for the EA model with Ising spins in two and three dimensions. A mean field assumption, which builds in the possibility of a transition from the start, is not made, and the replica technique is also avoided. Instead we use the rescaling, or real space renormalization group methods, which have proved very successful in reproducing known results for the two and three dimensional Ising model [10]. Consequently we feel that these

REAL SPACE RENORMALIZATION GROUP CALCULATIONS

techniques can now be used in situations where very little is known from other methods.

Our principal conclusion is that no transition occurs for the square lattice but that spin glass ordering does take place for the simple cubic lattice at about half the mean field transition temperature given in equation (2). For a Gaussian distribution of J our approximation gives $T_c/T_c^{mf} = 0.44$ which is in fair agreement with the Monto Carlo results considering that the three dimensional rescaling we employ is relatively crude. However the complete discrepancy between our results and the Monte Carlo studies in two dimensions is not understood. It is possible that the Monte Carlo estimate of T_c depends on the number of time steps allowed, because of the very slow relaxation to equilibrium, and that after an infinite time the estimated T_c would have decreased, perhaps to zero for $d = 2$. We feel that the more significant result is the presence of a transition in three dimensions confirming the Monte Carlo studies and mean field approximations. It remains to be seen whether the Heisenberg model would exhibit spin glass order for $d = 3$.

The rescaling technique employed involves performing a trace over a certain fraction of the spins, in two dimensions we use half the spins, a procedure which has become known as 'decimation' [11]. Correlations between the spins that remain can be described by a new effective Hamiltonian and the problem is to calculate the interactions in the new Hamiltonian in terms of the starting interactions. In one dimension this can be done exactly [12,13] but in higher dimension the new Hamiltonian contains interactions which were not present originally, and some approximation has to be made to truncate these extra terms beyond a certain point. One then iterates this procedure, in principle an infinite number of times, until a trace has been performed over all the spins. At each stage a spin independent term is generated and the free energy is just the sum of all these factors [14] because after all the traces have been performed these are the only terms left.

One can find out more conveniently whether or not a transition occurs by following the behaviour of the interactions between spins under iteration. For systems without disorder the effective couplings remain unchanged under iteration at the transition [10], which is a consequence of the infinite correlation length. This is the so called 'fixed point' of the recursion relation. In disordered systems the physics depends on the probability distribution for the interactions so a transition occurs when this probability distribution remains unchanged [15]. For $T > T_c$ the interactions decrease and go to a trivial fixed point with all interactions zero while for $T < T_c$, in the case of ferromagnets, the interactions grow and flow to a fixed point at infinity. For spin glasses the probability distribution is symmetric and so we expect that in the ordered spin glass phase, the interactions will flow to a fixed point where the interactions have values plus or minus infinity at random. This conjecture is confirmed for the one dimensional case [12,13].

In section 2 the rescaling transformation for the two dimensional square lattice is discussed within several approximations. The width of the distribution of J is found to go always to zero under iteration even down to very low temperatures, showing that correlations die off exponentially at large distances and that consequently there is no transition. For the simple cubic lattice discussed in section 3 the width of the distribution grows at low temperatures indicating spin glass order. A brief discussion of these results is given in section 4.

2. TWO DIMENSIONAL CALCULATIONS [12]

Except in one dimension the decimation transformation introduces interactions which were not present in the original problem, such as second neighbour couplings and four spin terms. Furthermore, in random systems, correlations between the distributions for different interactions are generated by the rescaling. In order to make the calculations tractable these extra complications have to be neglected beyond a certain point. A convenient way to do this is to perform the rescaling exactly on a finite cluster, so all interactions and correlations which can be fitted into the cluster are included and others are neglected. First of all we consider the simplest possible approximation obtained by working with the cluster of four sites shown in Fig. 1a. The trace over sites k and l is evaluated to generate a new effective interaction between i and j so only nearest neighbour couplings are considered at each stage and correlations are neglected. Building up a lattice out of these clusters one sees that the sites which are left still form a two dimensional square lattice but the lattice spacing is increased by a factor of $b = \sqrt{2}$. Denoting the interactions round the sides of the square in figure 1a by J_i, $i = 1...4$, and defining $t_i = \tanh\beta J_i$ then one finds

$$t' = \frac{t_1 t_2 + t_3 t_4}{1 + t_1 t_2 t_3 t_4} \qquad (4)$$

where $t' = \tanh\beta J'$ with J' equal to the new interaction between i and j.

Before considering the spin glass case the method will be illustrated for the case of a non random ferromagnet.

The distribution $P(t)$ is given by $P(t) = \delta(t-t_0)$ and clearly $P'(t')$ is also a single delta function at t_0'

REAL SPACE RENORMALIZATION GROUP CALCULATIONS

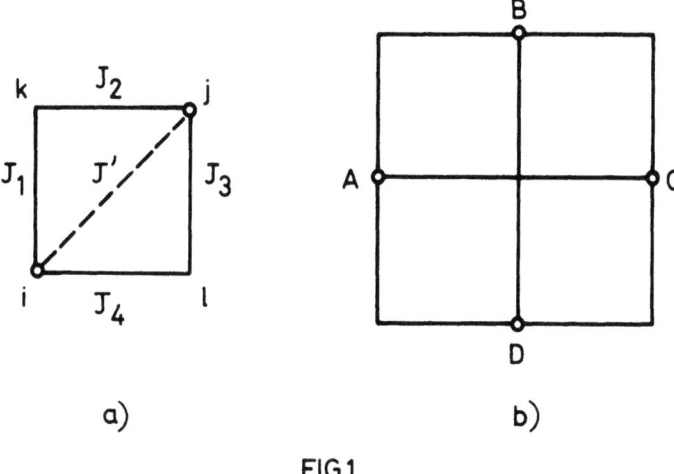

FIG.1

The two clusters used for the rescaling transformation on the square lattice.

where

$$t_o' = \frac{2t_o^2}{1 + t_o^4} \qquad (5)$$

In addition to the stable trivial fixed points at $t_o = 0$ and 1 there is another fixed point solution of (5) with $t_o' = t_o = t_o^* \simeq 0.544$ which describes the ferromagnetic transition [16]. If t_o is initially greater than t_o^* the trajectory always goes to the trivial fixed point at $t_o = 1$ and for $t_o < t_o^*$ the trajectory always goes to $t_o = 0$. Thus the fixed point describing the transition separates two regions of parameter space and the trajectories are very different from the one dimensional case where they always flow to $t_o = 0$ unless the temperature is precisely zero [12,13]. We stress that although the numbers one obtains are not very accurate, the overall picture is qualitatively correct. Replacing P(t) by a distribution centred at t_o^* but with a small width $\delta \ll t_o$, one finds that after rescaling the width of the new distribution δ' is given by $\delta' = y\delta$ with $y = 0.972$. Since $y < 1$ the effect of the randomness goes to zero after many iterations. The specific heat exponent α is zero for the two-dimensional Ising model, so one expects from arguments due to Harris [17] that the randomness should be "marginal" i.e.

y = 1. It is not surprising that our simple approximation does not exactly reproduce this result.

The appropriate distribution for the spin glass is one symmetric in t, i.e. $P(t) = P(-t)$, and this symmetry is preserved by the recursion relation (4). We find that there is no non trivial symmetric fixed point distribution for the recursion relation (4) and after many iterations all the weight of the distribution is at $t = 0$. This result implies that there is no singularity in the zero field free energy. Correlation functions like $<<S_i S_j>_T>_C$, where T denotes a thermal average and C denotes a configurational average, vanish for $i \neq j$ because of the symmetry of P(t). However quantities like $<<S_i S_j>_T^2>_C$ do not vanish identically but they fall off exponentially at large distances for any temperature because P(t) collapses to zero after many iterations and its width Δ transforms like $\Delta' \sim \Delta^2$ for $\Delta \ll 1$. This ensures that there are no divergences in the derivatives of the free energy with respect to an applied field.

The flow of the probability distribution of $\tanh\beta J$ was followed on a computer method suggested by D.J. Thouless (private communication). A starting set of N values of J was taken distributed according to a rectangular distribution, where typically N was set equal to 200 but some runs were performed for N = 1000 to check that random noise in the technique did not affect the results. Four out of these N values were chosen by a random number generator and a value of J' calculated from (4). This procedure was repeated until N values of J' had been obtained, which were used as the starting distribution for the next iteration. If we define the rectangular distribution to run from $-J_0$ to J_0 then the lowest temperature which could be treated without overflow problems was $kT \sim 0.1 J_0$. At all temperatures down to this value the width in the distribution for $\tanh\beta J$ went steadily to zero.

One can also show that the small deviations from a symmetrical distribution do not alter the results. For example, starting with a distribution of the form $p\delta(t-t_0) + (1-p)\delta(t+t_0)$ then for p = 1 we have the ferromagnetic fixed point discussed earlier and decreasing p one iterates to the same fixed point but the transition temperature goes to zero for $p \simeq 0.89$. For $0.11 \lesssim p \lesssim 0.89$ there is no transition and no long range order in the ground state.

The above calculations have neglected two spin interactions beyond first neighbour, four and higher spin terms and correlations between the distributions. For the non random ferromagnet the largest coupling, after nearest neighbour, is the second neighbour interaction [16]. We have therefore investigated the spin glass in an approximation which includes first and second neighbour couplings but neglects correlations [18]. A second neighbour coupling on the original lattice becomes a first neighbour interaction on the rescaled lattice and so the new first neighbour interaction J_1' is taken to be $(J'+L)$ where L is the second neighbour coupling and J' is the contribution from the nearest neighbour paths and is given by (4). Assuming that the dominant contribution to L', the value of L on the rescaled lattice, arises from the single direct path of two steps on the original lattice, we can write the recursion relations as

$$t' = \frac{r + t_1 t_2 + t_3 t_4 + r t_1 t_2 t_3 t_4}{1 + t_1 t_2 t_3 t_4 + r(t_1 t_2 + t_3 t_4)} \quad (6)$$

$$r' = t_5 t_6$$

where $r = \tanh\beta L$, and (6) is equivalent to equation (8) of reference 18. The recursion relations (6) have been studied numerically by the same method as equation (4). At low temperatures the width of the distribution went to zero much more slowly than with (4) and for $kT = 0.1\, J_0$ the width was practically static although it showed a slight tendency to decrease. Clearly this is a borderline situation and to clarify this issue we have carried through a better approximation where the rescaling is performed on the larger cluster shown in figure 1b, where four sites A, B, C and C are left after rescaling. First neighbour, second neighbour and four spin interactions can be accommodated as well as a variety of correlations. A full treatment would involve rescaling the joint probability distribution for all seven interactions which can be fitted on to a single square. If we neglect these correlations the problem is much more tractable and we find that the interactions iterate steadily to zero. It is difficult to estimate reliably the effect of correlations but we note that for the simpler case of the percolation problem the correlations are very small at the fixed point. Recently B.W. Southern (private communication) has investigated the recursion relations for this cluster by rescaling the joint distribution. Using a sample size of up to 2000 sets of the seven interactions the width of the distribution still went to zero. This was checked both for the rectangular and two delta function distributions.

3. CALCULATIONS IN THREE DIMENSIONS

It is not possible to carry out precisely the same transformation for a simple cubic lattice as was used in §2 for the square lattice because removing half the sites gives a different lattice structure. Suppose, however, that we perform the rescaling indicated in figure 1b and then rescale, as in figure 1a, the four sites that remain. A square of 2 × 2 lattice spacings has now been reduced to two sites in the centre of opposite edges with a single interaction between them. In this case the scale factor is b = 2. Our transformation for the cubic lattice is very similar and consists of reducing a cube of 2 x 2 x 2 lattice spacings down to two sites A and B in the centres of opposite faces, as shown in figure 2.

Evaluating all the traces exactly is very complicated and to make the calculations tractable one slight approximation is made. When the trace is taken over a site on an edge, such as C, the interaction between two adjacent corners, D and E, is neglected. Instead the trace is performed over spin C, neglecting the interaction CE, and new couplings between D, F and G are calculated. Then the trace is taken again over spin C neglecting interaction CD which gives interactions between E, F and G. In this way the coupling DE is represented approximately by an extra contribution to the other interactions. All the remaining traces are evaluated exactly. A value of the new interaction between A and B is calculated for a particular choice of the 54 interactions in the figure, chosen at random from a starting distribution of 1000 values. This procedure is repeated until 1000 new values of the interaction have been calculated, which then form the starting distribution for the next iteration. One such iteration took about 5 seconds on the CDC 7600 at Brookhaven National Laboratory.

At low temperatures it is found that the width in the distribution of $\tanh\beta J$ grows although the mean stays small, with a magnitude that decreases with increasing sample size. This is interpreted as spin glass order. The temperature below which the distribution grows depends on the form of the starting distribution and we find

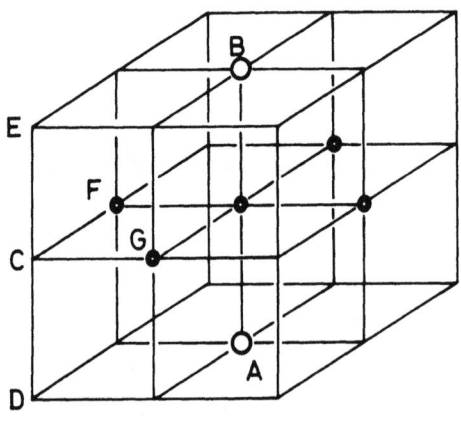

FIG.2

The cluster used for the rescaling transformation on the simple cubic lattice.

$$\frac{T_c}{T_c^{mf}} = \begin{cases} 0.44 \pm 0.02 & \text{Gaussian} \\ 0.50 \pm 0.02 & \text{Rectangular} \\ 0.55 \pm 0.02 & \text{Two delta function} \end{cases} \quad (6)$$

which is to be compared with the Monte Carlo results given by equation (3). The errors given are those caused by the finite sample size and the extra errors arising from the cluster approximation are unknown.

We have also checked how well the approximation fares in predicting the ferromagnetic transition temperature when there is no disorder. Series expansions [19] indicate that $T_c \simeq 0.75$ of the mean field value whereas the above approximation gives a numerical factor of 0.66 and so underestimates T_c by about 12%. It seems likely then that (6) is also an underestimate. The Monte Carlo calculations, on the other hand, may, if anything, overestimate T_c as discussed in the introduction. Consequently the difference between the results in (6) and (3) is not very surprising.

4. DISCUSSION

The rescaling calculations presented here indicate in a fairly convincing manner that a true phase transition occurs in the Edwards and Anderson spin glass model with Ising spins in three dimensions. No transition is predicted for a square lattice, which is in disagreement with Monte Carlo studies. The discrepancy is not properly understood but we would expect the two dimensional calculation to be more accurate because second neighbour and four spin couplings, as well as correlations between the distributions have been included.

We are not able to make any definite statements on whether the Heisenberg model has a transition in three dimensions. Rescaling calculations would be more difficult than for the Ising case because the angular dependence of the interaction becomes more complicated than just $\underline{S} \cdot \underline{S}$ after iteration [20]. The rescaling technique in real space is only applicable to models with short range forces so we do not feel that the RKKY interaction can be treated by this method.

ACKNOWLEDGEMENTS

The author would like to thank D.J. Thouless and B.W. Southern for stimulating comments. He is grateful to Brookhaven National Laboratory for hospitality and use of computing facilities during part of summer 1976 when the three dimensional calculations reported here were carried out.

REFERENCES

1. V. Cannella and J.A. Mydosh, Phys. Rev. $\underline{B6}$, 4220 (1972).

2. D.E. Murnick, A.T. Fiory and W.J. Kossler, Phys. Rev. Lett. $\underline{36}$, 100 (1976).

3. S.F. Edwards and P.W. Anderson, J. Phys. F $\underline{5}$, 965 (1975).

4. D. Sherrington and S.K. Kirkpatrick, Phys. Rev. Lett. $\underline{35}$, 1792 (1975).

5. K.H. Fischer, Phys. Rev. Lett. $\underline{34}$, 1438 (1975).

6. D.J. Thouless, P.W. Anderson and R.G. Palmer (to be published).

7. K. Binder and K. Schröder, Solid State Comm. $\underline{18}$, 1361 (1976).

8. K. Binder and D. Stauffer, Phys. Lett. $\underline{57A}$, 177 (1976).

9. J.L. Tholence and R. Tournier, J. de Physique $\underline{35}$, Colloque C4, 229 (1974).

10. T. Niemeyer and J.M.J. Van Leeuwen, Physica $\underline{71}$, 17 (1974); K.G. Wilson, Rev. Mod. Phys. $\underline{47}$, 773 (1975); L.P. Kadanoff, Phys. Rev. Lett. $\underline{34}$, 1005 (1975).

11. L.P. Kadanoff and A. Houghton, Phys. Rev. $\underline{B11}$, 377 (1975).

12. A.P. Young and R.B. Stinchcombe (to be published in J. Phys. C).

13. G. Grinstein, A.N. Berker, J. Chalupa and M. Wortis, Phys. Rev. Lett. $\underline{36}$, 1508 (1976).

14. M. Nauenberg and B. Nienhuis, Phys. Rev. Lett. $\underline{33}$, 1598 (1974).

15. A.B. Harris and T.C. Lubensky, Phys. Rev. Lett. $\underline{33}$, 1540 (1974).

16. M.N. Barber, J. Phys. C $\underline{8}$, L203 (1974).

17. A.B. Harris, J. Phys. C $\underline{7}$, 1671 (1974).

18. A.P. Young and R.B. Stinchcombe, J. Phys. C $\underline{8}$, L535 (1975).

19. C. Domb, J. Phys. C $\underline{3}$, 255 (1970).

20. T. Niemeyer and T.W. Ruijgrok, Physica $\underline{81A}$, 427 (1975).

THE REMANENT MAGNETIZATION OF SPIN GLASSES AND THE DIPOLAR COUPLING

F. Holtzberg[*], J.L. Tholence and R. Tournier

Centre de Recherches sur les Très Basses Températures

C.N.R.S., BP 166, 38042 Grenoble-Cedex, France

The specific heat and the magnetization of perfectly disordered solid solutions such as dilute $\underline{Cu}$ $\underline{Mn}$ alloys follow scaling laws[1] which depend only on the variables $\frac{T}{c}$ and $\frac{H}{c}$.[2] The saturated remanent magnetization[2] itself is only a function of $\frac{T}{c}$. In a solid solution out of equilibrium it is always possible to anneal the alloys and consequently to produce chemical clustering and changes in the amplitude of the irreversible properties.[3] We want here to limit the field of spin glasses to magnetic well disordered dilute alloys, containing impurities interacting only by a $\frac{1}{r^3}$ interaction, outside the mictomagnetism domain[3] which includes first neighbour direct interactions and sometimes chemical clusters.

The best region of concentration for the observation of scaling laws generally extends up to 1 %. In this case the first neighbour interaction is limited to a small number of pairs of impurities. The damping of the R.K.K.Y. interaction by the reduction of the electronic mean free path is negligible. For instance, the $\underline{Au}$ $\underline{Fe}$ alloy, because of large resistivity, does not follow[4] a scaling law for concentration of several percent. The scaling laws are a direct consequence of the $\frac{1}{r^3}$ character of the RKKY interaction between magnetic impurities in a well disordered solid solution. The existence of a scaling law for the remanent magnetization eliminates the assumption that it is due to clustering effects ; it is the sign of an intrinsic property directly related to a $\frac{1}{r^3}$ interaction creating the potential barriers which retain a large number of moments in the direction of the field previously applied. In a normal metal the most probable anisotropic interaction which

[*]Permanent adress : IBM T.J. Watson Research Center, Yorktown Heights, N.Y. 10598 USA

can create potential barriers is the dipolar one. We will examine this possibility in this paper.

NÉEL MODEL FOR INDEPENDENT GRAINS[5]

Different authors[6,7,8] have noted that the remanent magnetization of magnetic dilute alloys has the properties of fine grains as described by L. Néel.[5] This observation had led us to interpret[8,9] the irreversible properties in the Néel scheme in the sense that the perfectly disordered solid solution behaves like an assembly of independent regions. Each region has an anisotropy E_a depending on the number of spins n it contains.

$$E_a = \frac{1}{2} M_g H_a = (Q + \text{Ln } \tau) k T . \tag{1}$$

M_g is the uncompensated moment of the cloud, and is proportional to the difference between the number n^+ of up spins and the number n^- of down spins in an Ising model ($n = n^+ + n^-$). H_a is the anisotropy field of the region, τ its relaxation time at the temperature T. Equation 1 can be written:

$$\tau = \tau_o \, e^{E_a/kT} .$$

Q is equal to $\log 1/\tau_o$. For a time of measurement $t = \tau$ and a value of E_a, each region has a blocking temperature $T = T_b$. M_g is superparamagnetic above T_b and is frozen below this temperature.

As it has been previously shown[8,9], the remanent magnetization varies as log t. An example is given for Au Fe 8 % in Figure 1. The same property has been observed at lower concentration[9] in Au Fe 3 %, and in Cu Fe 1 %[10]. One can justify this property, if $p(E_a) \, dE_a$ is the number of clouds having an anisotropy energy between E_a and $E_a + dE_a$. The corresponding change of remanent magnetization is

$$d\sigma_r = - p(E_a) \frac{\overline{M_g}(E_a)}{2} d E_a .$$

$\overline{M_g}(E_a)$ is the mean moment of the clouds having their E_a between E_a and $E_a + dE_a$. The geometrical factor $\frac{1}{2}$ takes into account the mean value of the projection of Mg on the direction of the field initially applied. Using (1):

$$d E_a = \frac{\partial E_a}{\partial \log t} d \log t + \frac{\partial E_a}{\partial T} d T .$$

Consequently

$$\frac{\partial \sigma_r}{\partial \log t} = - p(E_a) \frac{\overline{M_g}(E_a)}{2} \frac{\partial E_a}{\partial \log t}$$

$$\frac{\partial \sigma_r}{\partial T} = - p(E_a) \frac{\overline{M_g}(E_a)}{2} \frac{\partial E_a}{\partial T} .$$

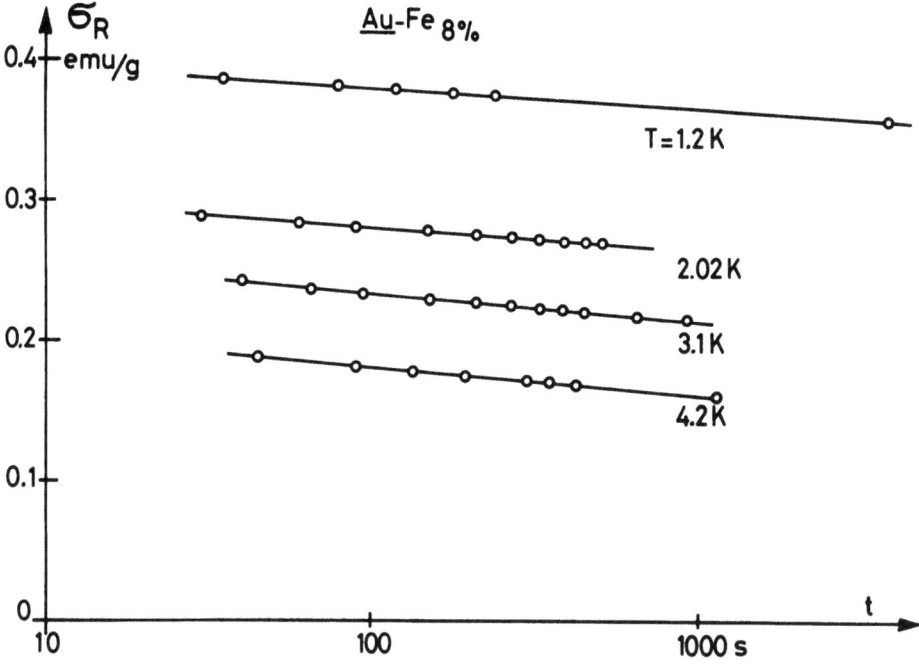

Fig. 1 : Log t dependence of the remanent magnetization of Au Fe 8 %.

Assuming the quantity $p(E_a) \overline{M_g}(E_a)$ varies slowly with E_a, we can deduce from these relations and from (1) that σ_r varies as log t at a fixed temperature T. The following ratio permits a determination of Q if Q is considered nearly independent of T and neglecting at low temperatures the thermal variation of the spontaneous magnetization of frozen clouds :

$$\left(\frac{\partial \sigma_r}{\partial T} / \frac{\partial \sigma_r}{\partial \log t}\right)_T = \frac{Q + \log \tau}{T} . \qquad (2)$$

Q has been determined some years ago by this method[8,9,10], by comparing the thermal variation to the time variation of the remanent magnetization of Au Fe and Cr Fe. Q is generally equal to 15 around T = 2 K and the relaxation time τ_o is about 3×10^{-7} s.

In agreement with Néel's model the thermoremanent magnetization (TRM)[6,8,9] is higher than the isothermal remanent magnetization (IRM) obtained in the same field h. In low fields, the TRM varies as h ; the IRM begins to vary as h^2 and as h in higher fields[9]. The IRM and TRM saturations are the same at each temperature.

THE REMANENT MAGNETIZATION OF A SPIN GLASS[9,12]

We want, now, to focus attention on the distribution of the uncompensated moment of clouds, which is assumed to be a gaussian law.

$$P|M_g| = \frac{2}{M_{g_o} \sqrt{2\pi}} e^{-\frac{M_g^2}{2M_{g_o}^2}}, \quad M_{g_o} = \mu\sqrt{n_o}. \qquad (3)$$

n_o is a mean number of spins in a cloud. We may determine n_o from the saturated remanent magnetization σ_{rs} at 0 K and from the total saturation of the alloy $\sigma_s = N_o n_o \mu_o$.

$$\sigma_{rs}(T=0) = \frac{N_o}{2} \left(\frac{2 n_o}{\pi}\right)^{1/2} \mu_o \; ; \; n_o = \frac{1}{2\pi}\left(\frac{\sigma_s}{\sigma_{rs}}\right)^2.$$

N_o is the number of clouds : $N_o = \frac{Nc}{n_o}$, n_o is independent of c when σ_{rs} and σ_s are proportional to c, i.e when the scaling laws are obeyed.

For Cu Mn, n_o is nearly independen of c and equal to 260. For Au Fe alloys $350 < n < 500$ when $0.01 > c > 0.001$ (See table 1).

Au Fe c	σ_s emu/g	σ_{rs} (T→0) emu/g	n_o	C_{cal}	C_{exp}
0.001	$5,7 \times 10^{-2}$	1×10^{-3}	512	$0,61 \times 10^{-5}$	$0,73 \times 10^{-5}$
0.002	$11,4 \times 10^{-2}$	2×10^{-3}	512	$1,22 \times 10^{-5}$	$1,5 \times 10^{-5}$
0.005	$28,4 \times 10^{-2}$	6×10^{-3}	355	$3,05 \times 10^{-5}$	$4,6 \times 10^{-5}$
0.01	$56,7 \times 10^{-2}$	$12,5 \times 10^{-3}$	330	$6,0 \times 10^{-5}$	$6,25 \times 10^{-5}$

Table 1 : Calculated values of n_o and C from the saturated remanent magnetization.

If the assumption of a gaussian distribution for M_g is respected, and if the model is correct, we have to find again the Curie constant of all impurities using the superparamagnetic Curie constant of the clouds, i.e. knowing only σ_{rs} (T=0 K).

$$C_{cal} = C_{superpara.} \times \left(\frac{\mu_{eff}}{\mu_o}\right)^2 = \left(\frac{N_o n \mu_o^2}{3k}\right)\left(\frac{\mu_{eff}}{\mu_o}\right)^2$$

$$= \frac{2\pi \sigma_{rs}^2}{N_o 3k}\left(\frac{\mu_{eff}}{\mu_o}\right)^2 = \frac{Nc(\mu_{eff})^2}{3k}. \qquad (4)$$

A good agreement between C_{cal} from $\sigma_{rs}(T=0)$ and C_{exp} is found for Cu Mn[9] and Au Fe (Table 1). Then the gaussian law is a good representation of the M_g distribution.

In order to determine the origin of the anisotropy in this system we calculate the width Δ_a of the distribution of the local anisotropy fields on each impurity. The anisotropy energy E_a of a cloud is <u>assumed to be proportional to the number n of spins in this cloud</u> :

$$E_a = 20 \, k \, T_b = n \, \mu \, \Delta_a \qquad (5)$$

$(Q + \text{Log } \tau = 20$, because $Q = 15$ and $\text{Log} \tau \simeq 5)$.

In order to respect the scaling law Δ_a will be proportional to c. Similarly, the blocking temperature T_b of an n-cloud is proportional to c.

The mean moment of the clouds which have their T_b between T and $T + dT$ increases with T_b. Assuming

$$\overline{M_g(E_a)} = \overline{M_g(T_b=T)} = \mu_o \sqrt{\frac{2n}{\pi}}$$

with a Gaussian model, we can write :

$$\frac{\overline{M_g(T)}^2}{2 M_{g_o}^2} = \frac{n}{\pi \, n_o} = \alpha \frac{T}{c} \qquad (6)$$

where α is independent of c. From (5) and (6) we deduce :

$$\Delta_a = \frac{20 \, kc}{\pi \, n_o \, \mu \, \alpha} \, . \qquad (7)$$

α has to be determined from the thermal variation of the saturated remanent magnetization. Using (3) we calculate

$$\sigma_{rs}(T) = \frac{N_o}{2} \int_{\overline{M_g(T)}}^{\infty} M_g \, P(M_g) dM_g \, .$$

$$\boxed{\sigma_{rs}(T) = \sigma_{rs}(T=0) \, e^{-\frac{\alpha T}{c}}} \qquad (8)$$

This relation is important to test our model. Figure II shows that this law is perfectly obeyed over a large range of temperatures for different alloys.

We are obliged to consider that this law is limited at a temperature of the order of the mean RKKY coupling in the alloy. At this temperature the remanent magnetization (given by (8)) is expected to be negligible compared to $\sigma_{rs}(T=0)$. Figure II permits the determination of Δ_a. For Au $Fe_{0.01}$, $n_o = 330$, and Δ_a the width of the distribution of local anisotropy fields is 135 Oe.

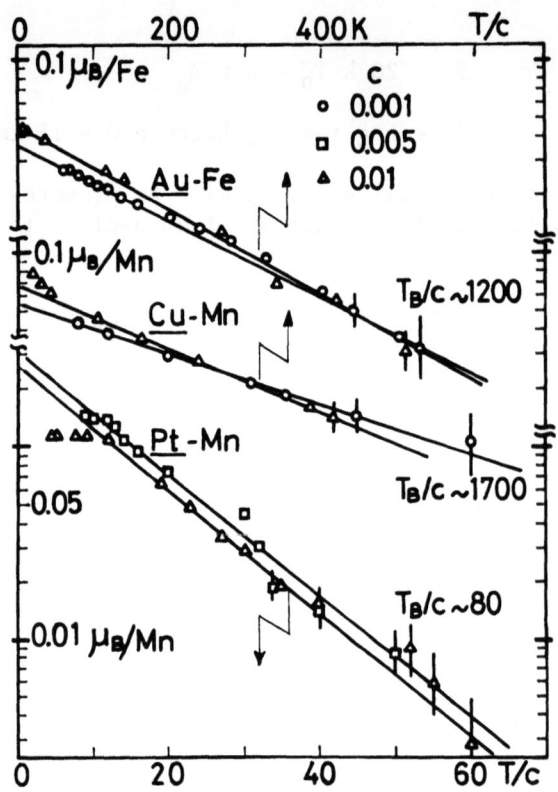

Figure II : Thermal dependence of the saturated remanent magnetization, in reduced units, for Au Fe[9], Cu Mn[10], Pt Mn[13].

ORIGIN OF THE ANISOTROPY : THE DIPOLAR COUPLING

In order to determine the origin of Δ_a, we calculate the width Δ_d of the distribution of dipolar fields, using the width Δ_{KB} of the distribution of RKKY molecular fields as calculated by Klein and Brout[14]. The RKKY field in an Ising model is written as $H_{KB} = A \mu \cos(2k_F r)/r^3$. The anisotropic part of the dipolar coupling between two sites i and j is $\mu h_{da} = 3\mu^2/r_{ij}^3 \cos^2\theta$, where θ is the angle between the axis of the moment and the line r_{ij} joining the sites i and j. Calculating the second moments of the distributions of h_{KB} and h_{da} taking into account the fact that $\overline{\cos^4\theta} = \frac{3}{8}$ and $\overline{\cos^2(2k_F r)} = \frac{1}{2}$ we determine :

$$\Delta_{KB} = 19.6 \frac{A \mu c}{d^3} \text{ and } \Delta_{da} = 29.4 \frac{\mu c}{d^3} \qquad (9)$$

where d is the lattice constant. We obtain Δ_{da} = 87 Oe a value which is in very good agreement with the experimental value Δ_a = 135 Oe for c = 0.01 in Au Fe (μ = 2.2 μ_B).

For Cu Mn and Au Fe, the anisotropic dipolar coupling is sufficient to justify the presence of potential barriers.

VOLUME OF INDEPENDENT CLOUDS

Up to now we have not tried to justify why the alloy can be divided into <u>independent</u> clouds containing n_o spins. We may have an idea of the phenomena which determines the number n_o. A n-cloud can freeze if its potential barrier is equal to or larger than the coupling of the n spins with the other spins outside the cloud. We assume the existence of an uniaxial anisotropy of the cloud. Its M_g has two orientations on this axis. The exchange energy of the cloud with the spins outside it, which is lost by changing the orientation of M_g on this axis, must be smaller than the potential barrier of the n-cloud. An order of magnitude of this energy is :

$$2 \mu \delta_{r>r_c} \sqrt{n} \leq E_a = n_o \mu \Delta_{da}$$

where $\delta_{r>r_c}$ is at least the width of the gaussian distribution of the molecular field created at the center of a sphere having the volume of an n-cloud ($r_c^3 = \beta d^3 c^{-1}$) by the spins located outside this sphere.

$$\delta_{r>r_c} = \frac{A c \mu}{d^3} \sqrt{\frac{8\pi}{3\beta}} , \beta = 0.054 \text{ n} .$$

We obtain n = 0.85 A, i.e. of the order of A, inside the rough approximation we have used. <u>The number of spins inside the clouds is of the order of the ratio of the amplitudes of the RKKY and</u>

dipolar couplings in agreement with the experimental results on Cu-Mn and Au-Fe alloys. From this property we can deduce the following consequences. The ratio σ_{rs}/σ_s will increase when n_o will decrease. Then the irreversible properties will be significantly enhanced in systems with a low RKKY interaction. It will be the case for Gadolinium diluted in normal metals. It could also be true for impurities diluted in an insulator and interacting by dipolar coupling. Because of the $\frac{1}{r^3}$ character of the interaction we expect scaling laws for dipolar spin glasses and large hysteresis phenomena.

EXPERIMENTAL RESULT ON A DIPOLAR SPIN GLASS : EuSrS

We have studied a specimen of SrS containing 1.7 % of Europium impurities. The Europium is divalent and is in a S state carrying a magnetic moment of 7 Bohr magnetons. The structure Sr S is fcc. Each Europium has 12 first neighbours of strontium. The lattice constant is = 6.02 Å. The compound Eu S has approximately the same lattice constant and is a ferromagnet. Then we expect that our specimen contains some first neighbour ferromagnetic pairs of europium which are present in a perfectly disordered solid solution.

From figure III, we see that the magnetization measured at T = 0.05 K is not saturated in a 3 kOe field. It indicates some negative interactions are present in high fields. An extrapolation of the magnetization versus $\frac{1}{h}$ gives the saturation magnetization. Its value (fig. III) corresponds approximately to the analyzed concentration 1.68 %.

The initial susceptibility follows a curie law (fig. IV) down to 0.05 K. A remanent magnetization is observed which is time-dependent.

In figure V the remanent magnetization obtained after suppressing different external applied fields is plotted as a function of the time. The specimen is in good thermal contact with the cold source which is an ammonium iron alun cooled by adiabatic demagnetization. However the relative stability of the remanent magnetization in the short times could be due to the competition between the after-effects of the sample and the tendency of it to cool after it has been quickly demagnetized. Then we look more at the general character of these phenomena which is a rapid decrease of the remanent magnetization versus time than its precise behaviour. Two remarks can be done :
- The relaxation time seems to be about the same for all the moments contributing to the remanent magnetization.
- The fact that the remanent magnetization obtained at short times is independent of the temperature below 0.15 K gives a new indication : the number of spins which contributes to the relaxation of the remanent magnetization does not change with temperature.

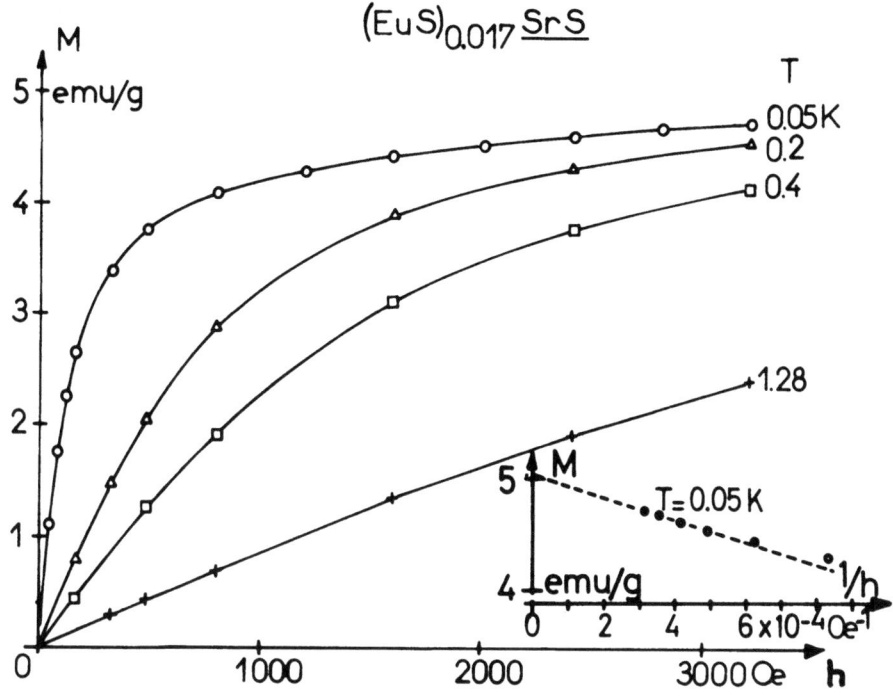

Fig. III : Magnetization curves of Sr S Eu

ORIGIN OF THE REMANENCE AND OF ITS RELAXATION

The Curie temperature of EuS is about 16 K (12 first neighbours in the lattice). So the first neighbour interaction is probably sufficient to produce ferromagnetic pairs below 1 K. But the dipolar coupling

$$E_d = \frac{\vec{\mu}_i \vec{\mu}_j}{r^3} - 3 \frac{(\vec{\mu}_i \vec{r}_{ij})(\vec{\mu}_j \vec{r}_{ij})}{r_{ij}^5}$$

plays an important role because for a ferromagnetic pair it is of the same order of magnitude ($\simeq 1.2$ K). The saturation of the remanent magnetization we observe is equal to 7.5 % of the saturated magnetization. If we multiply this quantity by the geometrical factor 2.12 to take into account the fact that the moments are alined on their anisotropy axis, and that we project them on the direction of measurement (the 001 axis) we approximately obtain the value we expect for the saturation of pairs of first neighbours

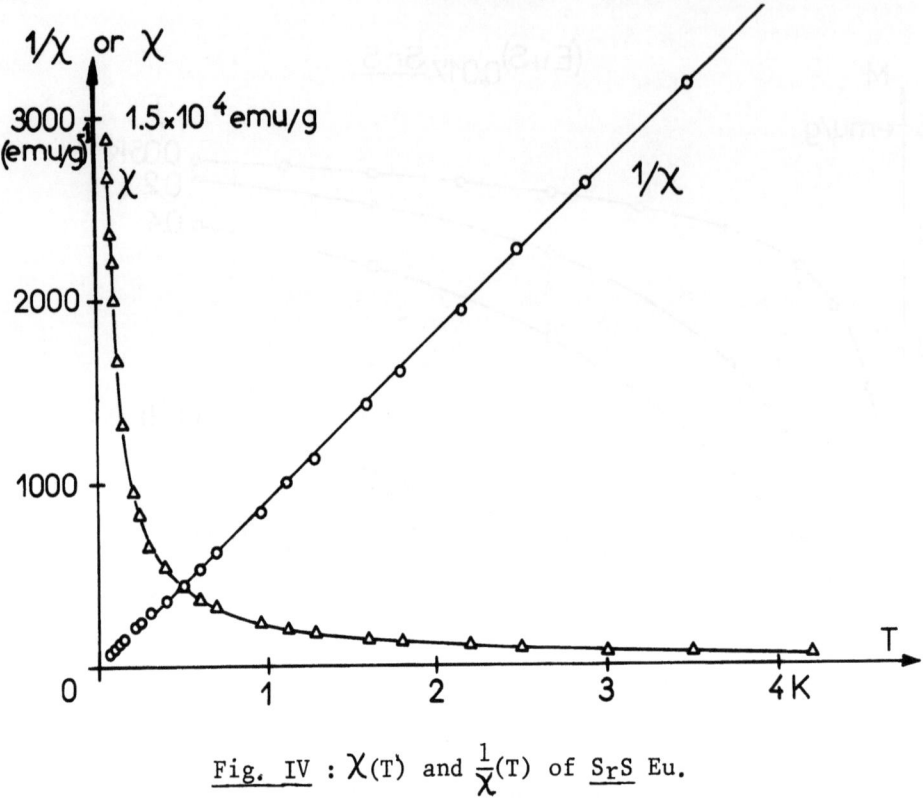

Fig. IV : $\chi(T)$ and $\frac{1}{\chi}(T)$ of <u>SrS</u> Eu.

(15 %). Then we conclude that the remanent magnetization down to 0.05 K, is essentially due to ferromagnetic pairs, and not to the freezing of all magnetic impurities.

In order to calculate the potential barrier which governs the relaxation of pairs we use the relation (1). The diagram (fig. VI) log τ versus $\frac{1}{T}$ gives E_a since :

$$\text{Log } \tau = \frac{E_a}{kT} + \text{Log } \tau_o .$$

At each temperature a value of τ is determined at the time t = τ for which $\sigma_r(t)$ is divided by 2.

We obtain $\frac{E_a}{k}$ = 0,19 ± 0.06 K. Assuming a rigid coupling between the 2 spins of a pair the potential barrier is then at the maximum equal to $3\mu^2/r_{ij}^3$ = 1.8 K. So the anisotropic dipolar coupling is sufficiently large to explain the observed value of the anisotropy energy.

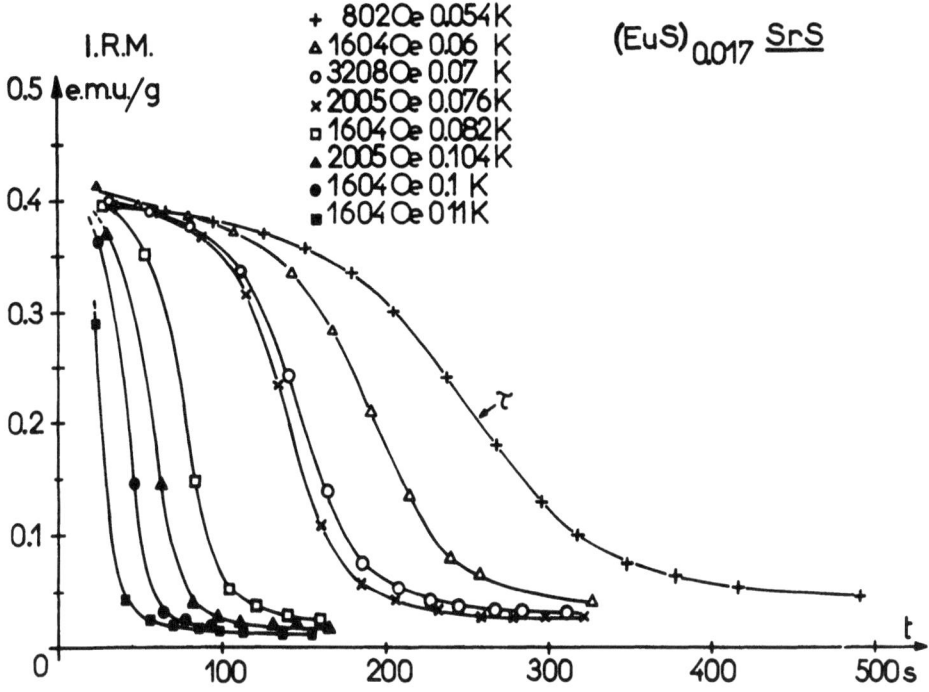

Fig. V : Time dependence of the saturated remanent magnetization of SrS Eu.

WIDTH OF THE DISTRIBUTION OF DIPOLAR FIELDS

The width Δ_d of the distribution of dipolar fields is calculated using the Klein and Brout method, when a great majority of spins are alined in a high external field. We obtain $\Delta_d = \frac{34}{d3} c \mu$, for $c = 0.017$, $\Delta_d = 170$ Oe. From the slope of the magnetization versus $\frac{1}{h}$ it is possible, in an Ising model, to have an idea of the width of the distribution of the molecular fields. If we assume a Lorentzian distribution :

$$P(H) = \frac{\Delta}{\pi} \cdot \frac{1}{H^2 + \Delta^2}.$$

The fraction of spins which are in a negative molecular field H, in a large applied field h is $\frac{\Delta}{\pi h}$. So the approach to the saturation is $\sigma = \sigma_s(1 - \frac{\Delta}{\pi h})$. From fig. 3 we determine $\Delta \simeq 700$ Oe, value enhanced by the thermal agitation. This value is in agreement with the estimated width of the distribution of the dipolar fields (~ 170 Oe).

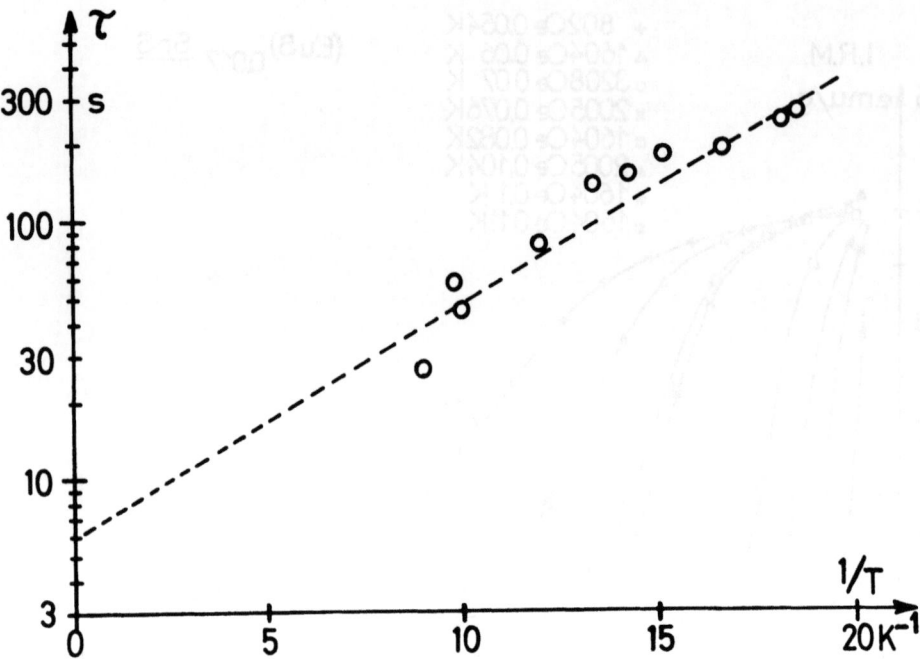

Fig. VI : Temperature dependence of the mean relaxation time τ in the diagram Log τ = f(1/T).

CONCLUSION

We have shown that a spin glass is at low temperatures an assembly of clouds containing n spins with uncompensated moments ($\sqrt{\overline{M^2}} = \mu\sqrt{n}$) each one frozen below its own blocking temperature. For Cu-Mn and Au-Fe alloys the potential barriers are due to the anisotropic dipolar coupling. The number of spins in the clouds is in mean of the order of the ratio of the amplitudes of the RKKY and dipolar couplings. We have shown for the first time the exponential decrease of the saturated remanent magnetization of different spin glasses. We have presented some previous results on a dipolar spin glass the Eu Sr S which is an insulator. The blocking temperature has not been attained but a remanent magnetization and relaxation effects are observed and due to ferromagnetic pairs of first neighbour europium atoms.

We are much indepted to Prof. R. Orbach for helpfull discussions. The help of R. Rammal and J. Souletie was greatly appreciated.

REFERENCES

1 - A. Blandin, Thesis, Paris (1961) unpublished.
2 - J. Souletie and R. Tournier, J. Low Temp. Phys. $\underline{1}$, 95 (1969).
3 - P.A. Beck, Met. Trans. $\underline{2}$, 2015 (1971)
   J. Less Common Met. $\underline{28}$, 193 (1972).
4 - J. Souletie, Thesis, Grenoble (1968) unpublished.
   O. Bethoux, J.A. Careaga, B. Dreyfus, K. Gobrecht, J. Souletie, R. Tournier and L. Weil, Proceedings of the 10th Int. Conf. on Low Temp. Phys. (Vimiti Publishing House, Moscow 1967, $\underline{4}$, 292.
5 - L. Neel ; Cours de Physique Theorique : Les Houches 1961, p. 412 (Presses Universitaires de France, Paris 1961).
   L. Neel, J. Phys. Soc. of Japan, $\underline{17}$ B-1, 676 (1962).
   L. Neel, Ann. Geoph. $\underline{5}$, 99 (1949).
6 - O.S. Lutes and J.L. Schmitt, Phys. Rev. $\underline{125}$, 433 (1962).
   "   " $\underline{134}$, A676 (1964).
7 - J.S. Kouvel, Phys. Chem. Solids $\underline{21}$, 57 (1961).
   "   "   " $\underline{24}$, 795 (1963).
8 - R. Tournier, Thesis, Grenoble (1965) unpublished.
   R. Tournier and L. Weil, Journal de Physique et le Radium $\underline{23}$, 522 (1962).
   R. Tournier and Y. Ishikawa, Phys. Letters $\underline{11}$, 280 (1964).
9 - J.L. Tholence, Thesis, Grenoble (1973).
   J.L. Tholence and R. Tournier, J. Phys. $\underline{35}$, C4-229 (1974).
10 - J.A. Careaga, Thesis, Grenoble (1967).
11 - Y. Ishikawa, R. Tournier and J. Philippi, J. Phys. Chem. Solids $\underline{26}$, 1727 (1965).
12 - J.L. Tholence and R. Tournier, Int. Conf. on Magnetism, Amsterdam (1976).
13 - J.L. Tholence and E.F. Wassermann, Int. Conf. on Magnetism, Amsterdam (1976) and to be published.
14 - M.W. Klein and R. Brout, Phys. Rev. $\underline{132}$, 2412 (1963).

The Magnetic Behaviour of Pd-Mn Alloys

C.N. Guy and W. Howarth

The Blackett Laboratory, Imperial College, London

Abstract

Low field magnetisation measurements are presented for Pd-Mn alloys with impurity concentrations in the range $1.3 \leqslant c \leqslant 5.8$ at%$\overline{\text{Mn}}$. Spin glass behaviour is well observed in alloys containing more than 4.5 at%Mn, and the variation of spin glass temperature T_g with concentration in this region is in good agreement with the maxima in $\Delta c/T$. Alloys with concentrations in the range $3.6 \leqslant c \leqslant 4.2$ at%Mn show a mixed magnetic behaviour, retaining a sharp, spin glass-like peak in the initial susceptibility but also exhibiting a small, reverse isothermal remanence at temperatures above the susceptibility peak. Two alloys with concentrations $c \leqslant 2$ at%Mn exhibited a more ferromagnetic like transition in the initial susceptibility but also exhibited reverse isothermal remanence below T_c.

The variation of isothermal remanence with external field shows an unusual waisted hysteresis loop for all alloys with $c \leqslant 4.2$ at%Mn. The variation of saturated reverse remanence with temperature has a maximum for the 3.8 and 4.2 at% alloys at nearly the same temperature as the susceptibility peak; the remanence is time dependent below this temperature in both alloys. No time dependence of remanence was observed in 1.3 or 2 at%Mn alloys and the saturated remanence increased monotonically with decreasing temperature to 1.7°K.

The existence of a reverse remanence and a waisted hysteresis loop suggest the existence of two magnetic phases in alloys with $c \leqslant 4.2$ at%Mn possibly caused by small impurity concentration fluctuations. Interactions between these regions may be responsible for the remanent behaviour in an analogous fashion to exchange anisotropy. Preliminary results on the effect of heat treating the alloys, which seem to confirm this, are discussed.

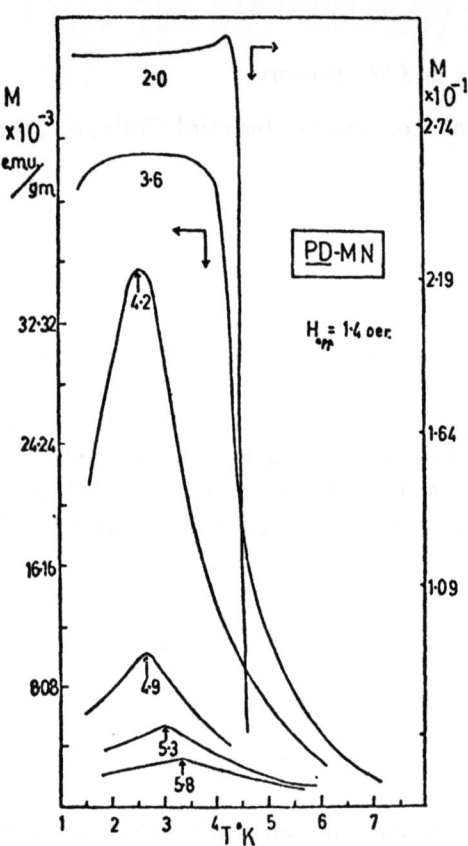

Fig. 1

The observed magnetisation in 1.8 Oers of Pd-Mn alloys without any demagnetisation correction. Samples with c $\geqslant$ 3.6 at%Mn were spherical. The 2 at%Mn was a disc edge on to the field.

1. Introduction

Disordered Pd-Mn in common with Pd-Fe and Pd-Co shows giant moment ferromagnetism at low solute concentrations arising from the long range d electron polarization around local magnetic moments on the impurity atoms. The magnetic behaviour of this system is however complicated by the existence of antiferromagnet Mn-Mn interactions which Moriya[1] has suggested can persist up to third neighbour distances. At low impurity concentrations ($c < 3$ at.%), specific heat data, Boerstoel et al.[2], resistivity, Williams et al.[3], and magnetisation Rault & Burger[4], Star et al[5], all show the existence of a sharp transition to some sort of long range ferromagnetism. As the impurity concentration is increased beyond 3 at.% Mn, the magnetic behaviour becomes extremely complicated, leading different authors to varying conclusions. See Niewenhuys[6], Coles et al[7]. The overall picture which has emerged so far is one of a complicated mixed magnetic behaviour between 3 and 4.5 at.%Mn giving way to spin glass ordering at higher concentrations, Figure 2.

The magnetisation work of Star et al[5] and the resistivity work of Williams et al[3] led both sets of authors to the conclusion that antiferromagnetic Mn-Mn couplings have a strong influence even at low (~1 at.%) impurity concentrations. The total magnetisation is saturated only in extremely high fields (~240 kOer.) and the spontaneous moment is much smaller than would be expected on the basis of uniform ferromagnetism with $\sim 7.5 \mu_B$/Mn atom, Star et al[5]. Alloys with Mn concentrations less than 2.45 at.% show extraordinarily sharp specific heat anomalies at T_c contrary to the normal behaviour of dilute random systems. Niewenhuys[6] has interpreted this as resulting from a narrow distribution of Mn-Mn distances. Alloys with concentrations higher than 2.5 at.% Mn show the more normal rounded specific heat maxima; with the temperature of the maximum occurring at much higher temperatures than the previously reported magnetisation and resistivity anomalies, Rault and Burger[4] Coles et al[7], Zweers[8], Zweers et al[9] have suggested that the significant thermal quantity for a mixed magnetic system is $\Delta c/T$, the temperature derivative of the entropy. This shows a maximum at a temperature which agrees with the maximum in low field susceptibility (Figure 2).

In this paper we present low field d.c. magnetisation measurements of Pd-Mn alloys in the concentration range $1.3 < c < 5.8$. Although the initial d.c. susceptibility of ferromagnetic and nearly ferromagnetic alloys is complicated by irreversible processes, it does provide a rather sensitive tool for such investigations. The use of very low fields ($H_{ext} < 2$ Oers) has proved particularly valuable in the investigation of the remanent state of ferromagnetic and nearly ferromagnetic Pd-Mn which has not previously been observed in the low concentration region of ferromagnetism.

2. Experimental

All the alloys were prepared by arc melting together pure Pd (3N) and pure Mn (5N5) on a water cooled copper hearth. (The ingots were re-melted four times to ensure homogeneity). Subsequently small pieces of the ingot were melted into spheres. Microprobe analyses showed that the nominal compositions were correct to within the analysis uncertainty and that the alloys were sufficiently homogeneous.

Magnetisation measurements were made using a vibrating sample magnetometer whose sensitivity is 10^{-6} emu. in magnetisation and which is calibrated using the room temperature susceptibility of pure Pd ($\chi = 5.333$ μ emu / gm). Temperature measurements are made with carbon and diode sensors calibrated against a standard Ge thermometer for $T > 4.2°K$ and using the vapour pressure of He^4 below 4.2°K.

3. Results

3.1 Initial Susceptibility

The variation of low field magnetisation with temperature for most of the alloys studied is shown in figure 1. In the concentration region $c > 4.2$ at%Mn we interpret the magnetisation peak as the spin glass freezing temperature T_g and plot this as a function of manganese content in figure 2. The inverse of the initial susceptibility, χ^{-1}, was plotted for the four most concentrated alloys. These plots were linear in T down to temperatures close to T_g. From this linear portion the paramagnetic Curie temperature, θ_p, and the effective moment, p_{eff}, were obtained by fitting the data to the formula

$$1/\chi = (T - \theta_p)/C$$

where the Curie constant $C = N \mu_B^2 p_{eff}^2 / 3k_B$. These results are given in table 1. Only the region upto 40°K was used in this fit so as to negate any temperature dependence of the matrix susceptibility. The increasing importance of the direct Mn-Mn interaction with increasing concentration is reflected by the decrease in θ_p. The effective moments obtained are in fair agreement with the giant moments reported by Star et al [5] for a series of more dilute Pd-Mn alloys with concentrations upto 2.45 at%Mn.

THE MAGNETIC BEHAVIOR OF Pd-Mn ALLOYS

Table 1. The paramagnetic Curie temperature, θ_p, and effective moment, p_{eff}, for the four more concentrated alloys (where demagnetisation effects are not important) obtained from inverse susceptibility plots.

Alloy (at. % Mn)	θ_p (°K)	p_{eff} (μ_B / Mn)
4.5	3.6 ± 0.1	8.4 ± 0.2
4.9	3.1 ± 0.1	7.6 ± 0.2
5.35	2.5 ± 0.1	7.3 ± 0.2
5.8	1.7 ± 0.15	7.4 ± 0.2

All samples with c $>$ 4.2 at%Mn had strongly time dependent magnetisation at temperatures below T_g even in the very low applied fields (H ~2 Oers) used here. The data presented in figure 1 represents the incremental reversible magnetisation in 1.8 Oers. In this composition range (c > 4.2 at%Mn) the absolute value of the susceptibility is sufficiently small for the demagnetising field to be ignored to first order. (All these samples were nearly spherical). Thus, for these alloys the data of figure 1 gives the variation of initial susceptibility with temperature. The extremely rapid increase in peak susceptibility with decreasing composition (c $\leq$ 4.2 at%Mn) makes the effect of the demagnetising field increasingly important. In figure 1 this effect is clearly seen for the 4.2 and 3.6 at%Mn data. The intrinsic volume susceptibility χ_o is related to the observed susceptibility χ' by the relation

$$\chi_o = \frac{\chi'}{1-N\chi'}$$

where N is the demagnetisation factor. The peak susceptibility of the 4.2 at%Mn alloy just fails to reach the limiting value of $1/N = .2387$ cgs for a perfect sphere and the peak is apparently broadened. The susceptibility of the 3.6 at%Mn is much larger than this value and the susceptibility is limited to the maximum value of $1/N$ producing an apparently different behaviour to the higher concentration. This is undoubtedly the cause of the apparent two significant changes observed in the low frequency a.c. susceptibility reported in the preliminary abstract. At present we attach no special significance to this a.c. behaviour.

Since our samples were not exactly spherical we have not attempted to correct for demagnetisation in this presentation of the data. When approximate corrections are made to the data for the 3.6 at%Mn alloy, the curve is very similar to those shown by the higher concentration alloys. The curve for the 2 at%Mn alloy shown in figure 1 was obtained using a very thin disc sample edge on to the field, giving an approximate demagnetisation factor of $N = 9.4 \times 10^{-2}$. Again the susceptibility is limited by this demagnetising factor at T_c but the variation with temperature is rather different in character to the higher concentration alloys, showing a more conventional ferromagnetic behaviour.

The variation of T_g with composition (figure 2) differs from that reported by Coles et al[7] particularly in the higher concentrations. This we feel is due to the very low fields used here, making the determination of the temperature of the susceptibility maximum much more precise. We note that there is quite good agreement between this work and that of Zweers et al[9] who use the peak in $\Delta c/T$ as the spin glass freezing temperature. Both sets of data show very clearly that rather than going to zero at ~5 at%Mn, the spin glass freezing temperature remains above 2°K well into the complicated region between 3 and 4 at%Mn. Perhaps coincidentally, the curve of T_g versus concentration apparently extrapolates to zero close to 0 at%Mn.

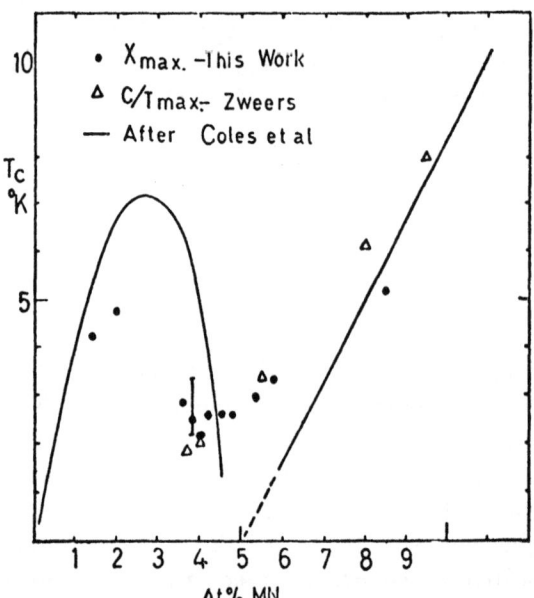

Fig. 2

The magnetic phase diagram of Pd-Mn. The solid lines due to Coles et al[7] are smooth curves drawn through data from several previous authors. Solid dots ● are from the present work. Δ are from Zweers et al[9].

3.2 Remanence and the Ferromagnetic Region

All the reports known to the present authors have suggested that the ferromagnetic Pd-Mn alloys exhibit no hysteresis and thus no remanence in zero applied field, Rault and Burger[4], Star et al[5]. In addition both these reports show a surprising magnetisation versus field variation for alloys with c below 2.5 at%Mn. Up to about 150 Oers the magnetisation is exactly linear in external field; at around 200 Oers there is an abrupt knee and this is followed by a very gradual increase of magnetisation with field which persists up to 240 KOer Star et al[5].

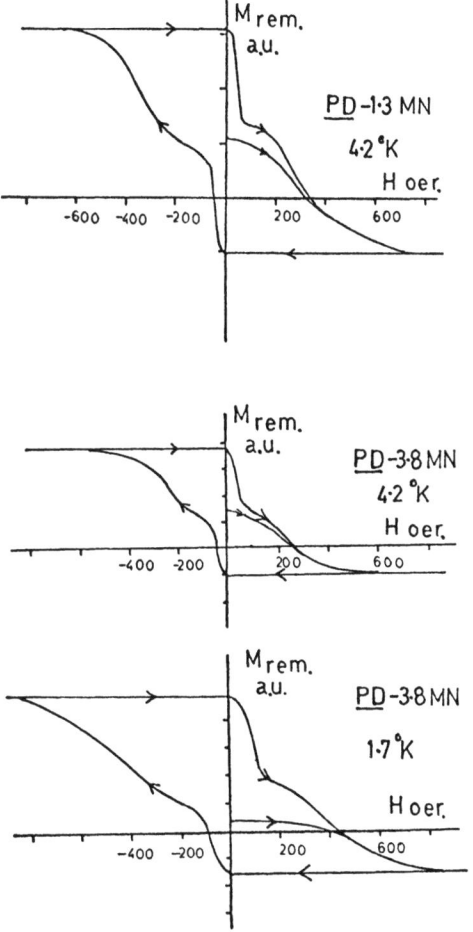

Fig. 3

The variation of isothermal remanence (Mrem) with the previously applied external field H, measured in 0.4 Oers. The asymmetry is due to the reversible response to the residual earth's field.

Our measurements show that there is a small isothermal remanence measured in the earth's field (0.40 Oers) for all the alloys studied having concentrations $c < 4.5$ at%Mn but it is in the opposite direction to, and has a very unusual dependence on, the previously applied external field (figure 3). This isothermal remanence appears to saturate in relatively low fields (600 Oers) with a value $10^{-2} - 10^{-3}$ emu/gm ; this is nearly three orders of magnitude smaller than the saturated magnetisation for the low concentration alloys. The temperature variation of this isothermal remanence is shown in figure 4 for 1.3 and 3.8 at%Mn.

In figure 3 we have plotted the variation of remanence with field to look like a reversed hysteresis loop. These measurements are made in an external field (earth's field) of 0.4 Oers : the reversible positive response to this field causes the asymmetry of the loops about the field axis (M = 0). These shifts are experimentally the same as the value of the magnetisation in 0.4 Oers obtained at 4.2°K after cooling in this residual field. Cooling in larger fields produces a negative thermoremanence. The data of figure 3 is typical of all the alloys studied with $c < 4.2$ at%Mn; we note that there are apparently two coercive forces, the higher one approximately matching the knee in the M versus H plot. For the low concentration alloys $c < 2$ at%Mn, preliminary measurements show that the width of the hysteresis curve is only weakly temperature dependent. In the higher concentration ; $c = 3.8$ at%Mn, however, the width apparently increases with decreasing temperature.

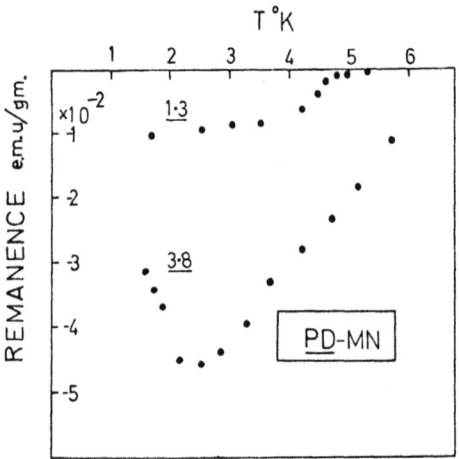

Fig. 4

The variation of saturated isothermal remanence with temperature for 1.3 and 3.8 at%Mn. Correction has been made for the reversible response to the residual earth's field (0.4 OERS). Time effects are observed below 2.5°K in the remanence of the 3.8 at%Mn sample. The peak in the low field susceptibility of 3.8 at%Mn is at 2.6 ± 0.3°K.

The temperature variation of reverse remance shown in figure 4 shows a clear difference between the low concentration, apparently ferromagnetic alloys, and those of the higher concentration mixed phase. In the low concentration alloys we found no time dependence and a monotonic rise in remanence to 1.7°K. Both 3.8 and 4.2 at%Mn alloys however show a maximum in the reverse remanence at nearly the same temperatures as the maximum in small field magnetization and time dependence of the remanence at temperatures below this maximum. It should be noted that remanence, albeit reversed, persists in these alloys to temperatures well above the susceptibility maximum. This is out of keeping with the general features of normal spin glass behaviour, Kouvel[10], Tholence and Tournier[11], Guy[12] and further points to the complexity of this intermediate concentration region.

Our results show that the sign of the isothermal remanence is unaffected by gross changes in the demagnetisation factor of a 2 at%Mn sample and that the thermoremanence of this alloy produced by cooling in a field of 160 Oers. is in the opposite direction to the cooling field. We have begun to study the effects of differing heat treatments on the reversible susceptibility and the sign of the remanence. Only qualitative results are at present available for 2 and 4.2 at%Mn alloys. In both cases we were unable to alter the temperature of the apparent transition, peak in $\chi(T)$ for 4.2 at%Mn, Curie temperature of 2 at%Mn. After a severe deformation and subsequent anneal at 650°C the 2 at% alloy showed a small positive isothermal remanence at 4.21°K ; the remanence of the 4.2 at%Mn alloy remained reversed at 4.2°K after a similar heat treatment. Chakrabati[13] studying the effects of differing heat treatments on 2.7 & 4.2 at%Mn samples, found that the magnetisation in 12.6 KOers is lowest in the deformed condition and highest in the aged condition. The remanence temperature construed from σ^2 versus Hi/σ plots for the 2.7 at%Mn was increased with ageing.

4. Discussion

The results described in the previous section strongly suggest that the transition from spin glass to ferromagnetism occurs gradually as the manganese concentration is decreased. Alloys with c between 3.6 and 4.5 at%Mn show some spin glass features, a sharp peak in $\chi(T)$ and magnetic viscosity at lower temperatures; at the same time these alloys exhibit a reverse remanence which persists at temperatures well beyond the susceptibility peak and is here apparently time independent. At a naive level this behaviour is suggestive of a magnetically two phase system possessing two different ordering temperatures, the higher one being some sort of ferromagnetic transition; the lower one being a spin glass freezing of the impurity Mn spins. Although this must remain conjectural at present it has some circumstantial support both from this data and that of other authors. We note that resistivity data in this concentration region shows a sharp change in slope at much higher temperatures than our susceptibility peaks; Coles et al[7], Williams et al[3], furthermore, the variation of isothermal remanence with applied field exhibits the unusual waisted characteristic and is of course in the opposite sense to the exciting field. The existence of a reverse remanence elsewhere is often associated with mixed systems close to the composition boundary between two magnetic and/or structural phases cf ilmenite - haematite solid solutions at 60% ilmenite Nagata[14]; disordered Ni-Mn alloys near the composition Ni_3Mn - Satoh et al[17].

It has long been apparent that the behaviour of Pd-Mn results from a competition between the ferromagnetism of indirect exchange interactions mediated by the highly exchange enhanced Pd hose and the antiferromagnetic direct exchange interactions between Mn spins. At very low concentrations the long range host polarization clearly dominates the latter more short range interaction. At intermediate concentrations the two effects must be of comparable magnitudes and thus the outcome of duel must depend rather critically on the precise distribution of Mn atoms. This is apparently borne out by the preliminary results of heat treatment effects on the remanence. However it is surprising that the remanence in a 2 at%Mn alloy can be reversed in sign without altering the Curie temperature.

The origin of the reverse remanence in Pd-Mn is at present unknown and thus we can only proceed by analogy with other self reversing systems. The hysteresis loops of figure 4 suggest a coercive force of about 200 Oers for the remanence. This would appear to rule out the magnetostatic model put forward originally by Neel, see Nagata[14]. Furthermore the sign of the remamence is independent of sample shape. Rather, some form of weak exchange interaction between predominantly ferromagnetic and predominantly antiferromagnetic regions seems more likely in an analogous fashion to the exchange anisotropy phenomena Kouvel[10].

The analogy cannot however be taken too far since Pd-Mn alloys do not apparently show the positive shifted hysteresis loops so characteristic of Cu-Mn alloys even though in this latter system long range Mn-Mn couplings are thought to be ferromagnetic and short range ones antiferromagnetic as in Pd-Mn.

Normally directed constricted hysteresis loops resulting from two magnetic phases in MnBi-Bi eutectics have been reported by Graham[15]. The well-known perminvar effect is also characterised by a constricted hysteresis loop but here the cause is thought to be due to directional ordering causing a stabilisation of some domain walls, Graham[16]. The perminvar effect is destroyed by cooling in moderate external fields (magnetic annealing) and exposure to a saturating field. The experiments with a 2 at%Mn alloy showed no such change in the remanence curve after cooling in 120 Oers. The effects of very high, saturating, fields on the remanence state of Pd-Mn are not known.

At present it is very difficult to find a critical concentration for the onset of true long ferromagnetism in Pd-Mn particularly in the light of the preliminary results of heat treatment studies. We note however the agreement between this work and that of Zweers et al[9] in the concentration region $c \gtrsim 3.6$ at%Mn; these two studies were carried out with rather different alloy heat treatments thus suggesting that the gross features that we observe are independent of alloy condition here. In this region it is fairly clear that no true long ferromagnetism exists.

The lower concentration alloys do show an apparently ferromagnetic transition close to the previously reported Curie temperatures but these alloys also showed reverse remanence in the arc melted state and a positive remanence in a deformed and heat treated state. This further points to the importance of antiferromagnetic Mn-Mn interactions and to the need for extreme metallurgical care even in these small concentrations.

Acknowledgements

The authors have great pleasure in acknowledging the help of Dr. H.E.N. Stone in sample preparation and the extremely useful discussions of this work with Professor B.R. Coles, Professor E.P. Wohlfarth and Dr. N. Rivier. One of us, W.H. thanks the SRC for a research studentship which supports this work.

1. Moriya T. Proc. Int. School of Physics, Enrico Fermi Course XXXVII New York Academic Press 1967
2. Boerstoel B.M. Physica 57, 397 (1972)
3. Williams G. Loram J.W. Sol. Stat. Comm. 7, 1261 (1969) Phys. Rev. 7B, 257 (1973)
4. Rault J., Burger J.P. C.R.Acad. Sci.(Paris) B269, 1085 (1969)
5. Star W.M., Foner S., Mcniff E.J., Phys. Rev. B12, 2690, (1975)
6. Nieuwenhuys G.J., Adv. in Physics 24, 515 (1975)
7. Coles B.R., Jamieson H., Taylor R.H., Tari A., J. Phys. F 5, 565 (1975)
8. Zweers H.A., PhD Thesis Leiden (1976)
9. Zweers H.A., Pelt W., Nieuwenhuys G.J., Mydosh J.A., Int. Mag. Conf. Amsterdam (1976)
10. Kouvel J.S., J. Phys. Chem. Sol. 21, 57 (1961)
11. Tholence J.L., Tournier R., Jnl. de Physique 35, C4-237 (1974)
12. Guy C.N., Int. Mag. Conf. Amsterdam (1976)
13. Chakrabarti D.J., Int. J. Magnetism 6, 305 (1974)
14. Nagata T., Rock Magnetism Chap. V, Maruzen, Tokyo (1961)
15. Graham C.D., Magnetic Properties of Metals and Alloys, A.S.M. Conf. Cleveland 1958
16. Graham C.D., Notin M.R., Int. Magnetism Conf. Moscow, Tom II p.186 (1973)
17. Satoh T., Patton C.E. and Goldfarb R.B., Int. Mag. Conf. Amsterdam (1976)

SOME CONTROVERSIAL ASPECTS OF ELECTRONIC AND MAGNETIC INTERACTIONS

IN THE AMORPHOUS METALLIC STATE*

C. C. Tsuei

IBM Thomas J. Watson Research Center
Yorktown Heights, New York 10598

INTRODUCTION

In recent years, considerable effort has been devoted to the study of electrical and magnetic properties of amorphous metallic alloys. Several important aspects of this field are still the subject of much controversy. The purpose of this article is to critically review the status of one such controversial problem, i.e. <u>the origin of the resistance minimum phenomenon</u>.

A large number of amorphous metallic conductors are characterized by a minimum in the electrical resistance versus temperature plot. The resistance minimum anomaly can occur at relatively high temperatures (e.g. $\sim$500°K) and even co-exists with ferromagnetism.[1-3] These findings have been attributed to either a s-d exchange interaction (i.e. a Kondo-type effect)[1] or to effects associated with a configurational indeterminacy in the atomic arrangement.[3] Recent experimental results will be discussed in terms of these two points of view.

BASIC PHENOMENA

The temperature dependence of the electrical resistivity $\rho(T)$, of various amorphous metals can be classified into the following three categories:

1) Like most of crystalline metals, the resistivity of amorphous elements or alloys in this category increases monotonically with increasing temperature. The resistivity is constant as a function of temperature for T $\gtrsim$ 15°K. At relatively high temperatures (T>100°K), resistivity depends linearly on temperature. The

*Invited paper.

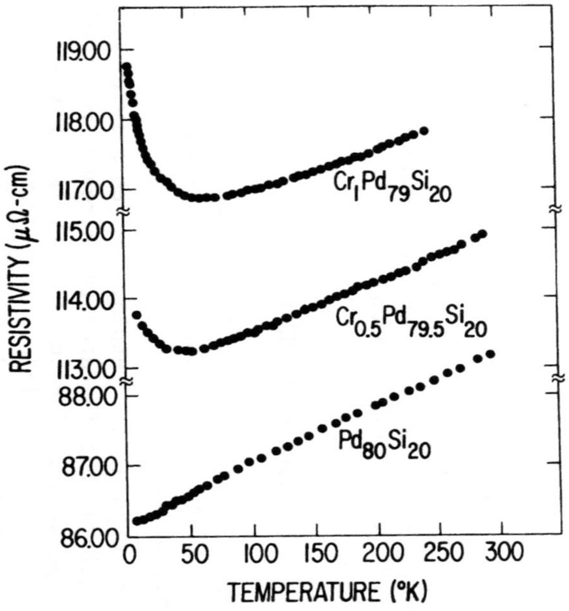

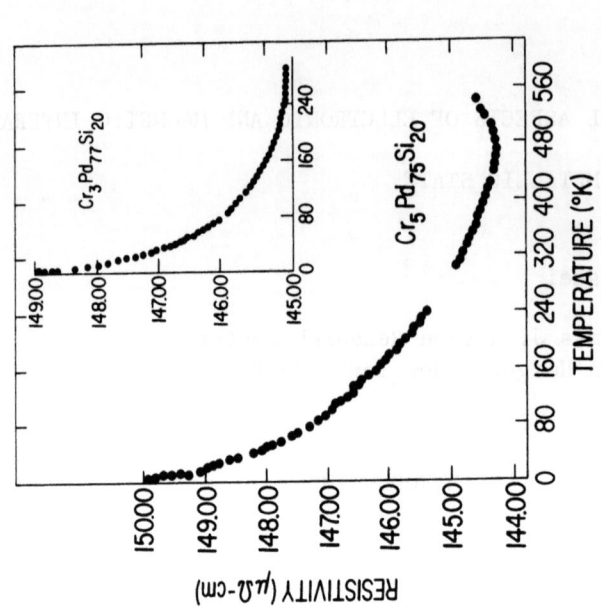

Fig. 1. Electrical resistivity as a function of temperature for amorphous $Pd_{80}Si_{20}$, and Cr-Pd-Si alloys obtained by liquid quenching.[1]

Fig. 2. Electrical resistivity as a function of temperature for amorphous Cr-Pd-Si alloys.[1]

resistivity between the two temperature regions is, in general, characterized by a power-law temperature dependence such as T^n with typically n=2.[4] Prototype examples of amorphous metals exhibiting this type resistivity behavior are pure amorphous Pd-Si alloys[1] (see Fig. 1).

2) There is a well-defined minimum in the resistivity versus temperature curve. The temperature at the resistance minimum (T_{min}) ranges from a few degress (°K) to as high as 500°K. The magnetic properties of the amorphous alloys with such a resistance anomaly ranges from weakly paramagnetic to strongly ferromagnetic. The temperature dependence of the amorphous Cr-Pd-Si alloys as depicted in Figs. 1 and 2 represents the typical resistivity behavior of the paramagnetic alloys. The resistivity of amorphous ferromagnets also shows a minimum at low temperature[2,3] (below the Curie temperature). Further analysis of the resistivity data indicates that the resistivity as a function of temperature for $T<T_{min}$ can be essentially described as a sum of a $\ln T$ term and residual resistivity (see Figs. 3 and 4).

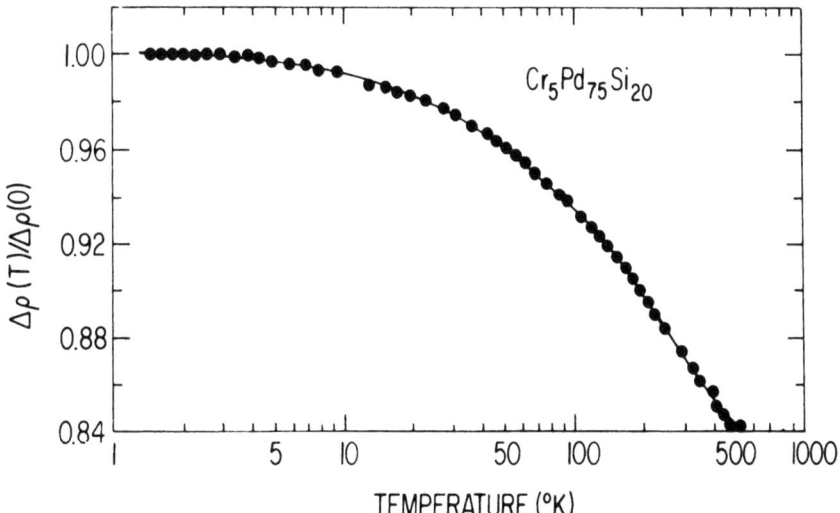

Fig. 3. Typical temperature dependence of differential resistivity for amorphous $Cr_5Pd_{75}Si_{20}$.[1]

$$[\Delta\rho(T) \equiv \rho_{alloy}(T) - \rho_{Pd_{80}Si_{20}}(T)]$$

3) The resistivity of alloys in this category decreases monotonically with increasing temperature. There is an upturn in the resistivity at low temperatures.[5,6] It has been found that the resistivity as a function of temperature in the upturn region also depends logarithmically on temperature. Typical examples are shown in Fig. 5. In addition to amorphous Ni-Pd-P alloys obtained by liquid quenching, amorphous Ni-P alloys prepared by electrodeposition also exhibit similar temperature dependence of resistivity.[7]

From the above summary of resistivity data for a variety of amorphous metallic alloys, several interesting features emerge:

a) Amorphous metals are basically characterized by a relatively high residual resistivity ρ_o and a small temperature-dependent term ρ_T (i.e. total resistivity $\rho(T)$ can be described as $\rho(T) = \rho_o \pm \rho_T$). The value of ρ_o is of the order of 10^2 μΩ cm and ρ_T, essentially the electron-phonon term, is only about one to two percent of ρ_o at room temperature. As a consequence, even a weakly-temperature dependent contribution to resistivity (other than the normal electron-phonon in origin) shows up conspicuously in a resistivity versus temperature plot.

b) A large number of amorphous metallic conductors exhibit the resistance minimum phenomena. It is, however, worth noting that this resistance anomaly is absent in a few amorphous metallic alloys. These alloys are either of high purity or of high valency. Ultra pure amorphous $Pd_{80}Si_{20}$ alloys are examples of the former,[1] Sn rich non-crystalline Cu-Sn alloys, the latter.[8]

c) Those alloys that do show a resistance minimum cover a wide spectrum of alloy systems and compositions and a variety of magnetic properties as mentioned before. Naturally, one would conclude that the resistive anomaly is not magnetic in origin. It is, however, important to point out that experimental results have not conclusively established that the resistance minimum anomaly is <u>not</u> associated with the existence of local moment in these alloys. In fact, there is no amorphous alloy that exhibits the resistance minimum phenomenon and contains no magnetic constituent.

ORIGIN OF THE RESISTANCE-MINIMA

To explore the origin of the resistance-minimum phenomenon in amorphous metals, two major approaches have been taken. These two theoretical models are practically identical in formalism, but attribute the minimum to quite a diametrically different origin. In the following, a brief account of these two theories will be given before the discussion of more experimental results.

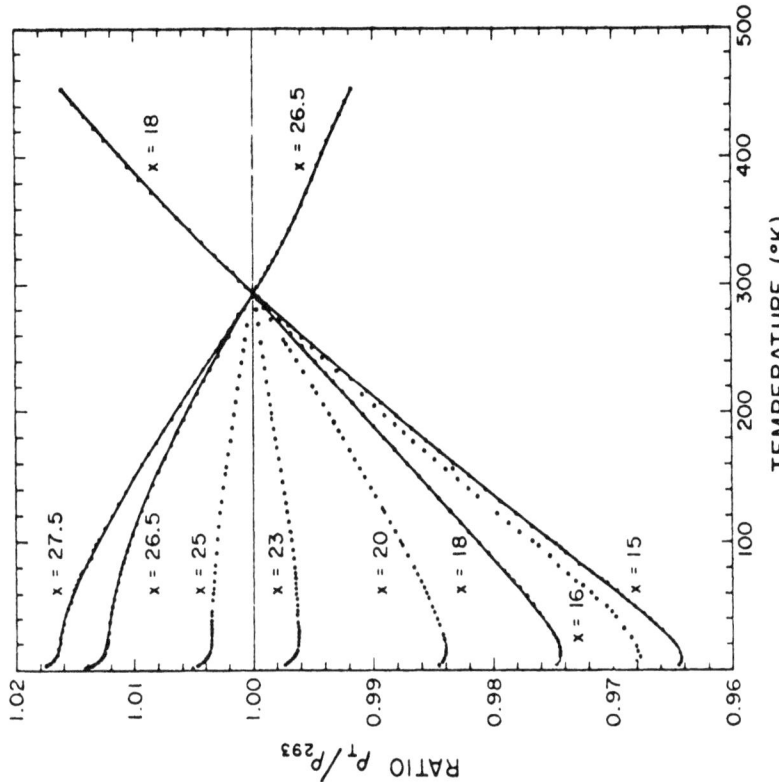

Fig. 5. Normalized resistivity as a function of temperature for amorphous $(Ni_{50}Pd_{50})_{100-x}P_x$ alloys prepared by quenching from the melt.

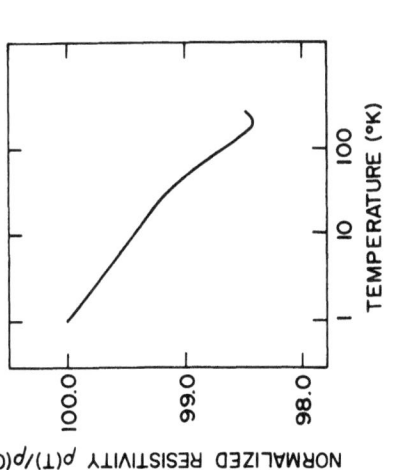

Fig. 4. Normalized resistivity $\rho(T)/\rho(0) \times 100$ as a function of temperature for Metglas 2826A $(Fe_{32}Ni_{36}Cr_{14}P_{12}B_6)$.[3]

A) Kondo-type effects:

Exchange interactions between conduction electrons and localized moments or equivalent can lead to a minimum in the ρ vs T plot. Extensive reviews of the theoretical[9] and experimental[10,11] results for the crystalline Kondo systems can be found in the literature. As Kondo theory is not explicitly restricted to deal with crystalline materials, it has been applied to the amorphous cases to explain the logarithmic temperature term in resistivity.[1] Basically, the s-d exchange scattering of a conduction electron in the state $\vec{K}$ by a magnetic impurity S_n into final state $\vec{K}'$ can be described by the following Hamiltonian:

$$H_{s-d} = -\frac{J_{s-d}}{2N} \sum_{n,\vec{K},\vec{K}'} \left\{ (a^+_{\vec{K}'\uparrow} a_{\vec{K}\uparrow} - a^+_{\vec{K}'\downarrow} a_{\vec{K}\downarrow}) S_{nz} + a^+_{\vec{K}'\uparrow} a_{\vec{K}\downarrow} S_{n-} + a^+_{\vec{K}'\downarrow} a_{\vec{K}\uparrow} S_{n+} \right\} \quad (1)$$

where J_{s-d} is the s-d exchange integral, a measure of the strength of the interaction. The second quantization operators and spin operators are defined conventionally. The arrows refer to the spin states of conduction electrons and n to the magnetic impurity. Using a perturbation approach, Kondo calculates the transition probability per unit time from the initial state a to the final state b:

$$W(a \to b) \simeq \frac{2\pi}{\hbar} \delta(E_a - E_b) \times \left\{ H'_{ab} H'_{ba} + \sum_{c \neq a} \frac{(H'_{ac} H'_{cb} H'_{ba} + c.c.)}{(E_a - E_c)} \right\}$$

where $H'_{ba} = \langle b | H_{s-d} | a \rangle$ \quad (2)

From $W(a \to b)$, one can calculate electrical resistivity in a fairly straightforward manner. The final expression for $\rho_{s-d}(T)$ obtained by Kondo is

$$\rho_{s-d} = c \left\{ A + B\, J_{s-d} \ln\left(\frac{k_B T}{D}\right) \right\} \quad (3)$$

where c is the concentration of magnetic impurity, A and B are constants, D the width of the conduction band. If the s-d interaction is antiferromagnetic ($J_{s-d} < 0$), ρ_{s-d} will increase with decreasing temperature.

To identify the scattering processes that are responsible for the $\ln T$ term in ρ_{s-d} (i.e. Eq. (3)), the probability per unit time for a transition from the state $\vec{K}\uparrow$ to the final state $\vec{K}'\uparrow$ will be considered in some details. Various possible first and second order scattering processes in the s-d interaction are schematically shown in Fig. 6. The first order (as shown in Fig. 6a) is a direct scattering of $\vec{K}\uparrow$ into $\vec{K}'\uparrow$ and its transition probability is

ELECTRONIC AND MAGNETIC INTERACTIONS

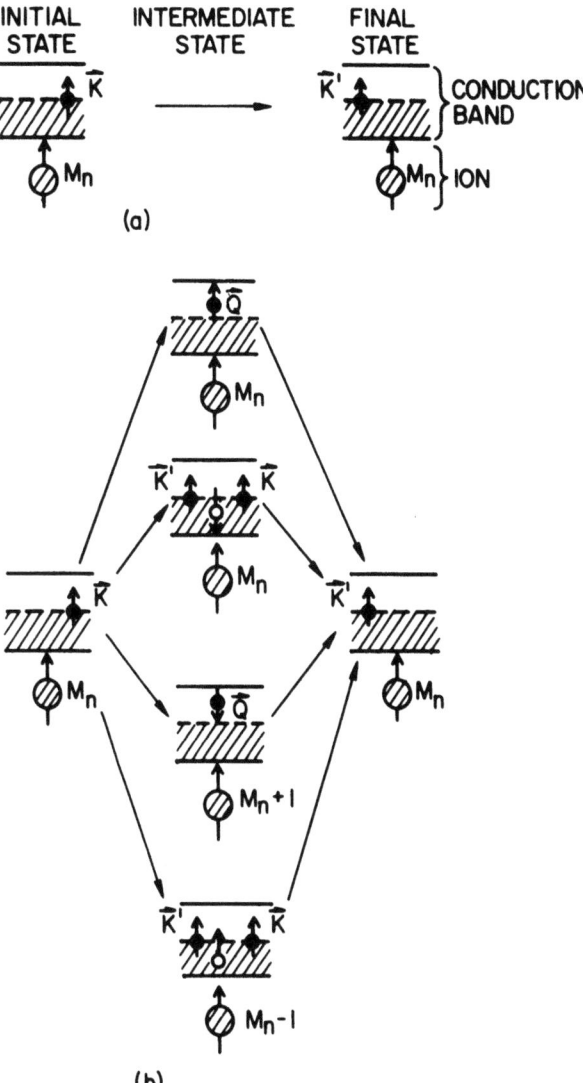

Fig. 6. Schematic diagrams showing the first and second order s-d scattering processes.

a constant of temperature. There are four possible second order processes arising from the variety of the intermediate states of virtual process. Of these four processes, the last two require that the z-component of the magnetic impurity spin (M_n) changes by one unit to conserve total angular momentum of the scattering system. And it is exactly these two processes that give rise to the ln T term in ρ_{s-d}. In terms of Kondo theory, a logarithmic increase of resistivity is then a manifestation of the dynamical nature of the magnetic spin system. In other words, the scattering system (e.g. the magnetic spins) has to possess some internal degree of freedom, e.g. spin-flipping for the case discussed here, to produce a Kondo-type resistive anomaly. This is consistent with the observation that resistance minimum disappears as the concentration of magnetic impurity increases (presumably this is due to the loss of spin-flip capability as a result of magnetic ordering) or under the influence of a strong magnetic field (say, H~200 KOe or higher) at low temperatures.

The formula for ρ_{s-d} as expressed by Eq. (3) predicts an unphysical logarithmic divergence at very low temperatures. This is, of course, a pitfall of the perturbation approach as used originally by Kondo.[12] This difficulty has been removed by Nagaoka[13] and others by using non-perturbational,, self-consistent treatments such as the technique of the double-time Green's function. They prove that the perturbational treatment breaks down below a characteristic temperature (called Kondo temperature, T_K) against the formation of a quasi-bound state between the conduction-electron spin and the localized spin. As a consequence of the spin-compensated states, resistivity is finite at T=0°K. Using Nagaoka's approach, Hamann derives an expression for ρ_{s-d} valid for all temperatures:[14]

$$\rho_{s-d} = \frac{2\pi c}{ne^2 K_F} \left\{ 1 - \frac{\ln(T/T_K)}{[\ln^2(T/T_K) + S(S+1)\pi^2]^{1/2}} \right\} \quad (4)$$

where S is the spin value of the localized spin, n the conduction electron density and K_F the Fermi wave vector.

B) Structural effects:

Recently, Cochrane et al.[3] suggest that the resistance-minimum phenomenon is <u>not</u> magnetic in nature. To account for the experimental results, they put forward a theory based on effects derived from the configurational indeterminacy in the atomic arrangement of amorphous alloys. This approach is essentially an analogy of the "tunneling model" proposed by Anderson, Halperin and Varma[15] for explaining the anomalous low temperature thermal properties of insulating glasses. The primary hypothesis of this model is that there are two (almost) equivalent atomic configura-

ELECTRONIC AND MAGNETIC INTERACTIONS

tions which a certain number of atoms (or clusters of atoms) can take. This is of course a manifestation of the structural (random) nature of the atomic arrangement in amorphous solids. Quantum mechanically, such structural characteristics can be represented by a distribution of localized "tunneling states" $|\pm\rangle$ and the corresponding electronic states with wave vector $\vec{K}$ by $|K_\pm\rangle$. The Hamiltonian that describes the amorphous alloy system with the interaction between conduction electrons and the tunneling configurations can be written as follows:

$$H = H_o + H_1$$

where $H_o = \sum_{\alpha=\pm} (\alpha\Delta) b_\alpha^\dagger b_\alpha + \sum_{\vec{K},\alpha=\pm} \varepsilon_{K\alpha} a_{K\alpha}^\dagger a_{K\alpha} b_\alpha^\dagger b_\alpha$

and

$$H_1 = \sum_{\substack{KK' \\ \alpha\alpha'=\pm}} V_{KK'}^{\alpha\alpha'} a_{K\alpha}^\dagger a_{K'\alpha'} b_\alpha^\dagger b_{\alpha'} \tag{5}$$

represents the scattering of conduction electrons from the tunneling states. The absence of spin-subscript for conduction electron merely reflects the a priori assumption that the origin of the resistance minimum anomaly is non-magnetic.

By defining the pseudo-spin operators[16] I_z and $I_\pm$ and by retaining only those terms involving the operators I_z, $I_\pm$, one can describe the interaction term of the total Hamiltonian H by:

$$H_{int} = V_c \sum_{KK'} [(a_{\vec{K}+}^\dagger a_{\vec{K}+} - a_{\vec{K}-}^\dagger a_{\vec{K}'-}) I_z$$

$$+ e^{-\lambda} (a_{\vec{K}+}^\dagger a_{\vec{K}'-} I_+ + a_{\vec{K}-}^\dagger a_{\vec{K}'+} I_-)] \tag{6}$$

where V_c is a measure of the strength of the exchange interaction, and λ is related to the energy difference of the tunneling states $|\pm\rangle$ by $2\Delta = \hbar\omega_o e^{-\lambda}$, $\hbar\omega_o$ is a zero-point energy.

At this point, it is important to note that there is a close resemblance between H_{s-d} (Eq. (1)) and H_{int} (Eq. (6)). Naturally, one would expect that an anisotropic exchange interaction as described by H_{int} would lead to a Kondo-type resistance behavior. This is indeed the case as Cochrane et al. have shown it. The resistivity as a function of temperature can be expressed by:

$$\rho_{int}(T) = g \ln (k_B^2 T^2 + \Delta^2) \tag{7}$$

where Δ is of the order 0.1 meV.

Based on this result, it was concluded that the resistance-minimum anomaly is only structural manifestation in amorphous metallic alloys and is not magnetic in character.

EXPERIMENTAL SUPPORTING EVIDENCE

From the above discussion, one can see that there are at least two plausible theoretical approaches to the problem of resistance-minimum in amorphous metals. Both of these models apparently are capable of producing a $\ln T$ contribution to the electrical resistivity as a consequence of some internal degree of freedom built intrinsically in each mechanism. For the Kondo-type effect, this comes from the spin-flip of the localized moment. For the structural effects, it stems from the random nature of the atomic arrangement - tunneling between two structural configurations. To make a choice from these two possible mechanisms, the following experimental findings are worthwhile considering:

1) From the structure analysis of many amorphous metallic alloys, it has been concluded that the radial distribution functions (RDF) of these alloys are quite similar if the differences in the atomic sizes are taken into account.[17] The RDF features can be explained in terms of a common model such as dense random packing of hard spheres (transition metals: Pd, Ni, Fe or Cr) with the metalloid (Si, P, C or B) filling the larger holes inherent in a Bernal structure. On the other hand, the occurence of resistance minimum has not been a common event for all the amorphous metallic alloys. Therefore, it is unlikely that the resistive anomaly is associated with the structure.

2) Resistance minimum with a $\ln T$ dependence below T_{min} has never been observed in any amorphous metallic element or alloy that is surely free of magnetic impurity such as Fe or Cr ($c <$ a few ppm). For instance, amorphous Pd-Si alloys obtained by liquid quenching usually exhibit a minimum in their ρ vs T plots. The anomaly was not found for alloys, of the same composition, made of ultra-high-purity constituents under the same quenching conditions (e.g. Fig. 1).

3) On the other hand, it has been well-established[1] experimentally that an addition of small amount of elements such as Cr, Fe or Co to an amorphous alloy such as $Pd_{80}Si_{20}$ results in a resistance-minimum and related phenomena characteristic of Kondo effect. The resistivity minimum temperature T_{min}, as a function of Cr concentration for Cr-Pd-Si alloys, varies as $c_{cr}^{1/2}$. This is consistant with the fact the electron-phonon resistivity has a T^2 temperature dependence in the T_{min} region. There is no observable Cr-effect on the structure of these amorphous alloys.

4) There are a number of amorphous metallic vapor-deposited films that do not exhibit the resistance minimum phenomenon.[8,18] Examples are: $Ga_{95}Ag_5$, $Sn_{84}Cu_{16}$ and $Pb_{80}Ag_{20}$. These alloys apparently were not prepared under the condition of ultra high purity. According to Anderson's theory of localized moments formation,[19] this absence of resistance minimum merely reflects the unlikelyhood of forming localized moments in high-valent metals such as Pb, Sn etc. It is interesting to note that a minimum was indeed observed in the $\rho(T)$ of an amorphous $Cu_{50}Ag_{50}$ vapor-deposited film.[8] And it is well-known that Fe, Cr or other transition-metal impurities do carry a magnetic moment in low-valent metals such as Cu and Ag.

5) The fact that a resistance-minimum anomaly co-exists with ferromagnetism in practically all amorphous ferromagnets is quite remarkable. There were several theoretical attempts to explain the phenomenon in terms of a Kondo-type effect.[20-22] Furthermore, it has been pointed out that weakly-coupled d-spins in a non-crystalline ferromagnetically ordered material can still contribute to the spin-flip-scattering process which would lead to a Kondo-type resistance minimum.[23] Recently, it has been shown that small amounts of $Cr(\leq 2at.\%)$ added to a strong amorphous ferromagnet such as $Fe_{75}P_{15}C_{10}$ result in a resistance minimum with a T_{min} up to 200°K, depending upon Cr concentration.[24] This experimental result indicates that a resistance minimum can indeed be produced by introducing Cr impurity into a magnetically ordered non-crystalline alloy. It also provides some insights as to interpreting the resistivity data for some of the Metglas alloys. For instance, there are two linear $\ln T$ regions in the ρ versus $\ln T$ plot for Metglas 2826A ($Fe_{32}Ni_{36}Cr_{14}P_{12}B_6$) as shown in Fig. 4. In view of the resistivity results for amorphous $Fe_{75}P_{15}C_{10}$ containing Cr just mentioned, the linear $\ln T$ region near T_{min} (∼210°K) in Fig. 4 is probably due to Cr contained in the alloy. It is of interest to note that other Metglas alloys containing no Cr show no resistance minimum at such high temperatures. The linear $\ln T$ that occurs at lower temperatures (T<15°K) is probably attributable to a Kondo-type scattering originated from some of the Fe spins in the alloy.

6) The effect of thermal annealing on the $\rho(T)$ of Metglas 2826A was studied by Cochrane and Strom-Olsen recently.[25] Their results indicate "No effect on the resistivity is observable until crystallization, whereupon the anomaly is reduced by about an order of magnitude, but does not disappear even when annealed at several hundred degrees above the crystallization temperature." The fact that the resistance minimum can not be eliminated by annealing at several hundred degrees above the amorphous-crystalline phase transition temperature strongly suggests that this anomaly is not a structural manifestation in amorphous alloys. On the other hand,

it is consistent with a Kondo-type effect caused by some magnetic constituents in the alloys. It is known that amorphous metallic alloys such as Metglas alloys crystallize into multi-phases, magnetic and non-magnetic. It is, therefore, of no surprise that some weak Kondo-type resistance minimum can be found in the well-annealed samples. Obviously, more work on the structure and magnetic properties of the crystallized sample is needed to substantiate the conclusions drawn here.

7) There are other related effects that can be of some help in clarifying the controversial situation centered around the resistance minimum phenomenon:

(a) The effect of high magnetic fields: Under the influence of high field (H $\gtrsim$ 200 kOe), the s-d exchange scattering should be greatly altered, while the structural interactions proposed by Cochrane et al. should not be effected. The $\ln T$ dependence of the resistance in several amorphous metals is not modified by a field of 45 kOe, aside from the regular magneto-resistance effect.[3] This result, however, can not be admitted as convincing evidence disproving the Kondo-effect, because the magnitude of the magnetic field effect hinges on the ratio of $(\mu \cdot H)/k_B(T+T_K)$ where μ is the magnitude of the localized moment. A straight forward estimate indicates that at low temperatures (say $T \sim 1°K$), an amorphous ferromagnet containing impurity with large moment such as Gd should exhibit an observable effect under a field of the order of 100 kOe or higher.

(b) Anomalous linear term in specific heat: This effect has been observed in insulating glasses by Pohl and his co-workers[26] and explained in terms of the structural tunneling model by Anderson et al.[15] In amorphous metallic alloys, this effect is likely tangled with the electronic linear-T specif heat.

(c) Anomalous accoustic attenuation at low temperatures: As predicted by Anderson et al.[15] the structural tunneling should result in an increase in ultrasonic attenuation at low temperatures. No experiment of this sort has been reported on amorphous metallic alloys with a resistance minimum. An increase of ultrasonic attenuation in insulating glasses with decreasing temperature at very low temperatures was observed.[27] And this was attributed to the effect of atoms tunneling between double-well potentials. Therefore, we suggest that acoustic measurements should be performed on amorphous materials such as Cr-Pd-Si, Fe-P-C, Metglas, Ni-P, Ni-Pd-P alloys etc. to provide additional experimental evidence in supporting (or disapproving) the structural model for the resistance minimum anomaly.

CONCLUDING REMARKS

The resistance-minimum phenomenon prevails in (almost) all

amorphous metallic conductors, except for the cases where the conditions of high-purity (e.g. c ≲ 1ppm of Fe) or high-valency (e.g. higher than di-valent), are satisfied. To account for the experimental findings associated with this anomaly, two theoretical models have been proposed. These two approaches are identical in theoretical formalism, yet quite different about the mechanism (the internal degrees of freedom) that is responsible for the $\ln T$ contribution in resistivity. The two mechanisms are: 1) Kondo-type s-d exchange scattering - a magnetic alloy-constituent effect; 2) configurational tunneling - a structural manifestation of the non-crystalline nature in the atomic arrangement. Available results on the resistance-minimum anomaly and its related effects are critically reviewed in this article. They are found to be in favor of the Kondo mechanism. Further experiments are suggested to pin down the origin of the resistance-minimum phenomenon on a more definitive basis.

ACKNOWLEDGEMENT

The author wishes to thank R. W. Cochrane for providing him with resistivity results prior to publication.

REFERENCES

1. C. C. Tsuei and R. Hasegawa, Solid State Commun. 7, 1581 (1969); R. Hasegawa and C. C. Tsuei, Phys. Rev. B 2, 1631 (1970).

2. S. C. H. Lin, J. Appl. Phys. 40, 2173 (1969).

3. R. W. Cochrane, R. Harris, J. O. Strom-Olson and M. J. Zuckermann, Phys. Rev. Lett. 35, 676 (1975).

4. R. Hasegawa, Phys. Lett. 36A, 425 (1971).

5. B. Y. Boucher, J. Non-cryst. Solids 7, 277 (1972).

6. A. K. Sinha, Phys. Rev. B1, 4541 (1970).

7. P. J. Cote, Solid State Commun. 18, 1311 (1976).

8. D. Korn, W. Mürer and G. Zilbold, Z. Physik 260, 351 (1973).

9. J. Kondo, in Solid State Physics, edited by F. Seitz, D. Turnbull and H. Ehrenreich (Academic Press, New York, 1969), Vol. 23, p. 183.

10. J. A. Heeger in Solid State Physics, edited by F. Seitz, D. Turnbull and H. Ehrenreich (Academic Press, New York, 1969), Vol. 23, p. 283.

11. G. Grüner, Advances in Phys. $\underline{23}$, 941 (1974).

12. J. Kondo, Prog. Theor. Phys. $\underline{32}$, 37 (1964).

13. Y. Nagaoka, Phys. Rev. $\underline{138}$, A1112 (1965).

14. D. R. Hammann, Phys. Rev. $\underline{158}$, 570 (1967).

15. P. W. Anderson, B. I. Halperin and C. M. Varma, Philos. Mag. $\underline{25}$, 1 (1972).

16. $b_\pm^\dagger b_\pm = \frac{1}{2} \pm I_z$ and $b_\pm^\dagger b_\mp = I_\pm$

17. G. S. Cargill III, in Solid State Physics, edited by F. Seitz, D. Turnbull and H. Ehrenreich (Academic Press, New York, 1975), Vol. 30, p. 225.

18. D. Korn, H. Pfeifle and G. Zibold, Z. Physik $\underline{270}$, 195 (1974).

19. P. W. Anderson, Phys. Rev. $\underline{124}$, 41 (1961).

20. R. N. Silver and T. C. McGill, Phys. Rev. $\underline{B9}$, 272 (1974).

21. A. Madhukar and R. Hasegawa, J. Phys. (Paris) $\underline{35}$, C4-291 (1974).

22. T. Kaneyoshi and R. Honmura, Phys. Lett. $\underline{A46}$, 1 (1973).

23. C. C. Tsuei and H. Lilienthal, Phys. Rev. B $\underline{13}$, 4899 (1976).

24. C. C. Tsuei (to be published).

25. R. W. Cochrane and J. O. Strom-Olsen, Proceedings of ICM 1976 (Amsterdam).

26. R. C. Zeller and R. O. Pohl, Phys. Rev. $\underline{B4}$, 2029 (1971);
V. Narayanamurti and R. O. Pohl, Rev. mod. Phys. $\underline{42}$, 201 (1970).

27. W. Arnold, S. Hunklinger, S. Stein and K. Dransfeld, J. Non-Cryst. Solids $\underline{14}$, 192 (1974).

DISCUSSION

J. S. Schilling: I would like to point out that there is a lot of pressure work on various systems containing transition metal impurities and what one finds in all cases is that the Kondo temperature increases with increasing pressure, so that the suggestion here is that one way to determine whether one has a Kondo behavior or not is to apply pressure and see the changes that occur in the system.

C. C. Tsuei: This sounds like a nice experiment which should be done.

R. Tournier: This may not be very easy to do because what we are dealing with here is a distribution of Kondo temperatures. You may have impurities in the system that will give you an increase in the resistivity with temperature and some others that will give a decrease in the resistivity. So I am not completely sure the pressure experiment would give any definite answer.

R. W. Cochrane: I think that an excellent probe of a magnetic system is the magnetic field, particularly in the case of systems for which the characteristic temperatures are in the liquid helium range, where you have accessible high fields. I think what really hung us up on these things being magnetic is the fact that you see no change in the minimum or the low temperature slope when you apply a magnetic field. Now the Kondo effect being the scattering from a free spin one should be able to split this level with a magnetic field. Since we never saw this, it led us into discussing the structure as part of a contribution for the low temperature minimum. As I would visualize it the structure and the Kondo effect contributions are very similar. You get electrons scattering from an internal degree of freedom. In one case the degree of freedom being the spin in the other case, the indeterminacy in position. You can have them both together so that the field effect is a negative criterion. If you can see a field effect you would say it might be one, or it might be the other, it might be a mixture of both, but when you see no field effect you rule out the fact that the characteristic magnetic temperature is low in which case you would not have a simple $\ln T$ term anyway.

C. C. Tsuei: As I mentioned in my talk, I would not rule out the possibility of structural effects. However, the fact that the minimum disappears when magnetic impurities are absent from the system tends to lead people to think the presence of magnetic impurities have a definite effect. I also think the 45 kG field that you used is probably not sufficient to observe such a small effect. One would have to go to fields of the order of 150 kG or to very low temperatures before anything can be seen.

J. A. Mydosh: In a way of a comment I would like to mention an experiment that we recently carried out. We looked at the resistivity of pure palladium with less than 1 ppm Fe that was in the form of a thin film. These films have a high degree of disorder. The films exhibited a resistivity minimum that disappears upon annealing to higher temperatures as the crystallinity becomes more pronounced. The conclusion at this time is that it has to be a structural effect because the Fe would not simply disappear.

C. C. Tsuei: It is very possible upon annealing Fe-Pd system forms a giant moment which could immediately kill a Kondo type scattering.

B. C. Giessen: What is the present status of the susceptibility of pure amorphous Pd-Si? It was diamagnetic when first reported, then in later publications it was taken to be paramagnetic.

C. C. Tsuei: The ultra pure amorphous Pd-Si is diamagnetic in behavior. It was a very small upturn at low temperatures.

MAGNETIC SUSCEPTIBILITY OF AN AMORPHOUS NON-TRANSITION METAL ALLOY: $Mg_{.70}Zn_{.30}$*

B.C. Giessen, A. Calka and R. Raman
Northeastern University
Department of Chemistry and
Institute of Chemical Analysis, Applications
and Forensic Science
Boston, MA 02115

D.J. Sellmyer
Behlen Laboratory of Physics
University of Nebraska
Lincoln, NE 68588

ABSTRACT

In the past, the preparation of amorphous metals by rapid quenching of the liquid has generally been limited to transition or noble metal alloys; the magnetic properties of amorphous metals containing only s-p electron (normal) metals have therefore not been studied. Recently, amorphous Mg-Zn alloy ribbons with compositions in a narrow range including $Mg_{.70}Zn_{.30}$ (the region of a deep eutectic) were prepared by melt spinning. Initial susceptibility measurements have shown that these alloys possess temperature-independent Pauli paramagnetism with susceptibilities ranging from about 5.0 emu/g-atom for $Mg_{.75}Zn_{.25}$ to about 3.0 emu/g-atom for $Mg_{.70}Zn_{.30}$. A treatment of these values analogous to that for the susceptibilities of liquid metals yields a valence electron contribution corresponding to a free-electron Pauli-Landau term, without a contribution from electron-electron interactions. It is possible that this agreement is fortuitous and due to the cancellation of corrections which are required to interpret the data for liquid Mg and Zn.

*Research supported at Northeastern University by the Army Materials and Mechanics Research Center and at the University of Nebraska by the National Science Foundation.
Communication No. 19 from the Institute of Chemical Analysis, Applications and Forensic Science, Northeastern University.

INTRODUCTION

Studies of the magnetism of amorphous metals have generally focussed on alloys exhibiting temperature-dependent paramagnetism or ferromagnetism. Temperature-independent conduction electron (Pauli) paramagnetism has been observed in several amorphous metals, e.g., Pd-Si,[1] Cu-Zr[2,3] and Nb-Ni.[3] ($Pd_{.80}Si_{.20}$ had been reported to be paramagnetic with $\chi \approx 1$ to $10 \cdot 10^{-6}$ emu/g-mole;[1] however, more recently, samples of very high purity were studied and found to be diamagnetic.[4]) All of these observations of Pauli paramagnetism were made on amorphous transition metal alloys; this reflects the fact that to date all reported amorphous metals prepared by rapid quenching from the melt[5,6] and available for magnetic studies contain at least one transition element.

The substantial interest in the properties of an amorphous metal containing only normal (s-p electron) metal constituents has prompted a search for such a material. This interest is partly due to the possibility of treating a normal metal theoretically, e.g. by pseudo-potential methods, and comparing its observed and calculated electronic and energetic properties. Another incentive for preparing such an alloy is to test the Nagel - Tauc theory on amorphous phase formation which predicts a minimum in the density of states $D(E_F)$ for compositions that readily form amorphous metals;[7,8] while this theory had been formulated to explain the composition of amorphous noble metal alloys, it should also apply to normal metals for which the pseudo-potential approach is more appropriate.

Recently a convenient amorphous alloy based on normal metals has been found; it was observed that Mg-Zn alloys with compositions near the eutectic at ~30 at. pct. Zn can readily be retained as glasses by rapid quenching, e.g. melt spinning, to room temperature.[9] The Mg-Zn phase diagram is given in Fig. 1; it can be seen that the eutectic located close to the phase Mg_7Zn_3 is quite low melting, with a melting point substantially below the value given by ideal solution theory; such a deviation favors the formation of amorphous metals.[10] Amorphous Mg-Zn alloys with between 25-32 at. pct. Zn have the typical broad-peak diffraction patterns, high-resolution transmission electron micrographs and fractographs typical of amorphous metals; rapid crystallization begins at 380 K for a $Mg_{.70}Zn_{.30}$ alloy.[9] Amorphous ternary alloys with additions of up to 2 - 8 at. pct. Ni, Cu, Cd, Si, and other elements (with the maximum amount depending on the solute) can also be obtained by rapid quenching.

Preliminary values for the magnetic susceptibilities of binary Mg-Zn glasses are reported in the following. The susceptibility of crystalline Mg_7Zn_3 was also determined.

EXPERIMENTAL

Samples were prepared as follows: master alloys were melted from high-purity elements (Johnson-Matthey Mg 99.999%, Zn 99.9999%; ferromagnetic impurities: 2 and .2ppm, respectively) in evacuated Pyrex capsules, using Ta foil buckets. Appropriate quantities (typically ∼500 mg) were then melt spun using a modified Pond-Maddin unit.[11] In this unit, the liquid is squirted in a continuous stream through a small orifice against the inside of a rotating copper drum, with simultaneous translation of the orifice to produce a ribbon about 50 μm thick, 1-2 mm wide, and 4-10 m long; operation takes place in inert gas or vacuum.[9,12]

The amorphous structure of the resulting ribbons was tested by X-ray diffraction; absence of perceptible crystalline peaks indicates absence of more than a few percent of crystalline material. Generally, the resulting ribbons are tough (i.e., do not show brittle fracture on bending), and can be bent into a chosen shape. 20-80 mg quantities of ribbons were then wound into coils, tied with Nylon thread and weighed with Faraday balance systems in fields of 2 to 5 KOe and 5 to 10 KOe, respectively.

Susceptibilities were evaluated by Owen-Honda least square fits for all compositions reported here. Temperatures were varied in several runs between 78 and 298 K and in one run between 4 and 298 K.

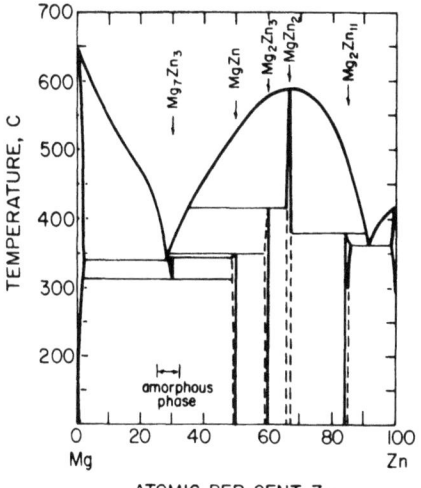

Figure 1. Equilibrium diagram for the Mg-Zn system with indicated composition range of amorphous phase formation by rapid quenching.

RESULTS

All amorphous Mg-Zn samples showed a positive, temperature-independent susceptibility χ of small magnitude, with a superimposed temperature-dependent susceptibility component. The room-temperature results for χ at 298 K are plotted against composition in Fig. 2; it is seen that there is a fairly wide scatter of $\pm 1 \cdot 10^{-6}$ emu/g-atom for samples of nominally identical compositions. This scatter may be due to the insufficient range of 1/H used for the Owen-Honda extrapolation in most measurements and failure to subtract the small paramagnetic component remaining at 298 K (see below); in the former case, it can be reduced by re-measurements in high fields, in the latter, by re-measurements at low temperatures.

Evaluation of the Owen-Honda plots revealed the presence of 1-6 ppm of ferromagnetic impurities in melt-spun samples, assuming $S = 3/2$. This impurity content is of the same order as that of the nominal materials analysis; it increases from the value for the as-received metals through the melted master alloys, and further to the melt-spun alloys.

While the variation of χ between 78 and 298 K were small, χ values for a $Mg_{.68}Zn_{.32}$ sample measured between 4.2 and 298 K could be fitted to a Curie-Weiss expression, $\chi - \chi_c = C/(T-\theta)$, with a Curie constant C corresponding to 130 ppm of a paramagnetic impurity with $S = 3/2$. While this is certainly due to contamination, further measurements are required to determine whether this was an individual contaminated sample or whether a batch effect is involved.

DISCUSSION

The principal results of the present work are the values of χ_c for amorphous Mg-Zn alloys. All values are shown in Fig. 2; two typical average values are listed in the Table together with those for the crystalline phases Mg, Mg_7Zn_3, $MgZn_2$, Zn and liquid Mg and Zn. They are seen to lie substantially below the average values of the elements (dashed line) and the crystalline intermediate phases. The following analysis of χ_c closely

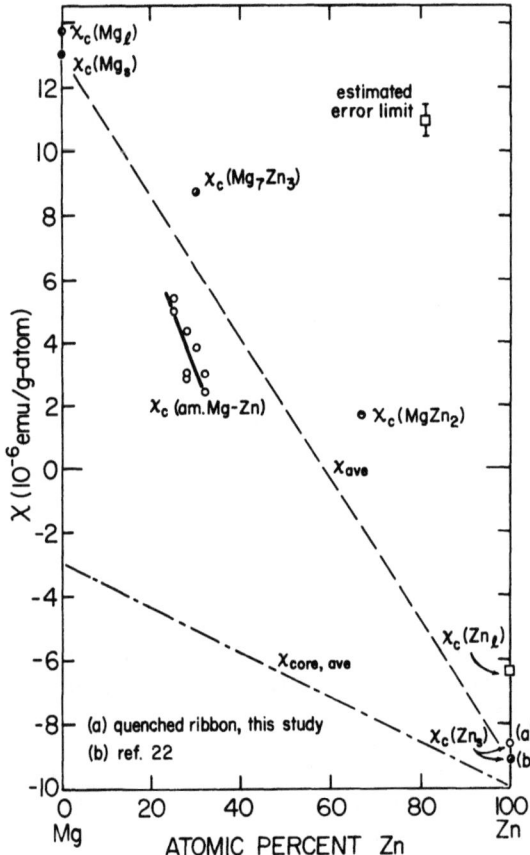

Figure 2. Magnetic susceptibilities (10^{-6} emu/g-atom) of amorphous Mg-Zn alloys, solid and liquid Mg and Zn, and other Mg-Zn phases.

follows that for liquid metals.[13,14] The customary assumption was made that

$$\chi_c = \chi_{ion} + \chi_{val} \tag{1a}$$

$$\chi_{val} = \chi_{Pauli} + \chi_{Landau} + \chi_{e-e} = \chi_{PL} + \chi_{e-e} \tag{1b}$$

where χ_{ion} is the diamagnetic contribution of the ion core electrons,
χ_{val} is the contribution of the valence electrons,
χ_{Pauli} is the Pauli conduction electron contribution,
χ_{Landau} is the contribution of the Landau conduction electron diamagnetism, and
χ_{e-e} is the contribution arising from electron-electron interaction (correlation).

χ_{ion} was obtained as a weighted average using Selwood's values[15] of -3 and $-10 \cdot 10^{-6}$ emu/g-atom (see Fig. 2); the use of these values has recently been reviewed concerning their application to liquid metals. For Mg, this value agrees closely with that from a theoretical calculation[16] ($-2.5 \cdot 10^{-6}$ emu/g-atom), while for Zn, the theoretical value[16] is considerably larger ($-15 \cdot 10^{-6}$ emu/g-atom). The valence electron susceptibilities χ_{Pauli} and χ_{Landau} (excluding electron-electron interactions but including electron-ion interaction via the effective mass approximation[13]) are combined in the expression[13,17]

$$\chi_{PL} = 64 \cdot 10^{-6} \frac{m^*}{m}(1 - \frac{1}{3} \cdot \frac{m^2}{m^{*2}})\underline{n}(E_F) \tag{2}$$

where $\underline{n}(E_F)$ is the density of states per eV atom (for one spin direction) and m, m^* are the electron and effective electron mass, respectively. In view of the absence of a band-structure and the spherical shape of the Fermi surface in the amorphous metal, $m^* \approx m$ as a first approximation, leading to the familiar factor 2/3 in (2).[18] The correction for electron-electron interaction has been discussed for the liquid metals;[13] it affects both χ_{Pauli} and χ_{Landau}. Its effect is to add between $3.8 \cdot 10^{-6}$ (for liquid Mg) and $2.8 \cdot 10^{-6}$ emu/g-atom (for liquid Zn) to the total χ_c. Again, a weighted average may be used for the present alloys.

Numerical values are given in the Table and Fig. 3. The derived valence electron contribution to the susceptibility χ_{val} and the calculated values for χ_{PL} are plotted in Fig. 3; it is seen that $\chi_{val} = \chi_{PL}$ near $Mg_{.71}Zn_{.29}$ although the χ values diverge quickly at other compositions. Within the limits of the assumptions made in this approximation (primarily, free electron Pauli and Landau terms and disregard of χ_{e-e}) amorphous Mg-Zn is well represented by

TABLE

Derived and Calculated Susceptibilities $\chi(10^{-6}$ emu/g-atom) at Temperature T and Densities-of-State $\underline{n}(E_F)$ (electrons of one spin/eV atom) for Mg, Zn and Mg-Zn Alloys (see Text)

(T = 298 K for solids and T = melting point for liquids)

Metal or Alloy	State and Structure	Measured Values χ_c(obs)	Derived Values			Calculated Values (free electrons)	
			χ_{ion}[a]	χ_{val}	$n(E_F)$	χ_{PL}	$\underline{n}(E_F)_{PL}$
Mg	Cryst., Mg type	13[b]	-3	16	(.37)	9.0	.21
Mg	Liquid	13.7[c]	-3	16.7	(.38)	~9.0	~.21
Mg$_{.75}$Zn$_{.25}$	Amorphous	5.0[d]	-4.8	9.8	(.23)	8.5	.20
Mg$_{.70}$Zn$_{.30}$	Amorphous	3.0[d]	-5.1	8.1	(.19)	8.3	.19$_5$
Mg$_7$Zn$_3$	Cryst.	8.5[d]	-5.1	13.6	(.32)	8.1	.19
MgZn$_2$	Cryst., MgZn$_2$ type	1.7[e]	-7.7	9.4	(.23)	7.9	.18$_5$
Zn	Cryst., Mg type	-9.15[f]	-10	.85	(.02)	6.8	.16
Zn	Liquid	-6.4[g]	-10	3.6	(.08)	6.9	.16

a) Ref. 15;
b) Ref. 19;
c) Refs. 13 and 20;
d) This study (average values for amorphous metals);
e) Ref. 21;
f) Ref. 22;
g) Average value of data in Ref. 14.

a free-electron description. The corresponding density-of-state functions are also compared in the Table.

χ_{val}(Mg) lies substantially above χ_{PL}(Mg) + χ_{e-e}(Mg), possibly due to an underestimate of χ_{e-e} [13] and liquid structure effects; χ_{val}(Zn) lies very much below χ_{PL}(Zn) + χ_{e-e}(Zn), possibly due to an underestimate of χ_{ion} and other effects. It is likely that the agreement between χ_{val} and χ_{PL} for the amorphous metals is in part fortuitous, as χ_{e-e} is neglected. It may also be due in part to the cancellation of opposing effects originating from the constituent elements. The difference between the χ_{val} for crystalline Mg_7Zn_3 and the amorphous metal with the same composition confirms that structure also has an effect on χ_{val}; it is of interest that χ_{val} for the amorphous solid is lower than χ_{val} for both the weighted average of the liquid elements as well as for the crystalline phases.

If an amorphous metal such as $Mg_{70}Zn_{30}$ is indeed free-electron-like, its structural stability is not due to a band structure effect but would derive solely from the volume-dependent part of the energy. The observed data are not sufficiently precise at this time to allow a good assessment of the change of χ with composition; however, it does not appear at present that there is a minimum of χ at the composition $Mg_{70}Zn_{30}$ and hence the prediction of Nagel-Tauc[7,8] is not fulfilled for this system.

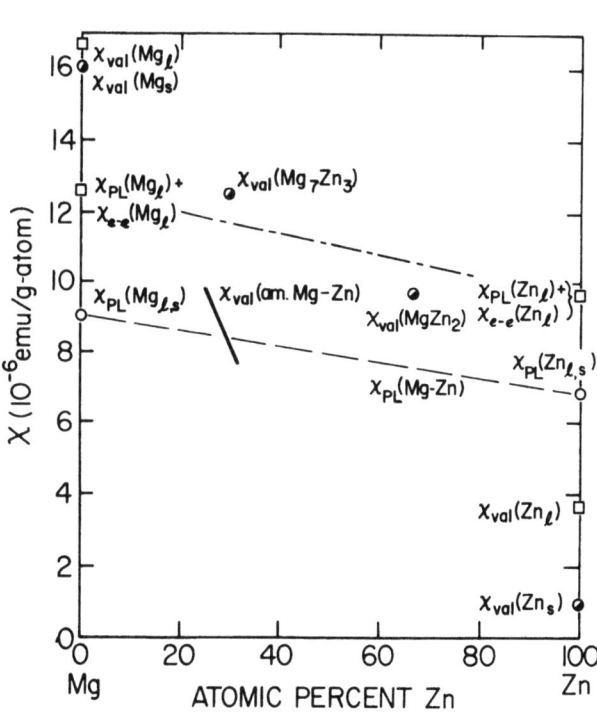

Figure 3. Valence electron contributions χ_{val} to the magnetic susceptibilities χ of solid and liquid Mg-Zn phases, after correction for diamagnetic ion contributions,[15] calculated Pauli-Landau valence electron susceptibilities χ_{PL} and electron-electron interaction contributions χ_{e-e}, (see text).

As the pair distribution function for amorphous $Mg_{70}Zn_{30}$ can be derived from the known diffraction pattern, $n(E_F)$ can be calculated directly by pseudo-potential methods for comparison with the present values, to improve on the approximation $m^* = m$ used here. Further susceptibility studies at high fields, other measurements of the conduction electron susceptibility, e.g. by EPR, or direct determinations of $\underline{n}(E)$, e.g. by photo-emission spectroscopy, would be desirable to confirm the present data. The effect of substitutions producing different valence electron concentrations should also be of interest and is under study.

ACKNOWLEDGEMENT

We acknowledge discussions with Profs. C. Foiles, D. Gelatt and H.J. Güntherodt and the experimental assistance of Messrs. A. Bogdan and G.R. Gruzalski. This work was supported at Northeastern University by the AMMRC (Contract DAAG 46-76-C-0021) and at the University of Nebraska by the NSF (Grant DMR 72-03208-A01).

REFERENCES

1. R. Hasegawa, J. Appl. Phys. $\underline{41}$, 4096 (1970).
2. G.R. Gruzalski, J.W. Weymouth, D.J. Sellmyer, and B.C. Giessen, Phys. Rev. B. Sept. (1976), in press.
3. G.R. Gruzalski, J.W. Weymouth, D.J. Sellmyer, and B.C. Giessen, in Proc. Second Internat. Conf. on Amorphous Magnetism (1976), R.A. Levy and R. Hasegawa, Eds., Plenum Press, New York.
4. C.C. Tsuei, in Proc. Second Internat. Conf. on Amrophous Magnetism (1976), R.A. Levy and R. Hasegawa, Eds., Plenum Press, New York.
5. H. Jones, Rep. Prog. Phys., $\underline{36}$, 1425 (1973).
6. A.K. Sinha, B.C. Giessen and D.E. Polk, in Treatise on Solid State Chemistry, N.B. Hannay, Ed., Vol. III, Plenum Press, New York, (1976), p. 1.
7. S.R. Nagel and J. Tauc, Proc. Second Internat. Conf. on Rapidly Quenched Metals, Section I, M.I.T. Press, Cambridge, Mass. 1976, p. 337.
8. S.R. Nagel and J. Tauc, Phys. Rev. Lett. $\underline{35}$, 380 (1975).
9. A. Calka, M. Madhava, D.E. Polk, B.C. Giessen, H. Matyja and J. Vander Sande, Scripta Met., submitted.
10. M. Marcus and D. Turnbull, Mat. Sci. Eng., $\underline{23}$, 211 (1976).
11. R. Pond and R. Maddin, Trans. Met. Soc. AIME, $\underline{245}$, 2457 (1969).
12. D.E. Polk and B.C. Giessen, unpublished.
13. R. Dupree and E.F.W. Seymour in Liquid Metals, (S.Z. Beer, Ed.) M. Dekker, New York (1972), p. 467.
14. G. Busch and J.H. Güntherodt, Solid State Physics, $\underline{29}$, (1974), 235.
15. P.W. Selwood, Magnetochemistry, Wiley - Interscience, New York (1956).
16. W.R. Angus, Proc. Roy. Soc., Ser. $\underline{A\ 136}$, 569 (1932).

17. A.H. Wilson, *The Theory of Metals*, Cambridge University Press, Cambridge, U.K. (1954), p. 155.
18. F.S. Ham, Phys. Rev. $\underline{128}$, 2524 (1962).
19. K. Honda, Sci. Rep. Tohoku Univ. $\underline{1}$, 1 (1912).
20. E. Wachtel, S. Woerner, and S. Steeb, Z. Metallk. $\underline{56}$, 776 (1965).
21. H. Klee and H. Witte, Z. Physik. Chem. $\underline{202}$, 352 (1954).
22. R.S. Tebble and D.J. Craik, *Magnetic Materials*, Wiley - Interscience, New York (1969).

DISCUSSION

N. Y. Rivier: Is there an obvious reason why it is much more difficult to prepare amorphous alloys out of non-transition metals?

B. C. Giessen: The fact that is has been so far quite difficult to obtain amorphous alloys out of non-transition metals by rapid quenching from the liquid has given rise to assumptions that there would be fundamental reasons. In the paper by Nagel and Tauc one finds the argument that there has to be a transition metal in there or an element which has d-electrons that would significantly contribute to binding such as for instance a nobel metal. However, at the present time, I would say there is no reason. If the phase diagram is right, as will be elaborated in Dr. Turnbull's talk, one can prepare amorphous non-transition metals just as well.

^{31}P NUCLEAR MAGNETIC RESONANCE STUDY IN THE METALLIC GLASS SYSTEMS $(Ni_yPt_{1-y})_{75}P_{25}$ AND $(Ni_{0.50}Pd_{0.50})_{100-x}P_x$*

William A. Hines and Lawrence T. Kabacoff

Univ. of Conn., Storrs, Connecticut 06268

Ryusuke Hasegawa

Allied Chemical Corp., Morristown, New Jersey 07960

Pol Duwez

Calif. Inst. of Tech., Pasadena, California 91109

I. INTRODUCTION

Recently, considerable attention has been focused on a class of materials known as metallic glasses.[1] This is due both to a desire for a re-examination of some heretofore fundamental concepts of solids as well as a possibility for a variety of technological applications. Metallic glasses have the general form $TM_{100-x}M_x$, where TM is a transition metal (or combination of transition metals) such as Fe, Ni, Pd or Pt and M is a high valence metalloid such as B, C, Si or P. For the most part, such alloys are prepared by rapid quenching from the liquid state and possess compositions typically ranging from x = 15 to 28 for the metalloid.

This work describes the application of nuclear magnetic resonance (NMR) in the study of the Ni-Pt-P and Ni-Pd-P metallic glass systems. Spectra obtained from both the transition metal and metalloid elements shed light on the applicability of the "dense random packing" (DRP) model, electronic structure, structural stability and nature of the short range order in these systems. This in turn leads to a better understanding of the electronic, magnetic and mechanical properties for metallic glasses.

In relation to the electronic structure, there is considerable interest in the nature of the density of states versus energy curve, N(E), for amorphous metals and how it is influenced by the lack of long range order. One point of view argues that due to the absence of a periodic structure, there are no sharp Brillouin zone boundaries imposed on the electronic structure. Consequently, the nearly free electron approximation might be expected to provide a reasonable description.[2] On the other hand, it has been suggested that a crystalline-like structure is retained on a small scale within the amorphous state.[3] Measurements of the NMR Knight shift are presented for the Ni-Pt-P and Ni-Pd-P systems as the compositions are varied which enable a preliminary description of N(E) for metallic glasses.

Several models have been proposed for the amorphous structure of metallic glasses. Among these models, the one proposed by Polk[4] seems promising particularly for systems in which the constituent metal and metalloid have a similar atomic size. A DRP structure is postulated in Polk's model for the transition metal atoms which is similar to that proposed for liquid metals by Bernal.[5] The metalloid atoms occupy the larger (interstitial) holes in the structure and, hence, are always surrounded by transition metal atoms as first nearest neighbors. One important feature of the DRP model proposed by Polk involves an "electron transfer" from the metalloid atoms to the transition metal atoms which results in a filling of the transition metal d-states. The occurrence of an electron transfer between the component atoms is said to stabilize the structure. This charge transfer mechanism offers a possible explanation for the magnetic behavior of certain metallic glasses. For example, $Pd_{80}Si_{20}$ possesses a much weaker paramagnetic susceptibility than crystalline Pd and this can be attributed to a filling of the Pd d-states.[6] We shall present NMR linewidth and Knight shift data which provide support for certain aspects of the DRP model and a transfer of charge from the metalloid to the transition metal atoms.

II. EXPERIMENTAL PROCEDURE

A first time observation of the ^{31}P and ^{195}Pt NMR in the Ni-Pt-P metallic glass system has recently been reported.[7] The present work describes additional, more detailed, measurements of the ^{31}P NMR for glasses of the form $(Ni_yPt_{1-y})_{75}P_{25}$, where y = 0.20, 0.30, 0.40, 0.50, 0.60, 0.64 and 0.68, and $(Ni_{0.50}Pd_{0.50})_{100-x}P_x$, where x = 16, 17, 20, 23, 25 and 26.5. All of the alloys were prepared by a rapid quenching process (piston and anvil technique) described in detail elsewhere.[8] The foils which resulted from this process were about 50 μm thick and 2.5 cm in diameter, and were checked by x-ray diffraction to verify their glassy structure.

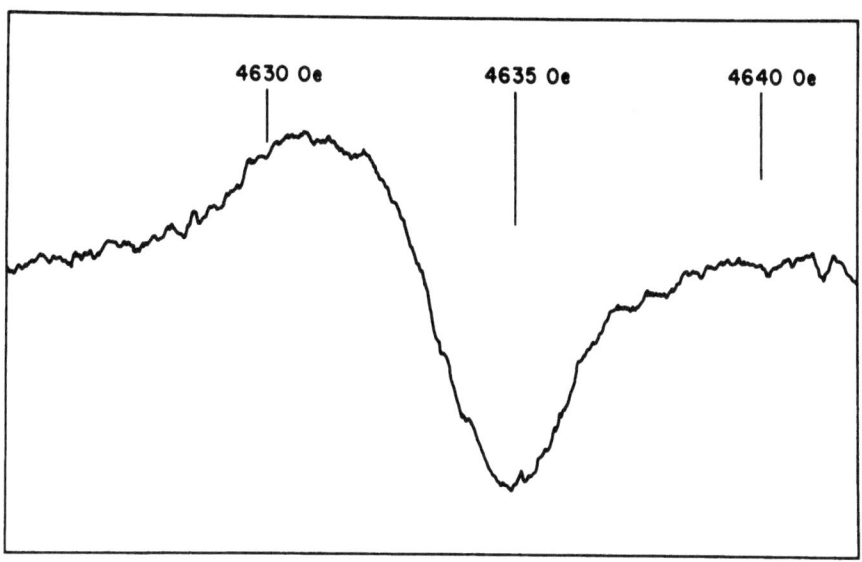

Fig. 1. Room temperature ^{31}P NMR absorption derivative (typical) at 8 MHz for the $(Ni_{0.20}Pt_{0.80})_{75}P_{25}$ alloy.

The NMR measurements were carried out at room temperature, for frequencies from 4 to 16 MHz, by utilizing a Varian wideline VF-16 cross coil spectrometer. Spectra were obtained from ^{31}P nuclei in all samples and ^{195}Pt nuclei in the $(Ni_{0.20}Pt_{0.80})_{75}P_{25}$ sample. Figure 1 shows a typical NMR spectrum (derivative of the absorption curve) for the ^{31}P nuclei in these samples. The lineshapes demonstrated some asymmetry in that the low field side was broader.

III. RESULTS AND ANALYSIS

A. NMR Knight Shift

Figure 2 shows the observed ^{31}P Knight shift (in %), K, (defined by the derivative crossover point) as a function of Ni concentration, y, for $(Ni_yPt_{1-y})_{75}P_{25}$. Data are shown for resonance frequencies, ν, of 8 and 16 MHz. It can be seen that the Knight shift remains unchanged over the entire range of Ni concentration (y = 0.20 to 0.68) within the error, however, it does possess a small but measurable frequency dependence. The detailed variation of the ^{31}P Knight shift with frequency for $(Ni_yPt_{1-y})_{75}P_{25}$ is shown in Figure 3. The solid vertical lines at the various fre-

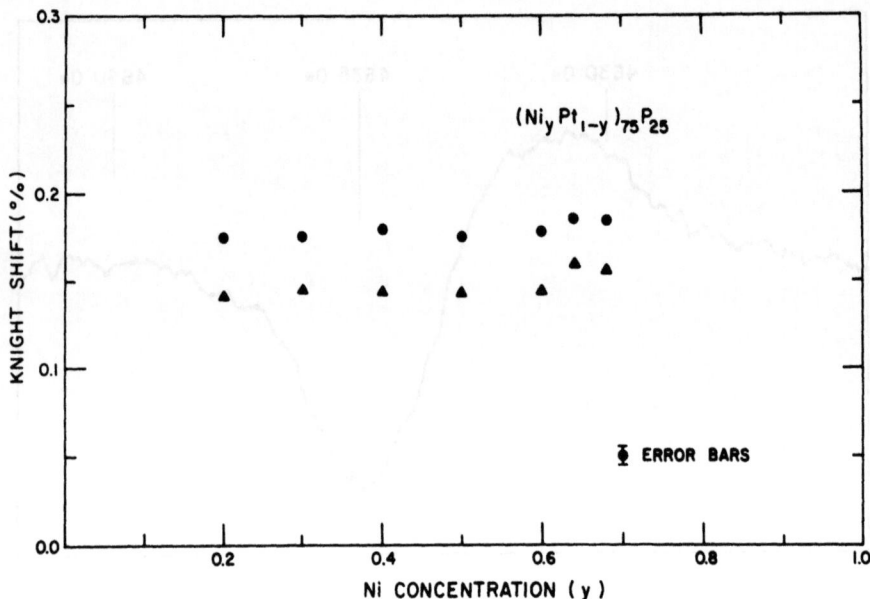

Fig. 2. ^{31}P Knight shift (in %) versus Ni concentration for the $(Ni_yPt_{1-y})_{75}P_{25}$ alloys: $\nu = 8$ MHz (▲) and $\nu = 16$ MHz (●).

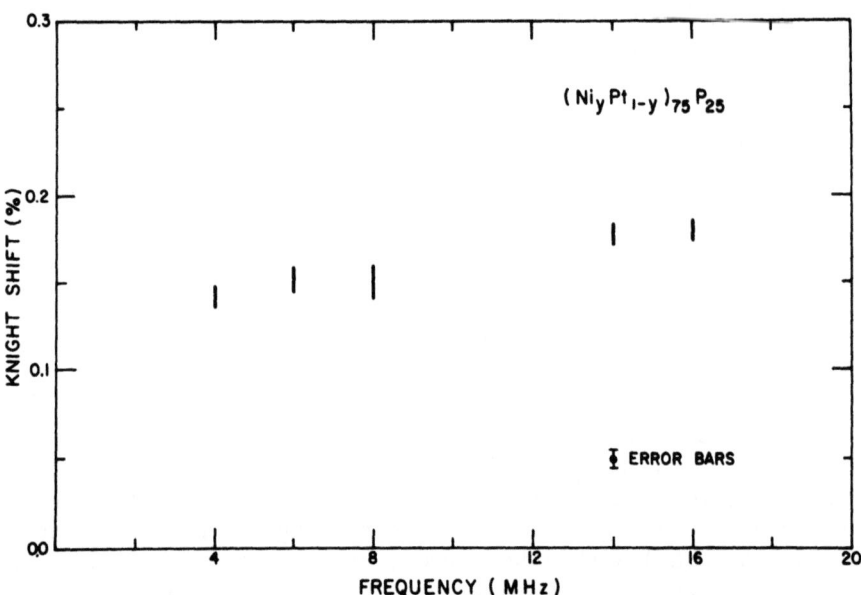

Fig. 3. ^{31}P Knight shift (in %) versus resonance frequency (in MHz) for the $(Ni_yPt_{1-y})_{75}P_{25}$ alloys; vertical lines represent various compositions.

quencies represent the range of shift values for the seven compositions studied; any one value would have the error indicated. The slight increase in Knight shift with frequency is clearly evident.

Figure 4 shows the observed ^{31}P Knight shift as a function of P concentration, x, for $(Ni_{0.50}Pd_{0.50})_{100-x}P_x$. Again, data are shown for ν = 8 and 16 MHz. In contrast to the $(Ni_yPt_{1-y})_{75}P_{25}$ results, $(Ni_{0.50}Pd_{0.50})_{100-x}P_x$ shows a strong decrease in Knight shift as the P concentration is increased from x = 16 to 26.5. This behavior might be expected as a change in P concentration would vary the average number of electrons per atom while a change in Ni relative to Pt would not. (It is unfortunate that the P composition is severely limited in Ni-Pt-P.) The variation of the ^{31}P Knight shift with frequency, for the respective $(Ni_{0.50}Pd_{0.50})_{100-x}P_x$ compositions, is shown in Figure 5 (closed symbols and dashed lines). It can be seen that all six compositions studied have essentially the same small increase in shift with frequency. For comparison, the $(Ni_yPt_{1-y})_{75}P_{25}$ results (vertical lines) from Figure 3 are included. We note that $(Ni_yPt_{1-y})_{75}P_{25}$ has the same amount of frequency dependence and falls between the x = 23 and x = 25 data for $(Ni_{0.50}Pd_{0.50})_{100-x}P_x$. Consequently, a certain degree of similarity in the electronic structure for the two systems is expected.

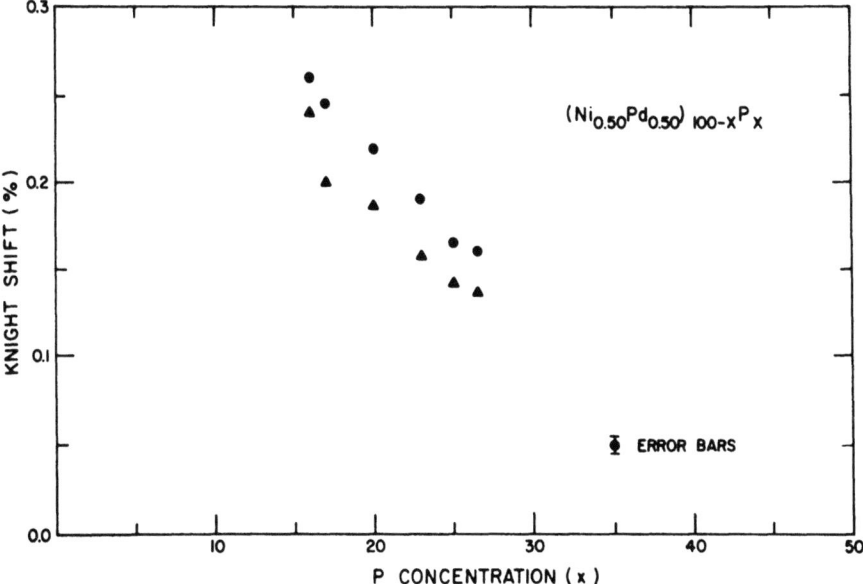

Fig. 4. ^{31}P Knight shift (in %) versus P concentration for the $(Ni_{0.50}Pd_{0.50})_{100-x}P_x$ alloys: ν = 8 MHz (▲) and ν = 16 MHz (●).

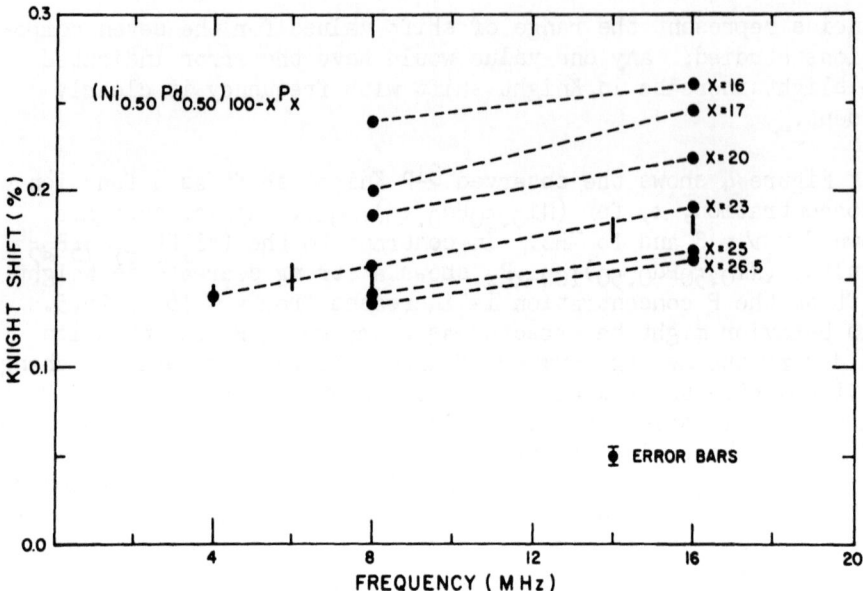

Fig. 5. ^{31}P Knight shift (in %) versus resonance frequency (in MHz) for the $(Ni_{0.50}Pd_{0.50})_{100-x}P_x$ alloys; vertical lines represent $(Ni_yPt_{1-y})_{75}P_{25}$.

B. NMR Linewidth

Figure 6 shows the variation of the NMR (peak-to-peak) linewidth, δ, with Ni concentration, y, for $(Ni_yPt_{1-y})_{75}P_{25}$. Data are given for the various resonance frequencies from $\nu = 4$ to 16 MHz. It can be seen that, for fixed frequency, the linewidth remains unchanged for $0.20 \leq y \leq 0.50$ and then decreases continuously for $0.50 < y \leq 0.68$. Also, Figure 6 shows that, for a given composition, the linewidth increases continuously with frequency from 4 to 16 MHz. The typical variation of the linewidth, δ, with frequency, ν, is illustrated in Figure 7. (Only the data for $y = 0.20$, 0.50 and 0.60 are shown). All $(Ni_yPt_{1-y})_{75}P_{25}$ samples show a frequency dependent linewidth which is indicative of a broadening mechanism resulting from a distribution of Knight shifts. As pointed out in reference 7, this is a consequence of the P atoms having a variety of environments in the glassy structure. All of the $(Ni_yPt_{1-y})_{75}P_{25}$ linewidth data (solid symbols, Figure 7) can be fitted extremely well by the form (solid curves, Figure 7)

$$\delta = \left[(\delta_1)^2 + (\delta_2)^2\right]^{1/2}, \qquad (1)$$

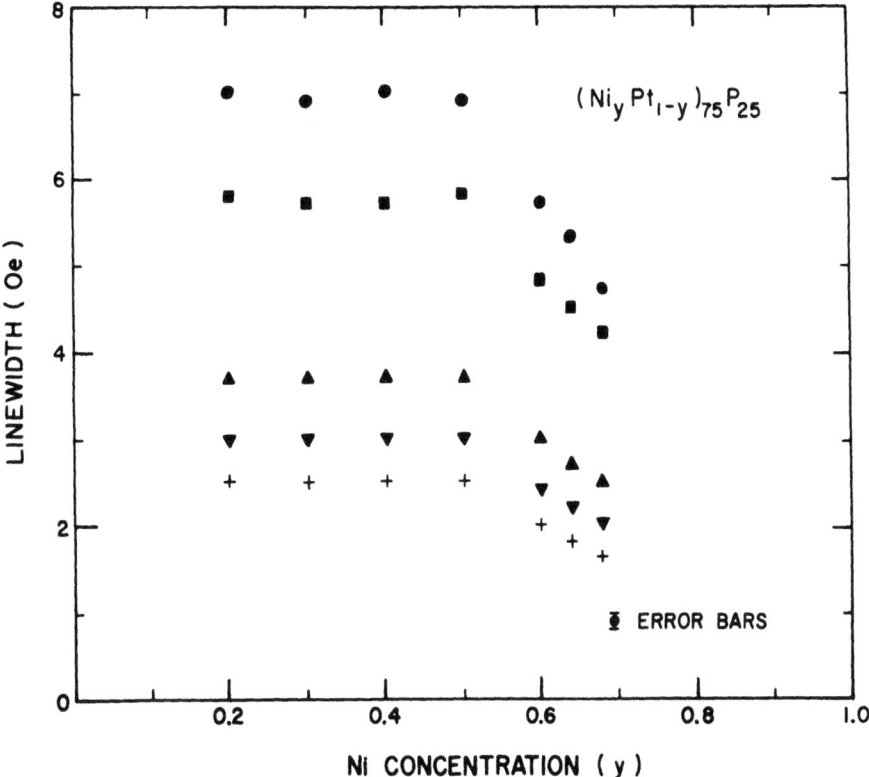

Fig. 6. ^{31}P peak-to-peak linewidth (in Oe) versus Ni concentration for the $(Ni_yPt_{1-y})_{75}P_{25}$ alloys: $\nu = 4$ MHz (+), $\nu = 6$ MHz (▼), $\nu = 8$ MHz (▲), $\nu = 14$ MHz (■) and $\nu = 16$ MHz (●).

where $\delta_1 = c\nu$ is the frequency dependent contribution and δ_2 represents the frequency independent contribution (e.g. dipolar broadening). Such a form is appropriate for combining uncorrelated broadening mechanisms. The values obtained for c and δ_2 associated with the respective compositions are listed in Table I. We note that the c value (as well as the δ_2 value) remains constant for $0.20 \leq y \leq 0.50$ and then decreases continuously for $0.50 < y \leq 0.68$. This behavior indicates that the distribution in Knight shifts, and correspondingly, the distribution in P sites is reduced for $y > 0.50$. The continuous nature of this reduction is characteristic of a "phasing out" of certain types of P sites above a particular concentration ($y = 0.50$).

The linewidth data for all six of the $(Ni_{0.50}Pd_{0.50})_{100-x}P_x$ compositions can also be described by Eq. (1) again indicating a

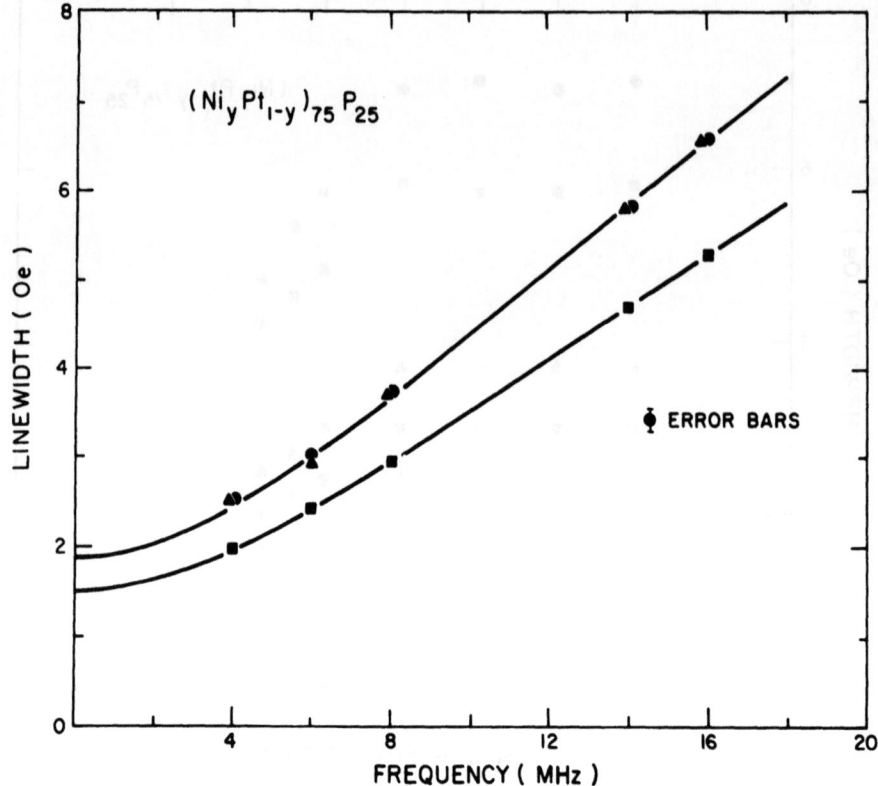

Fig. 7. ^{31}P peak-to-peak linewidth (in Oe) versus resonance frequency (in MHz) typical for the $(Ni_yPt_{1-y})_{75}P_{25}$ alloys: y = 0.20 (●), y = 0.50 (▲) and y = 0.60 (■). Curves represent Eq. (1).

distribution of shifts. The respective c and δ_2 values are listed in Table II.

IV. DISCUSSIONS AND CONCLUSIONS

A. Electronic Structure (Density of States)

A possible understanding of the origin and nature of the ^{31}P Knight shift in the $(Ni_yPt_{1-y})_{75}P_{25}$ and $(Ni_{0.50}Pd_{0.50})_{100-x}P_x$ metallic glass systems can be achieved by a consideration of the existing NMR work on crystalline transition metal alloys.[9] In particular, detailed measurements are available concerning the

TABLE I

y	c (Oe/MHz)	δ_2 (Oe)
0.20	0.41	1.8
0.30	0.41	1.8
0.40	0.41	1.8
0.50	0.41	1.8
0.60	0.34	1.4
0.64	0.32	0.93
0.68	0.29	0.76

Values of c and δ_2 obtained by fitting linewidth data for $(Ni_y Pt_{1-y})_{75} P_{25}$ alloys to Eq. (1).

TABLE II

x	c (Oe/MHz)	δ_2 (Oe)
16	0.77	1.7
17	0.67	1.7
20	0.43	1.6
23	0.36	1.4
25	0.34	1.4
26.5	0.32	1.4

Values of c and δ_2 obtained by fitting linewidth data for $(Ni_{0.50} Pd_{0.50})_{100-x} P_x$ alloys to Eq. (1).

hyperfine fields acting on noble metals dissolved in the transition metal hosts Ni, Pd and Pt.[10] These measurements lead us to believe that our ^{31}P Knight shift has two principal contributions. We write $K = K_s + K_d$, where K_s is the "direct contact shift" resulting from a polarization of the conduction s-electrons by the external magnetic field which is communicated to the ^{31}P nuclei via a contact hyperfine interaction and K_d is the "d polarization shift" resulting from a polarization of the transition metal d-electrons by the external magnetic field which is communicated to the ^{31}P nuclei via a s-d interaction that polarizes the conduction s-electrons (together with the contact hyperfine interaction). For Ni, Pd and Pt systems, the orbital (or Van Vleck) term is usually negligible. Using the customary two-band model, we can relate the two shift contributions to corresponding terms in the bulk magnetic susceptibility, χ, and density of states, $N(E)$, by

$$K = \alpha_s \chi_s + \alpha_d \chi_d = \alpha_s \mu_B^2 N_s(E_F) + \alpha_d \mu_B^2 N_d(E_F), \qquad (2)$$

where χ_s and χ_d are the paramagnetic spin susceptibilities, α_s and α_d are the coupling coefficients of the electron-nucleus interactions, and $N_s(E_F)$ and $N_d(E_F)$ are the density of states at the Fermi energy for the s- and d-bands respectively.

We have already noted that the $(Ni_yPt_{1-y})_{75}P_{25}$ system demonstrates a ^{31}P Knight shift which is independent of the relative Ni-to-Pt composition while the $(Ni_{0.50}Pd_{0.50})_{100-x}P_x$ system has a shift with a strong P concentration dependence. It was remarked that the former case does not constitute a variation of the average electron-to-atom ratio for the "band structure" while the later case does. In the context of a simple rigid band picture, changing the electron concentration would change the Fermi level and, consequently, the Knight shift will map out a portion of the density of states. For the two systems studied in this work, we have to be concerned with a possible variation of both terms (K_s and K_d) when the electron concentration changes. A detailed separation of terms can be achieved by plotting K versus χ using temperature as an implicit parameter.[11] (We are currently making the necessary susceptibility measurements.) However, a preliminary analysis of the Knight shift results leads us to believe that the K_d term dominates the K_s term by a factor of two or three and is primarily responsible for the observed decrease of the shift as the P concentration increases in $(Ni_{0.50}Pd_{0.50})_{100-x}P_x$. In other words, this decrease is a consequence of the transition metal d-states filling up because of a charge transfer from the P metalloid atoms.

Additional evidence for this idea comes from the room temperature observation of the ^{195}Pt resonance in $(Ni_{0.20}Pt_{0.80})_{75}P_{25}$. The resonance spectrum is quite broad with a linewidth of about

40 Oe and possessing a great deal of intensity in the tails. The most striking feature of the ^{195}Pt resonance in the metallic glass is a Knight shift value of approximately -0.2%. This is to be compared with a value of -3.5% for ^{195}Pt in pure crystalline Pt.[9] The large negative Knight shift for ^{195}Pt in fcc Pt metal has been attributed to a dominant core polarization contribution and the resulting positive increase in the Knight shift for ^{195}Pt in the metallic glass provides strong evidence that the transition metal d-states are being filled by a charge transfer from the metalloid atoms. Furthermore, the wide linewidth for the ^{195}Pt resonance suggests a broad distribution in the degree of d-state filling.

B. Electronic Structure (Bonding and Stability)

Three microscopic descriptions for the nature of the glass-forming tendency and stability in metallic glasses have been advanced. The first, based on the Bernal structure for the transition metal atoms with the smaller metalloid atoms filling the interstitial holes, suggests that the above-mentioned transfer of charge from the metalloid to the transition metal atoms results in a form of ionic-like bonding which stabilizes the structure.[4] This model has had some success, particularly in predicting the range of composition. However, objections to such an explanation have been raised because of detailed studies of the effects of alloying with elements of differing atomic radii.[12] The second model for these glasses suggests that it is chemical (covalent-like) bonding that stabilizes the structure.[12] From this point of view, we might expect the amorphous structure to have short range order similar to the crystalline state but no long range order. Recent x-ray photoemmission experiments on the metallic glass $Pd_{0.775}Cu_{0.060}Si_{0.165}$ are, however, inconsistent with such a bonding.[13] Finally, a model based on the nearly free electron (NFE) approach has received considerable attention recently.[2] This model suggests that the alloy is most stable when the composition is such that the Fermi level lies at a minimum in the density of states. This NFE model, which employs many of the concepts from Ziman's theory of liquid metals, gains support from measurements of the electrical resistivity and glass transition temperature, and also predicts the correct composition range. Of critical importance in this model is the relationship between $2k_F$ and k_p, where k_F is the Fermi wavevector and k_p is the value of the wavevector corresponding to the first peak in the structure factor, $a(k)$. Typically, $2k_F \sim k_p$ when the average electron to atom ratio is 1.7, and it is under this condition that the Fermi energy lies at the minimum of the density of states. Also, the larger $a(k_p)$, the deeper will be the minimum and, hence, the more stable will be the alloy.

Although we have not yet achieved a complete separation of the observed ^{31}P Knight shift into its components and, hence, have not obtained an exact measure of the density of states, we do note that our measurements show no evidence of any minimum in the density of states as required by the NFE electron model. For the $(Ni_yPt_{1-y})_{75}P_{25}$ system, the electron concentration does remain fixed over the entire composition range (y = 0.20 to 0.68). However, the structure factor, and hence $a(k_p)$, does change with composition and has been carefully determined.[14] According to the NFE model, some change in "band structure" and density of states would be expected. However, we do not observe any such change for $(Ni_yPt_{1-y})_{75}P_{25}$. Additionally, the electron concentration does vary over the range of P concentration (x = 16 to 26.5) for $(Ni_{0.50}Pd_{0.50})_{100-x}P_x$. The observed Knight shift appears to decrease monotonically with no indication of any minimum as the P content is increased.

In Section IV, part A above, we discussed the ^{31}P Knight shift results which lend support for the idea of a charge transfer from the metalloid to the transition metal atoms, an important feature of the DRP model proposed by Polk.[4] Below, in part C, we will describe the ^{31}P linewidth data which will also support the concept of a DRP structure.

C. Amorphous Structure

As indicated in Section III, part B, all of the $(Ni_yPt_{1-y})_{75}P_{25}$ and $(Ni_{0.50}Pd_{0.50})_{100-x}P_x$ alloys show a frequency dependent linewidth which is indicative of a broadening mechanism resulting from a distribution of Knight shifts. This is a consequence of the P atoms having a variety of environments in the glassy structure and, correspondingly, a distribution in the degree of transition metal d-state filling results. The distribution in Knight shifts enters through the K_d term.

For the $(Ni_yPt_{1-y})_{75}P_{25}$ alloys, we noted that the c value (Table I) remains constant for $0.20 \leq y \leq 0.50$ and then decreases continuously for $0.50 < y \leq 0.68$. The continuous nature of this reduction was attributed to a "phasing out" of certain types of P sites above the y = 0.50 concentration. This change in behavior for the linewidth between y = 0.50 and y = 0.60 correlates with other experimental measurements on the $(Ni_yPt_{1-y})_{75}P_{25}$ system. There is a significant increase in the glass temperature (T_g) and crystallization temperature (T_c) when the Ni content exceeds 0.50.[15] The increase of T_g may be considered as an increase in the structural stability, implying a decrease in configurational entropy. This is consistent with the smaller distribution in the Knight shift for y > 0.50. It means, further, that smaller atoms

(Ni) affect the local short range atomic arrangement to a greater degree that the larger ones (Pt). The manifestation of this aspect includes a larger rate of decrease (with increasing Ni content) of the first nearest neighbor distance for $y > 0.50$ compared with the case $y \leq 0.50$ and also a large change in the slope of the resistivity versus temperature curves in the vicinity of $y = 0.55$.[16]

These results provide support for certain aspects of the DRP structure and offer insight concerning computer structural simulations based on DRP for glassy systems. In such simulations, the size of the metalloid atoms must have a certain distribution as indicated by the present results.

REFERENCES

1. J. J. Gilman, Physics Today 28, No. 5, 46 (May, 1976).
2. S. R. Nagel and J. Tauc, Phys. Rev. Letters 35, 380 (1975).
3. B. G. Bagley, H. S. Chen and D. Turnbull, D. Mater. Res. Bull. 3, 159 (1968).
4. D. E. Polk, Scripta Metallurgica 4, 117 (1970).
5. J. D. Bernal, Nature 185, 68 (1960).
6. R. Hasegawa and C. C. Tsuei, Phys. Rev. B2, 1631 (1970).
7. R. Hasegawa, W. A. Hines, L. T. Kabacoff and Pol Duwez, submitted to Solid State Commun.
8. Pol Duwez, in Techniques of Metals Research, edited by R. F. Bunshah (Interscience, New York, 1968), Vol. I., part 1, Chap. 7, p. 347.
9. L. E. Drain, Metallurgical Reviews 12, 195 (1967).
10. A. Narath, J. Appl. Phys. 39, 553 (1968).
11. A. M. Clogston, A. C. Gossard, V. Jaccarino and Y. Yafet, Phys. Rev. Letters 9, 262 (1962).
12. H. S. Chen and B. K. Park, Acta Met. 21, 395 (1973).
13. S. R. Nagel, G. B. Fisher, J. Tauc and B. G. Bagley, Bull. Am. Phys. Soc. 20, 374 (1975).
14. A. K. Sinha and P. Duwez, J. Phys. Chem. Solids 32, 267 (1971).
15. H. S. Chen, Acta Met. 22, 1505 (1974).
16. A. K. Sinha, Phys. Rev. B2, 4541 (1970).

*Work at University of Connecticut supported in part by the University of Connecticut Research Foundation and at California Institute of Technology by the U. S. Energy Research and Development Administration under Contract No. AT(04-3)-822.

DISCUSSION

R. Tournier: Have you examined the Knight shift distribution for Pt also?

W. A. Hines: We have seen the Pt resonance in one sample of composition $(Ni_{0.20}Pt_{0.80})_{75}P_{25}$. The platinum resonance was very broad, about 40G wide and shifted in the positive sense. In the case of a pure platinum metal the Knight shift is -3.5%, the dominant contribution being attributed to the core polarization. In the metallic glass here it is shifted so it is about -0.20%, almost zero, so that the change in shift toward the positive sense is indicative of the fact that the core polarization contribution is definitely diminishing.

P. A. Cote: Are you aware of the heat capacity measurements made on NiP by Tyan and Toth which seem to support what you say in terms of your interpretation of how the density of states vary as a function of P composition?

W. A. Hines: I am not aware of that data but I thank you for bringing it up to my attention.

NMR AND MÖSSBAUER STUDIES OF THE AMORPHOUS SYSTEM $Fe_{79}P_{21-x}B_x$[+]

K. Raj[*]

Yale University, New Haven, Connecticut 06520

A. Amamou and J. Durand

Calif. Inst. of Tech., Pasadena, California 91125

J. I. Budnick

University of Connecticut, Storrs, Connecticut 06268

R. Hasegawa

Allied Chemical Corp., Morristown, New Jersey 07960

ABSTRACT

Combined NMR, spin-echo and Mössbauer experiments have been performed to obtain hyperfine field distributions of the transition metal and metalloid elements in splat-cooled amorphous $Fe_{79}P_{21-x}B_x$ alloys. These distributions are related to the local environments of the elements. The NMR signals are observed in the low frequency range 20 - 60 MHz and all the nuclei, i.e. Fe, P and B, may contribute to the spectral distribution. The resolution of the spectra into that due to Fe and (P + B) nuclei was made possible by using samples prepared with an Fe^{56} isotope. The Fe distribution thus obtained shows general agreement with the Mössbauer field distribution. From a careful analysis of the NMR data, the hyperfine field at the B nuclei in these amorphous alloys is found to range from 24 to 26 KG increasing with B content. An upper limit of 8 KG for the half-width is attributed to this distribution.

The Mössbauer spectra of the Fe^{57} nuclei resemble those for the crystalline $Fe_{75}P_{25-x}B_x$ alloys. A fit of our spectra shows a field distribution which suggests the presence of structure. Such a structure may correspond to various Fe sites, also seen in the crystalline alloys. The distributions, generally lie between about 160 and 330 KG, with a maximum at about 260 KG. These spectra do not show the presence of Fe nuclei with essentially zero hyperfine field as was obtained for amorphous Fe-Pd-P by Sharon et al. and for amorphous Fe-P by Logan et al. With increasing B content the center of gravity of the Fe distribution shifts to higher values. From a systematic study of the NMR lines and other considerations it is concluded that the P field distribution is broad and its hyperfine field is between 20 and 35 KG for the higher P concentration alloys.

INTRODUCTION

The determination and understanding of local atomic arrangements is of major interest in studies of amorphous materials. Microscopic probes such as x-ray, electron and neutron scattering have been used extensively to obtain this information. Mössbauer and NMR experiments, also probe the local environments around a given atom by measuring the hyperfine field distribution at the nucleus of the atom. For example, Sharon et al.[1] have used the Mössbauer technique to study local configurations around Fe sites in amorphous Fe-Pd-P. However, the applicability of the Mössbauer technique is limited to specific radioactive nuclei.[2] In the case of amorphous transition metal - metalloid systems - only alloys containing Fe^{57} have been extensively studied. On the other hand, NMR can be applied to a wide variety of nuclei including Fe^{57}. The only atoms to which NMR cannot be applied are the ones for which the nuclear spin is zero.

The ability to observe an NMR signal from a particular nucleus in a magnetically ordered state is a complicated function of several factors: one among them is isotopic abundance. Due to the low isotopic abundance of Fe^{57} (2.87%), it is not an especially favorable nuclei for NMR work, compared to, for example, Mn (100%). Thus in an amorphous alloy where Fe is present along with other elements with larger isotopic abundance, a combined approach of NMR and Mössbauer could be a useful one to provide accurate hyperfine field distributions for both the metal and metalloid atoms.

$Fe_{79}P_{21-x}B_x$ is an excellent system for the study of hyperfine field distributions. The Fe field distribution can be acquired by a Mössbauer study and can be compared with that obtained by NMR to check the fitting procedure. In addition, the large isotopic abundance of metalloid atoms, i.e. P (100%) and B^{11} (81.8%), makes them especially attractive for NMR measurements. It should be noted that metalloid hyperfine fields have not yet been reported in any amorphous alloy. This information is helpful for a wide variety of reasons - to understand conduction electron polarization and spin transfer effects, to shed light on the local environments of metalloid atoms and to compare these fields with their crystalline counterparts. The knowledge of these hyperfine field values is also necessary to extend our present theoretical calculations on crystalline systems into the area of amorphous materials.

THE Fe-P-B SYSTEM

The electronic and magnetic properties of splat-cooled amorphous Fe-P-B alloys have been reviewed by Durand in this conference proceedings as well as elsewhere.[3] We will summarize only some relevant features.

In $Fe_{79}P_{21-x}B_x$ alloys, B can be substituted for P to a large extent ($3 < x < 17$ at.%) in the amorphous phase. X-ray spectra on the $Fe_{79}B_{21}$ foils exhibited traces of microcrystalline phases. Both the value of $\bar{\mu}$ (the average magnetic moment per Fe atom) and T_c (the Curie temperature) increases as B replaces P. The value of $\bar{\mu}$ varies between about 1.96 and 2.01 μ_B and T_c between 670 to 770° K over the entire series. Furthermore, these variations in comparison with similar variations in the closest crystalline $Fe_{75}P_{25-x}B_x$ compounds suggest the presence of short range ordering. The NMR as well as Mössbauer experiments were performed on foils of these alloys. The central portions of the splat-cooled foils were removed due to their larger compositional inhomogenieties.

EXPERIMENTAL

<u>Spin-Echo</u>: The NMR measurements on amorphous $Fe_{79}P_{21-x}B_x$ alloys were made at 1.3 K and in zero external magnetic field. Nearly identical excitation conditions (pulses: 3 μs each, power: low) were employed to observe signals in these alloys.

The normalized spin-echo spectra corrected for a single power of frequency in some representative alloys are shown in Fig. 1. The important features of these spectra are: (i) a broad maximum in the spectral distribution, (ii) a shift of the maximum to higher frequencies with increasing B composition, (iii) an increase in asymmetry on the high frequency side with increasing P composition, and (iv) similar excitation conditions for the spin-echo, which suggests that enhancement factors are nearly the same for all the samples.

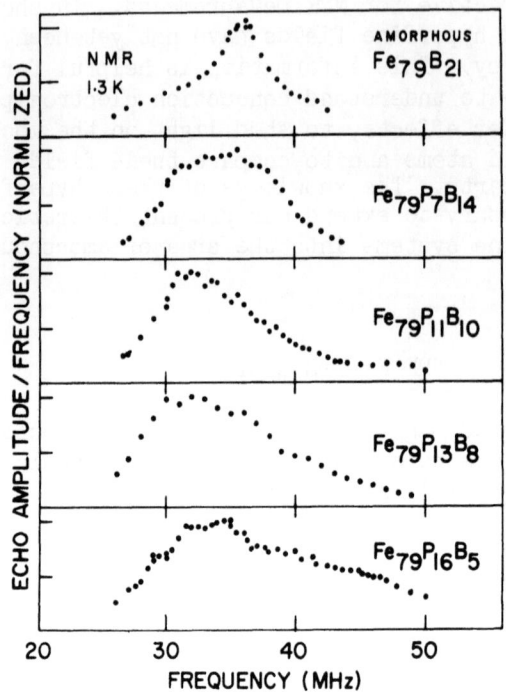

Fig. 1. Spin-echo spectra at 1.3 K of several amorphous $Fe_{79}P_{21-x}B_x$ alloys.

In principle the observed spectral intensity could arise from all of the elements since with the detection sensitivity available no signals were found in the low frequency range as well as at higher frequencies. In order to resolve this question two samples were prepared with an Fe^{56} isotope; namely $Fe^{56}_{79}P_6B_{15}$ and $Fe^{56}_{79}P_{16}B_5$. The nuclear spin for Fe^{56} is zero, thus the resonance due to Fe is not observed in samples containing only this isotope. The magnetic properties of the Fe^{56} samples, i.e. the magnetic moment and Curie temperature, on the other hand, are essentially the same as that for samples with the same composition containing natural iron.

The NMR signal strength depends on both the number and upon the properties of the nuclei present in the sample even if the enhancement factors for the nuclei are assumed to be the same. Thus in the sample $Fe_{79}P_5B_{16}$, for example, the contribution to the spectral distribution from each of the nuclei at a given frequency would be approximately: $Fe^{57}(173)$, $P^{31}(500)$, $B^{11}(1300)$ and $B^{10}(300)$. This means that the P^{31} and B^{11} signals are expected to be stronger than the Fe^{57} and B^{10}. The enhancement mechanism in vagnetically ordered materials is important and cannot be ignored. These mechanisms are not well understood and, among other factors, are proportional to the hyper-fine fields at the respective nuclei. The above numbers would be accordingly modified.

Figure 2 shows the actual intensity of both the natural Fe and the Fe^{56} samples. The echo amplitudes have been normalized with respect to the mass of the samples to obtain the relative magnitudes of the spectra. As expected, the Fe^{56} samples showed distributions with lower intensities than the natural Fe samples. The dotted lines represent the difference between the two spectra and correspond to the field distribution for the Fe. The relative height of the distribution should be treated with caution due to the assumptions involved but these difference spectra strongly suggest structure in the Fe hyperfine field distributions.

The variation of maximum echo amplitude, normalized to the

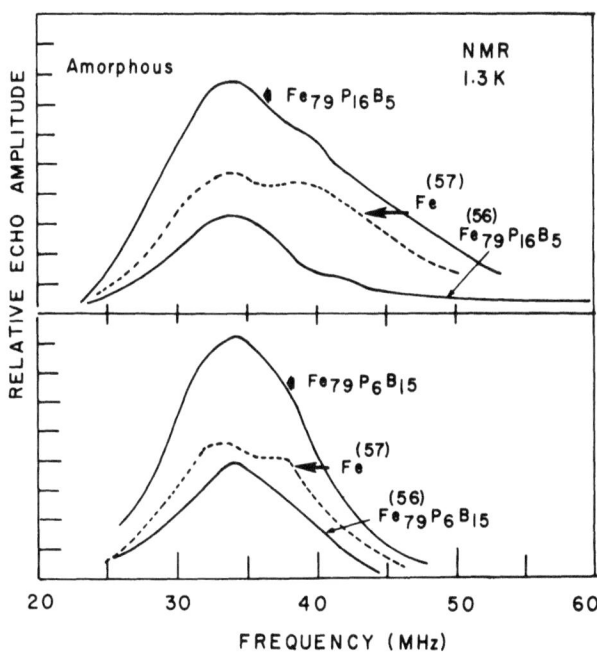

Fig. 2. Relative hyperfine field distributions of two amorphous $Fe_{79}P_{21-x}B_x$ (x = 5, 15) samples. Each sample is compared with the sample of same composition but containing Fe^{56} isotopes. The dotted spectra represent approximately the difference between the two distributions and are due to the Fe^{57} nuclei.

mass of the samples, with the B concentration is shown in Fig. 3. The amplitude increases with the B composition. This result indicates that the maximum in the NMR spectra arises predominately from the B nuclei since the Fe composition in the alloys remains constant and if the maximum had been produced by P the curve would show the opposite behavior. Since B^{11} is in greater abundance (82%) than B^{10} (18%), the maximum is associated with the B^{11} nuclei. The average B field in these amorphous alloys is about 25 KG with the maximum line-width of about 8 KG assuming a low intensity broad P distribution. The B hyperfine field increases slowly with x consistent with the trend observed in the crystalline alloys. The B field in Fe and in Fe_2B are < 4 [4] and 30 KG [5] respectively. This means that the B field increases with the B content in Fe as is found in the present amorphous alloys. The B enters with two less s-p electrons than P and thus reduces the electron concentration in the alloys, a situation similar to the crystalline alloys $Fe_{75}Si_{25-x}Al_x$ in which Al enters with one less electron than Si. It is interesting to note that the Al hyperfine field has also been shown to increase linearly with x.[6]

No clear sharp resonance corresponding to P is seen in these alloys. The appearance of a high frequency tail in the NMR spectra, however, suggests its presence.

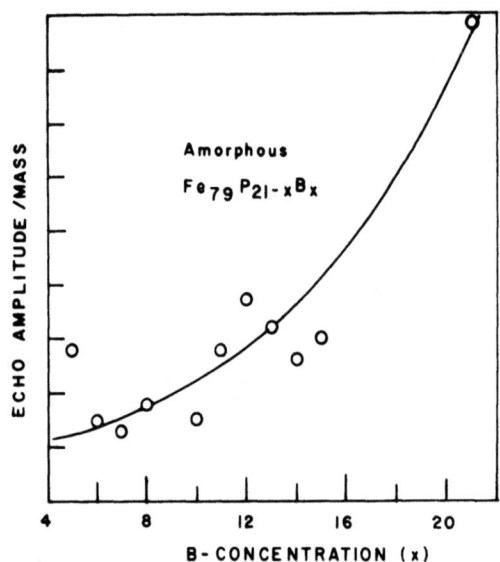

Fig. 3. Variation of peak echo amplitude per unit mass as a function of composition (x) in $Fe_{79}P_{21-x}B_x$ alloys. The behavior of the single point for the 5% Boron sample could reflect the growing dominance of phosphorous in the spectrum at lower x.

Mössbauer Absorption Measurements: The Mössbauer measurements were performed, at room temperature, with a standard spectrometer operating in a constant acceleration mode;[7] the source was Co^{57} in Cu and the specimens were about 40 microns in thickness. For all the $F_{79}P_{21-x}B_x$ amorphous alloys, the absorption spectrum as a function of the source velocity shows six overlapping but well defined peaks. The spectrum is roughly symmetrical with the symmetry axis close to that of the zero velocity. If we denote V_i as the velocity of the i th peak, the ratios V_6/V_1, V_5/V_2 and V_4/V_3 are approximately the same as in pure iron, corresponding to a hyperfine field of about 250 KG. Finally we note that no significant absorption is detected at the velocity of pure iron ($\pm$ 5.325 mm/s).

The experimental spectra as shown in Fig. 4, can likely be fitted by a continuous hyperfine field distribution assuming a given shape as a function of H. However we have chosen to deter-

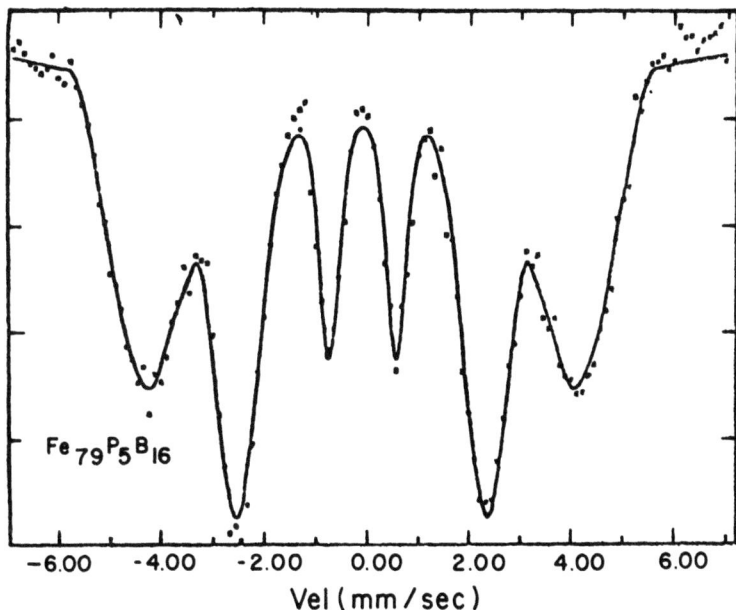

Fig. 4. Relative Mössbauer absorption versus the source velocity at T = 300 K for amorphous $Fe_{79}P_5B_{16}$. The full line represents the fit results.

mine the hyperfine field distribution P(H) by a method proposed by Hesse and Rubartsh;[8] this method does not presuppose a P(H) shape. In such a fit we assume that all the sextuplets have the same average shift, no quadrupolar splitting and the same linewidth as in pure iron (0.38 mm/s); the relative intensities of the lines being 3: R: 1 where R can be adjusted. A detailed study of the Mössbauer spectra and the fitting method will be presented in a further publication; in this paper we discuss the main features of our preliminary results. For all the investigated alloys a first fit was performed assuming the hyperfine field to lie between 0 and 400 KG. P(H) was determined for 40 values. The results showed that no significant probability occurred below 160 KG and above 330 KG. Therefore the investigated field range was reduced to 150 - 350 KG and the experimental spectrum could be fitted as well [Figs. 5(a) and 5(b)]. The obtained P(H) versus H curves are asym-

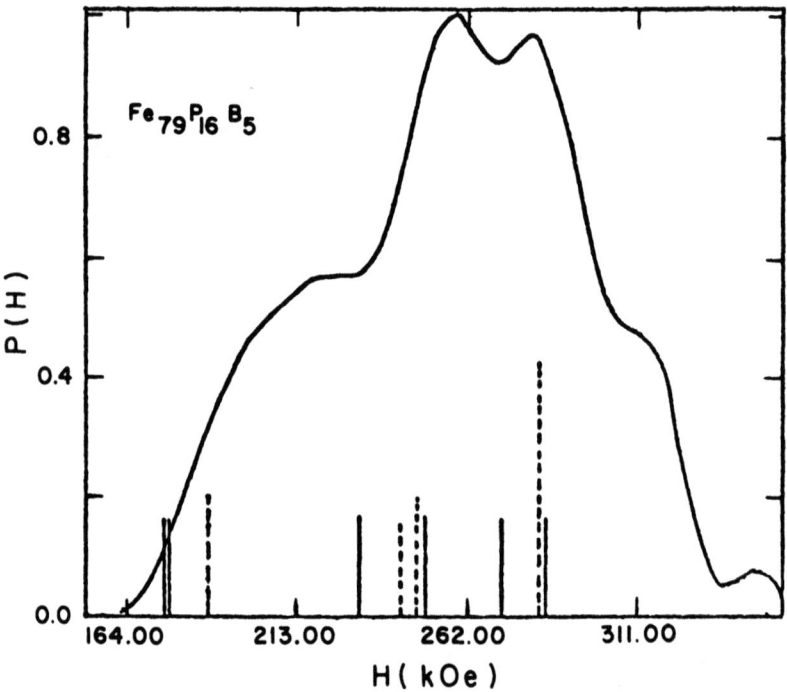

Fig. 5(a). Hyperfine field distribution P(H) vs. H, for amorphous $Fe_{79}P_{16}B_5$ at T = 300 K in the ϵ short range order, compared with values determined in crystalline Fe_3P (full lines) after ref.:E. J. Lisher et al: J. Phys. C7, 1344 (1974) and in crystalline $Fe_{3.03}P_{0.81}B_{0.19}$ (dashed lines) after ref. 10.

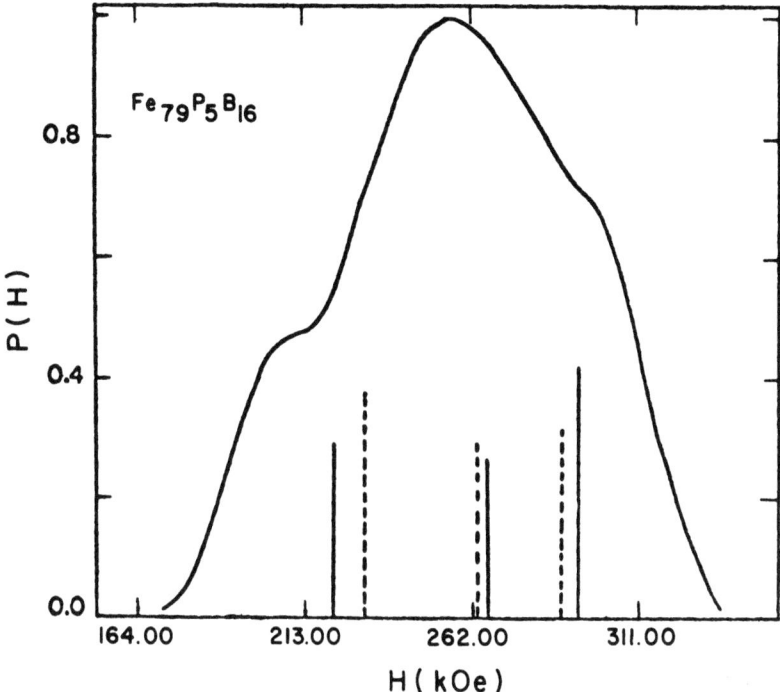

Fig. 5(b). Hyperfine field distribution P(H) vs. H, for amorphous $Fe_{79}P_5B_{16}$ at T=300 K in the ϵ_1 short range order, compared with values determined in crystalline $Fe_{3.01}P_{0.35}B_{0.65}$ (full lines) and $Fe_{2.85}P_{0.05}B_{0.95}$ (dashed lines) after ref. 10.

metrical for both samples with a rather broad tail below a maximum of about 250 KG [see Fig. 5 which refers to both Figs. 5(a) and 5(b)]; the contribution at low fields decreases slightly when the Boron content is increased. Moreover P(H) exhibits some bumps which suggest the existence of a structure in the hyperfine field distribution. It may be suggested that such bumps could arise from a microcrystalline structure in whole or in part of the alloy; however the samples studied can be considered as amorphous according to the current criteria: the x-ray pattern shows a broad first peak without any bump, the Curie transition is very sharp (1 to 3°K) and the behavior of the resistivity at high temperature exhibits an irreversible crystallization process similar to that observed in "good" amorphous alloys. Therefore the bumps in the hyperfine field distribution, more likely, have to be related to the amorphous state. Previous structural studies[9] on some amorphous transition metal alloys showed that the coordination number of each atom is the same as the crystalline counterpart. Thus, the P(H) shape may arise from the presence of iron atoms with various local environments which can be compared to those of the Fe-P-B crystalline alloys. Some previous studies on such alloys showed that hyperfine field values can be associated with each iron site; for $Fe_{75}P_{25-x}B_x$, the Mössbauer data is separated into four contributions for the ϵ phase and three contributions for ϵ_1 phase.[10] The hyperfine field values vary significantly with P and B content and the difference between the extreme values decreases when B is increased. The comparison between the results of the amorphous and the crystalline alloys have to be considered with caution since the compositions are not the same and the effect of iron concentration has not been determined. However, in our hyperfine field distribution the bumps occur at values which are close to those of the crystalline state (Fig. 5); these results suggest the presence of iron sites with first near neighbor configurations in the amorphous structure comparable to those found in the crystalline structure. So far, our results do not allow a clear separation between the two types of short range orders (SRO) ϵ and ϵ_1, as shown in magnetic studies.[3] Only some trends can be observed: (i) the line-width at half intensity is slightly larger for the ϵ-SRO region (51 ± 3 KG) than in ϵ_1-SRO region (45 ± 2 KG), (ii) the average hyperfine field H, defined as the center of gravity of P(H) ($H = \sum_i P(H_i) H_i / \sum_i P(H_i)$), is 248 ± 2 KG in the ϵ-SRO and 256 ± 2 in the ϵ_1-SRO. Finally let us note that in $Fe_{79}P_{21-x}B_x$ the ratio between the average magnetic field and the magnetic moment $\bar{H}/\mu$ is constant and equal to 126 KG/μ_B; this value is close to that found in the crystalline counterpart and in a large number of crystalline compounds of transition metals with s-p elements.[12,13]

It should be noted that the Fe Mössbauer field distributions (Fig. 5) are similar to the ones obtained by the NMR measurements (Fig. 2). The center of gravity of the two distributions is in agreement when corrected for the temperature difference. The analysis of our Mössbauer field distributions do not indicate the presence of nuclei with essentially zero hyperfine field as was found to be the case in amorphous Fe-P (Logan et al.[11]) and in Fe-Pd-P (Sharon et al.[1]).

SUMMARY AND CONCLUSIONS

The NMR measurements on the magnetically ordered splat cooled amorphous $Fe_{79}P_{21-x}B_x$ alloys show that the spectral distribution is a combined result of signals arising from the Fe, P, and B nuclei. A partial resolution of the NMR spectra into that due to Fe and (P+B) nuclei is made by using samples enriched in Fe^{56}. The Fe distributions thus obtained are not inconsistent with the Mössbauer results on the same samples indicating an overall consistency in the fitting of the Mössbauer data and the assumption of constancy of enhancement factor based on nearly identical excitation conditions employed in the excitation of the NMR spectra.

The center of gravity of the Fe hyperfine field distribution varies linearly with the B composition. This also implies a linear variation of the Fe field with the average Fe moment. The slope of the line is 126 KG/μ_B; a value reported for several of interstitial crystalline compounds. It is interesting to note that for amorphous Co-P alloys, the ratio of Co hyperfine field and its moment was also found to be constant.[14] The half width of Fe distribution remains practically constant around 80 KG. By comparing the variation of Fe hyperfine field distribution in the amorphous $Fe_{79}P_{21-x}B_x$ system with that of crystalline $Fe_{75}P_{25-x}B_x$, we believe that the short range atomic ordering, at least in the first neighbor shell in these amorphous alloys, is similar to that observed in their crystalline counterparts.

The broad maximum in the NMR spectra is associated with the B^{11} nuclei which are the more abundant of the two B-isotopes. This identification is supported by the variation of the echo amplitude with Boron composition (x) under constant excitation conditions. The B hyperfine field varies linearly with x just as does the Fe field. This observation implies scaling of the B hyperfine field with the average Fe moment and indicates that short range spin polarization effects are important. The average B hyperfine field in the amorphous alloys is 25 KG with a maximum line-width of ~8 KG, compared to an Fe hyperfine field of 260 KG and a line-width of ~80 KG. The small line-width observed for the B suggests a homogeneity in the nearest neighbor magnetic environments surrounding the metalloid atom.

The P resonance is not well resolved in the NMR spectra; however, its presence is clearly seen at higher frequencies when the P content is increased. We believe that the P resonance is spread out thus making it difficult to observe. The present investigation, along with a rough comparison of the NMR spectrum with a Mössbauer spectrum on a $Fe_{75}P_{15}C_{10}$ sample suggests that the P field lies in the range of 20 - 35 KG.[15] It is possible to make a rough prediction for the P line-width. The ratios of Fe and B hyperfine fields to their line-widths (in units of magnetic field) are approximately the same, i.e. ~3. If the same ratio is assumed for the P, then its line-width lies in the range 7 - 12 KG, or 12 - 22 MHz consistent with the observed broad distribution at higher frequencies in our spectra for high P composition.

If this reasoning is also applied to similarly coordinated amorphous Co-P alloys, in which one finds for the Co nuclei the ratio of field to line-width to be 1 - 1.5, the P line in Co-P could be as wide as its hyperfine field. This might be the reason that the P NMR has not been reported in amorphous Co-P alloys.[14]

In order to gain a more complete characterization of the P and B resonances, we have begun to prepare the binary amorphous $Fe^{56}_{79}P_{21}$ and $Fe^{56}_{79}B_{21}$ along with the samples $Fe^{56}_{79}P_{16}B_5^{10}$ and $Fe^{56}_{79}P_5B_{16}^{10}$. In addition, amorphous samples of Fe-P-B alloys enriched in Fe^{57} are being prepared in order to make further a detailed comparison of the Mössbauer fitting procedure with the NMR spectrum.

REFERENCES

+   Supported in part by The University of Connecticut Research Foundation.
*   Supported by the National Science Foundation.
1. T. E. Sharon and C. C. Tsuei, Phys. Rev. $\underline{B5}$, 1047 (1972).
2. For a general review of the applications to amorphous solids, see: J. M. D. Coey, Journal de Physique $\underline{C6}$, 89 (1974).
3. J. Durand, Joint MMM-Intermag. Conference, Pittsburg (1976). To be published in IEEE Transactions on Magnetics.
4. R. Avida, I. Benzyi, G. Goldring, S. S. Hanna, P. N. Tandon and Y. Wolfson, Nucl. Phys. $\underline{A182}$, 359 (1972).
5. H. Abe, H. Yasuoka, M. Matsuura, A. Hirai and T. Shinjo, J. Phys. Soc., Japan $\underline{19}$, 1491 (1964).
6. T. J. Burch, V. Niculescu, J. I. Budnick and K. Raj. To be published.
7. E. Kankeleit in: Mössbauer effect methodology, Vol. I, (Plenum Press, New York, 1965).
8. J. Hesse and A. Rubartsh, Journ. of Phys. E, Scientific Instruments $\underline{7}$, 526 (1974).
9. P. Duwez, "Annual Review of Material Science" $\underline{6}$, 83 (1976).
10. R. Wäppling, L. Haggstrom, S. Rundquist and E. Karlsson, J. of Solid State Chem. $\underline{3}$, 276 (1971).
11. J. Logan and E. Sun, J. Non-Crystall. Solids. To be published.
12. H. Bernas, I. A. Campbell and R. Fruchart, J. Phys. Chem. Solids $\underline{28}$, 17 (1967).
13. I. Vincze, M. C. Cadville, R. Jesser and L. Takacs, J. de Phys. (Paris) 35 C6 533 (1974).
14. K. Raj, J. I. Budnick, R. Alben, G. C. Chi and G. S. Cargill III, AIP Conf. Proc. $\underline{31}$, 390 (1976); A. Heidemann, Z. Physik $\underline{B20}$, 385 (1975); J. Durand and M. F. Lapierre, J. Phys. $\underline{F6}$, 1185 (1976).
15. To be published.

DISCUSSION

B. C. Giessen: Are you sure you are dealing exclusively with B^{10} and not a combination of two isotopes?

K. Raj: No, but we are considering doing experiments to determine the isotopic abundance of the resonant nuclei in our samples.

MAGNETISM IN AMORPHOUS Zr-Cu(Fe), Zr-Cu(Gd), AND Nb-Ni(Fe)*

G. R. Gruzalski, J. W. Weymouth, and D. J. Sellmyer

Behlen Laboratory of Physics
University of Nebraska
Lincoln, NE 68588

B. C. Giessen

Chemistry Department
Northeastern University
Boston, MA 02115

ABSTRACT

Magnetic properties of amorphous and crystalline $Zr_{40}Cu_{60-x}Fe_x$, amorphous $Zr_{40}Cu_{54}Gd_6$, and amorphous $Nb_{50}Ni_{50-x}Fe_x$ alloys are described, and the nature of the magnetic coupling in these alloys is discussed. None of these alloys exhibit spin-glass phenomena. There are localized magnetic moments on the Fe and Gd atoms and these lead to ferromagnetism in $Zr_{40}Cu_{48}Fe_{12}$ and $Zr_{40}Cu_{54}Gd_6$. The $Nb_{50}Ni_{50-x}Fe_x$ system shows some spin clustering effects but no magnetic order up to $x = 10$.

INTRODUCTION

Theoretical descriptions of magnetism in amorphous materials have been proceeding along two paths. In the first approach, it is assumed that the most important effect of the amorphous structure on the magnetism is that it introduces fluctuations in the Heisenberg coupling constant.[1] In the second approach, it is assumed that the crucial feature of an amorphous magnet is the existence of a local

---

*Research supported at the University of Nebraska by the National Science Foundation, and at Northeastern University by the Office of Naval Research.

crystal field whose direction is random in space.[2] An additional important aspect of the problem, at least in certain systems, is that the magnitude of the moment on a given type of atom may depend upon its local environment. In the following we refer to these three features as exchange fluctuations, random anisotropy and moment instabilities, respectively. In the case of metals, an additional problem is determining the nature of the dominant interaction between spins, i.e., is it a direct coupling or an indirect mechanism such as the Ruderman-Kittel-Kasuya-Yosida (RKKY) interaction. This problem is amplified in the case of amorphous metals, where one expects extremely short electron mean-free paths effectively eliminating interactions between all but the closest spins.

We have been studying the above questions in several magnetically dilute glassy transition-metal alloys. These include $Zr_{40}Cu_{60-x}Fe_x$ ($0 < x < 12$), $Zr_{40}Cu_{54}Gd_6$, and $Nb_{50}Ni_{50-x}Fe_x$ ($0 < x < 10$). In these alloys one expects predominantly local-moment magnetism associated with the Fe or Gd atoms, so that by varying the concentration from zero to several atomic percent, it is possible, in principle, to obtain information on the nature of the magnetic coupling.

RESULTS AND DISCUSSION

The alloys were produced by arc melting the pure materials several times. The amorphous samples were produced by rapidly propelling a plunger onto the melted alloy in a modified arc-melting furnace. All samples were checked by x-ray diffraction measurements to confirm that they had glassy structures.

$Zr_{40}Cu_{60-x}Fe_x$

Electronic states and localized magnetic moments and their interactions were studied in this system, for $0 < x < 12$, in both the amorphous and crystalline states. Since a comprehensive paper on the results for these alloys is being published elsewhere,[3] we summarize here only the major conclusions in order to compare the magnetic properties of this system with those of $Zr_{40}Cu_{54}Gd_6$ and $Nb_{50}Ni_{50-x}Fe_x$.

In the dilute ($0 < x < 6$) $Zr_{40}Cu_{60-x}Fe_x$ crystalline alloys, Curie-Weiss behavior is observed with localized moments in the neighborhood of 3 μ_B and with Kondo temperatures of about 25 K. At higher iron concentrations ferromagnetism is observed. Curie-Weiss behavior also is seen in the susceptibility of the dilute amorphous alloys; however, for these alloys the dependence on Fe concentration appears complex, the effective moments on the iron atoms are very small ($\mu_{eff} \simeq 0.7\ \mu_B$), and the paramagnetic Weiss temperature, though still negative, is only a few degrees in magnitude. The rather small values of μ_{eff} suggest that the moment on the Fe atoms may be unstable depending upon the local environment.[4] These moment instabilities may depend upon the number of Cu or Zr nearest neighbors, since it is known in crystalline alloys that Fe does not have a moment in Zr but does have one in Cu. Another possibility is that a given Fe atom may have a moment irrespective of its Cu and Zr nearest neighbors, only if it has a certain minimum number of Fe atoms as nearest or near neighbors. Such "cluster" models already have been applied to dilute crystalline alloys such as Au(Co) and Cu(Co). The resistivity data on amorphous $Zr_{40}Cu_{60}$ and the dilute Fe-doped alloys show a negative $d\rho/dT$ from 1.4 to 300 K. This is not to be associated with Kondo spin-flip scattering but it is consistent with several other mechanisms including localized-spin-fluctuation scattering,[5] s-d scattering in a nonmagnetic model due to Mott,[6] scattering from tunneling states in the amorphous alloy, as discussed by Cochrane et al.,[7] or quasi-liquid-metal-pseudo-potential scattering.[8] A recent theory due to Nagel and Tauc[9] on the nearly-free-electron approach to metallic glass alloys can be shown to be consistent with this last idea and also can account for other features exhibited by the amorphous Zr-Cu system.[3]

In the concentrated (x>6) amorphous alloys, resistance maxima and magnetic hysteresis are seen at low temperatures. These maxima do not appear to be connected with a spin-glass state since no maxima are seen in the susceptibility. Moreover, the temperature at which the resistivity maximum occurs does not increase with concentration as in typical spin glasses, but rather the maxima "turn on" suddenly at x = 10. This suggests the development of a random ferromagnetic state with T_0 = 30 K for x = 10 and 12, perhaps due to a percolation process. The ordering temperature for the x = 12 sample (30 K) is some five times smaller than the corresponding crystalline alloy. The saturation moment in the amorphous alloy (0.18 μ_B) is also considerably smaller than in the crystalline case (0.66 μ_B). This behavior is similar to other systems in which the crystalline-to-amorphous transition greatly weakens the magnetism. The lack of any evidence of spin-glass phenomena in Zr-Cu(Fe), as well as in the other systems discussed here, may be due to the extremely short electron-mean-free paths one expects to find in these amorphous alloys. We will discuss this further in the last section.

$$Zr_{40}Cu_{50}Gd_6$$

In this alloy the moment instability feature should be absent, so that exchange fluctuations and/or random anisotropy effects should predominate. Figure 1 shows the temperature dependence of the moment σ(T) measured at 0.23 kOe and the inverse of the local-moment contribution to the susceptibility. The data above 85K were fitted to the equation

$$\Delta\chi \equiv \chi - \chi_0 = x\mu_{eff}^2 [3k(T-\theta)]^{-1}. \qquad (1)$$

The resulting parameters are χ_0 = 3.33 μemu/g, μ_{eff} = 8.03 μ_B, and θ = 29.5 K. This moment value is very close to that expected for a Gd^{3+} ion (7.94 μ_B). From Fig. 1 and from preliminary σ(H) data at 4.2 K, it is clear that this alloy is ferromagnetic and a rough estimate of the ordering temperature is 15 K. Again, it is possible that extremely short electron-mean-free paths are responsible for the absence of spin-glass phenomena. Furthermore, in this system, an important aspect may be that large locally anisotropic crystal-field interactions do not exist; such interactions would greatly aid the random freezing of spins. Measurements on more dilute alloys are planned to determine whether a spin-glass state can be produced in this system.

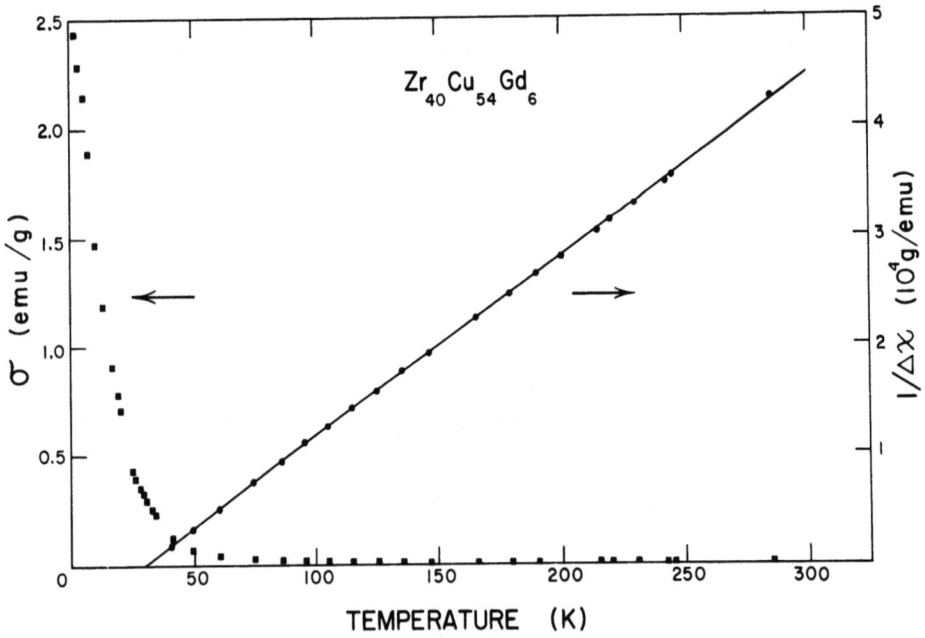

Fig. 1. σ(T) and 1/Δχ(T) for amorphous $Zr_{40}Cu_{54}Gd_6$

$Nb_{50}Ni_{50-x}Fe_x$

Figure 2 shows the susceptibility, $\chi \equiv \sigma/H$, of $Nb_{50}Ni_{50-x}Fe_x$ at H = 10.4 kOe for x = 0, 5, 10. The $Nb_{50}Ni_{50}$ data are clearly indicative of temperature-independent Pauli paramagnetism, and the x = 5 and 10 samples show a local-moment contribution. Figure 3 shows the field dependence of the moment at 4.2 K and 300 K. Apparently, there is a cluster contribution to the magnetization in that, even at 300 K, there is a significant nonlinearity in $\sigma(H)$. As an initial attempt to understand these data we have fitted the $\chi(T)$ data to Eq. (1) and have obtained the following parameters: For x = 5: χ_0 = 3.17 μemu/g, μ_{eff} = 1.69 μ_B, and θ = -134 K; for x = 10: χ_0 = 5.86 μemu/g, μ_{eff} = 1.65 μ_B, and θ = -79 K. The fits are shown in Fig. 4. It can be seen that there are systematic deviations from the simple Curie-Weiss law of Eq. (1), which may be due to the cluster contribution mentioned above.

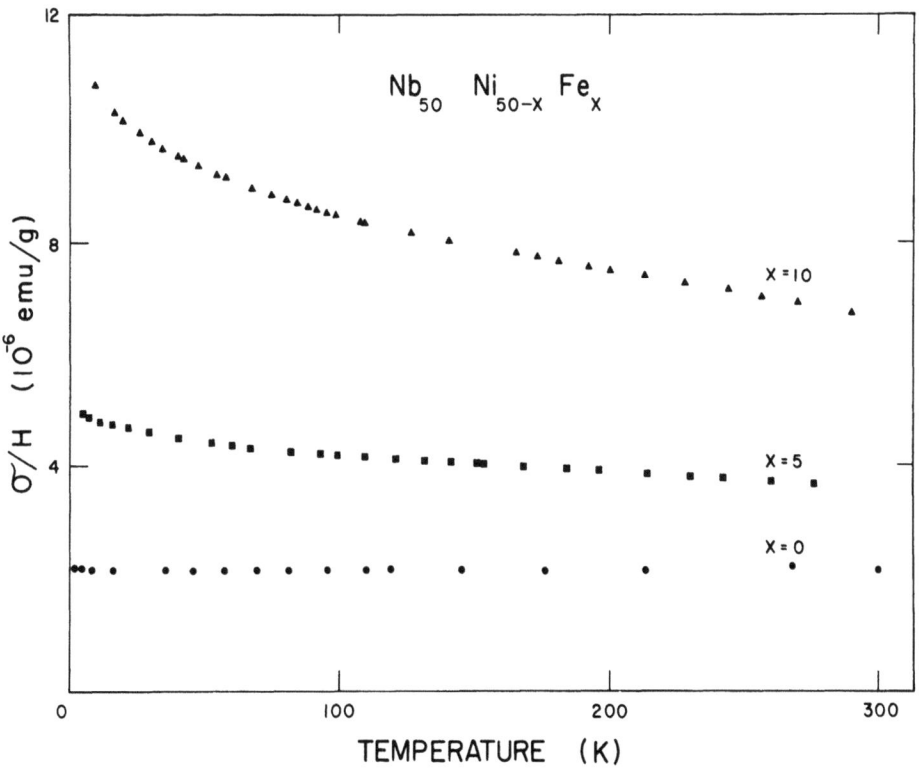

Fig. 2. Susceptibility for amorphous $Nb_{50}Ni_{50-x}Fe_x$

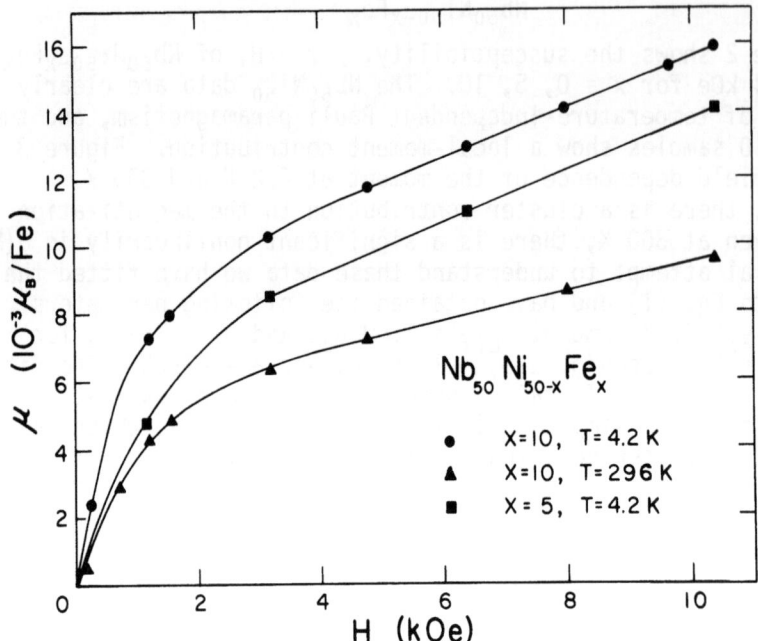

Fig. 3. Field dependence of average moment per Fe atom in amorphous $Nb_{50}Ni_{50-x}Fe_x$

Since μ_{eff} appears to be concentration independent, it seems to be the case that a major part of the temperature-dependent magnetism is due to "essentially isolated" Fe moments. If we hypothesize, then, that the local-moment magnetism can be separated approximately into "isolated" atom and "cluster" contributions, we can determine the temperature dependence of the susceptibility due to isolated Fe atoms, χ_i, as follows. Assuming that the cluster magnetization is saturated at 10 kOe, then χ_i is proportional to the high-field slope of $\sigma(H)$ in Fig. 3. The figure also shows that $\chi_i/(\mu/H)$ at $H = 10.4$ kOe is 0.47 for $x = 10$, $T = 4.2$ K, and 0.45 for $x = 10$, $T = 296$ K. This indicates that we can adjust the χ_0 values obtained in the fits mentioned above by multiplying them by the factor 0.46. Similarly, if we assume that the cluster contribution involves only a small fraction of the Fe atoms, the μ_{eff} values can be adjusted (by the factor $(.46)^{\frac{1}{2}}$) with the result that μ_{eff} ($x = 10$) = 1.1 μ_B, and χ_0 ($x = 10$) = 2.70 µemu/g. Following a similar procedure for x = 5 leads to the values μ_{eff} (x = 5) = 1.3 μ_B and χ_0 (x = 5) = 1.78 μemu/g. Naturally this adjustment procedure does nothing to the earlier determination of θ. It is possible to estimate the number of Fe atoms in clusters by making certain assumptions about the clustering itself. For example, if one assumes that all the clustered spins have a value of 3/2 and are ferromagnetically aligned, Fig. 3 leads to the estimate that only about 0.2% of the Fe atoms are clustered. It is reasonable to assume that this is a lower bound.

The results of this tentative analysis are, therefore, that there is in this system an inhomogeneous form of local-moment magnetism and no magnetic ordering down to 1.4 K, even with Fe concentrations as high as 10 at %. The effective moments determined for x = 5 and 10 are small and are similar to those seen in dilute $Zr_{40}Cu_{60-x}Fe_x$ alloys. Aside from the small cluster contribution involving perhaps as little as 0.2% of the Fe atoms, there is little evidence for magnetic interactions between the Fe atoms. The negative θ values (-134 K and -79 K), on the other hand, do indicate the presence of some type of antiferromagnetic interactions, and the obvious candidate is a conduction electron-magnetic impurity interaction of the Kondo type. If this

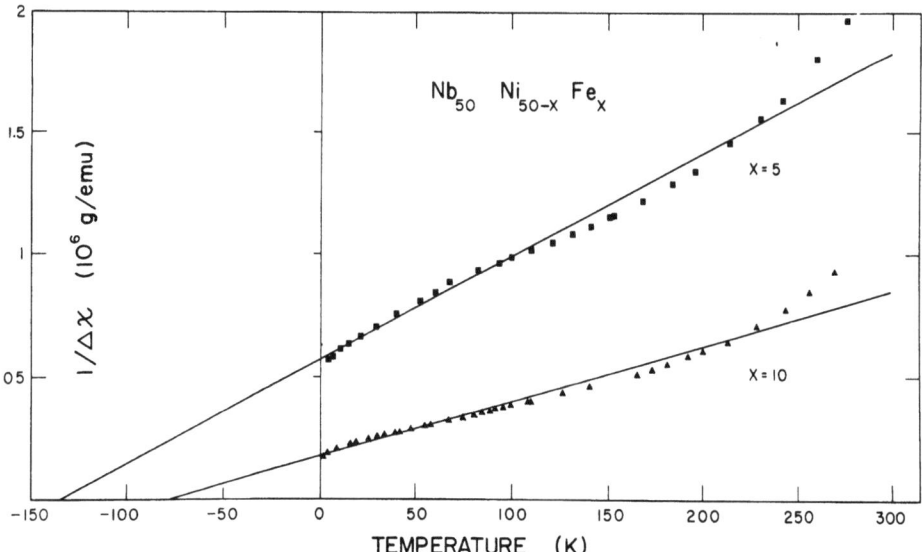

Fig. 4. Curie-Weiss fits for amorphous $Nb_{50}Ni_{50-x}Fe_x$

conjecture is true, and if θ for $x = 5$ is indicative of single impurity interactions, the data suggest a Kondo temperature of $T_K \simeq \theta/4.5 = 30$ K. Although it would be surprising if one had truly non-interacting magnetic impurities up to concentrations of 5 at %, it is also true that in glassy metals such as this, the restricted mean free paths attenuate any interimpurity interaction through the conduction electons. An additional example of this behavior is that of glassy $Ni_{41}Pd_{41}B_{18}$ containing Cr, in which isolated-atom Kondo phenomena were seen up to several atomic percent Cr.[10] Of course, it may be true in the $Nb_{50}Ni_{50-x}Fe_x$ system, as in the $Zr_{40}Cu_{60-x}Fe_x$ system, that the small effective moments result from moment instabilities caused by varying local environments. In addition to the restricted electron-mean-free paths, the predominance of a Kondo-type interaction may be responsible for the lack of any spin-glass phenomena in this system.

CONCLUSIONS

Many alloys of nonmagnetic metals with low to intermediate concentrations of randomly distributed magnetic impurities are characterized by a number of directly observable quantities, such as maxima in the magnetic susceptibility and electrical resistivity at a characteristic temperature which scales with concentration. Such alloys are called spin glasses, and the phenomena which they exhibit are believed to be due to the random freezing of the impurity moments at the characteristic temperature. Theoretical descriptions of such systems based upon exchange fluctuations rely upon the idea that these ions interact via an indirect, oscillatory exchange interaction, and it is believed that the competition between random ferromagnetic and antiferromagnetic interactions is the reason for the random freezing. The lack of spin-glass phenomena in the data discussed above does not necessarily mean the absence of an indirect, oscillatory exchange interaction. For example, an oscillatory exchange interaction may exist in $Zr_{40}Cu_{54}Gd_6$ provided that the dominant interaction is ferromagnetic. Such a situation may arise in an amorphous alloy where one expects extremely short electron-mean-free paths, effectively eliminating any interaction between all but the closest spins. One then has essentially a nearest-neighbor topology for the interacting spins with the exchange interaction depending quite strongly on the distance between spins. Under such circumstances, it is certainly possible that interactions of one sign will predominate. Another possibility for amorphous alloys doped with 3d magnetic impurities is the following. Due to restricted mean-free paths the direct interactions will predominate over indirect interactions for the more heavily concentrated alloys. In this case too exchange interactions of one sign may prevail. Whenever ferromagnetic exchange interactions predominate, the strength of the locally anisotropic crystal-field interactions may be crucial in determining whether such alloys

will be spin glasses.[2] It should also be pointed out here that
random systems can have two distinct kinds of thermodynamic behavior
depending upon whether they are (mathematically) quenched or
annealed.[11] Basically, an annealed system comes into thermal equilibrium at each temperature whereas in a quenched system the randomness does not change with temperature. There is a very different
mathematical structure in the quenched and annealed cases. For
example, if the systems discussed here are better described as
annealed random magnetic systems, then spin-glass phenomena should
not be expected. In the $Zr_{40}Cu_{48}Fe_{12}$ alloy there is a resistivity
maximum but no susceptibility maximum at the ordering temperature.
However, in amorphous $Pd_{75}Si_{20}Cr_5$ just the reverse is true.[12] A
possible explanation for this is that because of the reduced mean-
free-paths, direct d-d interactions will prevail at the expense of
long-range, RKKY-type interactions. If this is the case, it is
possible that the susceptibility maximum seen in the Pd-Si(Cr)
system, suggesting random antiferromagnetic order, results from the
predominance of antiferromagnetic near-neighbor Cr-Cr interactions.
However, in the Zr-Cu(Fe) system one would expect ferromagnetic
order on the basis of the likelihood of ferromagnetic Fe-Fe near-
neighbor interactions. This suggests that measurements on
$Zr_{40}Cu_{60-x}Mn_x$ and $Nb_{50}Ni_{50-x}Mn_x$ at high x values would be quite
valuable in assessing the importance of the sign of the d-d
interactions.

The $Nb_{50}Ni_{50-x}Fe_x$ system shows local-moment behavior but no
magnetic order for x up to 10 at % and temperatures down to 1.4 K.
There appear to be two contributions to the local-moment magnetism,
one due to isolated-Fe atoms and one due to Fe-atom clusters. It
is not clear whether the magnetic clusters are intrinsic or extrinsic, that is, due to magnetic interactions involving purely <u>random</u>
fluctuations in the Fe concentration, or to <u>atomic clustering
(segregation)</u> of the Fe atoms which naturally lead to magnetic
clustering. Annealing experiments and studies of alloys quenched
at various rates may be informative concerning this question.

Another interesting point is raised by our results on
$Zr_{40}Cu_{54}Gd_6$. This is that Harris and Zobin[2] have determined a
phase diagram, based on the random anisotropy model, which contains
paramagnetic, ferromagnetic, and spin-glass regions. The alloy is
increasingly driven from ferromagnetism towards spin-glass magnetism
as the uniaxial anisotropy coefficient is increased. Since Gd is an
S-state ion one would expect the anisotropy in $Zr_{40}Cu_{60-x}Gd_x$ to be
small compared to, say, an alloy such as $Zr_{40}Cu_{60-x}Tb_x$. The ferromagnetism of $Zr_{40}Cu_{54}Gd_6$ is consistent with Harris and Zobin's work,
but a significant test, which we are pursuing, is the possibility
of observing a spin-glass state in an alloy such as $Zr_{40}Cu_{60-x}Tb_x$,
in which anisotropy is expected to play a more significant role.

ACKNOWLEDGEMENTS

We thank Drs. E. R. Domb and F. R. Szofran for help with the measurements, and Dr. R. Ray for assistance in sample preparation. The support of the National Science Foundation at the University of Nebraska (Grant DMR72-03208 A01) and the Office of Naval Research at Northeastern University (Contract N14-68-A207-3) is deeply appreciated.

REFERENCES

1. See: D. Sherrington and S. Kirkpatrick, Phys. Rev. Letters $\underline{35}$, 1972 (1975), and references therein.
2. See: R. Harris and D. Zobin, AIP Conf. Proc. $\underline{29}$, 156 (1976), and references therein.
3. F. R. Szofran, G. R. Gruzalski, J. W. Weymouth, D. J. Sellmyer, and B. C. Giessen, Phys. Rev. B $\underline{14}$, 2160 (1976).
4. For preliminary accounts of this work on the Zr-Cu(Fe) alloys, see F. R. Szofran, J. W. Weymouth, G. R. Gruzalski, D. J. Sellmyer, R. Ray, and B. C. Giessen, AIP Conf. Proc. $\underline{18}$, 282 (1974); and J. W. Weymouth, F. R. Szofran, G. R. Gruzalski, D. J. Sellmyer, and B. C. Giessen, Bull. Am. Phys. Soc. $\underline{19}$, 253 (1974). See also T. Mizoguchi and T. Kudo, AIP Conf. Proc. $\underline{29}$, 167 (1976).
5. M. J. Zuckermann, J. Phys. F $\underline{2}$, L25 (1972).
6. See, e.g., F. J. Blatt, <u>Physics of Electronic Conduction in Solids</u>, (McGraw-Hill, New York, 1968) p. 185.
7. R. W. Cochrane, R. Harris, J. O. Strom-Olson, and M. J. Zuckerman, Phys. Rev. Letters $\underline{35}$, 676 (1975).
8. A. K. Sinha, Phys. Rev. B $\underline{1}$, 4541 (1970).
9. S. R. Nagel and J. Tauc, Phys. Rev. Letters $\underline{35}$, 380 (1975).
10. V. K. C. Liang and C. C. Tsuei, Phys. Rev. B $\underline{7}$, 3215 (1973).
11. M. F. Thorpe and D. Beeman, Phys. Rev. B $\underline{14}$, 188 (1976).
12. C. C. Tsuei and R. Hasegawa, Solid State Comm. $\underline{7}$, 1585 (1969).

DISCUSSION

U. Larsen: Does the resistance maximum occur at a lower temperature in the amorphous state compared to that in the crystalline state?

D. J. Sellmyer: We do not see any resistance maximum in the crystalline state.

L. N. Mulay: You have referred to intrinsic and extrinsic clustering. Can you clarify those concepts?

D. J. Sellmyer: By intrinsic clustering I am referring to a random alloy in which one expects normal fluctuations in the local concentration of the iron atoms. By extrinsic clustering I am referring in this case to a segregation of the iron atoms.

FROM SUPERCONDUCTIVITY TO FERROMAGNETISM IN AMORPHOUS Gd-La-Au ALLOYS.[*]

S. J. Poon and J. Durand[†]

California Institute of Technology

Pasadena, California 91125

I. ABSTRACT

Splat cooled alloys of composition $(La_{100-x}Gd_x)_{80}Au_{20}$ have been obtained by complete substitution of gadolinium for lanthanum in amorphous $La_{80}Au_{20}$ matrix. Results of high field magnetization (up to 70 kOe), ac and dc low field susceptibility, and resistivity measurements over temperature range of 1.7 to 300°K for these alloys are reported. The $La_{80}Au_{20}$ alloys are superconducting at 3.5°K. For $x \lesssim 1$, a suppression of T_c described by the relation $dT_c/dx \approx -4.0$°K per atomic per cent gadolinium is observed. For alloys within the concentration range $1 \lesssim x \lesssim 70$, maxima in low field susceptibility measurements are observed. The 'ordering' temperatures T_M are proportional to x for $1 \leq x \leq 16$, similar to those observed in crystalline spin-glass alloys. For $16 \lesssim x \lesssim 70$, T_M is increasing at a faster rate than in the low concentration region, and this intermediate type of ordering corresponds to a mictomagnetic regime. As $x \gtrsim 70$, a ferromagnetic regime emerges. The maximum Curie temperature is observed for $Gd_{80}Au_{20}$ at ~ 150°K. The moment per gadolinium atom is found to be constant and close to that of the crystalline value throughout the concentration range investigated. Results of resistivity measurements are correlated with the magnetic properties of different regimes in the magnetic phase diagram.

II. EXPERIMENTAL PROCEDURES

The sample foils were prepared in the usual way as discussed elsewhere.[1] X-ray scanning of the foils indicated patterns with broad maxima centered at $\sim 31.5°$ with full widths at half maxima

Table 1. Parameters derived from magnetization measurements for amorphous $Gd_{80}Au_{20}$ alloys.

$T_c(°K)$	β	γ	δ	$\mu_{Gd}(\mu_B)$	$\mu_{eff}^{(a)}(\mu_B)$	$\theta_p(°K)$	$J_n(°K)$ $(T>T_c)$	$J_n(°K)$ $(T<T_c)$
149.45	0.439	1.294	3.948	7.00	9.37	165	2.28	1.34

(a) $\mu_{eff} = g[J(J+1)]^{\frac{1}{2}}$, some authors use p_{eff}.

of ~ 4.5°, which were typical of a glassy metal. Magnetic ordering temperatures were observed using a standard ac inductance bridge technique. Magnetization measurements were made between 1.7°K and 290°K in fields up to 70 kOe using the Faraday method.[1] Electrical resistivity as a function of temperature was measured using a standard four-probe technique.

III. RESULTS AND DISCUSSION FOR $(La_{100-x}Gd_x)_{80}Au_{20}$ ALLOYS

A. $La_{80}Au_{20}$ Superconductors.

The critical behavior and transport properties of amorphous superconducting $La_{80}Au_{20}$ alloys has been discussed previously.[2,3] The alloys are ideal type II superconductors characterized by $T_c \simeq 3.5°K$, $H_{c2}(0) \simeq 60$ kOe, $J_c(0) \simeq 10^4 A/cm^2$, and a Ginzburg-Landau parameter κ of ~ 70. Spin-orbit scattering effects are found to be stronger in the amorphous samples than in disordered crystalline samples. Fluctuation conductivity in three dimensional amorphous superconductors has been investigated by Johnson and Tsuei.[4] We obtained magnetization results between 1.7°K and 290°K. The susceptibility is found to be temperature independent with a value of ~ 0.5×10^{-6} emu/g.

B. $Gd_{80}Au_{20}$ Ferromagnets.

The magnetic properties of this amorphous ferromagnet[1] are summarized in Table 1. The normalized $M(H,T)$ data are fitted to an equation of state derived for second order phase transition in fluid systems. The equation takes the form $h/m = f_\pm(m)$ where the ± signs stand for temperatures above and below the Curie temperature (T_c) respectively. This is illustrated graphically in figure 1. Together with the equality relation observed by the critical exponents, the results indicate clearly a second order phase transition in the amorphous state. The small deviations of the exponents from the Heisenberg values are probably due to the crystal-field anisotropy. The exchange integrals J_n for temperatures above and below T_c are found to be lower than those in crystalline Gd. The values equal 2.28 and 1.34°K as determined from the Rushbrooke-Wood formula and spin-wave theory respectively. The low temperature saturation magnetization follows the $T^{3/2}$ law from 0.13 to 0.80 T_c. Amorphousness is found to be more detrimental for T_c than for μ_{Gd}. Fluctuations in J_n due to structural randomness are under investigation.[1] The differences in the exchange integrals J_n, and that between the effective moment μ_{eff} and saturation moment μ_{Gd} at different temperature regimes can be attributed to the nearest-neighbors antiferro-

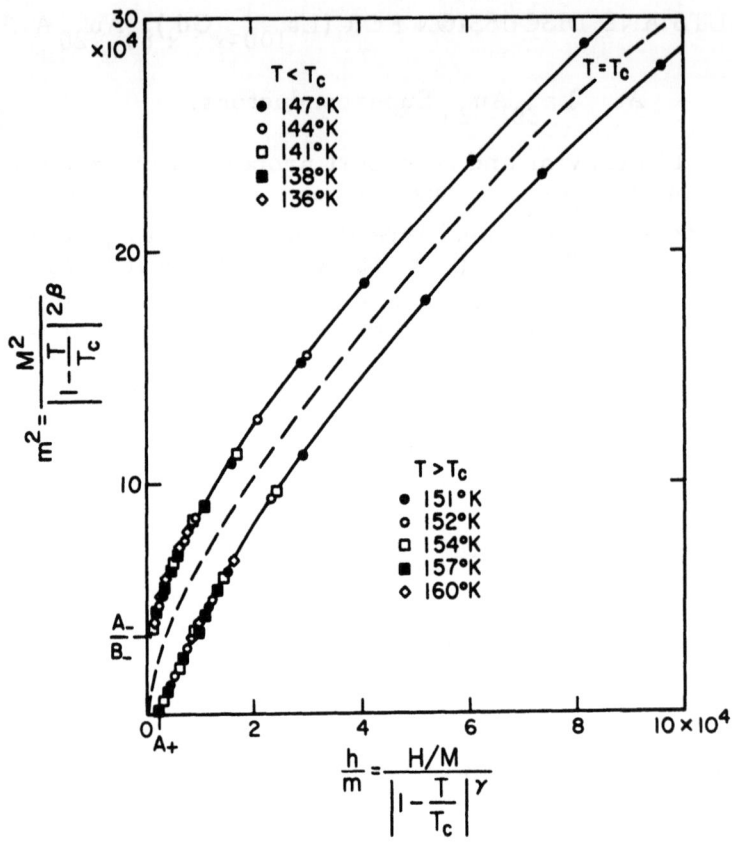

Fig. 1. The normalized magnetization m^2 versus normalized inverse susceptibility χ_o^{-1} for temperatures around T_c. The dashed line indicates asymptotic behavior of the two curves for large m above and below T_c.

magnetic couplings in the presence of Au below T_c. This is supported by case studies in stochiometric Gd-Au compound.[5]

C. x < 1, Coexistence of Magnetic Short-Range Ordering and Superconductivity Regime.

In figure 2, the suppression of superconducting transition temperature in the presence of magnetic impurity Gd follows

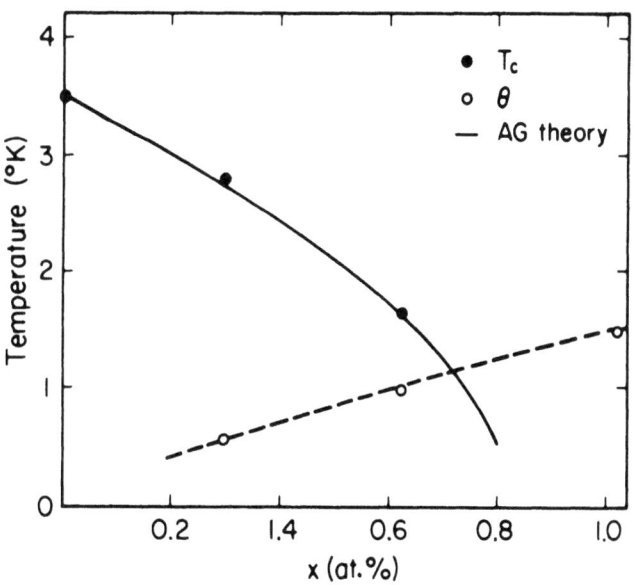

Fig. 2. The dependence of T_c and θ on Gd concentration in $La_{80}Au_{20}$ matrix.

closely the Abrikosov-Gor'kov theory[6] below the critical concentration x. The gradient dT_c/dx gives a value of $-4°K$ per atomic per cent Gd. Superconductivity in La-Gd dilute solid solutions has been investigated thoroughly by Matthias et al.[7,8] By employing specific heat results, Finnemore et al[9] demonstrated the coexistence of antiferromagnetic coupling and superconductivity in the dilute Gd limit. It might be more interesting to investigate the coexistence phenomena directly by performing magnetization measurements. Such results have been obtained recently.[1] The magnetizations, after correcting for the $La_{80}Au_{20}$ matrix contributions to the susceptibility, are found to satisfy a Brillouin equation of the form $M(H,T) = M(\infty, 0) B_S(H/T + \theta)$. The parameters S and θ (> 0) give the spin of the moments and the antiferromagnetic characteristic temperature respectively. For the samples with 0.24 and 0.5 per cent Gd investigated, it is found that $\theta \lesssim 1°K < T_c$ as shown in figure 2. The Gd atoms carry a moment of about 7 Bohr magnetons per atom which are close to the $^8S_{7/2}$ ionic value.

D. Mictomagnetic Regime (1 < x < 70).

Alloys in this regime are characterized by susceptibility maxima in low field measurements and thermomagnetic effects[10,11] (isothermal remanent magnetization and thermal remanent magnetization at least for high concentrations). The dependence of the 'ordering' temperature T_M on Gd concentration can be divided into two regimes

$$T_M(in\ °K) \simeq \begin{cases} 0.38\ x & 1 < x < 16 \\ 6 + f(x) & 16 < x < 70 \end{cases}$$

where $0 < f(x) < 48$ is a monotonic increasing function of x. The distribution of the regimes are evident from the magnetic phase diagram of figure 3. The values of the initial susceptibility $\chi_o(T)$ for x = 20 and 60 in "zero" field and in fields up to 500 Oe on zero-field cooled samples are shown in figure 4. It is clear that the peaks in $\chi_o(T)$ are reduced and rounded off in small applied fields as observed in the crystalline case. They disappear in samples cooled in fields greater than ~ 1 kOe. The paramagnetic Curie temperature θ_p (≃ 3 T_M) is found to increase with x indicating a trend towards stronger ferromagnetic couplings above the 'ordering' temperature. The large value of $\theta_p - T_M$ also measures the temperature range of inhomogeneous ferromagnetic interactions. Using the classical molecular field approach, the effective number of Bohr magnetons per Gd atom p_{eff} is found to remain constant at the value of 8 which corresponds to the ionic value of 7.94.

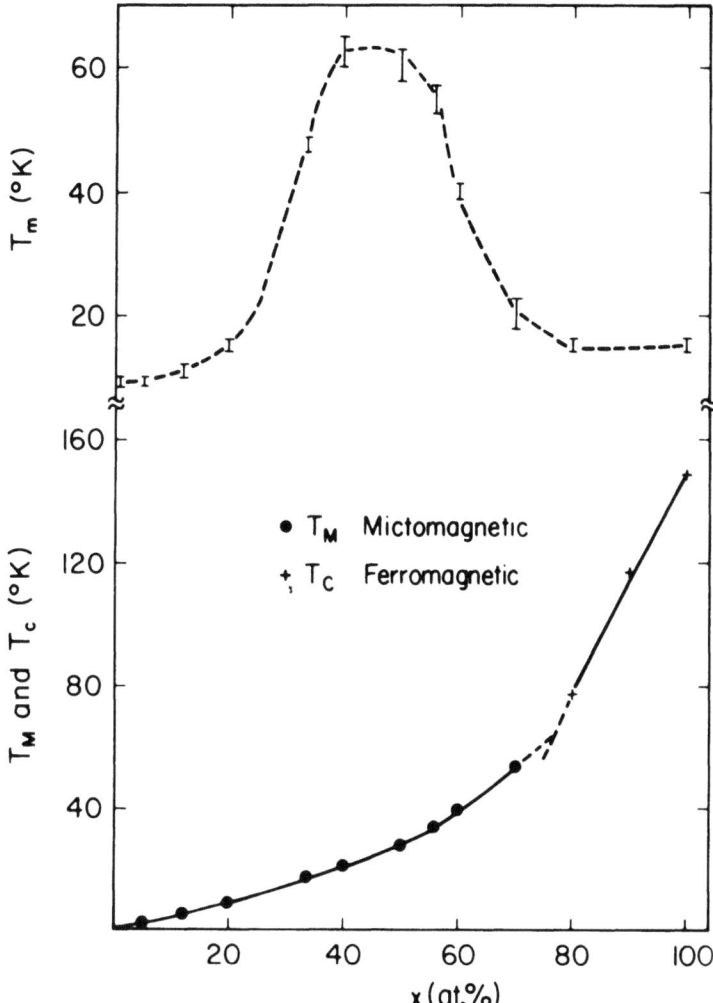

Fig. 3. Magnetic phase diagram of $(La_{100-x}Gd_x)_{80}Au_{20}$ indicating the transition temperatures T_M and T_c in the mictomagnetic and ferromagnetic regimes respectively as functions of x. The resistivity minima T_m are also included for comparison.

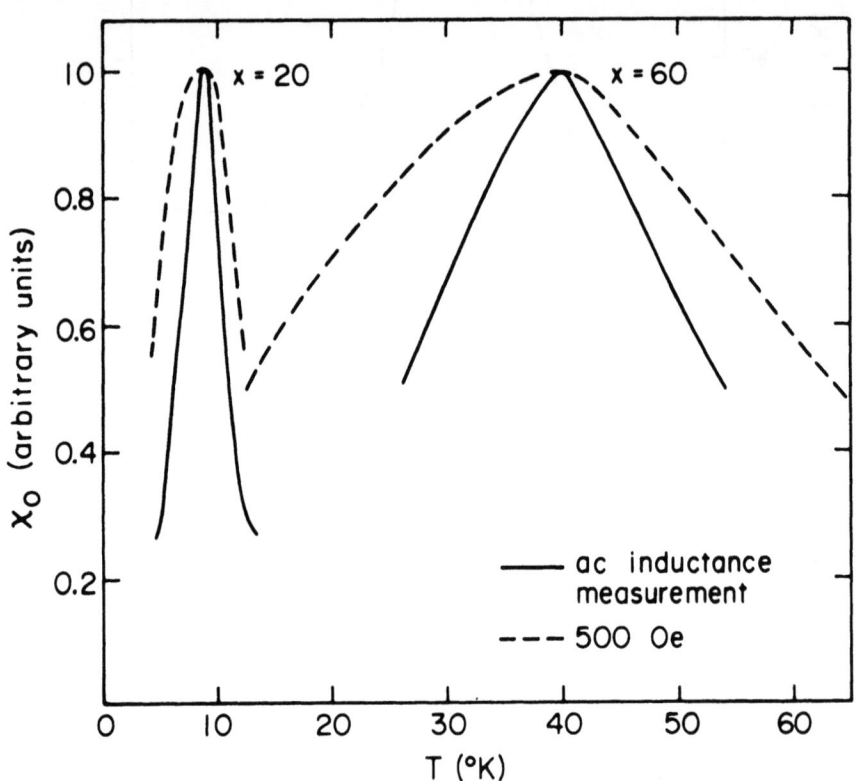

Fig. 4. Susceptibility (in arbitrary unit) as a function of temperature measured in "zero" field and in 500 Oe for x = 20 and 60 samples.

The linear dependence of T_M on concentration in the first regime resembles those observed in typical spin glass systems.[12] However, unlike the well studied spin glass systems, the present magnetization study does not indicate any scaling law relations and the concentration of magnetic impurities is far beyond those observed in the spin glass alloys in the sense of reference 12. Therefore, it might be more appropriate to call this region a mictomagnetic[13] regime without loss of generality.

The classical Arrott plots (M^2 versus H/M) in both regimes exhibit strong departures from linearity at small and high fields for all temperatures below θ_p, so that any spontaneous magnetization and Curie temperature cannot be defined by this method. However, the M^2(H/M) isotherms are observed to approach closer to the M^2-axis for higher Gd concentrations indicating a gradual onset of spontaneous magnetization for $x \gtrsim 70$. The absence of spontaneous magnetization from the Arrott plots for all $T \ll \theta_p$ also points towards the possibility of weak and inhomogeneous ferromagnetic interactions. For temperatures between T_M and θ_p, the superparamagnetic clusters break up gradually at increasing temperature to yield single magnetic atom moments above θ_p. The persistence of mictomagnetic regime at high Gd concentrations (up to $\sim$ 56 atomic per cent) is probably favored by two conditions. First, the structural randomness in the amorphous state introduces inhomogeneities in ferromagnetic couplings which depend strongly on local structural environment. Second, it has been mentioned beforehand[1] that Gd atoms tend to couple antiferromagnetically in the presence of Au and La. Even in crystalline Gd, there is already a trend towards this type of couplings, as the difference in the exchange interactions J_n at different temperatures can be accounted for by the RKKY interaction.[14]

E. Ferromagnetic Regime (70 < x < 100).

Alloys in this regime are characterized by a well defined Curie temperature. The magnetic phase transition determined from ac inductance bridge measurement gives a transition width of $\sim 10°K$. The Curie temperature is defined by the inflection point on the signal intensity versus temperature curve.[1] The spontaneous magnetization can be determined from the Arrott plots. Nonlinearity in M^2 versus H/M is observed even for $x > 70$ indicating inhomogeneities in ferromagnetic couplings. This is also supported by the fact that the inflection point on the M(H) versus T plots disappears in fields greater than $\sim$ 2 kOe.[1] The temperature domain over which ferromagnetic inhomogeneities dominate narrows as x increases until $Gd_{80}Au_{20}$,

$(\theta_p - T_c)/T_c \simeq 0.1$. The variation of the mean magnetic moment per atom when La substitutes for Gd obeys fairly well a dilution law, which means the possible polarization of La (or Au) atoms has to be rather small.[15] The effective moment p_{eff} determined at $T > \theta_p$ gives ~ 8 Bohr magnetons for Gd atom if the effects due to La and Au are ignored. The saturation moment μ_{Gd} gives approximately the same value. However, the detail trends in μ_{eff} and μ_{Gd} can be explained in terms of antiferromagnetic couplings, conduction electron polarizations, and crystal-field effects.[1] The suppression of T_c defined by $(1/T_c)(dT_c/dx)$ when La is substituted for Gd in crystalline Gd[15] and amorphous $Gd_{80}Au_{20}$ alloys are found to be 1.82×10^{-2} and 2.32×10^{-2} per La atom respectively.

F. Resistivity Results.

Resistivity minima have been observed over the whole concentration range for $x \gtrsim 0.6$. The variation of the resistivity minima (T_m) follows a bell shaped curve as shown in figure 3. The invariance in T_m (10-15°K) at both the low concentration range and ferromagnetic regime suggests a structural rather than a magnetic origin for this phenomena[16] based on the following reasons. For $x = 0.6$ dilute alloys, it is unlikely that the resistivity minimum results from the Kondo effect.[17] Recent magneto-resistivity measurements[18] indicate that the shape of the resistivity curve is unaltered except for the sign of magnetoresistance in fields up to 40 kOe. For the ferromagnetic regime, conjectures had been made to explain the occurrence of resistivity minima in terms of conduction electrons being scattered off low energy magnons.[19,20] It is not clear in which way the magnon dispersion spectrum is changed in a field of 40 kOe (~ 5°K). Present experiments give the same results for the ferromagnetic samples as in the low concentration limit in an applied field. However, the strong variation of T_m in the mictomagnetic regime might point towards a magnetic origin.[21] Thus, the resistivity minimum phenomena might be caused by either structural or magnetic mechanism, and the interplay of these two mechanisms on the shape of the resistivity curves is under investigation.

Acknowledgment - The authors wish to thank Professor Pol Duwez for his interest and support throughout this work.

REFERENCES

*Work supported by the Energy Research Development Agency, Contract No. AT(04-3)-822.
†On leave from Laboratoire de Structure Electronique des Solides, E.R.A. 100 4, rue Blaise Pascal, 67000 Strasbourg, France.

1. S. J. Poon and J. Durand, to be published.
2. W. L. Johnson, S. J. Poon, and P. Duwez, Phys. Rev. B11, 150 (1975).
3. W. L. Johnson and S. J. Poon, J. Appl. Phys. 46, 1787 (1975).
4. W. L. Johnson and C. C. Tsuei, Phys. Rev. B13, 4827 (1976).
5. C. C. Chao, Ph.D. Thesis, Caltech (1965).
6. M. B. Maple, in 'Magnetism' Vol. 2B, p. 289-325, edited by G. T. Rado and H. Suhl, Academic Press, New York (1966).
7. B. T. Matthias, H. Suhl, and E. Corenzwit, Phys. Rev. Lett. 1, 92 (1958).
8. R. A. Hein, R. L. Falge, Jr., B. T. Matthias, and C. Corenzwit, Phys. Rev. Lett. 2, 500 (1959).
9. D. K. Finnemore, D. C. Hopkins, and P. E. Palmer, Phys. Rev. Lett. 15, 891 (1965).
10. V. Cannella, in 'Amorphous Magnetism' p. 195-206, edited by H. O. Hooper and A. M. deGraaf, Plenum Press, New York (1973).
11. H. Claus and J. S. Kouvel, Solid State Comm. 17, 1553 (1975).
12. J. L. Tholence and R. Tournier, J. Phys. (Paris) 35, C4-229 (1973).
13. P. A. Beck, Trans. AIME 2, 2015 (1971).
14. D. A. Goodings, Phys. Rev. 127, 1532 (1962).
15. A. B. Beznosov and V. G. Nazarenk, Fiz. Tverd. Tela 16, 576 (1974). [Sov. Phys. Solid State 16, 372 (1974)].
16. R. W. Cochrane, R. Harris, J. O. Ström-Olson, and M. J. Zuckermann, Phys. Rev. Lett. 35, 676 (1975).
17. A. Blandin, in ref. 6, p. 57-88.
18. S. J. Poon, J. Durand, and M. Yung, to be published.
19. A. Madhukar and R. Hasegawa, J. Phys. (Paris) 35 C4-291 (1974).
20. R. N. Silver and T. C. McGill, Phys. Rev. B9, 272 (1974).
21. R. Hasegawa and C. C. Tsuei, Phys. Rev. B3, 214 (1971).

DISCUSSION

U. Larsen: It appears that your definition of a spin glass alloy is restricted to a rather narrow concentration range.

S. J. Poon: Yes, you are right, the difference between a spin glass and a mictomagnet alloy is never quite clear.

N. Y. Rivier: How do you establish your transition point from a mictomagnet to a ferromagnet?

S. J. Poon: It is not quite obvious. However, the persistance of the mictomagnet behavior up to very might concentrations of gadolinium might be due to two factors. The first arising from disorder or randomness considerations because the ferromagnetic and antiferromagnetic couplings depend strongly on the local environment. The second point is that gadolinium tends to couple antiferromagnetically in the presence of Au atoms, and this is supported by case studies of stoichmetric gadolinium-gold compounds.

THE HIGH TEMPERATURE ELECTRONIC AND MAGNETIC PROPERTIES OF Pd-ALLOYS IN THE GLASSY, CRYSTALLINE AND LIQUID STATE

H.-J. Güntherodt, H.U. Künzi, M. Liard, M. Müller and R. Müller

Institut für Physik, Universität, Basel, Switzerland

C.C. Tsuei

IBM Thomas J. Watson Research Center, Yorktown Heights, N.Y., USA

INTRODUCTION

At the first International Symposium on Amorphous Magnetism several questions have been raised concerning the comparison of electrical transport and magnetic properties in the amorphous, crystalline and liquid state. For more details see the discussions of the papers presented by Bagley and DiSalvo[1] and Tsuei[2]. We have tried to compare electrical resistivity, Hall coefficient and magnetic susceptibility in these three states for the binary $Pd_{81}Si_{19}$ and for the ternary $Pd_{77.5}Cu_6Si_{16.5}$ alloys. The difference in the glass-forming abilities of both alloys is remarkable. For the binary alloy a cooling rate of $10^{6}°C/sec$ and for the ternary alloy only a cooling rate of $10^{2}°C/sec$ is required. In view of the similar constituents, such a comparison is thought to provide information from the electronic structure point of view on the particularly enhanced stability against crystallization of the ternary alloy.

EXPERIMENTAL RESULTS

The binary alloy was rapidly quenched by a splat cooling technique. The ternary alloy was rapidly quenched into water.

Fig.1 shows the electrical resistivity of $Pd_{81}Si_{19}$ (empty dots) and $Pd_{77.5}Cu_6Si_{16.5}$ (full dots) as a function of temperature in the glassy (a), crystalline (c) and liquid (ℓ) state. At room temperature we measure an electrical resistivity of 80 µΩcm for glassy $Pd_{81}Si_{19}$ which is slightly larger than the observed resistivity value of $Pd_{77.5}Cu_6Si_{16.5}$. Our new values do not agree with previously published values. Duwez et al.[3] found for glassy $Pd_{83}Si_{17}$ a value of 60 µΩcm whereas in reference [1] a value of 95 µΩcm for $Pd_{77.5}Cu_6Si_{16.5}$ is mentioned.

Upon crystallization, the room temperature resistivities of both alloys decrease by a factor of three and become more temperature dependent.

At the melting points T_M the electrical resistivity of the binary alloy changes to a value of 95 µΩcm and for the ternary alloy to a value of 88 µΩcm. The temperature coefficient of the electrical resistivity is slightly larger in the liquid state than in the glassy state. The data in fig.1 indicate that the resistivity values of the

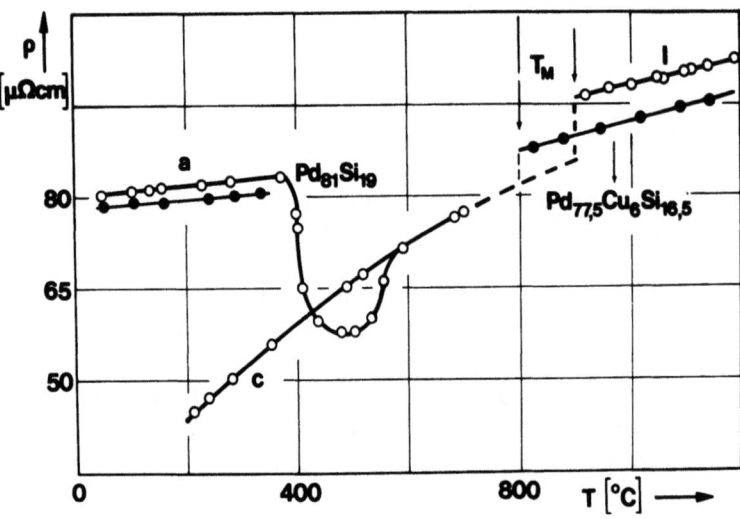

Fig. 1   Electrical resistivity

liquid state extrapolate to the resistivity curve of the glassy state. This is in good agreement with the unpublished work of Crewdson.[2,4] Furthermore, fig.1 shows a general trend that the extrapolated resistivities of amorphous, crystalline and liquid states tend to show nearly the same values at high temperatures.

The Hall coefficient of $Pd_{77.5}Cu_6Si_{16.5}$ (full dots) and $Pd_{81}Si_{19}$ (empty dots) in the glassy (a), crystalline (c) and liquid state are shown in fig.2. The Hall coefficient of the glassy ternary alloy shows a nearly temperature independent value of $-10.7 \cdot 10^{-11} m^3/As$ which is a larger negative value than observed for the $Pd_{81}Si_{19}$ alloy. The latter corresponds to $-9.6 \cdot 10^{-11} m^3/As$. These observed Hall coefficients are in good agreement with the values mentioned in reference 1 where a value of $-9.3 \cdot 10^{-11} m^3/As$ for $Pd_{80}Si_{20}$ and a value of $-10 \cdot 10^{-11} m^3/As$ for $Pd_{77.5}Cu_6Si_{16.5}$ have been observed.

The Hall coefficient in the crystalline state seems to show smaller negative values but do not differ very much from the glassy state data. Below 600°C, a different behavior of the Hall coefficients in different samples have been observed.

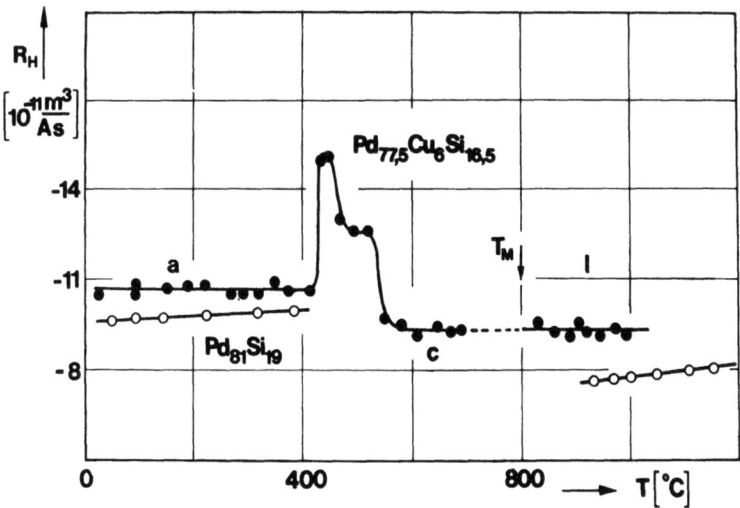

Fig. 2   Hall coefficient

The Hall coefficients of the liquid alloys show 15% smaller negative values compared to the glassy state. This difference of the Hall coefficients seems to be not very significant and it might be difficult to draw conclusions due to the lack of mass density data. The liquid state Hall coefficients indicate that Si provides approximately four and Pd less than one conduction electrons. Such values are in good agreement with measurements of the Hall coefficient in pure liquid Ge and in similar liquid transition metal alloys.[5] Hall coefficient measurements of pure liquid Si, Pd and the corresponding liquid alloys are under investigation.

The available magnetic susceptibility data are shown in fig.3. The magnetic susceptibilities of the liquid binary and ternary alloys are very small and paramagnetic at higher temperatures. We observe that the liquid state susceptibilities decrease linearly with decreasing temperature, becoming less paramagnetic.

The magnetic susceptibility of glassy $Pd_{77.5}Cu_6Si_{16.5}$[1] is diamagnetic and increases linearly with increasing temperature, becoming less diamagnetic. The magnitude of the diamagnetic susceptibility and its temperature coefficient correspond very closely to the magnetic susceptibility of liquid noble metals[5] such as Ag, Au and Cu. This

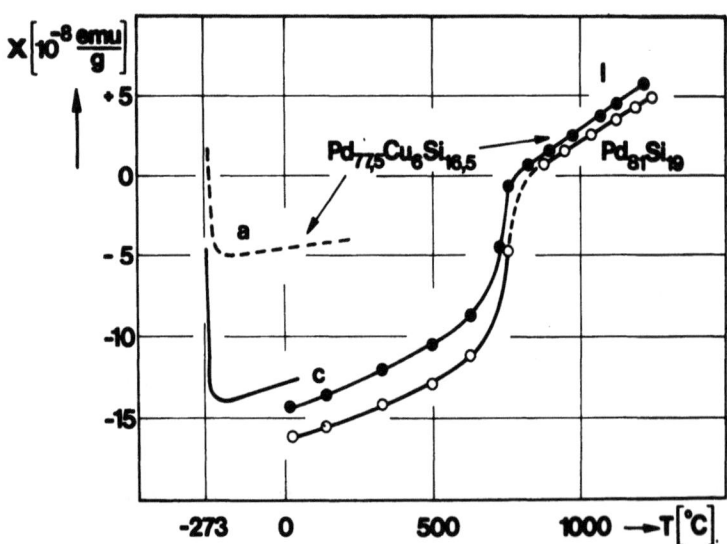

Fig. 3   Magnetic susceptibility

analogy suggests that the susceptibilities also show some correspondence in the glassy and liquid state. The measurements of the glassy binary alloy are still under investigation.

DISCUSSION

The main conclusions of our measurements are that the glassy state appears as a continuous extension of the liquid state and that the experimental results of the binary and ternary alloys do not differ very much.

The first observation might help to extend the liquid transition metal theory to the glassy state. It seems that electrical resistivity, Hall coefficient and magnetic susceptibility can be understood in terms of theoretical concepts developed for the liquid state.

The present electrical resistivity data of the glassy Pd alloys can be explained by an extension of the Faber-Ziman theory in the t-matrix approximation.[6] The electrical resistivity of 78μΩcm for pure liquid Pd[7] is mainly determined by resonant scattering of the conduction electrons by the d states lying in the conduction band. On alloying the contribution of the interference function increases due to the increasing $2k_F$ value of Si.

In view of the similarities of structure, the comparable magnitudes of the resistivity and its temperature coefficient we believe that a similar model for glassy transition metal alloys can explain the observed high-temperature electrical transport data in the glassy state.

The observed magnetic susceptibility data of the glassy Pd alloys can be understood by comparison with liquid transition and noble metal alloy behavior.[8,9] The measurements of the magnetic susceptibility in liquid transition metal alloys have indicated some models for the density of states in the liquid state.[5] But due to the experimental difficulties of photoemission experiments in the liquid transition metals such experiments have not been done and compared with model density of states. However, on glassy materials photoemission work is relatively easy and results of such measurements can be used to test the suggested models for the density of states in the liquid state.

The strong paramagnetism of pure Pd is drastically reduced in the glassy and liquid state by alloying of Si. The magnetic susceptibility of pure Pd is explained in terms of exchange and correlation enhancement and the density of states arising from d electrons. The smaller susceptibility observed by alloying Si can be attributed to a decrease of the enhancement factor and a smaller density of states at the Fermi energy in the alloys compared to pure Pd. In pure Pd the d band is almost filled. By adding extra electrons the Fermi energy is raised above the d bands resulting in a similar density of states as observed for the noble metals and the density of states at E_F is decreased.

X ray and UV photoemission data have been published[10] for $Pd_{77.5}Cu_6Si_{16.5}$ which are in agreement with this data. In fact the valence band looks very similar to a Cu density of states. Preliminary results[11] from XPS measurements of the valence band of glassy $Pd_{81}Si_{19}$ show a reduction in the density of states at E_F and what appears to be a narrowing of the Pd d band width compared to polycrystalline Pd. The reduction in the density of states at E_F is further supported by the absence of asymmetry of the Pd core levels in glassy $Pd_{81}Si_{19}$.

The second observation indicates that the electrical resistivity, the Hall coefficient and the magnetic susceptibility of the binary and ternary alloy do not differ very much. Therefore we cannot see any drastic difference between both alloys with respect to the investigated properties explaining the difference in glass-forming ability. But before final conclusions can be drawn we need more experimental data, particularly the magnetic susceptibility of the glassy $Pd_{81}Si_{19}$ alloy and the Hall coefficients in the liquid state. The only obvious difference seems to be the different melting temperatures of both alloys.

Certainly it would be more convincing to see the strong similarities of the glassy and liquid state for other alloys and in a larger concentration range.

Financial support of the "Schweizerische Nationalfonds" is gratefully acknowledged.

REFERENCES

1. B. G. Bagley and F. J. DiSalvo, in: Amorphous Magnetism, H. O. Hooper and A. M. de Graaf ed. (Plenum Press, N. Y. 1973) 143.
2. C. C. Tsuei, in: Amorphous Magnetism, H. O. Hooper and A. M. de Graaf ed. (Plenum Press, N. Y. 1973) 299.
3. P. Duwez and S. C. H. Lin, J. Appl. Phys. $\underline{38}$ (1967) 4096.
4. R. C. Crewdson, unpublished (1966).
5. G. Busch and H.-J. Güntherodt, Solid State Physics $\underline{29}$ (Academic Press, N. Y. 1974) 235.
6. O. Dreirach, R. Evans, H.-J. Güntherodt and H. U. Künzi, J. Phys. F $\underline{2}$ (1972) 709.
7. H.-J. Güntherodt, E. Hauser, H. U. Künzi and R. Müller, Phys. Letters $\underline{54A}$ (1975) 291.
8. K. H. Bennemann, J. Phys. F $\underline{6}$ (1976) 43.
9. R. Dupreee and C. A. Scholl, Z. Physik $\underline{B20}$ (1975) 275.
10. S. R. Nagel, G. B. Fischer, J. Tauc and B. G. Bagley, Phys. Rev. $\underline{B13}$ (1976) 3284.
11. M. Müller et al., Proceedings IMC (Amsterdam) 1976.

DISCUSSION

G. S. Cargill: In reference to your use of liquid metal theory to explain the magnitude of resistivity or the dependence of that on silicon content, I wonder whether an alternate explanation might be given in terms of impurity scattering.

H. J. Güntherodt: It is possible, but one has to realize that we are dealing here with highly concentrated alloys.

J. Wong: Are you dealing here with a single or multiphase system?

H. Güntherodt: I do not know, we have not investigated that aspect of the problem, our aim was primarily to compare the glassy and liquid state.

FERROMAGNETIC AND ANTIFERROMAGNETIC COUPLING IN AMORPHOUS $(Ni_{100-c}Mn_c)_{78}P_{14}B_8$*

A. Amamou[†]

California Institute of Technology

Pasadena, California 91125

ABSTRACT

The magnetic properties of amorphous alloys $(Ni_{100-c}Mn_c)_{78}P_{14}B_8$ with 0.7 at.% ≤ c ≤ 20 at.%, have been investigated for temperatures between 1.7°K and 270°K. Samples were prepared by the splat cooling method; the susceptibilities at zero field and the magnetizations in fields up to 70 kOe have been measured. $Ni_{78}P_{14}B_8$ is paramagnetic and Ni-Mn-P-B alloys exhibit different magnetic characteristics depending on the manganese concentration and the temperature range. At "high temperature" T ≥ 30°K the initial susceptibility has a Curie-Weiss behaviour; all the paramagnetic Curie temperatures θ are equal to zero or positive. The low temperature studies show that three concentration regimes can be determined;
i) for c ≤ 2 at.%, a dilute alloy behaviour is observed. For higher manganese concentrations the magnetization features show the existence of a mixing of ferromagnetic and antiferromagnetic coupling between atoms. ii) For 2 at.% < c ≤ 8 at.% the alloys present spin glass characteristics i.e. a random magnetic coupling occurs between magnetic moments. iii) For 8 at.% < c ≤ 20 at.% the alloys are mictomagnetic and show a trend toward an antiferromagnetic order; irreversible phenomena are observed. In this paper the experimental results are interpreted and discussed in relation with the spin glass and mictomagnetic models.

INTRODUCTION

Amorphous alloys of transition metals (Mn, Fe, Co, Ni) with metalloids (B, C, Si, P) exhibit several types of magnetic behaviours. For alloys containing one transition element magnetic properties are relatively simple to understand; Mn-P-C[1] was reported antiferromagnetic; Fe-P-B[2], Fe-P-C[3] and Co-P-B[4] are ferromagnetic, Ni-P-B[5] is paramagnetic although a trend toward a ferromagnetic transition is observed at high nickel concentration. In an amorphous alloy, when one transition element can be continuously substituted for another, a wide range of varying magnetic behaviours can be obtained. Magnetic properties of an Fe-Mn-P-B-Al[6] alloy exhibit irreversible phenomena which are interpreted as evidence for ferromagnetic-antiferromagnetic "exchange anisotropy". Fe-Mn-P-C[1] exhibits a transition from ferromagnetism to antiferromagnetism when the manganese concentration is increased. Studies on Ni-Co-P-B[4] alloys show the existence of a paramagnetic-ferromagnetic transition which can be understood in the same way as in crystalline concentrated alloys and compounds.

In this paper we study the magnetic properties of $(Ni_{100-c}Mn_c)_{78}P_{14}B_8$ amorphous alloys with $0.7 \le c \le 20$ at.% Mn. In this study we are mainly interested in the magnetic behaviour of an isolated manganese atom and the characteristics of interactions between magnetic atoms when the manganese concentration is increased. We show that for $c \lesssim 2$ at.% Mn, Ni-Mn-P-B has a dilute alloy behaviour; for higher manganese concentrations, the magnetic properties can be understood by the coexistence of ferromagnetic and antiferromagnetic coupling between magnetic atoms, these properties can be compared in some way with those observed in spin glass and mictomagnetic crystalline alloys. In order to do such a comparison and since definitions depend on the authors, let us review briefly what is meant by spin glass and mictomagnetism in crystalline alloys like Cu-Mn[7,8,9], Au-Fe[10], Mo-Fe[11], which are characterized by the existence of long range interactions between magnetic impurities. Such a definition was first proposed in reference 10: at low concentrations of the transition element Mn or Fe, where interactions are negligible the concentration range is called dilute regime. In the spin glass regime the impurty concentrations were such that the magnetic properties are mainly determined by the long range interactions. In the mictomagnetic concentration regime the magnetic properties are related to both long range and short range interactions between magnetic impurities.

EXPERIMENTAL PROCEDURE

Foils of $(Ni_{100-c}Mn_c)_{78}P_{14}B_8$ were obtained by quenching from the liquid state using the "piston and anvil" technique, full details of the alloys preparation can be found in reference 12; the concentrations given in this paper are nominal. The X-ray diffraction spectrum of each foil was checked with a Norelco diffractometer. These spectra showed that in the amorphous state manganese can be substituted for nickel for concentrations up to 20 at. % Mn. The variations of zero field susceptibilities were determined by an induction method using an ac bridge; the investigated temperature range was included between 1.7K and 270K. The magnetization measurements were performed by the Faraday method using an Oxford Instrument magnetometer described in reference 13; the magnetic field was varied between 0 and 70 kOe and the temperature between 1.7 and 270K.

EXPERIMENTAL RESULTS AND DISCUSSION

For manganese concentrations between 2 at. % and 8 at. % the zero field susceptibility S exhibits a cusp at a temperature T_M. This temperature T_M varies linearly with c at a rate of about 1K/at. % Mn (Fig. 1); extrapolation of the T_M vs c curve shows that if an S cusp exists for c ≤ 2 at. % it occurs below the investigated temperature range. For c ≥ 10 at. % no S cusp was detected; this could be due to a large decrease of the cusp magnitude when c is increased, as observed in Cu-Mn alloy.[8]

For all the investigated samples the initial susceptibility χ_o varies according to a Curie-Weiss law $\chi_o = C_{cw}/(T + \theta)$, over a wide temperature range. For instance for c = 20 at.%Mn, deviations from this law occur at T < 30 K and for lower c such deviations are observed at lower temperatures. The paramagnetic Curie temperatures θ are all equal to zero or positive (0 ≤ θ ≤ 6 K). θ increases with c until a concentration of about 8 at.% Mn, then it decreases smoothly with c (Fig. 1). The Curie constant increases linearly for c ≤ 2 at.%, then saturates progressively. The effective moment per manganese atom, determined from C_{cw}, is constant for c ≤ 2 at% Mn, then it decreases from $5.9\mu_B$ to $4.3\mu_B$. Amorphous alloys with c ≥ 5 at. % Mn exhibit a maximum of the initial susceptibility at a temperature close to that of the S cusp, when such a cusp exists; for samples for which no cusp is detected, a rather broad maximum is observed at a temperature which is increasing with c.

At low temperature T ≲ 30 K, the magnetization M(H, T) shows a continuous approach to saturation with increasing magnetic field. However this saturation approach becomes slower when the manganese concentration is increased. At constant field and temperature, M(H, T) as a function of c increases until about 8 at. % Mn, then it decreases smoothly (Fig. 2).

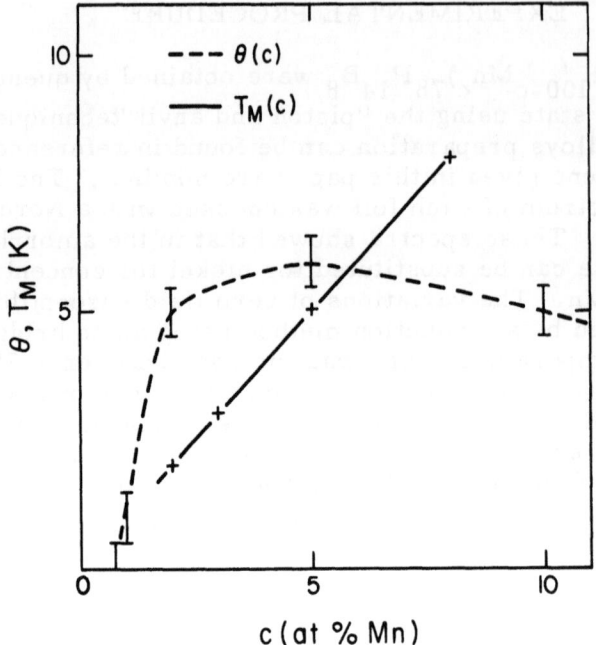

Fig. 1. $(Ni_{100-c}Mn_c)_{78}P_{14}B_8$: variation of the "ordering temperature" T_M and the paramagnetic Curie temperature θ as a function of manganese concentration.

For $c \geq 8$ at.% Mn, irreversible magnetic phenomena are observed below T_M or below the maximum of the susceptibility χ_0. For the sample containing 5 at.% Mn our measurements show that such phenomena exist but their magnitude is very small compared to the total magnetization and could not be determined accurately. At a given temperature, for a sample cooled at zero magnetic field, the first magnetization curve exhibits roughly an S shape; it starts increasing linearly with H until H $\simeq$ 3 kOe then it has a positive curvature and finally becomes concave downward at high field H $\gtrsim$ 7 kOe. At decreasing field, the magnetization curve is concave downward. If the magnetic field is again increased, the magnetization curve is linear in a rather wide field range (H $\lesssim$ 7 kOe) then it is concave downward (Fig. 3). As a result M(H, T) exhibits an hysteresis loop comparable to those observed in antiferromagnetic alloys, after applying a high magnetic field. When the alloy is cooled in a constant magnetic field, the magnetization increases until T_M then it is constant below T_M.

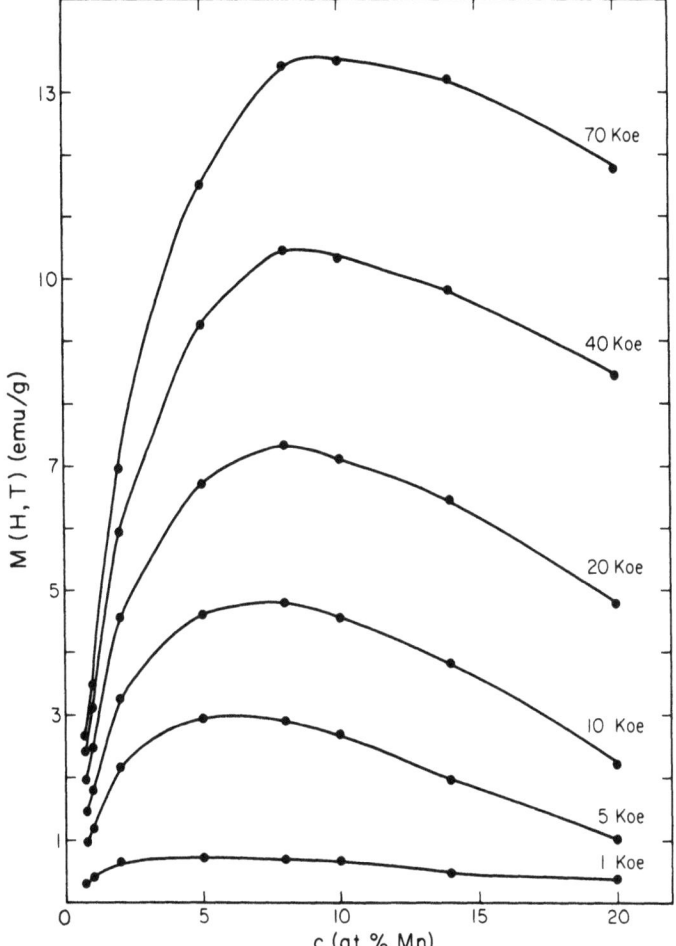

Fig. 2. $(Ni_{100-c}Mn_c)_{78}P_{14}B_8$, T = 1.7K: variation of the magnetization with manganese concentration, at constant field, for alloys cooled at zero field.

The previous experimental results on $(Ni_{100-c}Mn_c)_{78}P_{14}B_8$, in particular the initial susceptibility maxima, the variation of the magnetization as a function of c and H, show the coexistence of a ferromagnetic and an antiferromagnetic coupling between magnetic atoms. The small values of the paramagnetic Curie temperature may be related to the structure of amorphous alloys. It has been previously suggested[14] that the near zero θ values obtained in the amorphous state are due to the fluctuations of the interatomic distances. From our experimental results we can roughly determine three manganese concentration ranges: a dilute regime c ≲ 2 at.% where the manganese atoms behave as

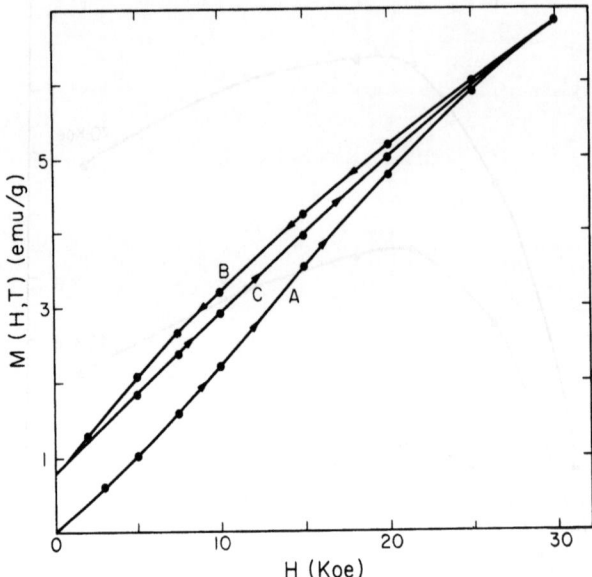

Fig. 3. $(Ni_{80}Mn_{20})_{78}P_{14}B_8$ T = 1.7K alloy cooled at zero field: magnetization M(H, T) vs magnetic field
A: first magnetization curve
B: magnetization at decreasing field
C: "   "   increasing   "

isolated atoms; a spin glass like regime 2 at. % ≲ c ≲ 8 at. % where a random magnetic coupling occurs between magnetic moments and finally a mictomagnetic regime c ≥ 8 at. % where a trend toward an antiferromagnetic order occurs. However let us note that this separation expresses only the fact that a certain type of magnetic characteristics is predominant in a given regime; the transition from one regime to another is not abrupt but progressive, therefore the boundary between them is only approximately defined. In the following we discuss in more detail the magnetic characteristics of each concentration regime.

Dilute regime c ≤ 2 at. % Mn: In this concentration range C_{cw} and M(H, T), at a given temperature and magnetic field, are increasing linearly with manganese concentration; for c = 2 at. % some departure from the dilute behaviour is observed at 1.7°K (Fig. 4); this can be attributed to the occurrence of magnetic interaction below 4.2°K. Previous studies on Mn-Pd-Si[15]

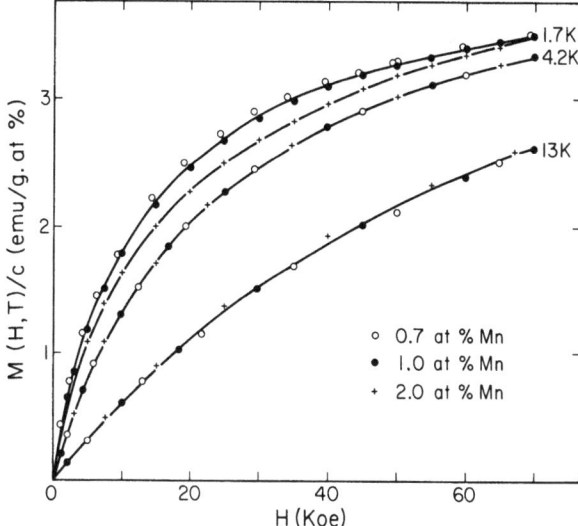

Fig. 4. $(Ni_{100-c}Mn_c)_{78}P_{14}B_8$: magnetization per impurity $M(H,T)/c$ as a function of magnetic field, for low manganese concentration.

amorphous alloys showed a limit of the dilute regime at about 1.5 at.%Mn. The effective moment per manganese atom determined from C_{cw} is $5.9\mu_B$; this value is close to those obtained in Mn-Pd-Si.[15] However at low temperatures the magnetization cannot be described by a single impurity contribution and the moment deduced from the saturation magnetization is only about $3\mu_B$. These features may be attributed to a polarization of nickel atoms surrounding the manganese atoms.

Spin glass like regime 2 at.% ≤ c ≤ 8 at.% Mn: in this concentration range, as mentioned previously, the zero field susceptibility exhibits a cusp at T_M and $M(H,T)$ is increasing with c. Morever for 3 at.% ≤ c ≤ 7 at.% the magnetization at low temperature is varying roughly according to a scaling law (Fig. 5) i.e.: the magnetization per manganese impurity $M(H,T)/c$ is a unique function of the reduced variables H/c and T/c. These experimental features suggest the occurrence of a randomly distributed ferromagnetic and antiferromagnetic coupling between manganese atoms. In crystalline alloys such as Cu-Mn[7,8,9], and Mo-Fe a similar behaviour has been

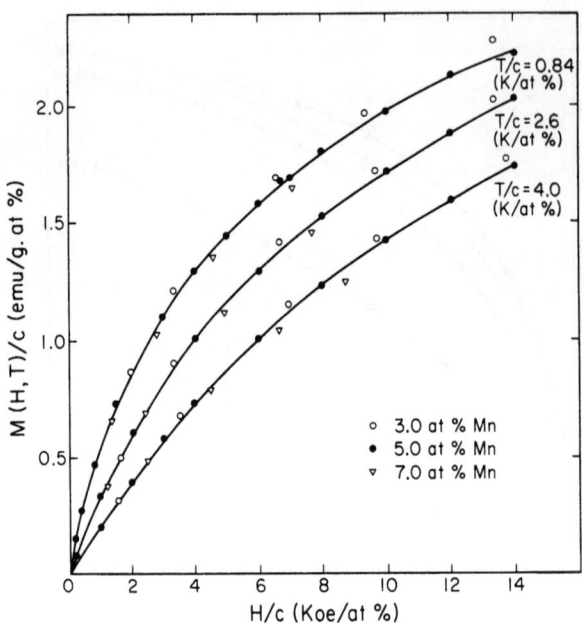

Fig. 5. $(Ni_{100-c}Mn_c)_{78}P_{14}B_8$: magnetization per impurity $M(H,T)/c$ as a function of H/c and T/c (scaling law).

observed, and some theoretical models [16, 17, 18, 19] have been proposed to account for the experimental results. The cusp of the zero field susceptibility is attributed to the occurrence, below T_M, of a magnetic ordering where the impurity moments are frozen in randomly distributed directions; this ordering arises from an oscillating interaction between magnetic moments. Assuming a RKKY interaction [17, 18, 20], it has been shown that the magnetization and the specific heat are varying according to a scaling law. However recent NMR results on CuMn [9] suggest that the interpretation of a cooperative freezing involving all the moments at T_M, is questionable. Such a process would be rather progressive with decreasing temperature, although an important freezing occurs at T_M. On the other hand for Ni-Mn-P-B amorphous alloys if we assume that a transition atom is surrounded by an average of 12 transition atoms, as suggested by structural studies [20], the spin glass properties occur in a concentration range where cluster effects cannot be neglected. As a matter of fact for these concentrations, the probability of having pairs, triplets etc. --- of manganese first neighbors is large. Therefore the magnetic properties of Ni-Mn-P-B cannot be unambiguously related to long range interactions of RKKY type. The origin of ferromagnetic and antiferromagnetic coupling may also be due to a short range effect; it may be related to the existence, for a given manganese atom, of neighboring manganese atoms at various distances; thus the nature of the magnetic coupling may depend on the distance between magnetic moments.

Mictomagnetic regimes c > 8 at.%: In this concentration range, at a given low temperature ($T \lesssim 30°K$) and magnetic field, the magnetization is smoothly decreasing with increasing c. This shows that antiferromagnetic coupling is becoming predominant as the manganese clusters become larger in the alloy. The low field measurements show that, for a sample cooled at zero field, the initial susceptibility deduced from M(H,T), measured at increasing field, is independent from any field previously applied. The high field magnetization shows the persistence of a strong antiferromagnetic coupling between magnetic atoms; for instance at 20 at.% Mn, the average magnetization is about $0.68\mu_B$ per manganese atom. The constant susceptibility and the characteristics of the irreversible effects suggest, as in Au-Fe[10] and Mo-Fe[11] alloys, the formation below the χ_o maximum, of magnetic domains of which resulting moments are interacting. When the alloy is cooled in zero field these resulting moments are progressively frozen in random directions. When a small magnetic field H is applied these directions are not affected; at a higher field the domains are oriented in the H direction. When the alloy is cooled in a magnetic field the resultant moments freeze in a preferential direction which is that of the applied field. A further study of the irreversible effect, especially the remanant magnetizations and the hysteresis loop, should provide more details on the domains and the origin of their coupling.

Acknowledgement: The author wishes to express his appreciation to Professor Pol Duwez for his advice and encouragement throughout this work. Thanks are also due to C. Geremia, S. Kotake, J. Wysocki and Mrs. Bressan for expert technical help.

REFERENCES

*Work supported by the Energy Research Development Agency, Contract No. AT(04-3)-822.

†On leave from Laboratoire de Structure Electronique des Solides, E.R.A. 100, 4, rue Blaise Pascal, 67000 Strasbourg, France.

1. A. K. Sinha: J. of Appl. Phys. 42, 338 (1971).
2. J. Durand: M.M.M. Intermag Conf. Pittsburgh, Pa (June 1976) to be published in IEEE Transaction on Magnetics (1976).
3. C. C. Tsuei, G. Longworth and S. G. H. Lin: Phys. Rev. 170, 603 (1968).
4. A. Amamou: M.M.M. Intermag Conf. Pittsburgh, Pa, (June 1976) to be published in IEEE Transaction on Magnetics (1976).
5. A. Amamou and J. Durand: Comm. on Phys. 1 (7) 191 (1976).
6. R. C. Sherwood, E. M. Gyorgy, H. J. Leamy and H. S. Chen: M.M.M. Intermag Conf. Pittsburgh, Pa. (June 1976).
7. H. Klaus and J. S. Kouvel: Solid State Comm. 17, 1553 (1975).
8. R. W. Tustison and P. A. Beck: to be published in Solid State Comm. (1976).
9. D. E. MacLaughin and H. Alloul: Phys. Rev. Lett. 36, 1158 (1976).
10. J. L. Tholence and R. Tournier: J. de Phys. Paris 35 C4, 229 (1974).
11. A. Amamou, R. Caudron, P. Costa, F. Gautier and B. Loegel: 14th Int. Conf. on Low Temp. Phys. L.T. 14 Otanami, Finland (1975), v. 3. J. of Phys. F to be published.
12. P. L. Maitrepierre: J. Appl. Phys. 40, 4826 (1969).
13. G. Tangonan Ph.D. Thesis Caltech (1975).
14. P. W. Anderson in "Amorphous Magnetism" Proc. of the Int. Symp. on Amorphous Magnetism, Detroit, Michigan (1972). Edt. H. O. Hooper and A. M. deGraaf.
15. R. Hasegawa and C. C. Tsuei: Phys. Rev. B 2, 1631 (1970).
16. A. Blandin: Ph.D. Thesis Paris (1961).
17. W. Marshall: Phys. Rev. 118, 1519(1960).
18. S. F. Edwards and P. W. Anderson: J. of Phys. F.: Metal Phy. 5, 965 (1975).
19. A. I. Larkin and Khmel'nitsskü: Sov. Phys. J.E.T.P. 31, 958 (1970).
20. P. Duwez: "Annual Review of Material Science" 6, 83 (1976).

DISCUSSION

P. W. Anderson: It is not really so much a question for this particular speaker but it appears that we still have not answered the problem whether mictomagnetism is different from spin glass. I would welcome anybody's contribution to this. I do not know of any genuine experiment that has established a clear distinction between the two.

A. Amamou: I do not know of any clear distinction either.

ELECTRONIC AND MAGNETIC PROPERTIES OF AMORPHOUS Fe-P-B ALLOYS.[*]

J. Durand[†] and M. Yung

California Institute of Technology

Pasadena, California 91125

I. ABSTRACT

The ternary diagram of the amorphous phase in the splat-cooled Fe-P-B system was investigated by X-ray diffraction measurements. Alloys were found to be amorphous within the Fe concentration range $75 \leq c_{Fe} \leq 83$ at.%. For a fixed Fe concentration, B substitutes for P over a large scale, the maximum substitution occurring for $c_{Fe} = 80$ at.%. Concentration dependence of the electronic and magnetic properties was systematically studied in alloys within the whole amorphous diagram by means of bulk magnetization $\bar{\mu}$ (4.2°K), Curie temperature T_c and electrical resistivity ρ measurements from 4.2°K up to the amorphous-crystalline transformation. The composition dependence of $\bar{\mu}$ and T_c when B substitutes for P at a constant c_{Fe} suggests the existence of two different short range orders in the amorphous alloys, corresponding to ϵ and ϵ_1 crystal structures in the $Fe_3P_{1-x}B_x$ compounds. The concentration dependence of $\bar{\mu}$ and T_c when P (or B) substitutes for Fe at a constant c_B (or c_P) is explained by a comparison between the amorphous Fe and the fcc Fe. A minimum in the variation of ρ as a function of temperature T occurs at T_m. The temperature dependence of ρ is expressed by a phenomenological law:

$$\rho(T) = \rho_o + A \log T + B T^2 + CT.$$

The concentration dependence of T_m and of the coefficients A and B is discussed in relation with the magnetic and structural properties of the amorphous Fe-P-B alloys.

II. INTRODUCTION

In an earlier report[1] it has been shown that the substitution of B for P in a series of amorphous $Fe_{79}P_{21-y}B_y$ alloys results in an increase of both the Curie temperature T_c and the mean magnetization $\bar{\mu}$ averaged over all atoms in the alloy. Moreover, a change of slope for the variation of T_c and $\bar{\mu}$ as a function of y occurs at a value of y which corresponds fairly well to a change of crystal structure ($\epsilon \to \epsilon_1$) in the ternary $Fe_3P_{1-c}B_c$ compounds.[2] In order to check whether this phenomenon was fortuitous or not, we studied the variation of T_c and $\bar{\mu}$ for as many as possible Fe-P-B alloys available in the amorphous phase. The ternary diagram of the amorphous phase in the splat-cooled Fe-P-B system is presented in section III. The variation of T_c and $\bar{\mu}$ as a function of y when B substitutes for P at a constant c_{Fe} ($75 \leq c_{Fe} \leq 83$ at.%) is discussed in section IV A. On the other hand, it has been found that T_c and $\bar{\mu}$ vary in an opposite way, i.e. T_c increases and $\bar{\mu}$ decreases as a function of x (or y) when P (or B) substitutes for Fe, the content of the other metalloid being kept constant. The previous measurements[1] were made on series of $Fe_{92-x}P_xB_8$ and $Fe_{87-y}P_{13}B_y$ alloys. In this paper, we report the results obtained for all the $Fe_{100-x-y}P_xB_y$ series available in the amorphous diagram (section IV B). Thus, we were able to obtain extrapolated values of T_c and $\bar{\mu}$ for the binary $Fe_{100-x}P_x$ and $Fe_{100-y}B_y$ amorphous systems. These values were found to be consistent with the results obtained for electrodeposited amorphous Fe-P alloys[3] and for some splat-cooled Fe-B alloys. Besides, some resistivity measurements as a function of temperature $\rho(T)$ were undertaken on samples over the whole amorphous phase. We report also (section IV C) some preliminary results on the concentration dependence of the different parameters involved in $\rho(T)$. Details on the experimental procedures may be found in ref. 1.

III. DIAGRAM OF THE AMORPHOUS PHASE

The ternary diagram of the amorphous phase in splat-cooled Fe-P-B system is shown on Fig. 1. As in many amorphous systems obtained by splat-cooling, the most favorable concentration of transition metal is around 80 at.%. We succeeded in obtaining some single-phase amorphous $Fe_{80}B_{20}$ alloys; but alloys with 19 and 21 at.% B exhibited some traces of microcrystalline phases. On the P side, we did not obtain any good amorphous binary alloy with $17 \leq x \leq 21$ at.%. Substitution of B for P is possible to a large extent for $c_{Fe} = 79$, 80 and 81 at.%.

ELECTRONIC AND MAGNETIC PROPERTIES OF Fe-P-B ALLOYS

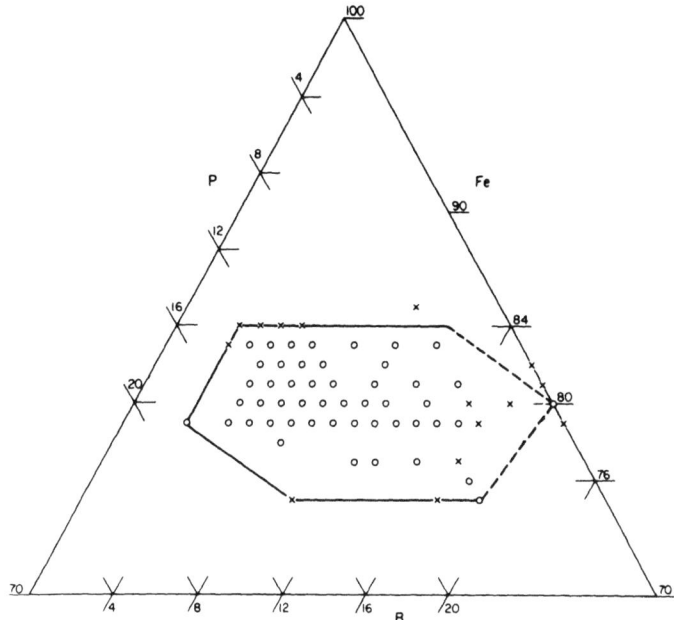

Fig. 1. Ternary diagram of the amorphous phase in splat-cooled Fe-P-B alloys. Alloys with traces of microcrystalline phases (X). Single-phase amorphous (O).

The limit defined by c_{Fe} = 83 at.% seems to be absolute. For $c_{Fe} < 79$ at.%, one still can obtain some single-phase amorphous, but the foils were found to be more and more brittle when c_{Fe} decreases down to 75 at.%, and the efficiency became poor (one amorphous foil out of a set of five). In the 75 at.% line, only the $Fe_{75}P_6B_{19}$ alloy was found to be fully amorphous.

In order to obtain an estimate of the influence of P and B on the stability of the amorphous phase, we compared the temperatures at which the resistivity drops down for samples of $Fe_{79}P_{21-y}B_y$ heated up at a rate of 3°K/min. These so called

Table 1. Critical concentration for the change of slope in the variation of T_c and $\bar{\mu}$ in the Fe-P-B amorphous alloys and in crystalline $Fe_3P_{1-c}B_c$.

	Cryst. Fe_3PB	Fe_{79}	Fe_{80}	Fe_{81}	Fe_{82}	Fe_{83}
T_c	c_{crit} = 0.50	0.57	0.55	0.58	0.55	0.59
$\bar{\mu}$	c_{crit} = 0.68	0.71	0.60	0.63	0.67	0.71

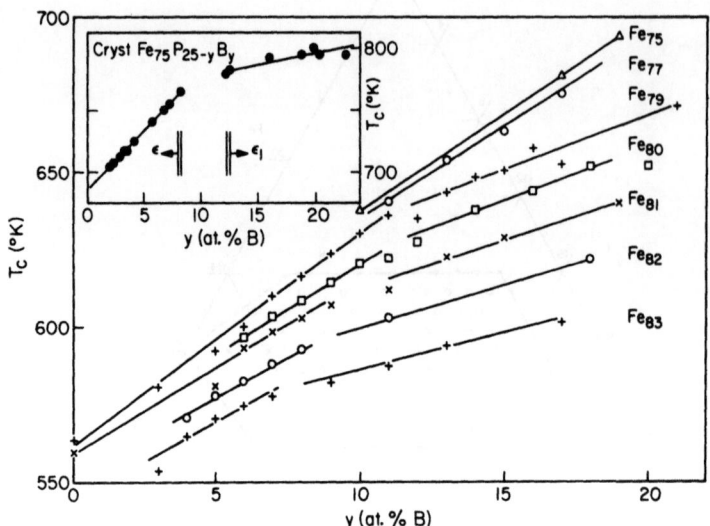

Fig. 2. Curie temperature T_c versus B content y when B substitutes for P in amorphous $Fe_c P_{100-c-y} B_y$ alloys with $75 \leq c_{Fe} \leq 83$ at.%. Insert: T_c versus y for crystalline $Fe_{75} P_{25-y} B_y$ compounds (ref. 5).

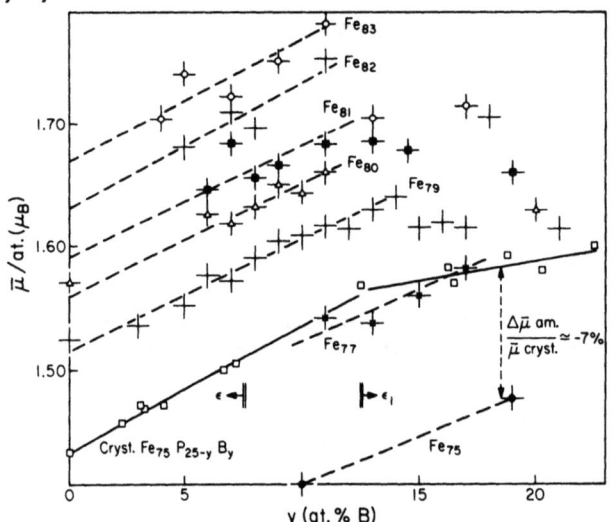

Fig. 3. Mean atomic moment per atom of alloy $\bar{\mu}$/at. as a function of B content y when B substitutes for P in amorphous $Fe_c P_{100-c-y} B_y$ alloys with $75 \leq c_{Fe} \leq 83$ at.% Fe. Values obtained for crystalline $Fe_{75} P_{25-y} B_y$ compounds (ref. 5) are also included for comparison (□).

"crystallization temperatures" defined for the same rate of heating were found to increase proportionately to y up to y = 13 at.% B. The slope is about 3°K/at.%B. For y > 13 at.% B, the "crystallization temperature" starts to decrease. These results are in qualitative agreement with the results obtained by Polk and Chen.[4]

IV. RESULTS AND DISCUSSION

A. Substitution of B for P at a constant c_{Fe}.

The results of our measurements of T_c (Fig. 2) and of $\bar{\mu}$ (Fig. 3) are plotted as a function of the B content y. The results reported before are confirmed. When B substitutes for P, the Fe content being kept constant, the linear variation of T_c as a function of y changes slope at some value of y. Below this concentration, T_c increases at a rate of 5 to 7°K/at.% B while above the critical concentration the slope is smaller and is between 2.5 and 4°K/at.%B. For comparison, we normalize the y critical concentration by writing: $c_{crit} = y_{crit}/x + y$ (see Table 1).

As shown in Fig. 3, the variation of $\bar{\mu}$ as a function of y changes drastically at some value of y. Normalized values of the critical concentration are listed on Table 1. This behavior has to be compared with a similar feature in the crystalline $Fe_{75}P_{25-y}B_y$ alloys.[5] These ternary compounds have a crystal structure $\epsilon - Fe_3P$ (tetragonal of the Ni_3P type) up to y = 8.2 at.% B. For 8.2 < y < 12.3 at.% B, there is a solubility gap. For 12.3 ≤ y ≤ 24.4 at.% B, the compounds crystallize in a new ternary phase called $\epsilon_1 - Fe_3P_{1-c}B_c$ (tetragonal also, but not isomorphous with Ni_3P).[2,5] The variation of the magnetic properties as a function of y were found to be different in the ϵ and ϵ_1 phases. The results of Fruchart et al.[5] are collected for comparison in Fig. 3 for $\bar{\mu}$ and in Fig. 2 (insert) for T_c. In the ϵ phase, T_c and $\bar{\mu}$ increase at a rate of 9.5°K/at.% B and 0.97 μ_B/B at., respectively. In the ϵ_1 phase, the slopes are definitely smaller and rather poorly defined; $d\bar{\mu}/dy$ is estimated to be about 0.30 μ_B/B at. The values of the critical concentration for the change of slope are not accurate because of the solubility gap. But, as in our amorphous alloys, this concentration is a little higher for $\bar{\mu}$ than for T_c (see Table 1). The rigid band model predicts an increase of 2 μ_B/B at., when B substitutes for P, from the difference of s p electrons between P and B. Actually, the experimental values of $d\bar{\mu}/dy$ are much smaller than 2 μ_B in both phases. Moreover, the rigid band model fails to explain that the variations of $\bar{\mu}$ and T_c are crystal structure dependent. This phenomenon was tentatively explained by a contraction of the $Fe_{III}-Fe_{III}$ distances in the ϵ_1 phase.[5]

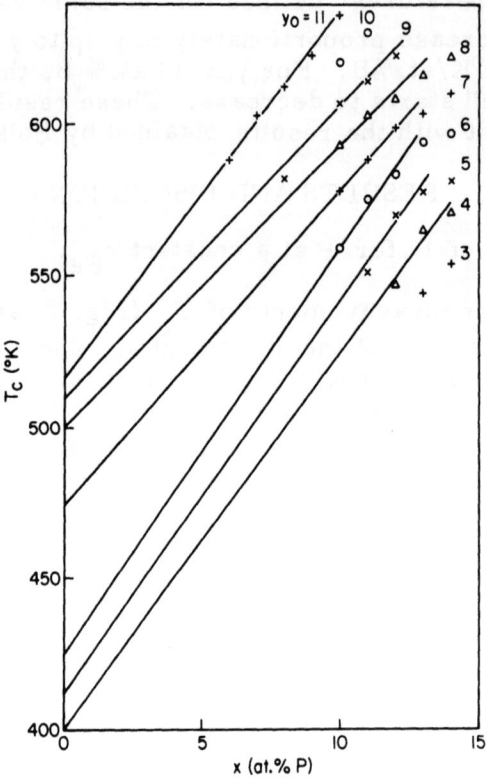

Fig. 4. Curie temperature versus P concentration when P substitutes for Fe, the B content being kept constant, in amorphous $Fe_{100-yo-x}P_xB_{yo}$ alloys, for some values of y_o ($3 \leq y_o \leq 11$ at.% B).

For the low values of y in the amorphous case, the slope of $\bar{\mu}(y)$ ($d\bar{\mu}/dy = 0.90$ μ_B/B at.) is quite similar to that for the ε-phase of the crystalline compounds. For high y values, $\bar{\mu}(y)$ seems to be roughly constant. On the other hand, the critical concentrations are roughly the same in both amorphous and crystalline systems, so that the same physical explanation has to hold in both cases.

Two conclusions may be drawn from our study. First, the magnetic properties of these crystalline compounds and of these amorphous alloys are mainly determined by short-range interactions. The short-range order (SRO) is basically the same in the amorphous alloys and in their crystalline counterpart. It is

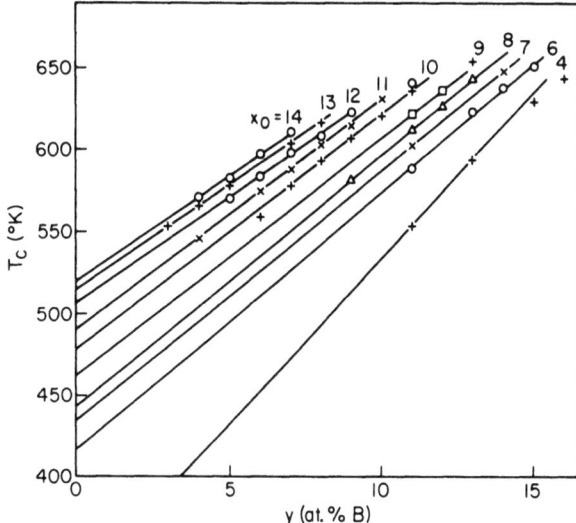

Fig. 5. T_c versus B concentration when B substitutes for Fe, the P content x_o being kept constant, in amorphous $Fe_{100-xo-y}P_{xo}B_y$ alloys, for some values of x_o ($4 \leq x_o \leq 14$ at. %P).

a well established result from X-ray and neutron diffraction measurements[6] that in the amorphous alloys of the transition-metal metalloid type, the coordination numbers (CN) for the metal and metalloid atoms are the same in the amorphous alloy and in the equivalent compound. In our case, the ϵ and ϵ_1 phases have the same CN but differ only by the interatomic distances. Our study suggests that the two different SRO, corresponding to the two ϵ and ϵ_1 phases, differ also by the mean interatomic distances.

A second conclusion is related to the effect of the long-range disorder on the magnetic properties of the amorphous Fe-P-B alloys. This effect is small in any case. It seems to be a little more important in the ϵ_1 region than in the ϵ region where the slopes were found to be about the same in both amorphous and crystalline cases. For the ϵ_1 region, this effect is evaluated by comparison of the values of T_c (694 and 794°K, respectively) and $\bar{\mu}$ (1.48 and 1.59 μ_B per atom, respectively) for the amorphous $Fe_{75}P_6B_{19}$ alloy and the compound of same composition (Fig. 3). The values in the amorphous alloy are lower by 7% for $\bar{\mu}$ and by 12% for T_c. This effect is somewhat larger than that

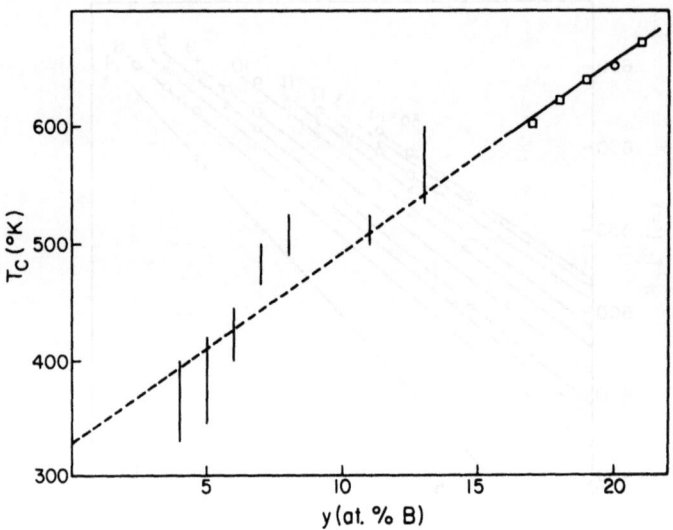

Fig. 6. T_c versus B concentration in amorphous Fe-B alloys. Extrapolated values obtained from Fig. 4 (bars). Values measured on binary splat-cooled Fe-B alloys (□).

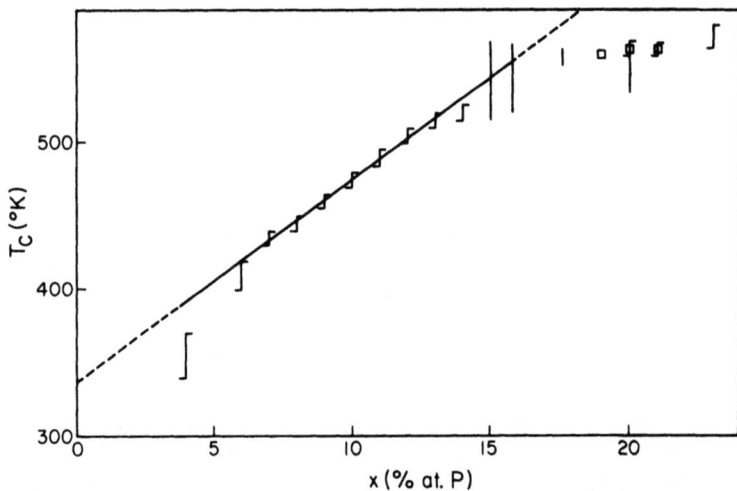

Fig. 7. T_c versus P content in amorphous Fe-P alloys. Values of T_c extrapolated from Fig. 5 (⌐). Experimental values obtained on electrodeposited Fe-P alloys (bars) (ref. 3) and on binary splat-cooled Fe-P alloys (x = 19, 20 and 21 at.% P) (□).

ELECTRONIC AND MAGNETIC PROPERTIES OF Fe-P-B ALLOYS

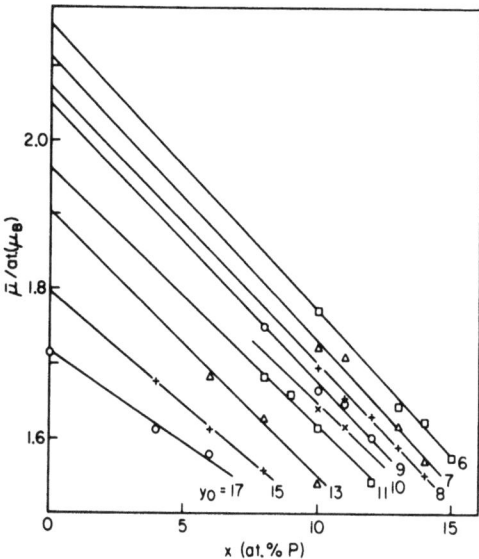

Fig. 8. Mean magnetic moment per atom of alloy as a function of P concentration when P substitutes for Fe, the B content being kept constant in amorphous $Fe_{100-y_0-x}P_x B_{y_0}$ alloys for some values of y_0 ($6 \leq y_0 \leq 17$ at.% B).

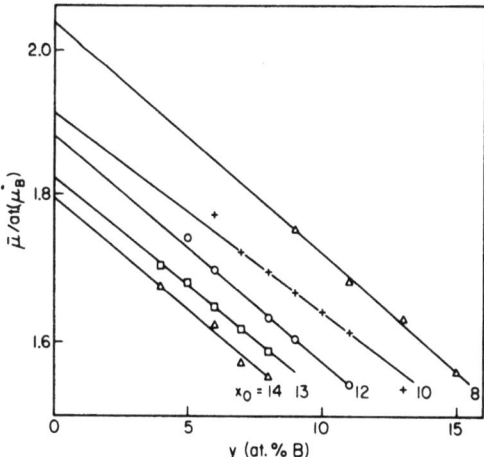

Fig. 9. Mean magnetic moment per atom of alloy as a function of B concentration when B substitutes for Fe, the P content x_0 being kept constant in amorphous $Fe_{100-x_0-y}P_{x_0}B_y$ alloys for some values of x_0 ($8 \leq x_0 \leq 14$ at.% P).

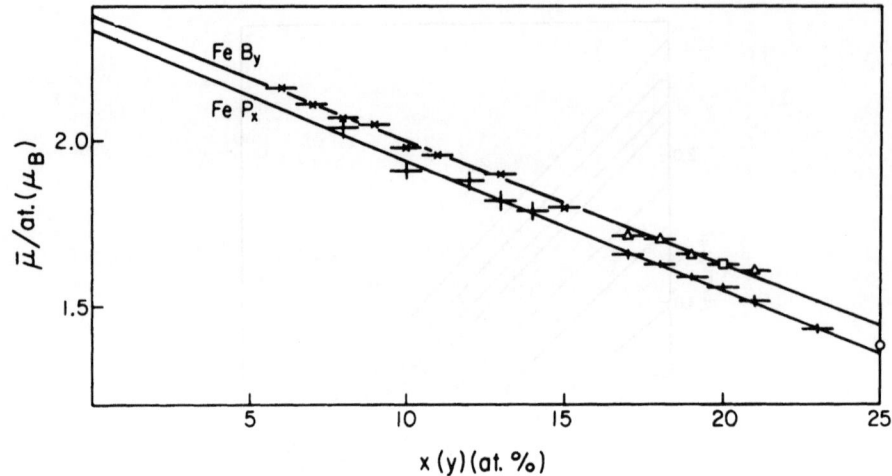

Fig. 10. Mean magnetic moment per atom of alloy as a function of P concentration (+, extrapolated values) and as a function of B content (*, extrapolated values - see text). Experimental values obtained on binary splat-cooled Fe-B alloys (△) are also included for comparison (only $Fe_{80}B_{20}$ being a single-phase amorphous) (□).

observed in Fe-P-C (5% for $\bar{\mu}$ and 4% for T_c).[7] On the other hand, as in amorphous $Gd_{80}Au_{20}$, the effect of amorphousness seems to be more detrimental for T_c than for $\bar{\mu}$.[8]

B. Substitution of P (or B) for Fe, the content of the other metalloid being kept constant.

The results reported in ref. 1 for the $Fe_{92-x}P_xB_8$ and the $Fe_{87-y}P_{13}B_y$ alloys are confirmed: T_c and $\bar{\mu}$ vary in an opposite way for all the sets of concentrations investigated (compare Fig. 4, 5 and Fig. 8, 9). The influence of the substitution of P for Fe on T_c at different constant values of y_o is shown in Fig. 4. The results for the substitution of B for Fe are plotted Fig. 5 for different constant values of x_o. Linear extrapolations to $x = 0$ in Fig. 4 and to $y = 0$ in Fig. 5 allow us to define the values of T_c for the concentrations y_o and x_o in the binary $Fe_{100-y}B_y$ and $Fe_{100-x}P_x$ amorphous systems. These extrapolated values are plotted in Fig. 6 and 7, respectively. In Fig. 6 are shown also

the measured values of T_c for some binary Fe-B alloys
($17 \leq y \leq 21$). Among these latter alloys, only $Fe_{80}B_{20}$ was
found to be a single phase amorphous. The other ones
(y = 17, 18, 19 and 21 at.% B) display some traces of microcrystalline phases, which do not greatly affect the value of T_c for the amorphous phase. This has been verified by measuring T_c on "good" and "bad" amorphous foils of the same composition. The extrapolated values obtained from Fig. 4 are in the form of a straight line defined by the binary Fe-B alloys. This line extrapolates for y = 0 to a value of about 320°K for T_c of "pure amorphous Fe".

In Fig. 7, together with the extrapolated values of T_c obtained from Fig. 5, are plotted the values of T_c measured on electrodeposited amorphous Fe-P alloys[3] and on microcrystalline splat-cooled alloys (x = 19, 20 and 21 at.% P). The variation of T_c as a function of x is fairly linear for $x \leq 16$ at.% P. It extrapolates at x = 0 to a value of T_c = 335°K for "pure amorphous Fe", which is in good agreement with the estimate obtained for the Fe-B system. Let us note that the straight line defined for $x \leq 16$ at.% P would extrapolate for x = 25 at.% P to a value of 685°K very close to the value of T_c for crystalline Fe_3P (716°K).[5] For x > 16 at.%, the variation of T_c as a function of x is considerably smoothened. The change in the variation of T_c occuring for $16 \leq x \leq 17.5$ at.% P may be related to the singularities already observed around the same concentration in the variation of the quadrupolar electrical moment, the isomer shift, the width of the Curie transition and the electrical resistivity parameters.[3,9] These phenomena may be due to some structural modifications occurring around the eutectic composition.

The concentration dependence of $\bar{\mu}$ was analyzed in the same way by taking into account the two regions defined in IV A. The variation of $\bar{\mu}$ as a function of x when P substitutes for Fe is shown in Fig. 8 for different constant values of y_o. Extrapolated values are plotted in Fig. 10. The results for the substitution of B for Fe are shown in Fig. 9 and the extrapolated values for the different sets of x_o are also plotted in Fig. 10. As for T_c, these extrapolated values are in good agreement with the measured values on splat-cooled binary Fe-B alloys, but they disagree with those obtained on electrodeposited Fe-P alloys.[10] From a least squares fit, we obtain for Fe-B and Fe-P the following variations of the mean moment:

$$\text{Fe-B:} \quad \bar{\mu}/at = (2.37 - 3.7\,y)\,\mu_B$$

$$\text{Fe-P:} \quad \bar{\mu}/at = (2.33 - 3.9\,x)\,\mu_B$$

The uncertainty in the values of the slopes and of the moment μ_{Fe} for pure amorphous iron is probably as large as $\pm 0.2 \mu_B$. The agreement obtained for μ_{Fe} for the two extrapolations is surprisingly good. Let us note that the least-squares line for Fe-P in Fig. 10 extrapolates for x = 25 at.% P to a value which is practically the experimental value for Fe_3P.[11] But, keeping in mind the dependence of $\bar{\mu}$ and T_c on the detail of the crystal structure (or of the SRO) as described in IV A and the particular concentration dependence observed[12] for the dilute limit in amorphous Fe-Si, Fe-Ge, Fe-O, one must be cautious about the results obtained from linear extrapolations between 0 and 25 at.% of metalloid. However, let us comment briefly on the values thus determined.

The concentration dependence of T_c as plotted in Fig. 6 and 7 gives for "pure amorphous Fe" $T_c \simeq 300°K$. Actually the trend observed for low concentrations of P and B shows that T_c is probably less than 300°K. This is in good agreement with the extrapolated values obtained from measurements on amorphous Fe-Tb (T_c = 200°K)[13] and RE-Fe (T_c = 270°K) alloys.[14] On the other hand, our value for μ_{Fe} is close to that assumed by Yamauchi et al.[15] (2.6 μ_B). The problem arising from the opposite variation of $\bar{\mu}$ and T_c in amorphous Fe-P and Fe-B may be discussed by different approaches: molecular field model[16], mixture of localized and itinerant magnetism. A comparison between amorphous Fe and crystalline Fe in a fcc environment was suggested in ref. 1. More experimental information is needed for solving this problem (variation of T_c as a function of pressure, hff studies).

It has been a custom to discuss the variation of the magnetic moment in transition-metal metalloid compounds or alloys in terms of electron transfer from metalloid s-p electrons to transition metal d electrons. Assuming that this mechanism is realistic and crystal structure (or SRO) independent, B would donate 1.4 ($\pm$ 0.2) and P, 1.6 ($\pm$ 0.2) electrons to the Fe d bands. These values agree more or less with those already proposed for B[15, 17], but are lower than those suggested for P. They are lower than the estimates in crystalline cases.[18]

C. Resistivity measurements.

Electrical resistivity measurements as a function of temperature $\rho(T)$ were undertaken in an effort to find a better understanding of the variation already observed for the magnetic and structural properties. A resistivity minimum was observed at a temperature T_m for all the samples investigated. The low temperature part ($4.2 \leq T \leq 60°K$) of $\rho(T)$ was found to follow the phenomenological law:

$$\rho(T) = \rho_0 + A \log T + B T^2$$

In order to avoid the uncertainties in the determination of the geometrical factor, we studied the normalized coefficients $a = A/\rho_0$ and $b = B/\rho_0$. We summarize our preliminary results on the concentration dependence of a and b in the series $Fe_{79}P_{21-y}B_y$, $Fe_{87-y}P_{13}B_y$ and $Fe_{92-x}P_xB_8$. The coefficients a and b were calculated by a non-linear least-squares computer program used in ref. 9. When B substitutes for P at c_{Fe} constant, both a and b increase when the B content increases. Both increase at about the same rate, so that T_m defined by $(-a/2b)^{\frac{1}{2}}$ is roughly constant. We hardly see a slight trend to higher values of T_m for the high y values. When B substitutes for Fe at a constant x, there is an increase of T_m proportional to y in the $Fe_{87-y}P_{13}B_y$ series. In the $Fe_{92-x}P_xB_8$ series, the effect due to the substitution of P for Fe seems to be smaller than in the substitution of B for Fe. But the data scattering is too large for any slope to be determined with accuracy. Measurements are needed on more samples before the concentration dependence of a, b and T_m can be discussed quantitatively. But one qualitative conclusion may already be drawn out: the concentration dependence of T_m resembles much more the variation of the Curie temperature than the variation of the magnetic moment. Whether the composition dependence of T_c and T_m would be related to the existence of different magnetic states of Fe in a compact environment or to the variation of the mean interatomic distances is still an open problem. Besides, the physical meaning of the coefficient b is not apparent in our case, although it can be attributed, in other amorphous systems[19], to a spin fluctuations mechanism. On the other hand, if the mechanism giving rise to a resistivity minimum seems to be magnetic in origin in our system, the explanation suggested[20] for this minimum in amorphous Fe-P-C does not look very likely in the amorphous Fe-P-B. From Mössbauer experiments performed on our samples[21], there is no experimental evidence for a distribution of the local field H_0 at Fe sites down to $H_0 = 0$. A complete report of our resistivity measurements will be published elsewhere.

Acknowledgments - It is a pleasure to thank Professor Pol Duwez for his interest and support throughout this work. We would like to thank also A. Bressan, C. Geremia, S. Kotake and J. Wysocki for technical assistance, and M. Ma, Y. Chan and C. Thompson for their help in resistivity measurements. We gratefully acknowledge Dr. J. Logan for the use of his computer program and Professor J. I. Budnick for stimulating discussions.

REFERENCES

*Work supported by the Energy Research Development Agency, Contract No. At(04-3)-822.
†On leave from Laboratoire de Structure Electronique des Solides, E.R.A. 100, 4, rue Blaise Pascal, 67000 Strasbourg, France.

1. J. Durand, Joint MMM-Intermag Conf., Pittsburgh, Pa., June 1976, to be published in IEEE Trans. Mag.
2. S. Rundqvist, Acta Chem. Scand. 16, 1 (1962).
3. J. Logan and E. Sun, J. Non-Cryst. Sol. 20, 285 (1976).
4. D. E. Polk and H. S. Chen, J. Non-Cryst. Sol. 15, 165 (1974).
5. E. Fruchart, A. M. Triquet and R. Fruchart, Ann. Chim. (Paris) 9, 323 (1964).
6. J. Bletry and J. F. Sadoc, J. Phys. F 5, L110 (1975); J. F. Sadoc and J. Dixmier, Conf. Rapid-Quenched Mat. MIT 1975 (Elsevier, Lausanne 1976), vol. 2, p. 187.
7. N. Kazama and M. Kameda, Joint MMM-Intermag Conf., Pittsburgh, Pa., June 1976, to be published.
8. S. J. Poon and J. Durand, to be published.
9. J. Logan and M. Yung, J. Non-Cryst. Sol. 21, 151 (1976).
10. J. Logan and J. Durand, to be published.
11. A. J. P. Meyer and M. C. Cadeville, J. Phys. Soc. Japan 17, Suppl. BI, 223 (1962).
12. W. Felsch, Z. Physik 219, 280 (1969).
13. H. A. Alperin, J. R. Cullen and A. E. Clark, AIP Conf. Proc. 29, 186 (1976).
14. N. Heiman, K. Lee and R. I. Potter, AIP Conf. Proc. 29, 130 (1976).
15. K. Yamauchi and T. Mizoguchi, J. Phys. Soc. Japan 39, 541 (1975).
16. N. Heiman, K. Lee, R. I. Potter and S. Kirkpatrick, J. Appl. Phys. 47, 2634 (1976).
17. R. Hasegawa, R. C. O'Handley, L. E. Tanner, R. Ray and S. Kavesh, Appl. Phys. Lett. 29, 219 (1976); R. Hasegawa, R. C. O'Handley and L. I. Mendelsohn, Joint MMM-Intermag Conf., Pittsburgh, Pa., June 1976, to be published.
18. M. C. Cadeville and E. Daniel, J. Phys. (Paris) 27, 449 (1966).
19. J. Durand, this conference.
20. C. C. Tsuei and H. Lilienthal, Phys. Rev. B 13, 4899 (1976).
21. A. Amamou et al., this conference.

MÖSSBAUER STUDY OF A GLASSY $Fe_{80}B_{20}$ FERROMAGNET

C.-L. Chien and R. Hasegawa

Physics Dept., The Johns Hopkins University
Baltimore, Maryland 21218
Materials Research Center, Allied Chemical Corporation
Morristown, New Jersey 07960

ABSTRACT

The newly synthesized glassy ferromagnetic $Fe_{80}B_{20}$ (METGLAS®
2605) has been studied by ^{57}Fe Mössbauer spectroscopy from 4.2 K
to 1100 K. The easy axis of magnetization of the "as-quenched"
sample lies in the sample plane at 300 K. Substantial tilting of
the easy axis out of the sample plane occurs at temperatures below
200 K. The hyperfine magnetic field decreases at low temperatures
with a dependence of $1 - BT^{3/2} - CT^{5/2}$, where $B = (22\pm1) \times 10^{-6}$
$deg^{-3/2}$ and $C = (1.4\pm0.5) \times 10^{-8}$ $deg^{-5/2}$. The magnetic ordering
temperature T_c of glassy $Fe_{80}B_{20}$ has been determined to be 685±3 K.
Unusual crystallization behavior occurs in $Fe_{80}B_{20}$. At high heating rate, rapid crystallization begins at $T_{CR} \simeq 715$ K. The first
crystalline phase nucleating in the glassy sample is identified as
$Fe_3B(T_c=820$ K$)$, which is a nonequilibrium compound. The Fe_3B phase
then transforms within ∼1 hour to stable $Fe_2B(T_c=1016$ K$)$ and α-Fe
$(T_c=1040$ K$)$.

INTRODUCTION

There has been much interest in ferromagnetic metallic glasses
in recent years.[1] These metals present opportunities for studying
the effects of structural disorder as well as potential for technological applications.[2] Most of the reported metal-metalloid
glassy systems contain two or more metalloids. The newly synthesized glassy alloy $Fe_{80}B_{20}$, however contains only one metalloid.
The binary combination undoubtedly clarifies the role of the metalloid in the present and more complex glassy systems. Some magnetic
and related properties of $Fe_{80}B_{20}$ have been discussed elsewhere.[3-5]

We report in this paper the magnetic properties and the unusual crystallization behavior of the metallic glass as studied microscopically by Mössbauer spectroscopy.

EXPERIMENTAL

Samples of $Fe_{80}B_{20}$ were prepared by rapid quenching from the melt and were in the form of long ribbons 2 mm wide and 40 μm thick. X-ray analysis showed a diffraction pattern typical of glassy materials. A few of the ribbons about 2.5 cm long were placed parallel to each other and used as an absorber. Conventional Fe^{57} Mössbauer spectroscopy in a transmission geometry was employed at temperatures from 4.2 to 1100 K. At high temperatures the heating took place in a vacuum of 10^{-5} torr.

RESULTS AND DISCUSSION

Typical spectra of $Fe_{80}B_{20}$ show broad but well-defined six-line patterns as shown in Fig. 1. The broad lines are caused by hyperfine field distribution commonly observed in glassy magnetic solids. The distribution in the hyperfine field for $Fe_{80}B_{20}$ is narrower than that of other glassy systems (e.g., $Fe_{75}P_{15}C_{10}$,[6] $Fe_{40}Ni_{40}P_{14}B_{6}$[7]) and will be discussed elsewhere.

It is well known[8] that for Fe^{57} hyperfine magnetic spectra the intensities of the six lines have an area ratio of 3:b:1:1:b:3, where b varies from 0 (magnetization axis stays perpendicular to the sample plane) to 4 (magnetization axis stays in the sample plane). An intermediate value of b indicates either a unique magnetization direction other than parallel or perpendicular to the plane, or more likely a spatially averaged value for the entire sample.[9] It is apparent from Fig. 1 that the spectrum of the "as-prepared" sample at 300 K and those at higher temperatures (not shown) show an area ratio close to 3:4:1:1:4:3. This indicates that the magnetization axis stays nearly completely in the ribbon plane. The same conclusion has also been reached by ferromagnetic resonance and scanning electron microscopy measurements.[3]

However, upon cooling the "as-prepared" sample, the area ratio of the spectral lines changes drastically between 200 K and 100 K. The average magnetization direction begins to tilt out of the ribbon plane at about 200 K and tilt back toward the ribbon plane at about 100 K. It should be pointed out that this type of directional change in magnetization is not exclusive to $Fe_{80}B_{20}$. Similar behavior has been observed in $Fe_{40}Ni_{40}P_{14}B_6$ (#2826),[9] $Fe_{80}P_{16}B_1C_3$ (#2615) and $Fe_{29}Ni_{49}P_{14}B_6Si_2$ (#2826B). It seems that this is a general behavior for these samples. The causes of the unexpected directional change are not well-understood at present, although strains which are frozen-in during the rapid quenching process have been suspected.

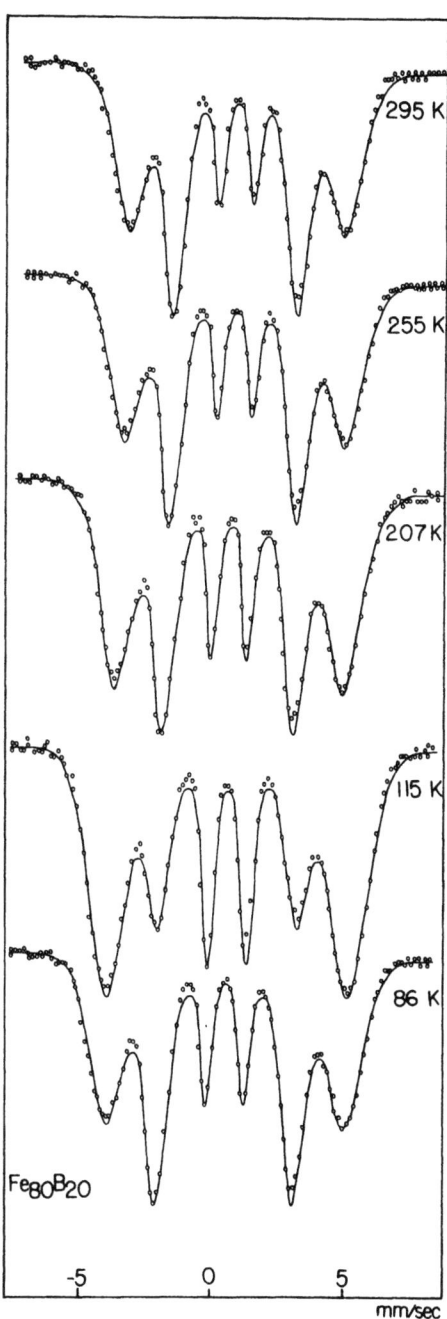

Fig. 1. Mössbauer spectra of $Fe_{80}B_{20}$.

The hyperfine field $H_{eff}(T)$ which is assumed to be proportional to the magnetization $M(T)$ has a temperature dependence as shown in Fig. 2. It decreases with temperature much more rapidly than that of crystalline Fe or Ni (Dashed curve in Fig. 2). Furthermore, at low temperatures the decrease of H_{eff} or $M(T)$ has a dependence of

$$\frac{M(0)-M(T)}{M(0)} = \frac{\Delta M}{M(0)} = B\, T^{3/2} + CT^{5/2} + \ldots \quad (1)$$

$$= B_{3/2}\left(\frac{T}{T_c}\right)^{3/2} + C_{5/2}\left(\frac{T}{T_c}\right)^{5/2} \ldots \quad (2)$$

which is a signature of the excitation of long wavelength spin waves. The dominant $T^{3/2}$ dependence can be clearly seen in Fig. 3. The coefficients B and C in the present case can be conveniently determined by plotting $\Delta M/M(0)\, 1/T^{3/2}$ vs. T as shown in Fig. 4. From the intercept and the slope of the straight line, the values $B = (22\pm1) \times 10^{-6}\ \deg^{3/2}$ and $C = (1.4\pm0.5) \times 10^{-8}\ \deg^{-5/2}$ have been determined. From the measured value of $T_c = 685$ K, $B_{3/2} = 0.4\pm0.05$ and $C_{5/2} = 0.17\pm0.08$ have been determined. These values, compared with those[10] of crystalline Fe and Ni are shown in Table 1. The following points should be made: (1) The values of B and $B_{3/2}$ are <u>anomalously large</u> in comparison with crystalline ferromagnets.

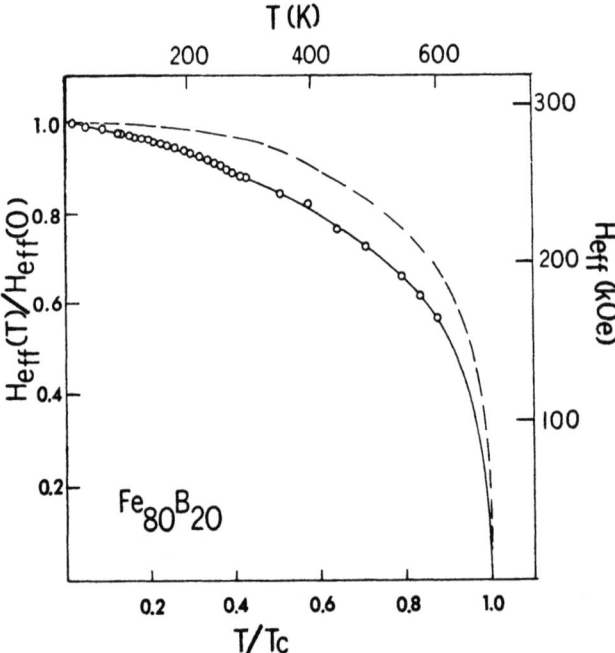

Fig. 2. Temperature dependence of $H_{eff}(T)$ of $Fe_{80}B_{20}$ and crystalline Fe and Ni (dashed curve).

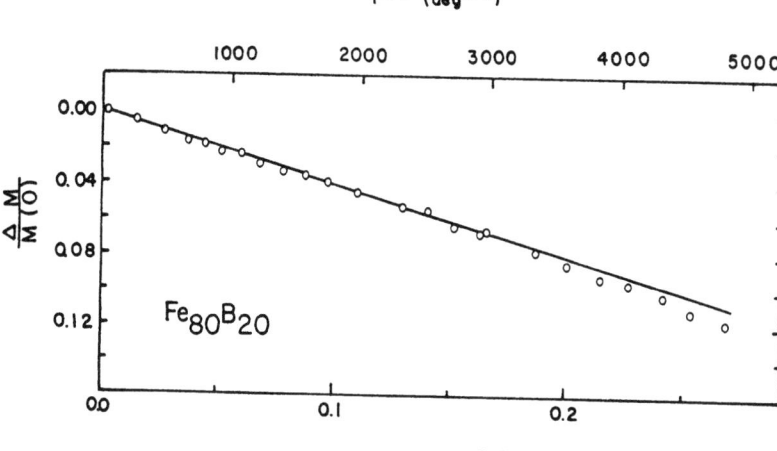

Fig. 3. Fractional decrease of M(T) versus $T^{3/2}$ in $Fe_{80}B_{20}$.

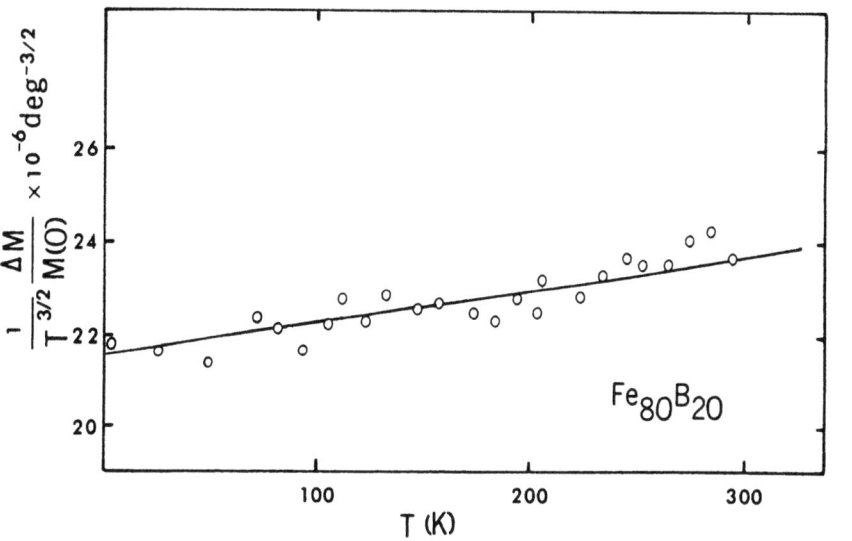

Fig. 4. $\Delta M/M(0)$ $1/T^{3/2}$ versus T for $Fe_{80}B_{20}$.

Similarly, large values have been observed in Co-P,[11] $Fe_{75}P_{15}C_{10}$,[6,12] $Fe_{40}Ni_{40}P_{14}B_6$,[12] $Fe_{80}P_{16}B_1C_3$,[12] etc. The quantity $B_{3/2}$ of glassy ferromagnets is typically 3 to 5 times larger. (2) The values, C and $C_{5/2}$, on the other hand, are not anomalously large. The values for glassy ferromagnets are, in fact, comparable to those of crystalline ferromagnets. (3) In crystalline ferromagnets, the B values as measured by magnetization, Mössbauer and FMR agree very well with those measured by neutron diffraction. In glassy ferromagnets, from the limited number of systems reported so far, neutron diffraction[13,14] always gives a smaller B value than other

TABLE I

Sample Composition, Curie Temperature T_c, and Coefficients B, $B_{3/2}$, C, $C_{5/2}$ for Fe, Ni and Glassy $Fe_{80}B_{20}$

	T_c(K)	$B(10^{-6}$ $deg^{-3/2})$	$C(10^{-8}$ $deg^{-5/2})$	$B_{3/2}$	$C_{5/2}$
Fe[a]	1042	3.4±0.2	0.1±0.1	0.114±0.007	0.04±0.04
Ni[a]	637	7.5±0.2	1.5±0.2	0.117±0.003	0.15±0.02
$Fe_{80}B_{20}$	685±3	22±1	1.4±0.5	0.4±0.05	0.17±0.08

[a]These values are taken from Ref. 10.

techniques. The discrepancy has been as large as a factor of 2.

Although the discrepancy between the neutron diffraction data and others for the values of B is not well understood, the larger values of B generally observed for the glassy ferromagnets can be interpreted as a considerable softening of the exchange interactions in these materials. Furthermore, the relatively small value of C/B ∿ 6 x 10⁻⁴ implies that the exchange interactions damp out considerably beyond the first nearest neighbors.

As the sample of $Fe_{80}B_{20}$ is very slowly (∿2°/min.) heated above room temperature, it begins to show the crystalline phase α-Fe at 615 K, which is below the T_c of $Fe_{80}B_{20}$. This unfortunate situation (often observed in glassy ferromagnetic samples)[12] prevents the conducting of prolonged sutdies near T_c necessary for the exploration of questions such as sharpness of T_c and critical phenomenon. However, T_c has subsequently been determined with a faster heating rate to be 685±3 K.

The crystallization behavior of glassy $Fe_{80}B_{20}$ is both complex and unusual. First of all, 615 K is not necessarily the crystallization temperature T_{CR}.[15] At a heating rate of 20 K/min, rapid crystallization occurred at 715 K. No noticeable change in this temperature was found for a heating rate of 40 K/min. Therefore, T_{CR} is taken to be approximately 715 K. After crystallization, $Fe_{80}B_{20}$ is transformed predominantly into a crystalline phase identified recently as an orthorhombic Fe_3B based on x-ray diffraction and magnetic measurements.[3,5] The Fe_3B phase shows T_c = 820 K in the present measurement, which is close to the previously reported value (T_c = 793 K).[3,5] As discussed in Ref. 16, this compound was originally postulated because of its structural instability. This nature is indeed reflected in the following observation: After heating at high temperatures ($\sim$1000 K) for less than an hour, the Fe_3B phase completely disappears, and Fe_2B (T_c = 1016 K) and α-Fe (T_c = 1040 K) are the resultant equilibrium crystalline phases. The fact that the transformation from Fe_3B into Fe_2B and α-Fe was not observed in the previous study[3,5] may indicate that the stability of the Fe_3B phase is largely controlled by the impurities.

CONCLUSION

The new metallic glass $Fe_{80}B_{20}$ is ferromagnetic below 685 K. Like many other METGLAS samples, the magnetization direction is in the ribbon plane at room temperature, and tilts substantially out of the ribbon plane at low temperatures. The decrease of the magnetization as measured by the hyperfine magnetic field shows characteristics of spin wave excitations. Unusual crystallization behavior is observed above the crystallization temperature of T_{CR} = 715 K. The glassy sample of $Fe_{80}B_{20}$ first transforms predominantly into an unstable Fe_3B phase and then transforms into stable Fe_2B and α-Fe.

ACKNOWLEDGEMENT

The authors would like to thank R. Ray for synthesizing the $Fe_{80}B_{20}$ metallic glass used in the present study.

REFERENCES

1. J. J. Gilman, Physics Today 28, 46 (1975).
2. T. Egami, P. J. Flanders and C. D. Graham, A.I.P. Conf. Proc. 24, 697 (1975).
3. R. Hasegawa, R. C. O'Handley, L. E. Tanner, R. Ray and S. Kavesh, Appl. Phys. Lett. 29, 219 (1976).
4. R. C. O'Handley, L. I. Mendelsohn, R. Hasegawa, R. Ray and S. Kavesh, J. Appl. Phys. 47, 4660 (1976).
5. R. Hasegawa, R. C. O'Handley and L. I. Mendelsohn, A.I.P. Conference Proc. for the Joint MMM-Intermag Conf. (Pittsburgh 1976).
6. C. C. Tsuei and H. Lilienthal, Phys. Rev. B13, 4899 (1976).
7. C.-L. Chien and R. Hasegawa, AIP Conf. Proc. 29, 214 (1976).
8. See e.g., G. K. Wertheim, Mössbauer Effect: Principles and Applications (Academic Press, New York, 1964).
9. C.-L. Chien and R. Hasegawa, J. Appl. Phys. 47, 2234 (1976); R. Hasegawa and C.-L. Chien, Solid State Comm. 18, 913 (1976).
10. B. E. Argyle, S. H. Charap and E. W. Pugh, Phys. Rev. 132, 2051 (1963).
11. R. W. Cochrane and G. S. Cargill, Phys. Rev. Lett. 32, 476 (1974).

12. C.-L. Chien and R. Hasegawa, AIP Conf. Proc. No. 31, 366 (1976).
13. H. A. Mook, N. Wakabayashi and D. Pan, Phys. Rev. Lett. <u>34</u>, 104 (1975).
14. J. D. Axe, L. Passel and C. C. Tsuei, AIP Conf. Proc. <u>24</u>, 119 (1975).
15. P. K. Rastogi and P. Duwez, J. Non-Cryst. Sol. <u>5</u>, 1 (1970).
16. R. Fruchart and A. Michel, <u>Mem. Soc. Chim.</u>, p. 422 (1959).

DISCUSSION

R. A. Levy: What is your value of the saturation field for this alloy?

C. L. Chien: The saturation field for this alloy is at around 290 kOe.

R. A. Levy: From your hyperfine field data were you able to estimate a value for the spin associated with the Fe atom?

C. L. Chien: No, we estimated the value of the spin from the magnetization data only.

K. J. Kim: Did you observe any evidence of oxidation in the sample at high temperatures?

C. L. Chien: It is possible that we might have some sort of an iron oxide in the sample but the Mössbauer data does not give any evidence of that.

THE RESISTIVITY OF AMORPHOUS FERROMAGNETS*

M. Baibich, R.W. Cochrane, W.B. Muir,
and J.O. Strom-Olsen

Eaton Electronics Laboratory, McGill University
P.O. Box 6070, Montreal, Quebec, Canada H3C 3G1

Abstract

We have measured the resistivity of several amorphous ferromagnets from 0.5 to 300K in magnetic fields up to 45 kOe. All the "as deposited" samples show the characteristic resistivity minimum with a log T region on the low temperature side. Moreover, the magnetic field does not affect this temperature dependence. For the alloy with an accessible ordering temperature $\rho(T)$ varies smoothly through T_c. Annealing at temperatures up to the crystallization point does not alter this behaviour. These results reinforce the view that the resistivity anomalies are structural rather than magnetic in origin.

Two unifying features of amorphous metals are their structure, which may be closely modelled by the dense random packing of hard spheres,[1] and their resistivity,[2,3] which at low temperatures varies as $-\ln T$ [3,4] and at higher temperatures shows a minimum. In a recent letter[4] two of the authors have contended that these features are in fact related: that the anomaly in the resistivity is caused by electrons scattering from structural degrees of freedom inherent in the random packing. In our model we considered an ion able to occupy either of two equivalent sites; scattering off this ion lead to a term in the resistivity

$$\rho_3 \sim V^3 \ln(T^2 + \Delta^2)$$

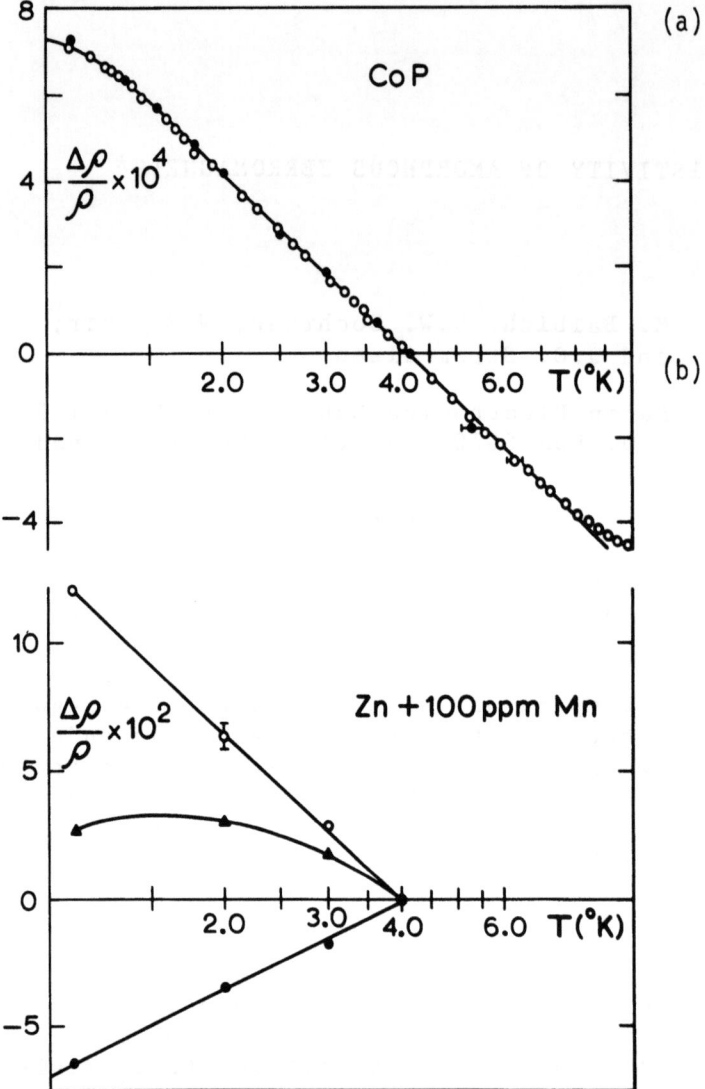

Fig. 1(a) $\frac{\Delta\rho}{\rho} = \frac{\rho(H,T) - \rho(H, 4.2)}{\rho(H, 4.2)}$ as a function of temperature for $Co_{76}P_{24}$. o, H=0 ; •, H=45kOe. The solid line is a fit to equation (1) with Δ = 0.6°K.

(b) $\frac{\Delta\rho}{\rho}$ as a function of temperature for ZnMn o, H=0 ; ▲, H=15kOe ; •, H=45kOe.

where Δ is a characteristic energy splitting and V the coulomb attraction between the ion and the electron. V being negative automatically ensures the correct sign of the anomaly.

In the present paper we wish to present some results of further studies of the resistivity of a number of amorphous ferromagnets including one (Metglas 2826A) whose T_c was low enough (~250K) to allow a very detailed study through the transition. The measurements were made from 1.1K to 300K in fields of up to 45kOe: in zero field we have extended measurements down to 0.5K. The resistivity was measured by a four-terminal AC method[5] with a resolution down to a few parts in 10^6.

Fig. 1(a) shows the behaviour for a typical amorphous metal, $Co_{76}P_{24}$. Between 1.3K and 8K the behaviour is strictly logarithmic, deviating upward at higher temperatures to a minimum at 12K and downward at lower temperatures towards saturation. This behaviour is of course characteristic of the Kondo effect and can be closely fitted to a Haman expression[6] with a T_k of about 3K, even though one would not expect an ordered ferromagnet to show the characteristics of a free spin system. That this is not in fact a Kondo effect, nor indeed any magnetic effect, is shown conclusively by the application of a magnetic field of 45kOe (open circles Fig. 1(a)). Within experimental error there is no effect on the temperature dependence of ρ, only a small temperature independent magnetoresistance, which has been subtracted. On the other hand Fig. 1(b) shows the effect of a field on a bona-fide Kondo system, ZnMn[7] with a T_k of around 1°K. Not only does a field of 45kOe destroy the lnT dependence but the sign of $\frac{\partial \rho}{\partial T}$ is actually reversed, while a change in the temperature dependence of ρ is observed in a field of only 2.5kOe. Such a result is exactly what one would expect: the resistivity must be affected when μH approaches the larger of $k_B T$ or $k_B T_k$. Even for CuFe[8] which has a much higher T_k of 15°K, large changes in the temperature dependence of ρ are visible in 20kOe.

The conclusion that the anomaly is non-magnetic is reinforced when one investigates the behaviour of ρ through the ferromagnetic transition temperature. This is illustrated in Fig. 2 for a sample of Metglas 2826A. It is clear that magnetic ordering has no significant effect on ρ or $\frac{\partial \rho}{\partial T}$.

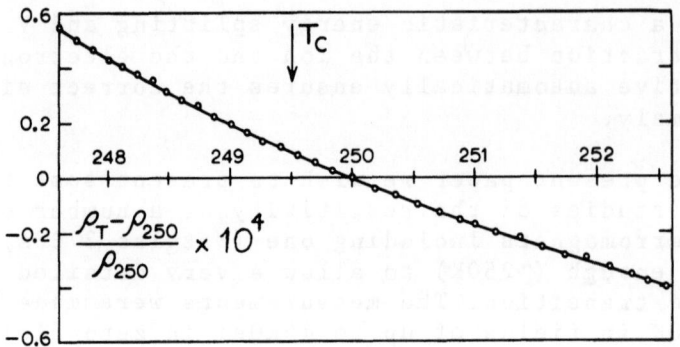

Fig. 2 The resistivity of Allied Chemical Metglas 2826A through the ferromagnetic tranistion temperature

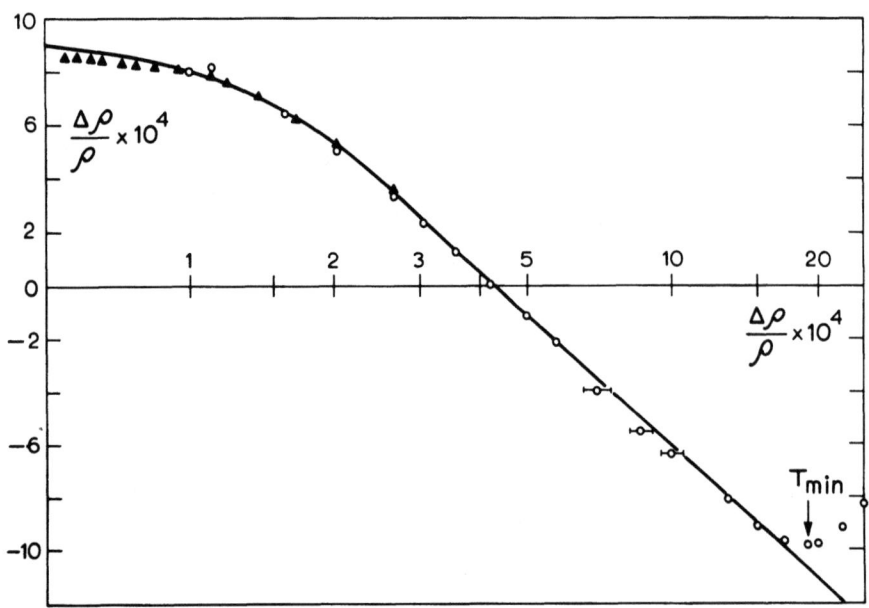

Fig. 3 The resistivity of Metglas 2862, ▲ taken in a He^3 run; ○ taken in a He^4 run. The solid line is a fit to equation (1) with $\Delta = 0.9°K$.

According to our model the resistivity caused by scattering from a single site should vary as $-\ln(T^2+\Delta^2)$. It is therefore interesting to see how closely the data may be described by a single value of Δ. This is shown in Fig.1(a) by the solid line where a value of Δ of 0.6°K fits the data very well below about 8°K. A more stringent test is obtained by going to temperatures below Δ, which we have done for a sample of Metglas 2826 as shown in Fig.3. Once again a single Δ (the solid line) fits the data quite well, while the resistivity shows clear signs of saturation. However above 7°K there is evidence that the resistivity deviates <u>downward</u> from lnT. We have seen such deviations in all our Metglas samples, and they are especially pronounced in those with a high T_{min}. But such behaviour is quite consistent with our model, since at higher temperatures sites with higher values of Δ must come into play, thereby increasing the coefficient of the $\ln(T^2+\Delta^2)$ term.

Finally we should mention that we have examined the effect of annealing at progressively higher temperatures. These results are reported in detail elsewhere[9] but may be summarised by saying that no effect on the resistivity is observable until crystallisation, whereupon the anomaly is reduced by about an order of magnitude, but does not disappear even when annealed at several hundred degrees above the crystallisation temperature.

In conclusion our data clearly support the view that resistivity anomalies are structural rather than magnetic in origin. Further the data are consistent with our simple model of scattering from atomic states.

REFERENCES

[1] G.S. Cargill III in Solid State Physics, ed. H. Ehrenreich et al (Academic Press, N.Y. 1975) 30, 277

[2] S.C.H. Lin, J. Appl. Phys. 40 (1969), 2173

[3] R. Hasegawa and J.A. Dermon, Phys. Lett. 42A (1973), 407

[4] R. W. Cochrane, R. Harris, J.O. Strom-Olsen, and M.J. Zuckermann, Phys. Rev. Lett. 35 (1975), 676

[5] W.B. Muir and J.O. Strom-Olsen, J. Phys. E 9, 163 (1976)

[6] D.R. Haman, Phys. Rev. **158**, 570 (1967)

[7] J.O. Strom-Olsen, unpublished

[8] P. Monod, Phys. Rev. Lett. **19**, 1113 (1967)

[9] R.W. Cochrane and J.O. Strom-Olsen, Proceedings ICM 76, Amsterdam (1976).

* Research supported by the National Research Council of Canada.

DISCUSSION

C. C. Tsuei: Did you examine the structural changes occurring in the Metglas alloy after annealing?

R. W. Cochrane: We have not done any x-ray studies so that we cannot draw specific conclusions on the structural properties per se. After you anneal the Metglas alloy and look at the field dependence of resistivity you can now in fact see a very small field dependence. In that case there may well be coexistence of ferromagnetic and paramagnetic phases.

C. L. Chien: In reference to Metglas 2826A I would like to point out that we have had difficulties characterizing the properties of this alloy through Mössbauer studies, in particular with regards to the precise determination of the value of T_c.

R. W. Cochrane: Would you attribute this to composition fluctuations?

C. L. Chien: We do not know the cause, but the Mössbauer data for this alloy do not seem to follow the general characteristics of the other amorphous Metglas alloys.

R. W. Cochrane: If you compare the slope of the low temperature resistivity, i.e., the log T term, the value of that slope for the Metglas 2826A alloy is within a factor of 2 or 3 similar to what you see in the non-magnetic Ni-P or the ferromagnetic Co-P alloys. So it is not anomalous in its magnitude in that temperature range. I would say that the peculiar behavior of that alloy may be due to the presence of chromium. One seems to get much larger temperature changes when chromium is added.

R. A. Levy: In a way of comment regarding your resistivity measurements of amorphous alloys, we have looked at a series of

Metglas alloys including Metglas 2826, 2826A, 2826B, 2615, 2605 and 2605A in the temperature range of 1.2 to 1000°K. All these samples appear to exhibit a minimum in the resistivity at a temperature depending on the composition of the particular alloy. Below the minimum one observes a logarithmic temperature dependence while way above the minimum a linear dependence sets in. Upon crystallization of these samples the resistivity exhibits a sharp discontinuity followed by a steep increase in the slope of the resistivity.

R. W. Cochrane: In the summary of your results shown in that first slide the location of the resistivity minimum for Metglas 2826A sticks out compared to all the other Metglas alloys; that seems to be consistent with the fact that chromium, which is predominant in this alloy, has a quite noticeable effect on the properties of the system.

MAGNETIC REGIMES IN AMORPHOUS Ni-Fe-P-B ALLOYS[*]

J. Durand[†]

California Institute of Technology

Pasadena, California 91125

I. ABSTRACT

A complete substitution of iron for nickel has been obtained by splat-cooling in amorphous alloys of composition $(Ni_{100-y}Fe_y)_{79}P_{13}B_8$. Results of high field magnetization (up to 70 kOe), ac and dc low field susceptibility, Curie temperature and resistivity measurements over a temperature range of 1.7 to 300°K are reported. The $Ni_{79}P_{13}B_8$ alloy is not ferromagnetic, but the magnetization behavior as a function of field and temperature is typically that of alloys in the critical concentration range for ferromagnetism. The $Fe_{79}P_{13}B_8$ alloy is ferromagnetic with a Curie temperature T_c of 616°K. For y = 1 at.%, the Fe atoms are magnetic. The variation of the moment per Fe atom as a function of y is discussed. When y is increased, the Ni atoms are likely to be polarized progressively and the moment per Ni atom would be roughly constant for y ≥ 30 at.%. Various magnetic regimes were defined as a function of the Fe content. The value of T_c reaches a maximum for y ≃ 90 at.%, and extrapolates to zero for y ≃ 7 at.%. Alloys within the range 1 ≤ y ≤ 10 at.% did not exhibit well-defined Curie transition, but sharp maxima in low field susceptibility measurements were observed at T_M. The value of T_M is proportional to y for 1 ≤ y ≤ 4 at.%, like in classical spin-glass regimes. For 4 < y ≤ 10 at.%, the variation of T_M as a function of y implies a more complicated type of magnetic ordering (mictomagnetism or superparamagnetism). Homogeneous ferromagnetic ordering emerges only for y > 10 at.%. Results of resistivity measurements are discussed in relation to the magnetic properties of different regimes in the magnetic phase diagram.

II. INTRODUCTION

Recent studies on crystalline concentrated alloys and compounds have emphasized the complex nature of the transition from paramagnetic to ferromagnetic state.[1] Most of the systems investigated undergo various intermediate magnetic regimes before reaching a long-range homogeneous ferromagnetic ordering. Experimentally, the main characteristics of these regimes are fairly well defined, although the physical mechanisms involved in each regime are not always clearly understood.[2] For example, in the classical cases, like Au-Fe, Mo-Fe....[3,4], one can distinguish a number of progressive steps of magnetic ordering between the Kondo concentration and the critical concentration for the onset of ferromagnetism: spin-glass regime (scaling laws, ordering temperature varying linearly with concentrations, $T^{3/2}$ dependence of the resistivity...), mictomagnetic regime (strong magnetic short-range order, irreversible processes, thermomagnetic historic effects, displaced hysteresis curves...), superparamagnetic regime (different ordering temperatures as determined by local or by bulk magnetic measurements...). In such systems, there is some evidence of a mixture of local antiferromagnetism and ferromagnetism attributed to long-range interactions of the RKKY type.[5] On the other hand, such phenomena as susceptibility maxima, irreversible processes...were observed in completely different systems where the dilute impurity has no localized moment (Ni in Cu, Fe in V...)[6,7], but, at higher concentrations, the magnetic moment resides in large polarization clouds. An interpretation of the magnetic "anomalies" in this case was suggested in terms of a random orientation of the anisotropy axes of the spin clusters.[6] Since the same phenomena may occur in physically different contexts, some authors suggested to give all the intermediate magnetic regimes the same name: mictomagnetic[1] or spin-glass regime.[8]

The onset of ferromagnetic order was already investigated by local and bulk magnetic measurements in some amorphous systems, like Pd-Si, Fe-Pd-P, Cu-Zr...[9-14], and the complexity of the para-ferromagnetic transition in amorphous systems was pointed out: analogy between amorphous Fe-Pd-P and crystalline Au-Fe systems suggested by Sharon and Tsuei[11], some irreversible phenomena mentioned by Hasegawa[9], susceptibility maxima in amorphous Pd-Si alloys with Fe and Co...[15] Recent progress in the understanding of the intermediate regimes in the crystalline case encouraged us to undertake a detail study of the electrical and magnetic properties at low (less than 500 Oe) and high (up to

70 kOe) fields in amorphous (Ni-Fe)-P-B alloys. The $(Ni, Fe)_{79}P_{13}B_8$ alloys seemed to be a good candidate for such an investigation, since a complete substitution of Fe for Ni is possible in the amorphous phase.

III. EXPERIMENTAL PROCEDURES

The samples were prepared by splat-cooling from the melt. Each foil was checked by a Norelco X-ray diffractometer. Concentrations are nominal. Magnetic ordering temperatures were observed using a standard ac inductance bridge. Magnetization measurements were made between 1.7 and 290°K in fields ranging from 0.1 to 70 kOe using the Faraday method. Electrical resistivity as a function of temperature was measured using a standard four probe technique. More detail on the experimental procedures may be found in ref. 16.

IV. RESULTS AND DISCUSSION

The magnetic properties of the $Ni_{79}P_{13}B_8$ alloys have been discussed previously.[17] The initial susceptibility exhibits a temperature independent term of about 2.10^{-6} emu/g and a small

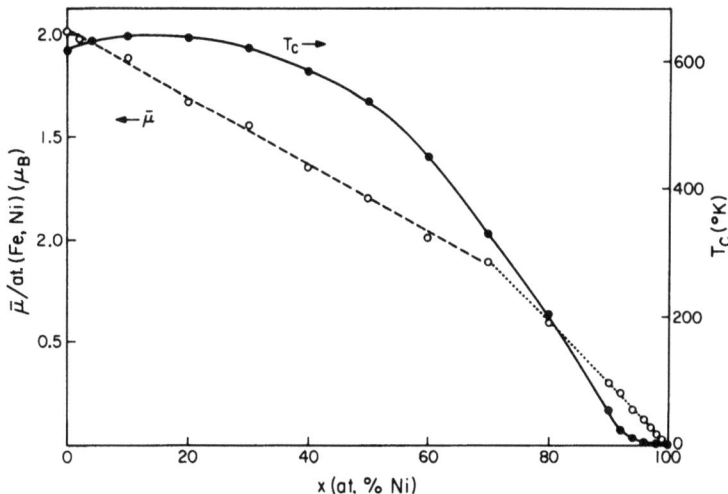

Fig. 1. Variation of the Curie temperature T_c and of the mean magnetic moment per transition metal atom $\bar{\mu}$/at. (Fe, Ni) as a function of the Ni content x in amorphous $(Fe_{100-x}Ni_x)_{79}P_{13}B_8$ alloys.

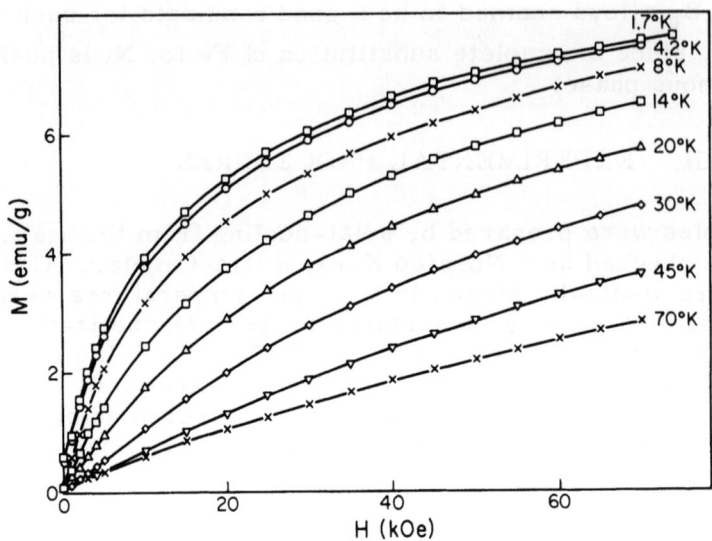

Fig. 2. Magnetization versus applied field at low temperature in amorphous $(Ni_{98}Fe_2)_{79}P_{13}B_8$ alloy.

magnetic contribution which is strongly temperature dependent ($\sigma_0(0°K) = 0.78$ emu/g). From a Brillouin function analysis, the magnetic part was found to arise from polarization clouds of Ni. These clouds are antiferromagnetically coupled and they have two different sizes: $\mu_1 = 5\mu_B$ and $\mu_2 \simeq 9\mu_B$, with characteristic temperatures $\theta_1 = 1.0°K$ and $0 < \theta_2 < 1°K$. The contribution of the $Ni_{79}P_{13}B_8$ matrix will be subtracted from the magnetization data of the $(Ni_{100-y}Fe_y)_{79}P_{13}B_8$ alloys for $1 \le y \le 10$ at.%. For $y > 10$, this contribution was found negligible. The variation of the Curie temperature T_c and of the mean magnetic moment per transition metal atom at $4.2°K$ μ/at.(Fe, Ni) as a function of the Ni content $x = 100-y$ is shown on Fig. 1. We first comment on the low Fe content part of the magnetic phase diagram (section A); then we discuss the magnetic properties for the ferromagnetic region (section B); finally, we summarize the results of electrical resistivity measurements performed over the whole concentration range.

A. $1 \le y \le 10$. Intermediate Magnetic Regimes.

The magnetic parameters of the alloys in the concentration range $1 \le y \le 10$ are listed in Table 1. Fig. 2 shows some low temperature isotherms $M = f(H)$ for a sample with 2 at.% Fe.

Table 1. Magnetic properties of amorphous $(Ni_{100-y}Fe_y)_{79}P_{13}B_8$ alloys ($1 \leq y \leq 10$ at.%).

y (at.% Fe)	T_M (°K)	θ (°K)	$\sigma_o^{(a)}$ (emu/g)	$C_{cw}^{(a)}$ (10^{-4} °K. cgs)	μ/Fe at.$^{(a)}$ (μ_B)
1	1.80	30	2.39	6.0	2.77
2	3.35	45	4.87	15.1	2.83
3	5.45	56	7.06	20.0	2.73
4	7.35	85	10.70	29.0	3.10
5	9.85	—	—	—	—
6	12.2	121	14.27	52.2	2.76
7	15.6	—	—	—	—
8	23.5	133	21.15	60.0	3.06
10	54	152	25.70	78.0	2.98

(a) Values obtained after correcting for the $Ni_{79}P_{13}B_8$ matrix contribution.

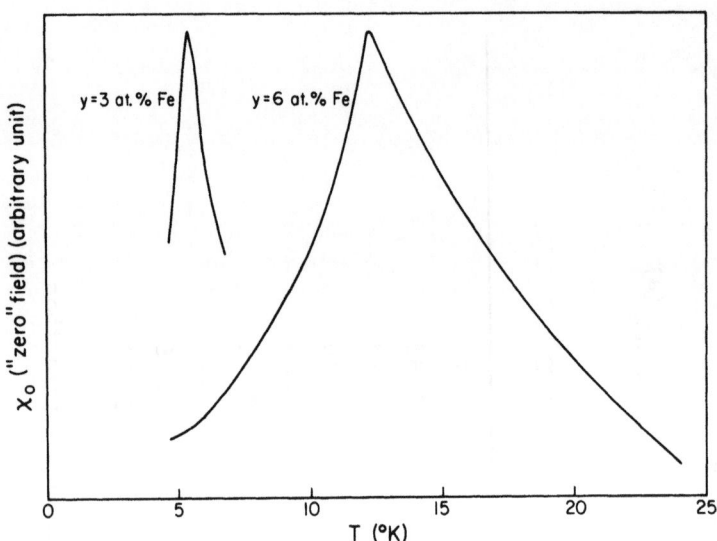

Fig. 3. Magnetic susceptibility (in arbitrary unit) as a function of temperature in "zero" field for y = 3 and 6 at.% Fe in amorphous $(Ni_{100-y}Fe_y)_{79}P_{13}B_8$ alloys.

The saturation becomes progressively easier when y increases. Even for 1 at.% Fe, the magnetization data at low temperature cannot be fitted to a unique Brillouin function. The isolated impurity regime in this system could be reached only for a much lower concentration of Fe. Our results were analyzed in the classical way used for crystalline concentrated alloys.[18,19] (see Table 1). Assuming no polarization of the Ni atoms (a possible polarization will be discussed in the following section) the Fe atoms would carry a moment of about $3\mu_B$, roughly constant over this concentration range. This moment is not carried by individual Fe atoms, but by some magnetic polarization clouds. The average size μ^* of these magnetic clusters is fairly constant and equal to 12-13 μ_B, and their concentration c^* increases with the Fe content.

The paramagnetic Curie temperature θ, defined by the variation at high temperature of the initial susceptibility $\chi_o(T) = C_{cw}/(T-\theta)$, increases monotonically as a function of y. The large differences between θ and the ordering temperature T_M are a measure of the inhomogeneous short-range ferromagnetic interactions. The value of $(\theta-T_M)/T_M$ decreases from a value of about 15 for y = 1 down to 1.8 for y = 10. This variation characterizes the critical range for the onset of long-range ferromagnetic order.

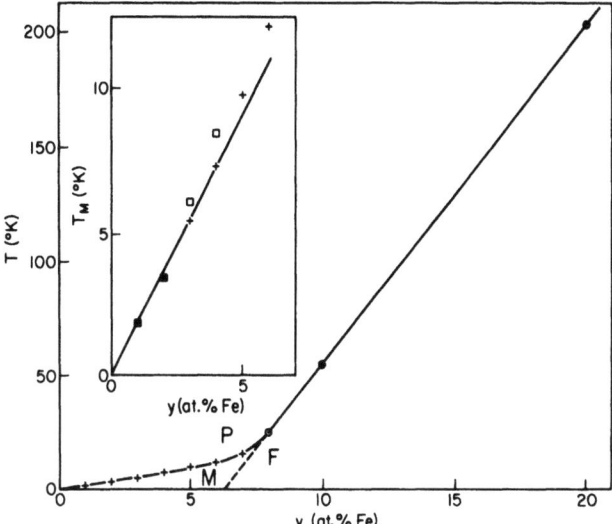

Fig. 4. Ordering temperatures (see text) as a function of Fe concentration in amorphous $(Ni_{100-y}Fe_y)_{79}P_{13}B_8$ alloys.
Insert: variation of T_M as a function of y (y ≤ 6 at.% Fe) in $(Ni_{100-y}Fe_y)_{79}P_{13}B_8$ (+) and in
$(Ni_{100-y}Fe_y)_{80}P_{12}B_8$ (□) amorphous alloys.

Measurements of the initial susceptibility at "zero" field (ac bridge) exhibit sharp peaks at temperature T_M (Fig.3). These peaks are rounded off when the susceptibility is measured at low dc fields (H < 500 Oe) as observed in crystalline cases.[20] A detail study of the concentration dependence of T_M (Fig. 4) and of the magnetic behavior of the different samples allows us to distinguish three different intermediate regions:

1. 1 ≤ y ≤ 4. T_M is proportional to y (insert of Fig. 4), like in Au-Fe in the spin-glass regime (y < 1 at.%) in the sense of ref. 3. The analogy is only formal, since in our case the scaling laws are not obeyed and there is no evidence for a substantial long-range RKKY coupling. In this region, the Fe-Ni interactions are negligible. This point was verified by measuring T_M for y = 1 to 4 at.% Fe in a $Ni_{80}P_{12}B_8$ matrix (insert of Fig. 4). We found exactly the same values for y = 1 and 2, but the departure from linearity starts at y = 3 when the alloy is richer in Ni.

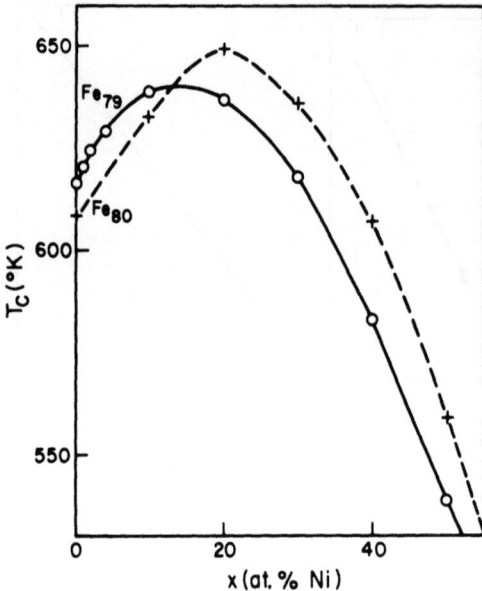

Fig. 5. Curie temperature as a function of Ni content in amorphous $(Fe_{100-x}Ni_x)_{79}P_{13}B_8$ (0) and $(Fe_{100-x}Ni_x)_{80}P_{12}B_8$ (+) alloys.

2. $4 \leqslant y < 8$. T_M as a function of y departs slightly from linearity. The Fe-Ni interactions become visible. Looking at the Arrott plots ($M^2 = f(H/M)$), one cannot define any Curie temperature, even for the low field part (about 100 Oe) of the curve.

3. $8 \leqslant y \leqslant 10$. For these alloys, a Curie temperature may be defined from the low values of H in the Arrott plots, and this T_c agrees fairly well with the values of the susceptibility maxima. On the other hand, a drastic change is observed at y = 8 on the variation of T_M as a function of y, and the values of T_M for y = 8, 10 and 20 increase linearly as a function of y. But

the observed temperature variation of the initial susceptibility is certainly not characteristic of a good ferromagnetic transition, as compared with the 20 at.% Fe sample where no peak is observed. This region resembles that called superparamagnetic in the Fe-Cr, Fe-Mo, Au-Fe systems. It is expected for our amorphous alloys that the ordering temperature observed by local measurements (Mössbauer, neutrons) will not coincide with the values of T_M. This problem is under investigation. Some measurements are planned also for samples with y between 10 and 20 at.% Fe to determine the "real" critical concentration for the onset of long-range ferromagnetic order.

In conclusion, we found, between the paramagnetic and the ferromagnetic states, some intermediate regions, whose characteristics resemble those defined in the Au-Fe system, although the mechanisms involved in the two cases are likely to be different.

B. $y \geq 20$. Ferromagnetic Regime.

1. Variation of the Curie temperature (Fig. 1, 4, 5). At the beginning of the ferromagnetic regime (Fig. 4), T_c increases with an initial slope of about 15°K/at.% Fe. The curve of T_c as a function of y is rounded off and T_c reaches a maximum value of 639°K for $y = 90$ at.%. The T_c decreases down to 616°K for $Fe_{79}P_{13}B_8$. Over the same concentration range, the magnetic moment $\bar{\mu}$ was found to decrease continuously from its value in $Fe_{79}P_{13}B_8$ (Fig. 1). When a small amount of Ni is alloyed with bcc Fe, T_c goes also through a maximum, but the variation of $\bar{\mu}$ follows that of T_c.[21] An opposite variation of $\bar{\mu}$ and T_c was observed in dilute Fe-Cr, Fe-V and Fe-Ti alloys in the bcc phase.[22] As far as we know, this anomaly has never been explained in a satisfactory way. We think that this effect has to be correlated to the crossing variation of $\bar{\mu}$ and T_c observed in amorphous Fe-P and Fe-B alloys.[16, 23] When Ni is added to a matrix having a lower T_c (amorphous $Fe_{80}P_{12}B_8$ - Fig. 5, or crystalline Fe_2P - ref. 24), the amplitude of the peak is increased and the Ni concentration for the maximum is shifted to a higher value.

2. Variation of the magnetic moment. The variation of the mean magnetic moment per transition metal atom $\bar{\mu}$/at. (Fe, Ni) as a function of x is shown on Fig. 1. The curve is divided into two regions. In each region, the variation is linear. According to a least squares fit:

$$0 \leq x \leq 70 : \bar{\mu}/\text{at. (Fe, Ni)} = (2.02 - 1.64x) \, \mu_B \tag{1}$$

$$70 \leq x \leq 99 : \bar{\mu}/\text{at. (Fe, Ni)} = 2.99 (1 - x) \, \mu_B \tag{2}$$

We first comment on equation (1), which is valid down to the dilute limit (2 at.% Ni in $Fe_{79}P_{13}B_8$), so that we can discuss the electronic structure of the $Fe_{79}P_{13}B_8$ alloy and of the transition metal impurities in this matrix. Second, we estimate the individual moments on the atoms of Fe and Ni.

2. 1- Electronic structure of $Fe_{79}P_{13}B_8$ and of the dilute impurities of transition metals in this matrix.

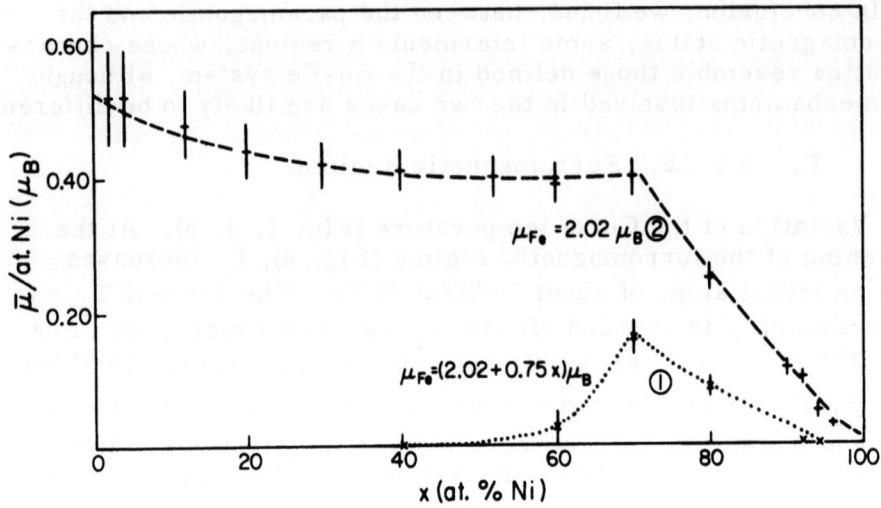

Fig. 6. Estimate of the magnetic moment per Ni atom as a function of Ni content in amorphous $(Fe_{100-x}Ni_x)_{79}P_{13}B_8$ alloys. Hypothesis 1: $\mu_{Fe} = (2.02 + 0.75\,x)\,\mu_B$. Hypothesis 2: $\mu_{Fe} = \text{constant} = 2.02\,\mu_B$.

As observed before[25], equation (1) gives for $\bar{\mu}$ a gradient of $-0.82\,\mu_B/e/a$, which is very close to the value of $-1\,\mu_B/e/a$ predicted by the Slater-Pauling curve. But, the Slater-Pauling approach based on a rigid band model fails to account for the behavior of $\bar{\mu}$ for impurities like Mn, Cr, V in the same matrix. The variation of $\bar{\mu}$ as a function of x is found to be linear from x = 2 to x = 70 at.% Ni (no impurity pair effect). So, this result, together with the values of the initial slopes $d\bar{\mu}/dc$ for Co, Mn, Cr, V[26] in $Fe_{79}P_{13}B_8$ may easily be interpreted in terms of electronic structure.[27] Let us call Z the impurity excess charge as compared with Fe, and Z_σ the displaced charge by one impurity in the bands of spin σ (↑ or ↓). From the Friedel's sum rule, $Z_\uparrow$ and $Z_\downarrow$ are related to Z by:

$$Z = Z_\uparrow + Z_\downarrow \tag{3}$$

The variation of the mean magnetic moment $\bar{\mu}$ (c) in the dilute limit is given by: (neglecting the spin-orbit coupling)

$$\bar{\mu}(c) = \mu_{Fe} + c\,\mu_B (Z_\uparrow - Z_\downarrow) \qquad (4)$$

From equations (3) and (4), the experimental values of $d\bar{\mu}/dc$ allow us to calculate the screening in each spin sub-band of $Fe_{79}P_{13}B_8$ when alloyed with impurities of the first transition series.

Table 2 - Displaced charge by one impurity in the bands ↑ or ↓ of $Fe_{79}P_{13}B_8$.

Impurity		V	Cr	Mn	Co	Ni
Displaced	$Z_\uparrow$	-4.4	-4	-2.6	+0.1	+0.2
charge	$Z_\downarrow$	1.4	2	1.6	0.9	1.8

Some interesting conclusions may be drawn from this analysis. First, the $Fe_{79}P_{13}B_8$ matrix is a strong ferromagnet. For Co and Ni in $Fe_{79}P_{13}B_8$, the screening in the ↑ bands is about zero. So, the rigid band model (and hence the Slater-Pauling curve) seems to be roughly valid for these impurities and the variations of $\bar{\mu}(c)$ may be approximated by:

$$\bar{\mu}(c) = \mu_{Fe} - cZ\mu_B \qquad (5)$$

The fact that the screening in the ↑ bands is very small for Co and Ni implies that the $d_\uparrow$ bands of the matrix are filled up (small density of states at the Fermi level, like in the crystalline Ni and Co metals). This conclusion may be justified as follows. Neglecting the s-d hybridization, the total displaced charge in a completely filled $d_\uparrow$ band is zero. Moreover, in a tight-binding description of the d-bands, the local density in each cell remains the same (and equal to 5) if all the states are occupied.[28] Thus, the band structure of amorphous $Fe_{79}P_{13}B_8$ is basically the same as that of compounds like Fe_2B[29], Fe_3P[30], Fe_3C[31], as suggested before.[32]

On the other hand, the negative values of $Z_\uparrow$ for Mn, Cr, V in $Fe_{79}P_{13}B_8$ are a classical result obtained for the same impurities in a strong ferromagnet. According to the Friedel's model[27], repulsive potentials are large enough to repell d bound states from the $d_\uparrow$ bands. These d bound states are hybridized with the s bands through the impurity potential. The occupation $n_{d\ell}$ of the corresponding virtual bound states (v.b.s.) is related to the total displaced charge by $Z_\uparrow = n_{d\ell} - 5$. Thus, the v.b.s. are half-emptied for Mn and almost emptied for V in $Fe_{79}P_{13}B_8$.

2.2 - Estimate of the invidual moments μ_{Fe} and μ_{Ni}. We try to give a consistent picture of equations (1) and (2) by means of reasonable assumptions on the respective variation of μ_{Fe} and μ_{Ni}. We assume that the variation of μ_{Fe} and μ_{Ni} in the amorphous

(Ni, Fe) PB alloys has to be roughly the same as that observed by neutron diffraction[33] and obtained by calculation[34] in crystalline Fe-Ni alloys. The fact that $Ni_{79}P_{13}B_8$ is "nearly magnetic" (like crystalline Pd) suggests also an anology with crystalline Pd-Fe alloys. Thus, we make several hypotheses on the variation of $\bar{\mu}_{Fe}$ and μ_{Ni} as Crangle did for the Pd-Fe system.[35] First, if μ_{Ni} remains constant and equal to zero, as in $Ni_{79}P_{13}B_8$, μ_{Fe} would increase slightly from 2.8 at y = 1 up to $3\mu_B$ at y = 30, and then decrease down to $2.02\mu_B$ for $Fe_{79}P_{13}B_8$. Such a variation of μ_{Fe} is unlikely as compared with the case of crystalline Fe-Ni.[33,34] Second, if μ_{Fe} increases with x according to:

$\mu_{Fe} = (2.02 + 0.75\ x)\mu_B$, then the value of μ_{Ni} would increase from y = 1 to y = 30 at.% Fe up to a maximum of $0.15\mu_B$; for y > 30, μ_{Ni} would be zero (hypothesis 1 on Fig. 6). Such a variation of μ_{Ni} is unlikely according to ref. 33, 34. Finally, we assume that μ_{Fe} remains constant and equal to $2.02\mu_B$. Then μ_{Ni} increases when y increases from 1 to 30 and remains roughly constant and equal to $0.40\ \mu_B$ for y ≥ 30 (hypothesis 2 on Fig. 6). Such a picture seems to be most likely. Thus, the Ni atoms would be progressively polarized when the Fe content increases like Pd in the crystalline Pd-Fe system.

C. Resistivity Measurements.

The temperature dependence of the electrical resistivity $\rho(T)$ for all the samples may be expressed at low temperature (4.2 < T < 60°K) by:

$$\rho(T) = \rho_o + A \log T + B T^2 \qquad (6)$$

At higher temperature (T > 100°K), $\rho(T)$ becomes proportional to T. The determination of the residual resistivity ρ_o is very inaccurate because of the difficulty in measuring the geometrical factor of the foils. This problem is avoided by using the normalized coefficients: $a = A/\rho_o$ and $b = B/\rho_o$. The coefficients a and b were calculated by a non-linear least-squares computer program used in ref. 36. The coefficient b displays a well-pronounced maximum between 10 and 20 at.% Fe, i.e. around the critical concentration for ferromagnetism. Similar dependence on concentration was found for many concentrated alloys in the critical region: Ni based alloys with Pd[37], Rh[38], Cr, V, Ru, Mo[19], and Fe based alloys with Cr[39], Ru.[40] This behavior is commonly explained by a spin fluctuation scattering especially important in this region of strong local enhancement. The coefficient a and the temperature of the resistivity minimum have a maximum value for samples with 4 to 6 at.% Fe. The fact that this maximum occurs in a region where the magnetic inhomogeneities are particularly important may suggest a magnetic origin for the resistivity minimum. Magnetoresistivity measurements are planned for clarifying this point.

Acknowledgments - It is a pleasure to thank Professor Pol Duwez for his interest and support throughout this work. Thanks also are due to A. Bressan, C. Geremia, S. Kotake and J. Wysocki for technical assistance, and to M. Yung, Y. Chan, and C. Thompson for their help in resistivity measurements. Interesting conversations with Professors P. A. Beck, H. Claus and J. S. Kouvel are gratefully acknowledged. We had many stimulating discussions with Dr. A. Amamou.

REFERENCES

*Work supported by the Energy Research Development Agency, Contract No. AT(04-3)-822.
†On leave from Laboratoire de Structure Electronique des Solides, E.R.A. 100, 4, rue Blaise Pascal, 67000 Strasbourg, France.

1. See the recent reviews of P. A. Beck, Met. Trans. $\underline{2}$, 2015 (1971); R. Tournier, L.T.13, Boulder, Colorado, 1972, edit. K. D. Timmerhaus, W. J. O'Sullivan and E. F. Hammel, (Plenum, New York 1974), vol. 2, p. 257; J. A. Mydosh, AIP Conf. Proc. $\underline{24}$, 131 (1974); J. P. Perrier and J. L. Tholence, J. Phys. (Paris) $\underline{35}$, C4, 163 (1974).
2. See for example N. Rivier and K. Adkins, Amorphous Magnetism, edit. H. O. Hooper and A. M. de Graaf (Plenum, New York 1973), p. 215; S. F. Edwards and P. W. Anderson, J. Phys. F $\underline{5}$, 966 (1975).
3. J. L. Tholence and R. Tournier, J. Phys. (Paris), $\underline{35}$, C4, 229 (1974).
4. A. Amamou, R. Caudron, P. Costa, J. M. Friedt, F. Gautier and B. Loegel, L. T. 14, Otaniemi, Finland, 1975, edit. M. Krusius and M. Vuorio, (North Holland, Amsterdam, (1975), vol. 3, p. 254, and to be published in J. Phys. F.
5. J. Souletie and R. Tournier, J. Low Temp. Phys. $\underline{1}$, 95 (1969) and ref. therein.
6. H. Claus, Phys. Lett. $\underline{51A}$, 283 (1975); Phys. Rev. Lett. $\underline{34}$, 26 (1975).
7. R. Kuentzler and J. P. Kappler, J. Less Com. Met. $\underline{47}$, 203 (1976).
8. B. V. Sarkissian and B. R. Coles, Commun. Phys. $\underline{1}$, 17 (1976).
9. R. Hasegawa and C. C. Tsuei, Phys. Rev. B$\underline{3}$, 214 (1971).
10. R. Hasegawa, J. Phys. Chem. Solids $\underline{32}$, 2487 (1971); J. Appl. Phys. $\underline{41}$, 4096 (1970).

11. T. E. Sharon and C. C. Tsuei, Sol.State Commun. 9, 1923 (1971); Phys. Rev. B. 5, 1047 (1972).
12. F. R. Szofran, G. R. Gruzalski, J. W. Weymouth, D. J. Sellmyer and B. C. Giessen, Phys. Rev. B 14, 2160 (1976), and AIP Conf. Proc. 18, 282 (1974).
13. P. Duhaj, J. Sitek, M. Prejsa and P. Butoin, Phys. Stat. Sol. (a) 35, 223 (1976).
14. A. Amamou, Joint MMM. Intermag Conf. Pittsburgh 1976 (to be published in IEEE Trans. Mag.)
15. A. Zentko, P. Duhaj, L. Potocki, T. Tima and J. Bansky, Phys. Stat. Sol(a) 31, K41 (1975).
16. J. Durand, Joint MMM. Intermag Conf. Pittsburgh 1976 (to be published in IEEE Trans. Mag.)
17. A. Amamou and J. Durand, Commun. Phys. 1, 191 (1976).
18. W. C. Muellner and J. S. Kouvel, Phys. Rev. B 11, 4552 (1975) and ref. therein.
19. A. Amamou, F. Gautier and B. Loegel, J. Phys. F 5, 1342 (1975).
20. V. Cannella and J. A. Mydosh, Phys. Rev. B 6, 4220 (1972).
21. J. Crangle and G. C. Hallam, Proc. Roy. Soc. 272, 119 (1963).
22. S. J. Stoelinga, A. J. T. Grimberg, R. Gersdorf and G. de Vries, J. Phys. (Paris) 32, C1, 330 (1971).
23. J. Durand and M. Yung, this conference.
24. R. Fruchart, A. Roger and J. P. Senateur, J. Appl. Phys. 40, 1250 (1969).
25. T. Mizoguchi. K. Yamaushi and H. Miyajima, in Amorphous Magnetism, edit. H. O. Hooper and A. M. de Graaf, (Plenum, New York 1973), p. 325.
26. J. Durand, to be published.
27. J. Friedel, Nuovo Cimento 7, Sup. 287 (1958).
28. F. Gautier, Ann. Phys. (Paris) 8, 284 and sq. (1973-74).
29. M. C. Cadeville and E. Daniel, J. Phys. (Paris) 27, 449 (1966).
30. A. Blanc, E. Fruchart and R. Fruchart, Ann. Chim. 2, 251 (1967).
31. A. Rouault and R. Fruchart, Ann. Chim. 5, 335 (1970).
32. J. Durand and M. F. Lapierre, J. Phys. F 6, 1185 (1976).
33. M. F. Collins and J. B. Forsyth, Phil. Mag. 8, 401 (1963).
34. H. Hasegawa and J. Kanamori, J. Phys. Soc. Japan 31, 382 (1971); id. 33, 1599 (1972).
35. J. Crangle, Phil. Mag. 5, 335 (1960).
36. J. Logan and M. Yung, J. Non-Cryst. Sol. 21, 151 (1976).
37. A. Tari and B. R. Coles. J. Phys. F 1, L69 (1971).
38. R. W. Houghton, M. P. Sarachik and J. S. Kouvel, Sol. State Commun. 10, 369 (1972).
39. B. Loegel, J. Phys. F 5, 497 (1975).
40. B. V. B. Sarkissian and B. R. Coles, J. Less.Com. Met. 43, 83 (1975).

RESISTIVITY OF METGLAS ALLOYS FROM 1.5 K to 800 K*

J. A. Rayne, Carnegie-Mellon University
Pittsburgh, Pennsylvania 15213

R. A. Levy, Rensselaer Polytechnic Institute
Troy, New York 12181

There is considerable interest in amorphous metal systems, particularly those of the Metglas[†] type. In this paper, we report resistivity measurements from 1.5 K to 800 K on a number of the latter alloys. The effects of recrystallisation on the resistivity have also been investigated.

The specimens used in this work were in the form of uniform ribbons kindly supplied by the Allied Chemical Corporation, Morristown, N.J. Nominal compositions of the alloys are given in Table 1. Resistivity measurements were made by an automated four-probe technique, using a PDP-11 minicomputer, at typical intervals of 1 K at low temperatures and 4 K above room temperature. In all cases the sample temperature was maintained electronically to an accuracy of better than 0.1 K using a germanium thermometer below 80 K, a platinum resistance thermometer between 80 K and 300 K and an iron-constantan thermocouple above 300 K.

Figure 1 shows the resistivity of Metglas 2826 up to 800 K in both the amorphous and recrystallised states. There is a clear minimum in the resistivity for the amorphous state at a temperature T_m approximately equal to 26 K. From Figure 2 it can be seen that below T_m there is an approximately logarithmic temperature dependence of the resistivity, normalised to the value ρ_{min} at the minimum. Except for Metglas 2605A all alloys exhibit similar behaviour, the temperature T_m and magnitude of the minimum depending sensitively on composition, as can be seen both from Table 1 and Figures 2 and 3. This sensitivity is particularly evident from the latter, which shows that both T_m and the minimum for Metglas 2826A are approximately an order of magnitude larger than those observed for the other alloys. Table 1 shows that there is no correspondence between T_m and the Curie temperature T_c, which in a number of cases also differ by an order of magnitude.

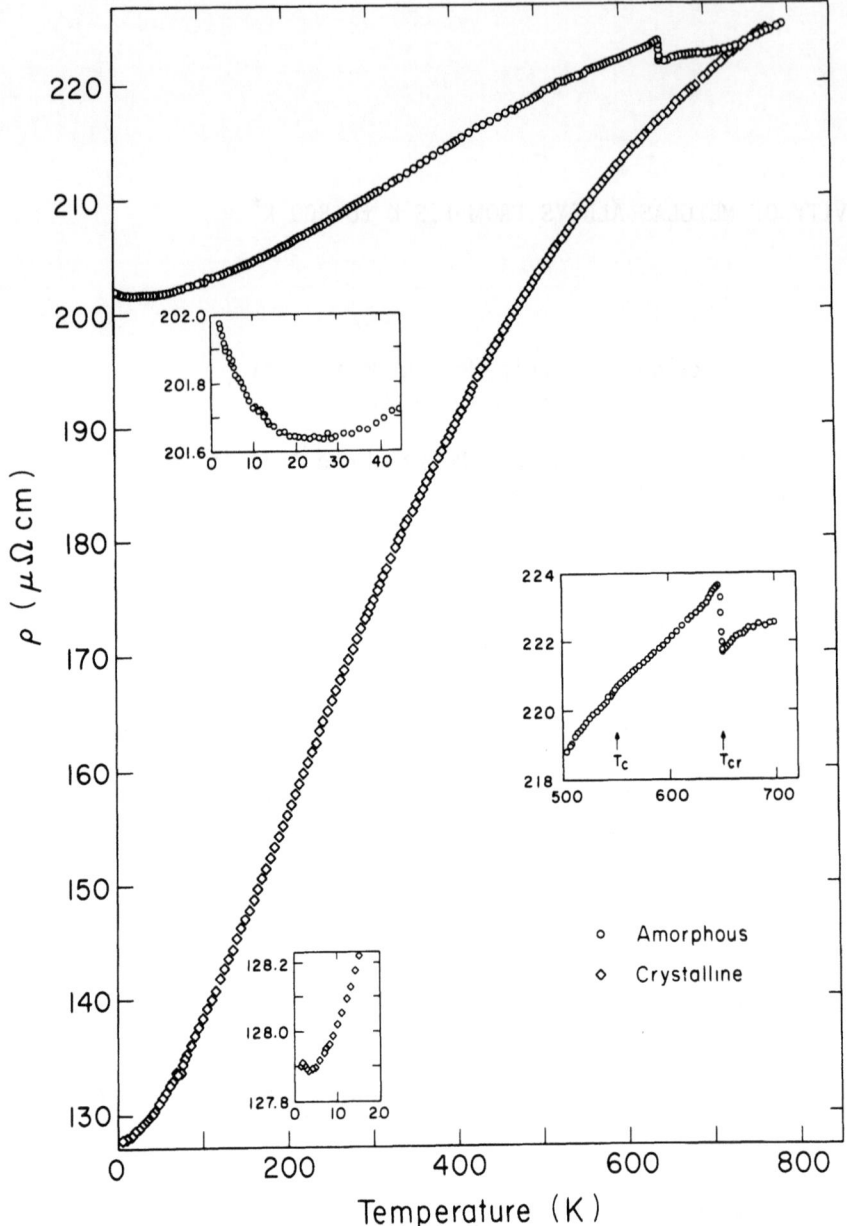

Figure 1. Temperature dependence of resistivity for Metglas 2826 in amorphous and crystalline states. The insets show the behaviour near the resistance minima and in the vicinity of the crystallisation temperature T_{cr} on expanded scales. Specimen recrystallised by annealing one hour at 800 K.

RESISTIVITY OF METGLAS ALLOYS FROM 1.5 K TO 800 K

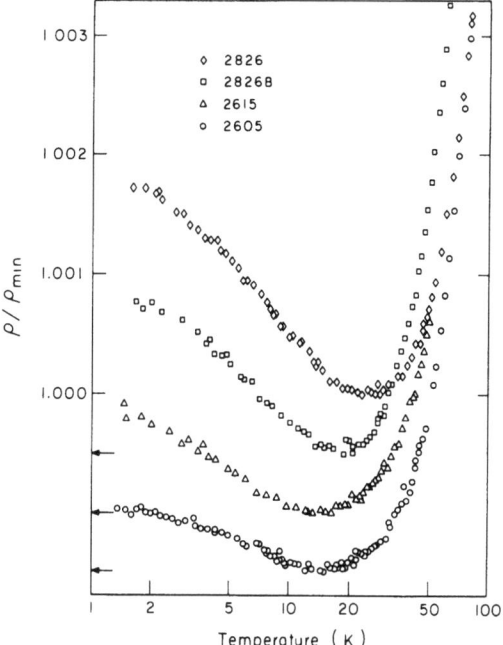

Figure 2. Resistance minima for Metglas alloys. Arrows show scale shift for each specimen.

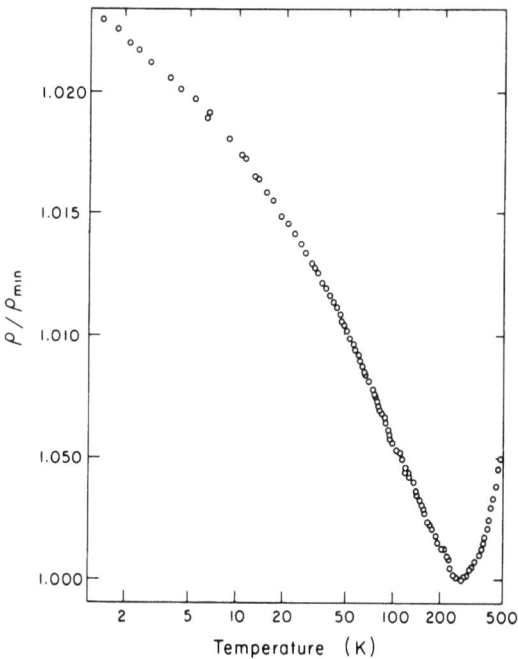

Figure 3. Normalised resistivity ρ/ρ_{min} versus temperature for Metglas 2826A in the amorphous state.

Table 1

Summary of Resistance Data for Metglas Alloys

Alloy[a]	T_c[b] (K)	T_m (K)	ρ_{300} (μΩcm)	$\rho_{4.2}/\rho_{300}$	$\rho_{1.5}/\rho_{min}$
2826	537	26	211±3	0.9617	1.0019
2826 A	250	270	183±3	1.0195	1.0216
2826 B	408	17	173±3	0.9479	1.0013
2615	565	12	168±3	0.9676	1.0009
2605	647	14	139±3	0.9596	1.0005
2605 A	595	80(7)[c]	163±3	0.9916	1.0003

[a]Nominal compositions in atom percent: 2826: Fe-40, Ni-40, P-14, B-6; 2826A: Fe-32, Ni-35, Cr-15, P-12, B-6; 2826B: Fe-29, Ni-49, P-14, B-6, Si-2; 2615: Fe-80, P-16, C-3, B-1; 2605: Fe-80, B-20; 2605A: Fe-78, Mo-2, B-20.

[b]See reference 7.

[c]Complex minimum

The existence of a resistance minimum and an approximate logarithmic temperature dependence below the minimum has been previously reported in a number of amorphous metal systems.[1,2] A probable explanation of this temperature dependence has recently been given by Cochrane et al.[2] in terms of a simple tunnelling model. The tunnelling involves states corresponding to low energy degrees of freedom, which presumably result from the indeterminacy of the local atomic configurations in amorphous alloys. From third-order perturbation theory it can be shown that there is a corresponding contribution to the resistivity

$$\Delta\rho_1 \sim - A \ln [(k_B T)^2 + \Delta^2] \quad , \tag{1}$$

where Δ is a measure of the energy separation between tunnelling eigenstates and A is a constant depending on the number of scattering sites. The crucial negative sign is a consequence of the attractive interaction between the ions and the conduction electrons. Clearly the present data are consistent with the saturation behaviour at low temperatures predicted by equation (1), the relevant values of Δ being of the order of 1 K. Further evidence for the correctness of this model will be presented later in the paper.

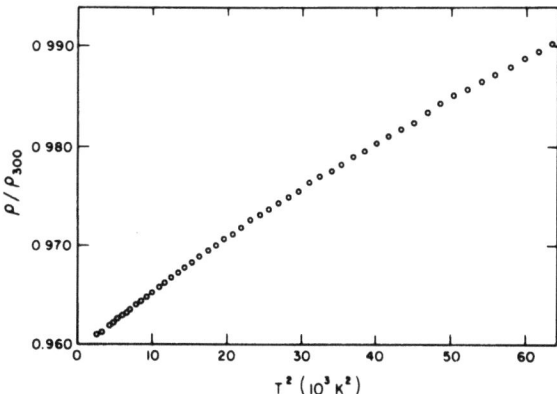

Figure 4. Plot of normalised resistivity ρ/ρ_{300} versus T^2 for Metglas 2826 in amorphous state.

Figure 1 shows that the resistivity of Metglas 2826 in the amorphous state exhibits an approximately linear temperature dependence well above T_m. At lower temperatures there is an approximately quadratic dependence, as can be seen from Figure 4. Except for Metglas 2605A, similar behaviour is observed in the other alloys studied. The existence of a quadratic term in the temperature dependence of the resistivity for other amorphous systems has previously been reported.[3] It is possible that the presence of this term arises from electron-electron scattering as in the case of pure transition metals.[4] However, it is more probable that the quadratic term is due to spin-density fluctuations.[5]

From the inset in Figure 1, it is clear that there is a small but significant change in slope of the resistivity versus temperature for Metglas 2826 near the reported Curie temperature. An extrapolation of the linear sections of the resistivity curve above and below the transition gives T_c = 550±10 K, in satisfactory agreement with the value of T_c = 537 K obtained from susceptibility data. Reference to Figure 5 shows that there is no detectable anomaly near T_c for Metglas 2826A. No reason for the apparent difference in behaviour of the two alloys can be suggested at this time.

Figure 1 shows that resistivity of Metglas 2826 changes discontinuously at the recrystallisation temperature T_{cr} = 650±2 K. Above T_{cr} the resistivity anomaly continues to approximately 750 K, at which temperature the data merge with those obtained for the crystalline state. Therefore, it is clear that the recrystallisation process involves both a sharp initial increase followed by a much more gradual growth of the crystalline phase. Figure 5 shows the corresponding effects of crystallisation on the resistivity of Metglas 2826A. In both cases it can be seen that the difference in resistivity between the amorphous and crystalline phases near T_{cr} is relatively small, showing that the principal contribution to the resistivity of these alloys arises from chemical rather than

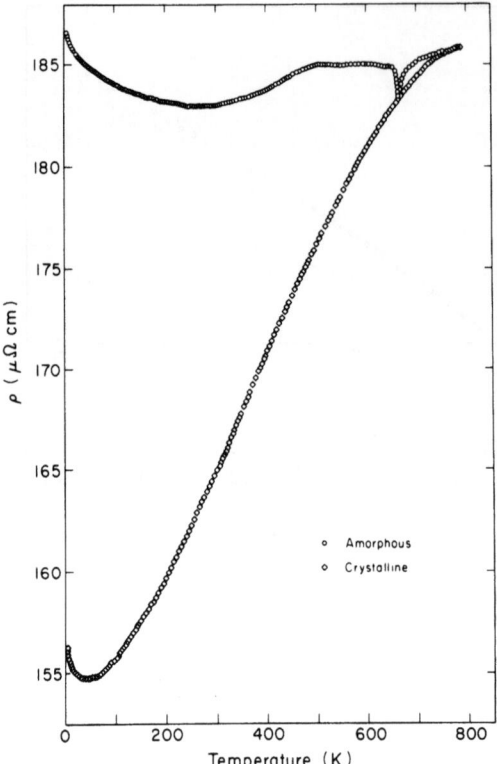

Figure 5. Temperature dependence of resistivity for Metglas 2826A in amorphous and crystalline state.

structural disorder. Similar effects have previously been reported in amorphous Ni-Pd-P alloys.[5] It is of interest to note that different specimens of Metglas 2826A show different changes in resistivity upon recrystallisation. However, the general shape of the ρ versus T curve for the crystalline state remains unchanged.

As can be seen from Figures 1 and 5, the ρ versus T curves for these alloys is profoundly modified by recrystallisation. Near 300 K, the value of dρ/dT is increased by roughly a factor of ten. Since the phonon spectrum and electronic structure are not expected to depend grossly on structural disorder, it is believed that the temperature dependence of the phonon contribution to the resistivity for the amorphous state should also be represented by that in the crystalline state. Assuming that the effects of structural disorder on the residual resistivity can be represented by a constant term $\Delta\rho_0$, it is clear that the difference $\Delta\rho = \rho_{amor}(T) - \rho_{cr}(T)$ can then be written in the form

$$\Delta\rho = \Delta\rho_0 + \Delta\rho_1 + \Delta\rho_2 \quad . \qquad (2)$$

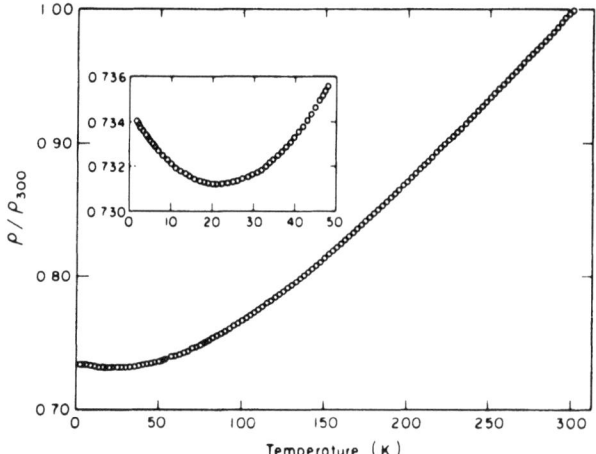

Figure 6. Temperature dependence of normalised resistivity ρ/ρ_{300} for Metglas 2826A in crystalline state. Recrystallised by annealing at 950 K for one hour. Inset shows behaviour at low temperatures on an expanded scale.

Here, $\Delta\rho_1$ is of the form given by equation (1) and $\Delta\rho_2$ represents the spin-disorder contribution, which rises to a constant value above the Curie temperature T_c. Evidently equation (2) is consistent with the observed behaviour and strongly suggests that the T^2 term in the resistivity for the amorphous state is in fact of magnetic origin.

In the crystalline states of both Metglas 2826 and 2826A, $d\rho/dT$ decreases with increasing temperature above 300 K. It can be shown that the resistivity of a metal at high temperatures can be expressed in the form[6]

$$\rho/T = \text{const}(1+2\alpha\gamma T)(1+AT^2..) , \qquad (3)$$

where α is the thermal expansion coefficient and γ is the Grüneisen constant. The coefficient A can be large and negative for transition metals, where the density-of-states can vary rapidly with energy near the Fermi level. Equation (3) appears to represent the data for the crystalline states of both alloys quite well, suggesting that observed departure from linearity in ρ versus T is indeed electronic in nature.

From Figures 1 and 5 it is clear that recrystallisation produces a substantial change in nature of the resistance minimum observed in Metglas alloys. The minimum is clearly reduced in magnitude and is moved to a lower temperature. Since the maximum annealing temperature for both alloys was 800 K, it is believed that the observed minima after heat treatment are due to incomplete recrystallisation. This view is supported by Figure 6, which shows the effects of annealing a specimen of Metglas 2826A at 950 K for

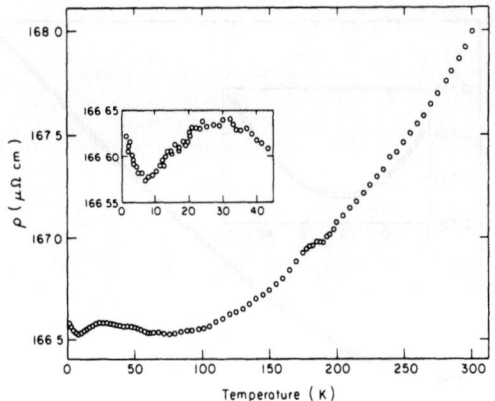

Figure 7. Temperature dependence of resistivity for Metglas 2605A in amorphous state. Inset shows the behaviour at low temperatures on an expanded scale.

one hour. The minimum is further reduced in magnitude and shifted to a still lower temperature. Therefore, it does not seem unreasonable to suppose that the observed resistance minima in the amorphous state of these alloys are of structural origin and that the tunneling model of Cochrane et al. is indeed correct.

As noted previously, the behaviour of Metglas 2605A is rather unusual. Figure 7 shows that the ρ versus T curve exhibits a complex structure of shallow minima at 7 K and 80 K, respectively. In addition, there appears to be further structure near 180 K. These anomalies disappear on recrystallisation and therefore are also believed to be of structural origin.

*Work supported by National Science Foundation.
†Trade name of Allied Chemical Corporation, Morristown, N.J.

REFERENCES

1. See for example, R. Hasegawa and C. C. Tsuei, Phys. Rev. $\underline{2}$, 1631 (1970).
2. R. W. Cochrane, R. Harris, J. O. Ström-Olson and M. J. Zuckerman, Phys. Rev. Letters $\underline{35}$, 676 (1975).
3. R. Hasegawa and C. C. Tsuei, Phys. Rev. $\underline{B3}$, 214 (1971).
4. N. V. Volkenshtein, V. A. Novoselov and V. E. Startsev, Sov. Phys. JETP 33, 584 (1971).
5. Phillipe Maitrepierre, J. Appl. Phys. $\underline{41}$, 498 (1970).
6. N. F. Mott and H. Jones, The Theory of the Properties of Metals and Alloys. Clarendon Press, Oxford (1936), p. 270.
7. R. Hasegawa, private communication.

ELECTRICAL RESISTIVITY AND CRYSTALLIZATION OF AMORPHOUS METGLAS 2826 AND METGLAS 2826A

W. Teoh, N. Teoh, and Sigurds Arajs

Department of Physics, Clarkson College of Technology

Potsdam, New York   13676

ABSTRACT

Electrical resistivity (ρ) of amorphous METGLAS 2826 and METGLAS 2826A has been studied as a function of temperature (T) between 78K and 900K. The ρ vs. T curves exhibit anomalies at elevated temperatures where these materials undergo a transition from the amorphous to the crystalline state. This study shows that the electrical resistivity is a sensitive probe for investigations of the crystallization process.

INTRODUCTION

Recently[1] Allied Chemical Corporation has developed new amorphous metallic (METGLAS) alloys which promise to be excellent engineering materials because of their unique strengths, hardness, magnetic and electrical properties, and outstanding corrosion resistances. These materials in their amorphous state are single phase systems with glass-like structures, and are obtained by very rapid quenching from the liquid state. At elevated temperatures these amorphous substances undergo a crystallization giving, in general, a multiphase structure. The crystallization process, including the detailed description of the formation and identification of the precipitated phases, is a relatively unexplored area of amorphous materials. Since the electrical resistivity (ρ) is a structure sensitive property, it is reasonable to expect that ρ measurements could give useful information about the crystallization processes in amorphous metallic alloys. The purpose of the present investigation is to apply this technique to two ferromagnetic amorphous alloys METGLAS 2826 ($Fe_{40}Ni_{40}P_{14}B_{6}$) and METGLAS

2826A ($Fe_{32}Ni_{36}Cr_{14}P_{12}B_{6}$). Our experimental results and their significance are presented in this paper.

EXPERIMENTAL CONSIDERATIONS

Ribbons (width ∼0.2 cm, thickness ∼0.005 cm) of the above amorphous alloys were purchased from Allied Chemical Corporation. They were studied in "as received" condition. Samples of the length of about 2 cm were cut from the received spools and were mounted in a cryogenic system described before.[2] This system was used for ρ studies between 78K and 350K. The behavior of ρ above 350K up to 1000K was explored using a vacuum platinum furnace. The electrical resistivity of the samples was determined by the standard four probe technique. For low temperature work spot-welded copper wires (diameter 2.9×10^{-2} cm) were used as current and potential leads. The high temperature investigations were done in a similar fashion with spotwelded molybdenum leads (diameter 3.8×10^{-2} cm). The electrical potentials in both regimes were measured with a Guildline Type 9176-G nanovoltpotentiometer capable of detecting voltage changes of 0.001 μV. Temperatures above 350K were determined with calibrated platinum and platinum 10% rhodium thermocouples. The low temperature thermometry used in this investigation has already been described elsewhere.[2]

EXPERIMENTAL RESULTS AND DISCUSSION

Figure 1 shows the electrical resistivity of METGLAS 2826 as a function of the absolute temperature (T). This graph represents two independent experimental runs using different samples cut from the same original spool. Run 1 (closed circles) was started at 78K and ρ measurements (up to ∼350K) were made with increasing temperatures. Above 350K this run was continued in the platinum furnace mentioned above. Since the rate of temperature change may play a role in the crystallization process these rates for the specified temperature intervals are given in Table 1.
Run 2 (open circular points) was started at ∼300K using a different sample of METGLAS 2826 cut from the same original spool. The rates of T changes used during the ρ measurements for this sample are also listed in Table I.

The electrical resistivity of METGLAS 2826A was measured in a manner similar to Run 1 of METGLAS 2826. Table I gives the corresponding rates of T changes. The ρ vs. T curve of this run in shown in Fig. 2.

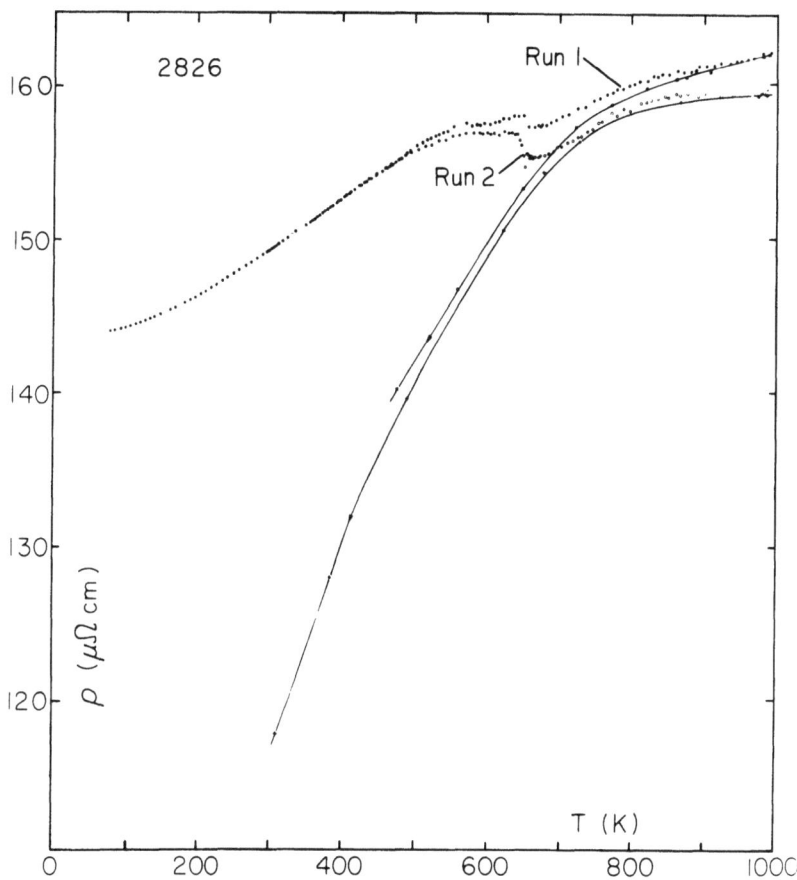

Figure 1. Electrical Resistivity of METGLAS 2826

Table I Cooling and Warming Rates Used in the Electrical Resistivity Studies

METGLAS 2826

Temperature Interval (K)	Warming Rate (K/hr)		Cooling Rate (K/hr)	
	Run 1	Run 2	Run 1	Run 2
300-350	17	13		
350-400	25	14		
400-450	34	14		
450-500	23	17		
500-550	23	29		
550-600	28	27		
600-650	30	31		
650-700	30	29		
700-750	41	20		
750-800	32	34		
800-850	28	31		
850-900	35	33		
900-950	33	34		
950-1000	40	21		
1000-300			57	60

METGLAS 2826A

Temperature Interval (K)	Warming Rate (K/hr)	Cooling Rate (K/hr)
300-400	8	
400-450	2	
450-500	8	
500-600	7	
600-700	17	
700-900	4	
900-300		10

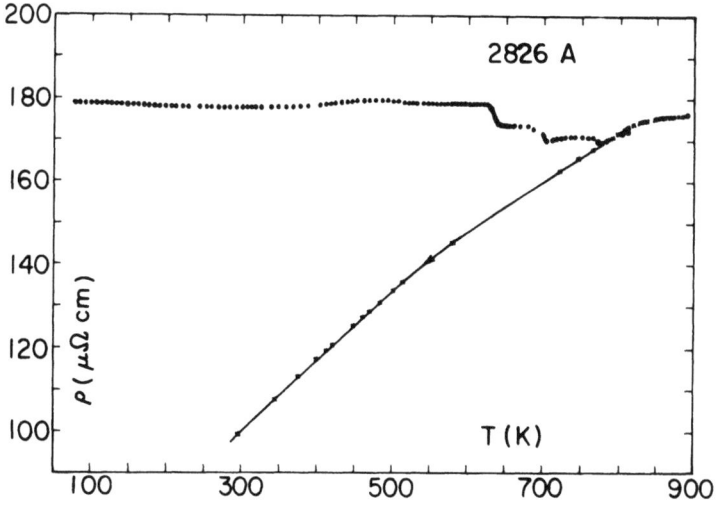

Figure 2. Electrical Resistivity of METGLAS 2826A

A second run was started at ~300K. The ρ data were obtained with increasing T. Unfortunately the sample broke at ~750K. This run, made with essentially the same heating rates, reproduced the corresponding structure in the ρ vs. T curve shown in Fig. 2.

According to the magnetization studies done by Hasegawa and O'Handley[3,4] the crystallization process in METGLAS 2826 takes place at 655K. Our ρ vs. T curves clearly show a step-type anomaly also in the neighborhood of this temperature. However, the decrease in ρ is surprisingly small (~1 μΩ cm) in comparison with those observed in other types of METGLAS materials[5] such as 2605 (~29 μΩ cm) and 2605A (~26 μΩ cm). After this step the electrical resistivity continues to increase monotonically with T up to 1000K without any additional anomalies. The ρ vs. T curve for the crystallized material is considerably below that of the amorphous sample. According to our studies the quantity ρ at 300K for the amorphous METGLAS 2826 is 149 μΩ cm while for the crystallized sample ρ (300K) = 117 μΩ cm (according to Run 2). Figure 1 clearly indicates that two independent runs using slightly different rates of T changes give very similar (although not completely identical) ρ vs. T curves. It also should be remarked that the slight scatter in the ρ data above ~550K results from the amorphous METGLAS 2826 samples and definitely are not due to some other source.

Figure 2 indicates that the crystallization in METGLAS 2826A consists of at least two steps. The initial crystallization starts at ~625K at which the electrical resistivity begins to decrease by ~5 μΩ cm. The second stage of crystallization takes place near 690K and is characterized by ~3 μΩ cm drop in ρ. The third small decrease (~1 μΩ cm) is noticeable at ~765K. Above this temperature ρ increases monotonically with T up to ~900K. The ρ vs. T curve for the crystallized METGLAS 2826A looks like that of a typical crystalline material with very high electrical resistivity. The quantity ρ (300) for the crystallized state is 100 μΩ cm. In the amorphous state ρ (300) = 178 μΩ cm. According to our best knowledge, the crystallization in this glassy alloy has not been studied neither by the x-ray diffraction or magnetic techniques. However, our ρ data suggest that the details of crystallization processes in METGLAS 2826A are more complicated than in METGLAS 2826.

According to Drierach et al[6] the values of ρ and their temperature dependence of glassy alloys is determined, primarily, by the value of the interference function at $2k_F$, $a(2k_F)$, where k_F is the magnitude of the Fermi vector. It seems reasonable to assume that the interference functions for METGLAS 2826 and METGLAS 2826A are similar. The present ρ vs. T data then suggest that $2k_F$ is closer to the first maximum of the interference function in METGLAS 2826A than in METGLAS 2826.

The experimental studies of the magnetic properties of METGLAS 2826 by Hasegawa et al[3,4] indicate that the average ferromagnetic Curie temperature is $\sim$530K. Our ρ measurements do not show any well-defined anomaly in the neighborhood of this temperature. This observation is consistent with the suggestion that $a(2k_F)$ dominates the ρ vs. T behavior of glassy alloys. The value of T_C for METGLAS 2826A is $\sim$250K.[7] We find a small minimum in the ρ vs. T curve at this temperature similar to that seen by Cochrane et al.[7] However, it appears that this minimum (not shown in this paper), which is likely a manifestation of the amorphous state, has nothing to do with T_C.

In summary, the ρ studies as a function of T can be very useful investigations for exploring the crystallization process in amorphous materials especially in combination with the x-ray diffraction and, if applicable, the magnetic work.

ACKNOWLEDGMENTS

The authors would like to express their appreciation to P.B. Harwood for his experimental assistance and K.V. Rao for his helpful comments.

REFERENCES

1. J.J. Gilman, Physics Today 28, No. 5 (1975).

2. S. Arajs, Canad. J. Phys. 47, 1005 (1969).

3. R. Hasegawa and R.C. O'Handley, "Magnetization in Some Iron-Base Glassy Alloys" (preprint).

4. C.L. Chien and R. Hasegawa, AIP Conf. Proc. 29, 214 (1976).

5. N. Teoh, W. Teoh, and S. Arajs, "Electrical Resistivity of Amorphous Metallic Glasses $Fe_{80}B_{20}$ and $Fe_{78}Mo_2B_{20}$ Between 78K and 1000K" (to be published).

6. O. Drierach, R. Evans, H.J. Guntherodt, and H.V. Kunzi, J. Phys. F2, 709 (1972).

7. R.W. Cochrane, R. Harris, J.O. Ström-Olson, and M.J. Zuckermann, Phys. Rev. Letters 35, 676 (1975).

TRANSFORMATION OF SOME AMORPHOUS FePC ALLOYS DURING ISOTHERMAL AGING

A.S. Schaafsma and F. van der Woude

Solid State Physics Laboratory, Materials Science Center

University of Groningen, Groningen, The Netherlands

ABSTRACT

A Mössbauer effect study of the liquid quenched alloys $Fe_{75}P_{18}C_7$, $Fe_{80}P_{13}C_7$ and $Fe_{83}P_{11}C_6$ as a function of aging time at 350 °C is reported. Crystallization occurs after a certain aging time which depends on the alloy composition; $Fe_{80}P_{13}C_7$ being the most stable. The crystallization process starts probably with the formation of clusters of iron atoms. The most important crystalline phases in the final decomposed state are identified. During the amorphous to crystalline transition the mean f-factor of the alloy remains constant to within about five percent.

INTRODUCTION

The amorphous to crystalline transformation is usually investigated by means of differential thermal analysis, resistance measurements, X-ray diffraction and electron microscopic methods [1,2]. According to Rastogi and Duwez [3] the most appropriate way to study this transformation are isothermal aging experiments. Of the above mentioned techniques only X-ray diffraction can be used for this kind of experiments. Electron diffraction and microscopic methods are also used but it has to be noted that in these methods the transformation is promoted by local heating of the sample through the electron beam.

We investigated the transformation upon isothermal aging of some liquid quenched FePC alloys of a composition near to $Fe_{80}P_{13}C_7$ by means of the Mössbauer effect. These alloys are ferromagnetic - at room temperature - which is favorable for the use of the Mössbauer effect [4,5]. Moreover an X-ray diffraction study of $Fe_{80}P_{13}C_7$ during isothermal aging has already been carried out by Waseda and Masumoto [6], and also electron diffraction and transmission electron microscope measurements were reported for this alloy [7].

EXPERIMENTAL

The alloys $Fe_{75}P_{18}C_7$, $Fe_{80}P_{13}C_7$ and $Fe_{83}P_{11}C_6$ were prepared by rapid quenching from the liquid state using the twin roller technique [8]. The foils obtained had a thickness of about 25µ (uniformity about 5%) and a width of 2 to 4 mm. Before using these foils as a Mössbauer absorber they were checked for the absence of crystalline phases by X-ray diffraction. The aging was carried out in evacuated quartz capsules.

RESULTS AND DISCUSSION

In Table I (second column) the X-ray diffraction results of the as-quenched materials prepared by us are listed together with some of the values given by Sinha and Duwez [9]. The domain size parameter L was calculated from the width of the first broad band using the Scherrer formula. Waseda and Masumoto [6] find a value of 15.0 Å for $Fe_{80}P_{13}C_7$ for a comparable parameter defined as to indicate the distance beyond which long range order disappears.

In Table I (third and fourth column) the glass transition temperature T_g and the crystallization temperature T_{cm} (corresponding to the maximum heat evolution) obtained from differential specific heat measurements are shown. The values for $Fe_{80}P_{13}C_7$ are not far from reported values [7,10]. $Fe_{80}P_{13}C_7$ and $Fe_{83}P_{11}C_6$ show a sharp glass transition. $Fe_{83}P_{11}C_6$ has two crystallization peaks. For $Fe_{75}P_{18}C_7$ no sharp glass transition was found. From these results as summarized in Table I we conclude that $Fe_{75}P_{18}C_7$ is microcrystalline, the other two alloys being amorphous.

Composition	L [Å]	T_g [°C]	T_{cm} [°C]	
$Fe_{80}P_{13}C_7$	15.7	430	452	
$Fe_{83}P_{11}C_6$	17.7	412	452	457
$Fe_{75}P_{18}C_7$	19.3		454	
$Fe_{75}P_{15}C_{10}$ [9]	15.7			
$Fe_{44}Pd_{36}P_{20}$ [9]	11.7			
$Pd_{80}Si_{20}$ [9]	14.6			

Table I: Results of X-ray diffraction (CuK_α) and differential specific heat measurements for the as-quenched materials. L was calculated with the Scherrer formula. The glass transition temperature T_g and the crystallization temperature T_{cm} (corresponding to the maximum heat evolution rate) were obtained at a scanning rate of 16 °C/min.

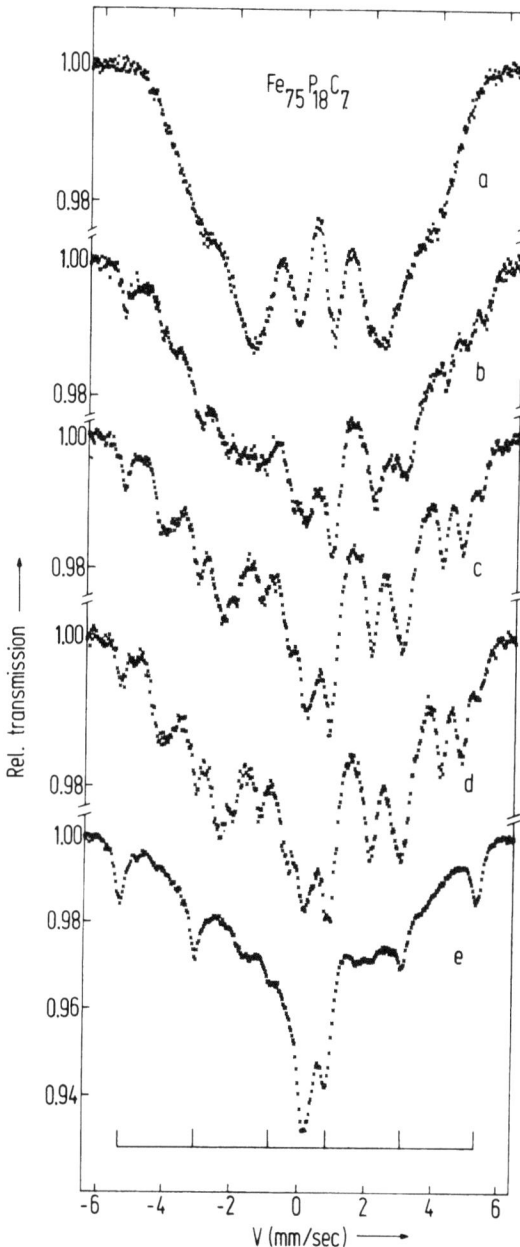

Fig. 1. Mössbauer spectra of $Fe_{75}P_{18}C_7$ at RT of: a) as-quenched material, and after aging at 350°C for b) 500 min, c) 1300 min, d) 1800 min, e) one week. The positions of the six lines of pure α-iron are indicated above the velocity scale.

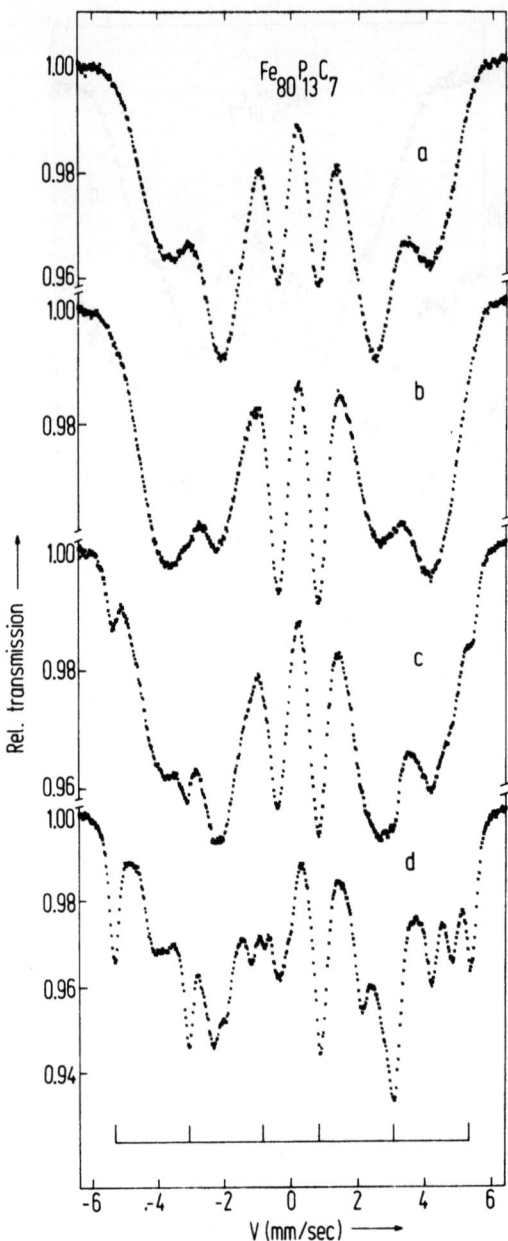

Fig. 2. Mössbauer spectra of $Fe_{75}P_{13}C_7$ at RT of: a) as-quenched material, and after aging at 350°C for b) 1300 min, c) 1800 min, d) one week. The positions of the six lines of pure α-iron are indicated above the velocity scale.

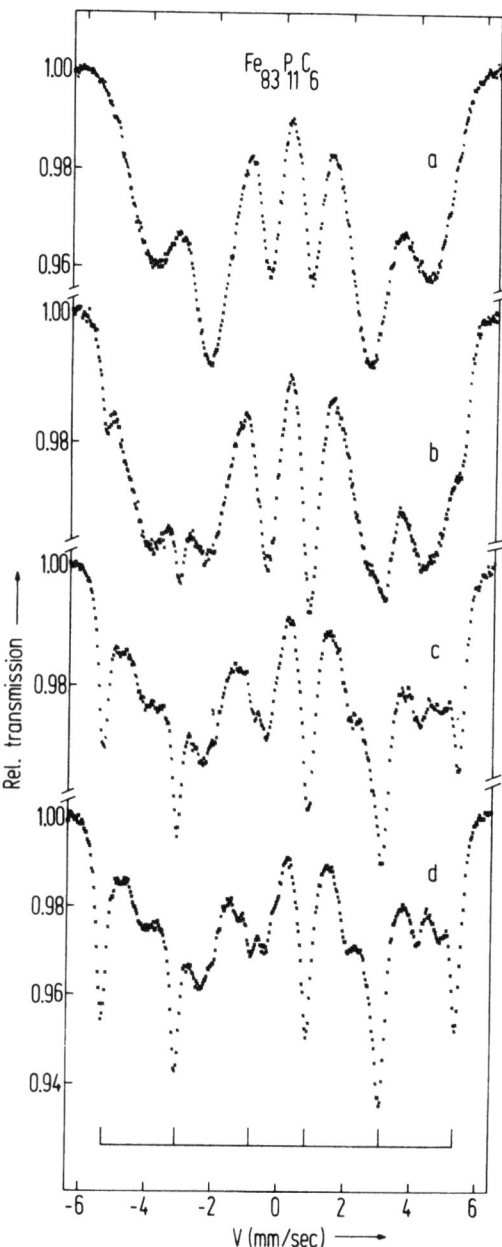

Fig. 3. Mössbauer spectra of $Fe_{83}P_{11}C_6$ at RT of: a) as-quenched material, and after aging at 350°C for b) 850 min., c) 1350 min., d) one week. The positions of the six lines of pure α-iron are indicated above the velocity scale.

The isothermal aging was done at 350 °C which is well below the glass transition temperatures. After aging for a certain time at 350 °C the Mössbauer spectrum was recorded at room temperature. Then the next aging treatment was given to the sample and so on, In the following t_a denotes the total (cumulative) aging time.

The spectra of the as-quenched materials are shown in fig. 1, 2 and 3 (upper spectrum). The mean magnetic splitting, obtained from the splitting of the outermost broad peaks, increases with increasing iron content of the alloys and is estimated to be about 210, 245 and 255 kOe respectively. The mean isomer shift, relative to pure iron, is 0.3 to 0.4 mm/sec for $Fe_{75}P_{18}C_7$ and 0.2 mm/sec for the other two alloys. These values are in agreement with published data for $Fe_{80}P_{12.5}C_{7.5}$ [11] and for some FeP amorphous alloys [12].

We will now discuss the general features of the spectra corresponding to different aging time t_a.

The series of spectra corresponding to the microcrystalline alloy $Fe_{75}P_{18}C_7$ are shown in fig. 1. After the first aging period (t_a = 500 min.) the typical six line pattern of α-Fe is weakly present indicating the formation of clusters of iron atoms. Also some other crystalline phases are present. The absorption lines corresponding to these phases sharpen somewhat when going from b (t_a = 500 min.) to d (t_a = 1800 min.), the amount of α-Fe remaining constant. The phases which can be identified after t_a = 1800 min. are Fe_3P and Fe_2P by comparison with published spectra of these compounds [13]. After t_a = one week (fig. 1e) the sharp lines corresponding to Fe_3P almost disappeared, in the middle of the spectrum the intensity of the asymmetric doublet corresponding to Fe_2P [13] increased significantly together and consistent with an increase in intensity of the α-Fe lines. So after one week aging we find Fe_2P and not Fe_3P, which is the equilibrium phase for the alloy composition, to be present which is a not expected but reproducable result.

Some of the $Fe_{80}P_{13}C_7$ spectra are shown in fig. 2. The spectrum after t_a = 300 min. is still the same as that of the as-quenched material (fig. 2a) and after t_a = 800 min. only the relative intensities of the broad lines have changed. This can be interpreted as a decrease of initial preferred orientation. After t_a = 1300 min. (fig. 2b) we can see the beginning of the formation of iron clusters as reflected by the appearance of weak shoulders at the outer limbs of the outer peaks. This process continues upon further aging. The spectrum after t_a = one week (fig. 2b) also clearly reveals the presence of Fe_3P [13].

Comparing to $Fe_{80}P_{13}C_7$ we see that in the case of $Fe_{83}P_{11}C_6$ (fig. 3) the formation of iron clusters starts earlier (after about 850 min.: fig. 3b)) and that the final relative intensity of the iron lines is greater. Besides this difference the spectra with t_a = one week are much the same, so also for this composition Fe_3P is one of the crystalline phases finally formed upon aging.

The samples of all the three compositions aged for one week were also examined by X-ray diffraction. The observations are the same as concluded from the Mössbauer spectra. The sharpness of the observed diffraction peaks indicates a crystallite size of at least 150 Å. We would expect also some Fe$_3$C to be present but this phase could not definitely be identified.

From the transformation behaviour of Fe$_{80}$P$_{13}$C$_7$ and Fe$_{83}$P$_{11}$C$_6$ as a function of aging time we arrive at the following qualitative picture. Firstly there is a stage in which the only change taking place can be interpreted as a decrease of initial preferred orientation. Secondly, after a time which depends on alloy composition, clusters of α-iron are formed in the amorphous matrix. The iron clusters grow upon further aging and also other crystalline phases begin to develop, but this process seems to proceed more slowly. Finally the stable decomposed state in reached.

Comparing the thermal stability of these two alloys against crystallization, at the aging temperature, we find that Fe$_{80}$P$_{13}$C$_7$ is the most stable one.

Of particular interest in amorphous alloys is the behaviour of the mean f-factor upon aging. For example in SnCu alloys an appreciable difference was found between the crystalline and amorphous state, the f-factor being smaller in the latter case [14]. The present measurements allow us to conclude that within an experimental uncertainty of about five percent, the mean f-factor at room temperature, remains constant during the amorphous-crystalline transformation.

The structural changes as a function of aging time at 330 °C were studied by Waseda and Masumoto [6] by means of X-ray diffraction for the composition Fe$_{80}$P$_{13}$C$_7$. They find only minor changes for samples aged for about 1300 minutes. This stage is interpreted as a rearrangement in the initial amorphous structure to relax its stress or its anisotropic configuration. They place the start of the crystallization at about 1500 min. aging time (at 330 °C). These results agree well with our Mössbauer measurements as we find the start of the crystallization for Fe$_{80}$P$_{13}$C$_7$ after about 1300 min. aging time at a slightly higher temperature (350 °C).

A temperature-time transformation diagram for Fe$_{80}$P$_{13}$C$_7$ of which the essential features are indicated in fig. 4, is proposed by Masumoto and Maddin [7], on the basis of results obtained by X-ray and electron diffraction and transmission electron microscopy. MS-I denotes a metastable phase in which bcc crystallites of α-Fe have separated from the amorphous matrix and MS-II stands for an unknown single metastable phase. MS-II will always decompose into the stable phases α-Fe, Fe$_3$P and Fe$_3$C upon further aging. This diagram (fig. 4) tells us in which state an initially amorphous sample will be after aging isothermally during a certain time. The diagram indicates that there is a critical aging temperature of about 350 °C, below and above this temperature the crystallization sequence and the final stable crystallization product being markedly different. Our aging experiments are done at a temperature of 350°C which is just above this critical temperature meaning that our results are in good agreement with fig. 4.

It may be mentioned that Waseda and Masumoto [6] and Masumoto and Maddin [7] used a different technique for rapid quenching from the liquid state as we did, which could lead to differences in properties of the amorphous material. However the isotherm we measured fits well into the temperature time transformation diagram (fig. 4) proposed by the latter authors.

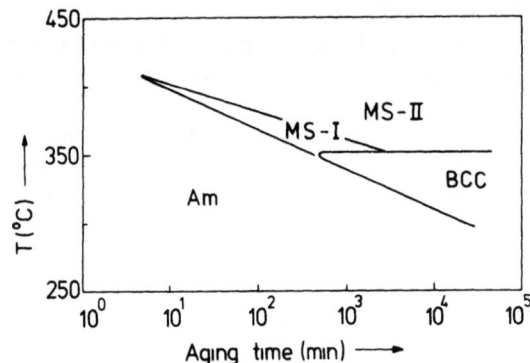

Fig. 4. Temperature-time transformation diagram of amorphous $Fe_{80}P_{13}C_7$ alloy on aging. (Taken from ref. [7]).

The present study deals with a transformation sequence in which the final state is decomposed. Experiments at a somewhat lower aging temperature, where according to Masumoto and Maddin [7], the material transforms into a single phase crystalline alloy of bcc structure, are now in progress.

In conclusion this study shows that the M.E.-technique is very helpfull for obtaining temperature time transformation diagrams due to the feature that each metal or alloy has a characteristic Mössbauer spectrum.

ACKNOWLEDGEMENTS

We thank our colleagues of the Physical Metallurgy Laboratory of our University for their help and their advice in preparing the amorphous materials.

This work forms part of the research program of the "Stichting voor Fundamenteel Onderzoek der Materie" (Foundation for Fundamental Research on Matter - F.O.M.) and was made possible by financial support from the "Nederlandse Organisatie voor Zuiver Wetenschappelijk Onderzoek" (Netherlands Organization for the Advancement of Pure Research - Z.W.O.).

REFERENCES

[1] Jones, H., Rep. Progr. Phys. 36 (1973) 1425.
[2] Takayama, S., J. Mater. Sci. 11 (1976) 164.
[3] Rastogi, P.K. and Duwez, P., J. Non-Cryst. Solids 5 (1970) 1.
[4] Litterst, P.J. and Kalvius, G.M. Proc. Int. Conf. on Mössbauer Spectroscopy, Cracow, 1975, 2, 189.
[5] Coey, J.M.D., J. Physique 35, C6 (1974) 89.
[6] Waseda, Y. and Masumoto, T., Z. Physik B 22 (1975) 121.
[7] Masumoto, T. and Maddin, R., Mater. Sci. Eng. 19 (1975) 1.
[8] Chen, H.S. and Miller, C.E., Rev. Sci. Instrum. 41 (1970) 1237.
[9] Sinha, A.K. and Duwez, P., J. Phys. Chem. Solids 32 (1971) 267.
[10] Jones, H. and Suryanarayana, S., J. Mater. Sci. 8 (1973) 705.
[11] Tsuei, C.C., Longworth, G. and Lin, S.C.H., Phys. Rev. 170 (1968) 603.
[12] Logan, J. and Sun, E., J. Non-Cryst. Solids 20 (1976) 285.
[13] Wäppling, R., Häggström, L., Rundqvist, S. and Karlsson, E., J. Solid State Chem. 3 (1971) 276.
[14] Bolz, J. and Pobell, F., Z. Physik B 20 (1975) 95.

PERSPECTIVE ON APPLICATION OF

AMORPHOUS ALLOYS IN MAGNETIC DEVICES[*]

F. E. Luborsky

General Electric Company
Corporate Research and Development
Schenectady, NY 12301

ABSTRACT

The soft magnetic material characteristics of iron-nickel, iron-cobalt, and iron-silicon are compared to a variety of amorphous alloys. The properties of the amorphous alloy toroids, as-wound, improve with decreasing magnetostriction as expected. However, even after annealing to completely stress-relieve the toroid, the properties still depend strongly on magnetostriction. To date, the amorphous alloys have somewhat higher losses and lower permeabilities than the same thickness Fe-Ni alloys have but the amorphous alloys are significantly superior to the Fe-Co and Fe-Si alloys. Applications of the amorphous alloys in small electronic devices appear to be justified where the design optimization can make use of (1) the lower cost expected from the amorphous alloys, (2) the higher induction of some of the amorphous alloys compared to the Fe-Ni alloys, or (3) their lower losses and higher permeabilities compared to the crystalline Fe-Co and Fe-Si alloys. The high saturation amorphous alloys of Fe-B as thin tapes have about one-fourth the losses of the best grain-oriented Fe-3.2% Si sheet steel measured with sine flux, but the saturation magnetization is 20% lower. The design implications of these differences for power devices is not immediately clear. The temperature dependencies of properties are equivalent to those of the crystalline alloys. The limiting metallurgical life, defined as the start of crystallization, is extrapolated to be 550 years at 175 °C and 25 years at 200 °C for the least stable, possibly useful alloy tested so far: the $Fe_{80}B_{20}$. The magnetically induced anisotropy changes direction and magnitude much more rapidly. This characteristic may be significant in only some specialized application conditions.

[*] Invited paper.

INTRODUCTION

The objective of this paper is to place in perspective the available information on the magnetic properties significant to the application of amorphous metallic alloys as soft magnetic materials in magnetic devices. The particular device and its use determine which properties are significant. These include not only the room temperature magnetic properties but also the temperature coefficients of these properties, thermal stability, environmental stability, and fabrication requirements. Results for these characteristics will be described.

The possible market for a new soft magnetic material may be judged by looking at the market of presently used materials. A survey [1] of the market in 1968 gives the annual dollar volume of electrical steels at $180 million, soft ferrites at $110 million, and iron-nickel alloys at $25 million. For a device or system, however, the impact of improved or different material greatly exceeds that of the materials market. Nevertheless, it should be noted that a new material will probably not impact the entire market in any of these three soft materials areas. Finally, in comparing the "quality" of new soft materials with existing materials, it must be remembered that existing materials are continually being both improved in quality and reduced in cost [1]. Thus, the quality and cost of existing materials must be projected out to the expected time of introduction of any new material.

Previous reviews [2-5] of amorphous alloys as soft magnetic materials presented an excellent view of the fundamental origins of the properties of interest. In brief, it was concluded that the macroscopic dynamic properties could be described by the same models as those used for their crystalline counterparts. We shall not attempt, in this paper, to discuss the properties in terms of their theoretical behavior.

In the following sections, rather, we shall summarize the application characteristics of typical materials and then summarize all of the available information on losses, permeability, and stability for amorphous metallic alloys. Some of these results have been previously reported in the literature, but much will be reported for the first time. Comparison with ferrites will be omitted because of lack of space.

CONVENTIONAL SOFT MATERIALS

Table I summarizes the properties and trade names of many of the alloys currently in use. Details follow here.

Table I Typical Characteristics of Some Commercial Alloys

Trade Names	Composition (wt%) and Description	$4\pi M_s$ kG	Curie Temp °C	Magnetostriction $\lambda \times 10^6$	Density g/cm³	ρ $\mu\Omega$-cm	H_c* Oe	M_r/M_s*
4-79 Mo-Permalloy Super Perm 80 HyMu "80" Mumetal Hipernom	80 Ni, 16 Fe, 4 Mo processed for high initial permeability	7.8	460	~0	8.74	55	0.025	–
Supermalloy Hymu "800"	80 Ni, 20 Fe processed for highest initial permeability and lowest H_c	8.2	400	~0	8.77	65	0.005	–
Square Permalloy Square 80 Hy-Ra "80"	80 Ni, 16 Fe, 4 Mo processed for B-H loop squareness	8.2	460	~0	8.74	55	0.028	0.72
Deltamax Square 50 Hy-Ra "49" Orthonol Hipernik V	50 Ni, 50 Fe grain-oriented, processed for maximum B-H loop squareness	16.0	480	40	8.25	45	0.10	.85
Supermendur	49 Fe, 49 Co, 2 V processed for maximum B-H loop squareness	23.0	940	70	8.15	26	0.18	0.87
Silectron Microsil Oriented T-S	96.8 Fe, 3.2 Si	20.3	730	4	7.65	50	0.50 0.30†	0.71 0.71†

*50 μm thickness unless noted
†305 mm thickness

Fabrication

Conventional Fe-Ni and Fe-Si alloys are prepared from large cast ingots by a sequence of roll reductions with intermediate anneals to control their crystallographic texture and, thus their properties. All of these materials are sensitive to mechanical strains to varying degrees, and therefore must be stress-relieved after fabrication to their final shape, to achieve optimum properties. They must then be housed in a stress-free environment— for example, housed in a case or encapsulated in a polymer. In the case of Fe-Si in power devices, a stressed coating is used to develop desired properties.

Application Areas of Various Materials

The material 4-79 Mo-Permalloy is processed to achieve low coercive force and high initial permeability. Its high material resistivity makes it valuable for higher frequency components. It is well adapted for use in current transformers, coupling transformers, and high-frequency power transformers, as well as in low-level signal transformers, modulators and magnetic amplifiers. It is available in cores wound from tapes of 25 µm, 50 µm, or 100 µm thicknesses, with 125 µm material available in many core sizes. The chemical composition is similar to that of both Square Permalloy and Supermalloy. This material, 4-79 Mo-Permalloy, can be made with different permeability vs temperature characteristics providing positive, zero, or negative temperature coefficients.

Supermalloy has the highest initial permeability of this group of metallic alloys. This property, coupled with a low coercive force, makes Supermalloy one of the most useful core materials. It is widely used in precision current transformers, standard ratio transformers, and low-level signal components. Supermalloy cores are usually prepared from 25 µm, 50 µm, and 100 µm strip thicknesses. The chemical composition is similar to that of 4-79 Mo-Permalloy, but Supermalloy is somewhat more expensive.

Square Permalloy combines the properties of high permeability, good squareness, and low core loss. Its high resistivity makes it useful in higher frequency applications. The maximum induction is only half that of Deltamax. Its low coercive force makes it particularly adaptable to high-gain magnetic amplifiers and high-frequency, low-power processing components, such as inverter transformers, low-level and high-frequency magnetic amplifiers, magnetic modulators, and pulse transformers. The chemical composition is the same as that of 4-79 Mo-Permalloy. The higher squareness ratio (B_r/B_m) is obtained by special processing.

Deltamax combines low enough core loss and a very square hysteresis loop to make it one of the most useful alloys in the power and audio frequency range. It has found extensive use in magnetic amplifiers and for dc inverter and converter transformers in higher frequency applications than the silicon steels. Other applications include its use in bistable switching devices, timing devices, and driver transformers. It is normally available in cores wound from 25 µm, 50 µm, or 100 µm strip thicknesses. Deltamax is a grain-oriented, medium cost material, being more costly than Silectron but less expensive than 4-79 Mo-Permalloy.

Supermendur has the highest available induction of any alloy normally used in tape-wound cores. This makes it useful for transformers and other components where small size is important. Special annealing techniques result in a very square hysteresis loop, making it adaptable for use in power magnetic amplifiers and dc converters and inverters, and wherever size and weight are a major design consideration. Magnetic properties are guaranteed for 100 µm thickness only, a fact tending to restrict usage to the lower audio frequency range.

Silectron is a grain-oriented alloy with 3% silicon, balance iron. For tape core uses, it is available in cores wound from 25 µm, 50 µm, 100 µm, or 305 µm thick tape. Its resistivity is similar to Deltamax, but the hysteresis loop is somewhat less square and the coercive force is higher. This difference makes the core loss higher than that of the nickel-iron alloys. Silectron is the least expensive alloy and is well adapted for use in power transformers, magnetic amplifiers, current transformers, saturable reactors, and power magnetic amplifiers. The thinner gauges are often used in pulse transformers.

Properties

For electronic device applications we are mainly concerned with higher frequencies; thus we show typical loss per unit volume vs induction for thin tapes at 1 to 50 kHz in Figs. 1-3. The upper, heavier curve in each case is for 50 µm thickness; the thinner weight curve is for 25 µm thickness. The permeabilities vs frequencies of some of these alloys, where the permeability is significant in its application, are shown in Fig. 4. Other characteristics of these alloys are listed in Table I.

For power device applications we are concerned principally with the FeSi alloys. The loss per kilogram vs induction up to 50 kHz is shown in Fig. 5 for a variety of tape thicknesses. Permeabilities vs frequencies are shown in Fig. 4. The major use, however, is at 60 Hz as relatively thick sheet steel in ballast, distribution, and power transformers. The properties available in sheet steels have

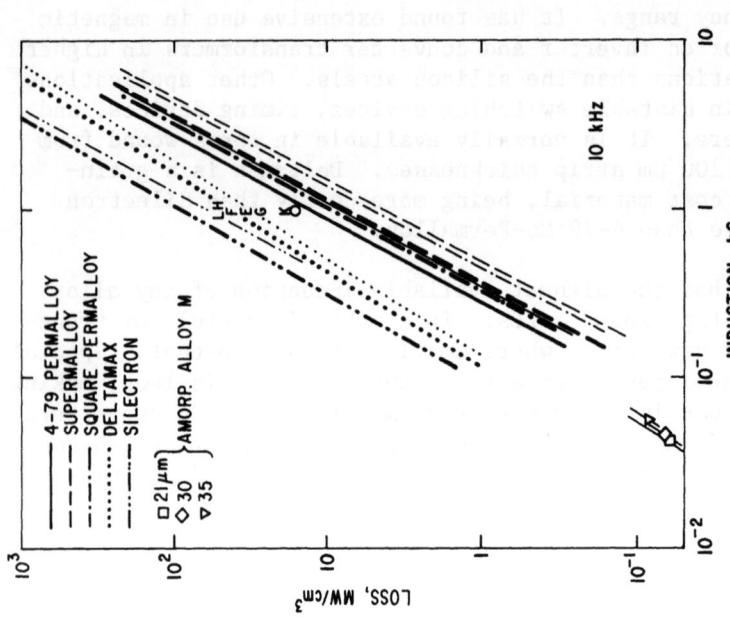
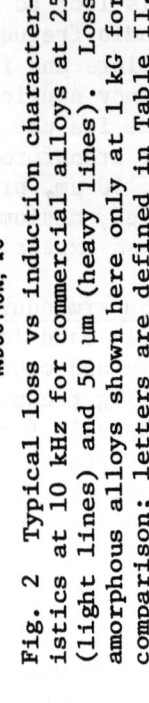

Fig. 2 Typical loss vs induction characteristics at 10 kHz for commercial alloys at 25 μm (light lines) and 50 μm (heavy lines). Loss of amorphous alloys shown here only at 1 kG for comparison; letters are defined in Table II.

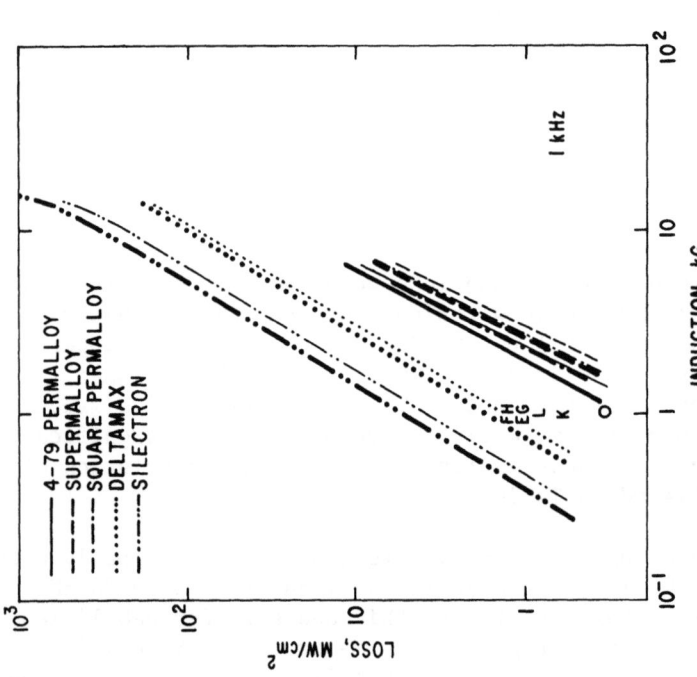

Fig. 1 Typical loss vs induction characteristics at 1 kHz for commercial alloys at 25 μm (light lines) and 50 μm (heavy lines). Loss of amorphous alloys shown here only at 1 kG for comparison; letters are defined in Table II.

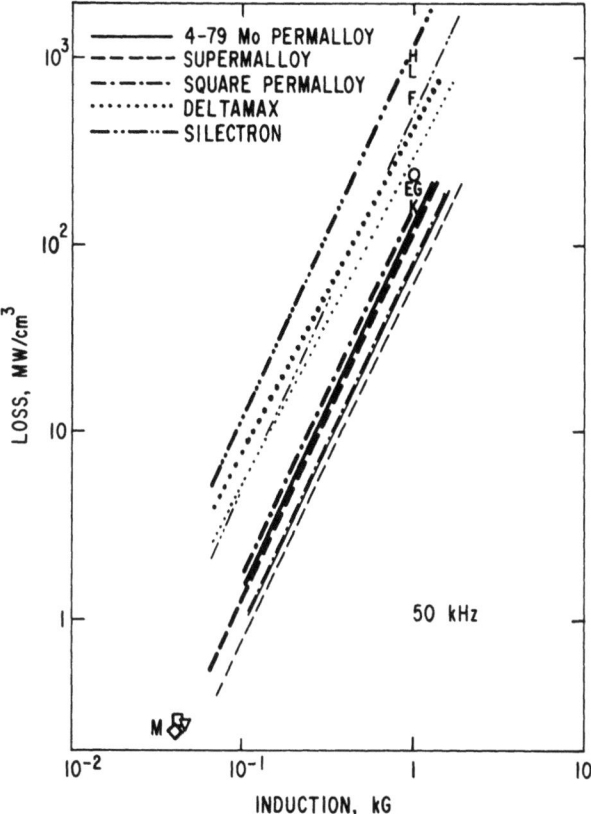

Fig. 3 Typical loss vs induction characteristics at 50 kHz for commercial alloys at 25 μm (light lines) and 50 μm (heavy lines). Loss of amorphous alloys shown here only at 1 kG for comparison; letters are defined in Table II.

been reviewed extensively [6]. The typical, high-quality-oriented Fe-3.2 Si steel is shown by the dashed curve in Fig. 5 for 300 μm thickness. The recently improved variety, "Orientcore Hi-B" [7], is also shown for 305 μm thickness, labeled "Hi-B" in the figure. In Fig. 6 we show these typical characteristics at 60 Hz for a variety of thicknesses in more detail. These characteristics will now be compared to the available data on amorphous alloys.

THE AMORPHOUS ALLOYS

Preparation and Testing

The amorphous alloy ribbons are usually prepared by quenching a molten stream of an alloy directly onto the surface of a spinning

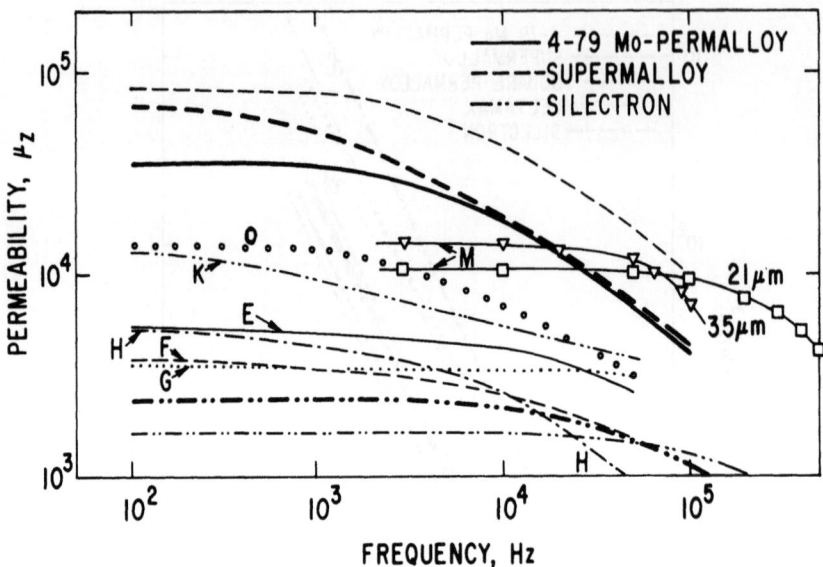

Fig. 4 Typical "initial" impedance permeability vs frequency for amorphous alloys and commercial alloys. Amorphous alloys in this work measured at $\Delta B = 100G$; $\triangle$, $\square$ at 45G; commercial alloys at 50G. Letters are defined in Table II.

drum [8]. The relatively low melting point of these alloys makes this direct casting easier to do than for pure Fe-Ni or Fe-Si. These alloys are also stress-sensitive, must be annealed in the final configuration, and must be housed in a stress-free environment.

The amorphous alloys evaluated in this paper were prepared for testing as previously described [9, 10] by winding about 15 turns of the amorphous ribbon into a 1.4 cm diameter toroid. Dynamic characteristics were obtained at frequencies up to 50 kHz using conventional [11] techniques, usually with a sine current drive. Sine flux tests were made on a few alloys using electronic feedback to maintain sine B at high induction levels.

Losses and Permeability

Amorphous alloy ribbons have been prepared with saturation magnetization, M_s, at room temperature, from very low values to a maximum (so far reported) of 16,500 G [12], which approaches the value of Fe-3.2 Si of 20,300 G. Similarly, Curie temperatures have been found from below room temperature to as high as 477 °C [13]-- temperatures comparable to those of the Permalloys but lower than those

AMORPHOUS ALLOYS IN MAGNETIC DEVICES

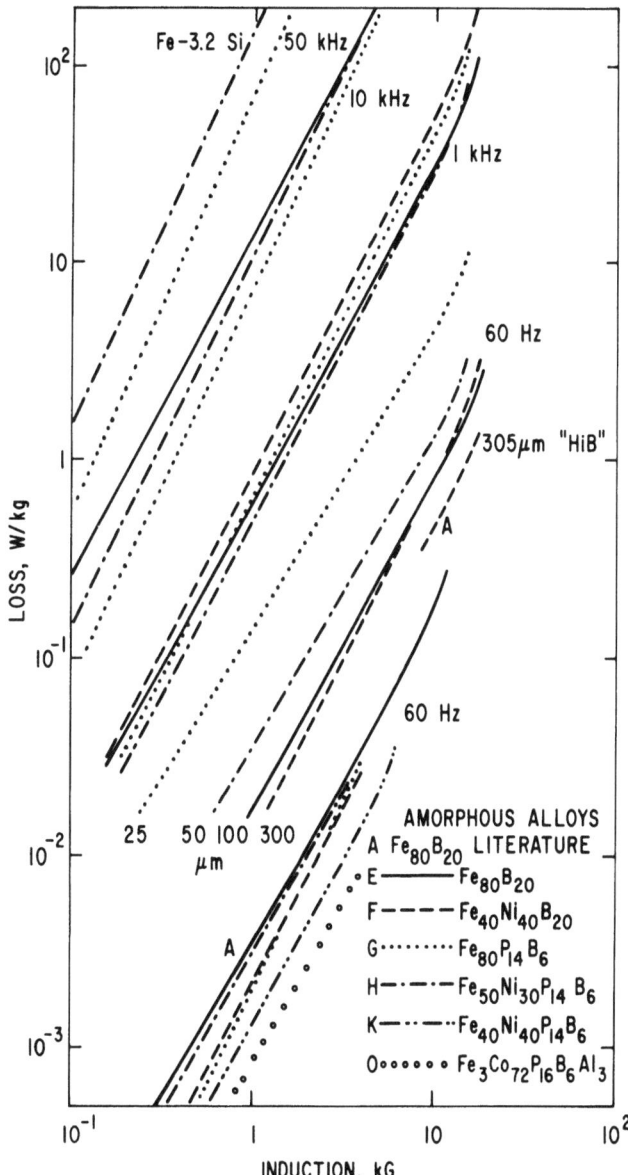

Fig. 5 Loss per kilogram vs induction. Typical characteristics for Fe-3.2 Si tapes at various frequencies and thicknesses. Conventional Fe-3.2 Si sheet steel at 300 μm is compared to the new "Hi-B" sheet steel and the amorphous alloys all at 60 Hz. The A symbols are located at the values given in the literature (Table IV) for annealed amorphous $Fe_{80}B_{20}$.

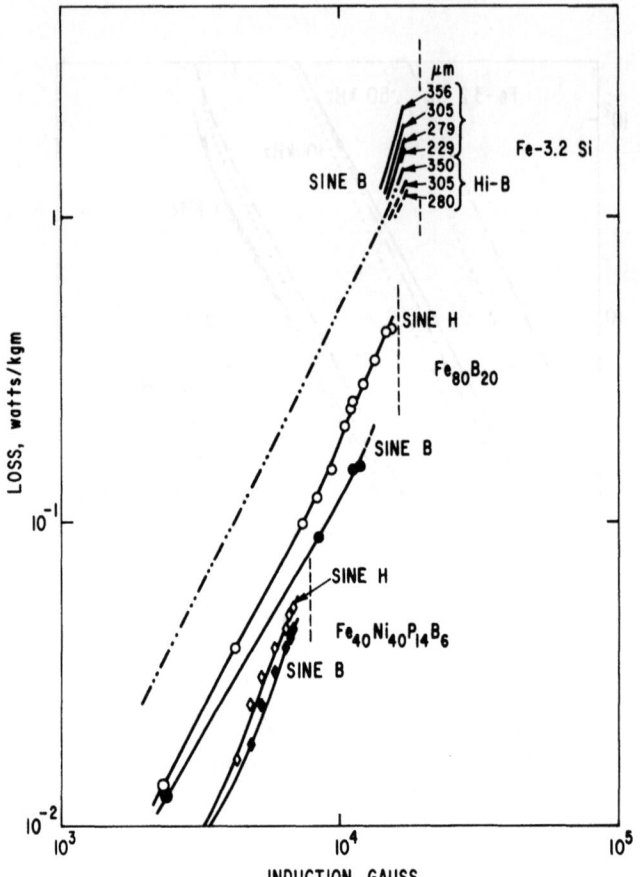

Fig. 6 Loss per kilogram for grain-oriented conventional and "HiB" Fe-3.2 Si at various thicknesses, both tested with sine B, compared to two amorphous alloys tested with sine B and sine H. The vertical dashed lines are located at $4\pi M_s$.

of the Fe-Co and Fe-Si alloys. Coercive forces and loop squareness measured on toroidal specimens have been equivalent to the best values found in their cousins, the Fe-Ni crystalline alloys [9, 10]. It is not surprising, then, that the losses and permeabilites so far found [9, 10] have been roughly equivalent to those of the Fe-Ni alloys.

Loss results and other pertinent properties reported in the literature [9, 10, 12, 14-19] and from the present work, all for stress-relieved toroids, are summarized in Table II at a fixed drive flux, B_m. Some of these are shown in Fig. 7 as a function of B_m and at various frequencies. The characteristics of these alloys are listed in Table III. These were all tested with a sine current

AMORPHOUS ALLOYS IN MAGNETIC DEVICES

Table II As-Cast vs Annealed Losses in Amorphous Alloys

Nominal Composition	Designation	Core Loss @ B_m = 1 kG, mW/cm^3									Tape Thick.	Annealed		REF.
		As Cast				Annealed						H_c	$\frac{M_r}{M_s}$	
		Frequency, kHz												
		0.060	1	10	50	0.060	1	10	50		µm	Oe		
$Fe_{80}B_{20}$ 2605*	A	0.071	2.8	65	—	0.030	1.1	18	—		30–35^p	0.040	0.78	12
$Fe_{80}P_{16}C_3B_1$ 2615	B	—	8.5	310	—	—	6.3	160†	—		40^p	0.050	0.42	15
$Fe_{40}Ni_{40}P_{14}B_6$ 2826	C	—	13	250	—	—	1.8	42	—		50^p	0.019	0.58	15
$Fe_{29}Ni_{49}P_{14}B_6Si_2$ 2826B	D	—	10	160	—	—	0.75	20	—		70^{p††}	0.011	0.70	15
$Fe_5Co_{70}S_{15}B_{10}$	M	—	—	—	—	—	—	—	—		—	0.010	0.85	16
$Fe_{80}B_{20}$ 2605	E	0.10	5	350	1000	0.025	1.2	35	200		30	0.075	0.46	t
$Fe_{40}Ni_{40}B_{20}$	F	0.07	4	450	2300	0.016	1.2	45	600		30	0.090	0.68	t
$Fe_{80}P_{14}B_6$	G	0.04	2.5	80	950	0.014	1.1	28	200		25	0.10	0.37	t
$Fe_{50}Ni_{30}P_{14}B_6$	H	0.09	5	380	1700	0.022	1.2	50	1000		30	0.050	0.84	t
$Fe_{40}Ni_{40}P_{14}B_6$ 2826	K	0.07	4	400	3000	0.010	0.60	18	180		50	0.020	0.70	9,10
$Fe_{40}Ni_{40}P_{14}B_6$	L	0.06	5.5	550	1300	0.014	0.92	49	820		36	0.035	0.85	t
$Fe_3Co_{72}P_{16}B_6Al_3$	O	0.025	1	28	190	0.006	0.35	16	230		20	0.015	0.82	t

* Allied Chemical Co. METGLAS® alloy designation
† Could not be annealed at a high enough temperature to fully stress-relieve
†† D-shaped cross section-maximum thickness
t This work
p Private communication from R.C. O'Handley

Table III Some Pertinent Characteristics of Amorphous Alloys

Nominal Composition	Designation	$4\pi M_s$ R.T. Gauss	Curie Temp. °C	Magneto-striction** $\lambda_s \times 10^6$	Density g/cm³	T_g °C	DSC Properties T_{cr} °C	T_{cm} °C	ΔH cal/g	2-hr Anneal T_{sr} °C	T_{cr} °C	Resistivity ρ μΩ-cm	REF.
$Fe_{80}B_{20}$ 2605*	A	16,000	374	30	7.4					310	389†	140	12,19
$Fe_{80}P_{16}C_3B_1$ 2615	B	14,900	292	30							327†	150	15
$Fe_{40}Ni_{40}P_{14}B_6$ 2826	C	8,200	247	11	7.7	390	400	427	19.9	300	—	180	15,18,19
$Fe_{29}Ni_{49}P_{14}B_4Si_2$ 2826B	D	4,200	382	5								140	15
$Fe_5Co_{70}Si_{15}B_{10}$	M	6,700	430	-0.1									
$Fe_{80}B_{20}$ 2605	E	16,100	378	29	7.07c 7.05	~441	451	473	34.8	330 345	340	134	14,16
$Fe_{40}Ni_{40}B_{20}$	F	10,300	396	13.5	7.48c 7.14	442	451	463	19.2	350	355		t
$Fe_{80}P_{14}B_6$	G	13,600	344	26	7.13c 6.86	418	428	451	30.0	325	—		t
$Fe_{50}Ni_{30}P_{14}B_6$	H	10,400	334	17.5	7.42c 7.21	418	423	432	23.0	305	350		t
$Fe_{40}Ni_{40}P_{14}B_6$	L	8,300	250	12	7.52	408	418	427	20.5	310	355		9,10
$Fe_3Co_{72}P_{16}B_6Al_3$	O	6,300	260	~0	7.60c	—	—	—	—	350	—		t

* Allied Chemical Co. METGLAS® alloy designation
**Magnetostriction measurements on our alloys were made by P. Flanders, Univ. Penn
† Temperature for 1/2 of total transformation
ΔH = heat of crystallization
T_g = glass
T_{cr} = initiation of crystallization
T_{sr} = stress-relief temperatures
t This work

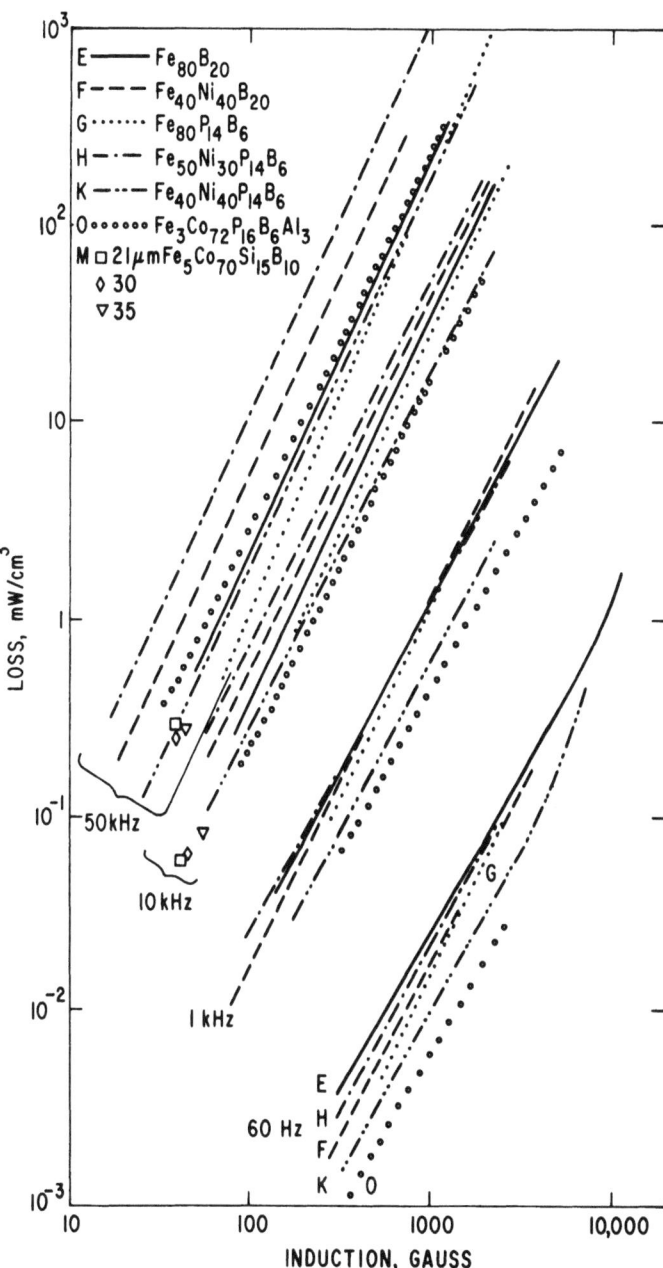

Fig. 7 Loss vs induction at various frequencies for amorphous alloys.

drive— some of them up into saturation, as indicated by the upward curving loss characteristic. For ease in comparing these results to the properties of conventional alloys, the 1 kG loss for each alloy is spotted on Figs. 1-3 using the letter designation given in Table II and in Fig. 7. In general, the losses are somewhat higher than those for most Permalloys, but lower than those for the Fe-Si and Deltamax, at all frequencies. Considering that the surfaces of these ribbons are somewhat rough compared to those of conventional alloy ribbons, which contributes to losses, these results are concluded to be quite good.

For power frequencies we have summarized the losses for the various amorphous alloys in Table IV, and we have compared the amorphous $Fe_{80}B_{20}$ to the sheet steels in Fig. 6. Since these large devices operate with sine B, this is the more meaningful comparison. Although alloys with lower saturation flux densities than those of $Fe_{80}B_{20}$ have lower losses, these are not of interest in power device applications where high flux-carrying capabilities are required. Thus, we concentrate on comparing the Fe-B with the conventional Fe-Si. It is clear that at the same induction the amorphous Fe-B has about one-fourth the losses, a very significant difference if we assume equivalent costs. However, it is not yet clear what design trade-offs will do to the cost of the entire device— that is, the cost not only of the magnetic core but of the conductors, insulation, and case— in view of the lower maximum induction ($\sim$25%) and the thinner gage available with the Fe-B tested here. Conventional transformers are designed to operate up to about 15 kG; and with the newer steels, up to about 17 kG. In sheet steels minimum losses are developed at thicknesses in the range of 250 to 350 µm as the result of careful control of grain size and texture. Sheets of this thickness are economical to handle and stack or wind into large transformer structures.

An alternative approach is to consider laminating a number of thin amorphous sheets: for example, by passing them through rolls to produce a 250 to 300 µm sheet for subsequent fabrication into a transformer. The questions to be considered here are the effect of the rolling and the increase in thickness on the properties. Rolling has been shown [20] to increase catastrophically the coercive force and thus, presumably, the losses of amorphous alloys. However, the subsequent stress-relief anneal appears to be almost as effective as that on the original as-cast alloy. Thus, a laminating procedure to achieve thicker sheets may be satisfactory. However, it is not clear what the increased thickness per se will do to the losses. There is evidence [21] that the losses will go down in agreement with the surface pinning model. The effect of thickness was examined at higher frequencies and low inductions for $Fe_5Co_{70}S_{15}B_{10}$, shown by the symbols in Fig. 7. This is the only data available for amorphous alloys on the effect of thickness.

Table IV Losses of Amorphous Alloys at 60 Hz

Nominal Composition	Designation	Core Loss[+], W/kG B_m, kG					REF.
		1	4	6	13	15	
$Fe_{80}B_{20}$ 2605*	A	0.0043			0.53		12
$Fe_{80}B_{20}$ 2605	E	0.0036	.035 (.028)	.070 .053	.34 .19	.44 (—)	t
$Fe_{40}Ni_{40}B_{20}$	F	0.0021	.027				t
$Fe_{80}P_{14}B_6$	G	0.0020	.029				t
$Fe_{50}Ni_{30}P_{14}B_6$	H	0.0030	.031				t
$Fe_{40}Ni_{40}P_{14}B_6$ 2826	K	0.0013	.014	.033			t
$Fe_{40}Ni_{40}P_{14}B_6$	L	0.0019	.017				t
$Fe_3Co_{72}P_{16}B_6Al_3$	O	0.00079	.0080				t

* Allied Chemical Co. METGLAS® alloy designation
+ sine H measurements
t This work
( ) sine B measurements

Furthermore, since the surfaces of the as-cast tape has an oxide on them, the electrical contact between layers will probably be poor. Thus, it may act more like the individual thin tapes, even at higher frequencies, where eddy current losses normally would become evident.

The very poor losses observed for the toroids before they were annealed (given in Table II) were expected to be a function of the strain-magnetostriction anisotropy, K_s. Assuming that the internal strains were all about the same for the toroids prepared in this work, we expect, then, that the losses in the as-wound toroid would be some function of both the magnetostriction and the tape thickness— that is, $W = f(\lambda \cdot t)$. This appears to be the case, as shown in Fig. 8. The ratio of the loss after the stress-relief anneal to the loss before the anneal should decrease with increase in magnetostriction. This also is observed but with considerable scatter, as shown in Fig. 9. After fully stress-relieving the toroid, we expect $K_s = 0$ and, thus, the losses should be independent of $\lambda \cdot t$. The lower curve in Fig. 8, however, shows that this is not true. The variation in thickness of the samples is random and amounts to approximately ±50%— too small to account for the trend shown. The dependence of loss on magnetostriction suggests, perhaps, that residual internal strains are still present or that the magnetostriction is contributing directly to the losses.

The absence of the magnetocrystalline anisotropy in amorphous alloys has changed the dependence of the properties on composition. In the crystalline Fe-Ni alloys, for example, the coercivity decreases in annealed stress-free samples as Ni is added, reflecting the decrease in crystal anisotropy. Although we have found a composition dependence in the amorphous alloys, its origin is not clear.

Temperature Coefficient of Properties

After their stress-relief anneal, the amorphous alloys have only a directional order anisotropy [22]. This anisotropy is proportional to M_s^x, where x is theoretically equal to 2 but in practice varies upward. Thus, we do not expect the properties to vary significantly with temperature except when approaching the Curie temperature, since there are no other significant contributions to the anisotropy. This is confirmed by the results shown in Figs. 10-12.

Metallurgical Stability

We have shown that at the onset of crystallization the coercivity of amorphous alloys, and thus their loss and permeability, change rapidly. We have thus defined the end of life as the onset of crystallization. The available data from the literature [19,23-26]

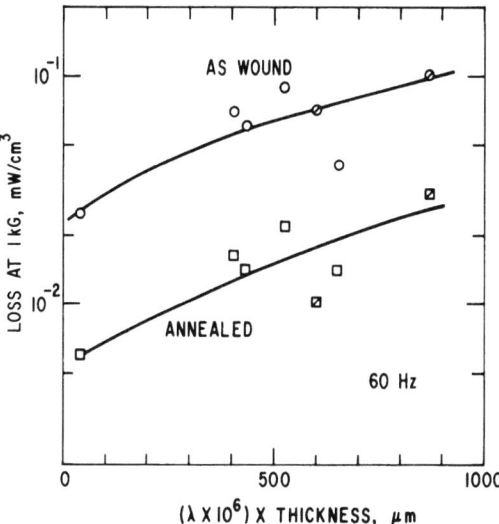

Fig. 8 Loss as a function of the product of magnetostriction and thickness for as-wound and annealed toroids. Toroids all wound, annealed, and tested in this work. Ribbons prepared in this work, open symbols; ribbons prepared by Allied Chemical Co., slashed symbols.

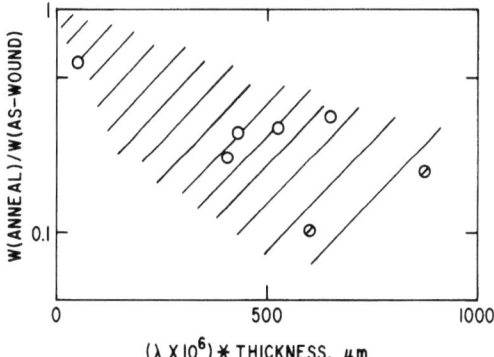

Fig. 9 Ratio of loss after anneal to before anneal vs magnetostriction times ribbon thickness. Ratio obtained by averaging ratio at 60 Hz, 1 kHz, 10 kHz, and 50 kHz.

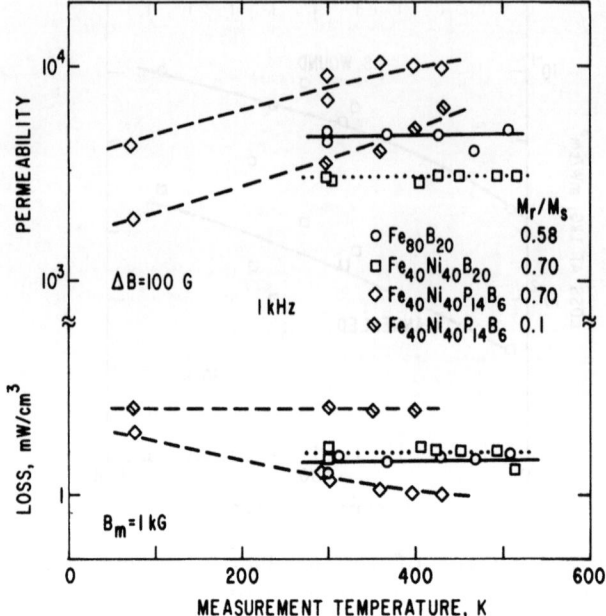

Fig. 10 Loss and permeability at 1 kHz of some amorphous alloys vs measurement temperature.

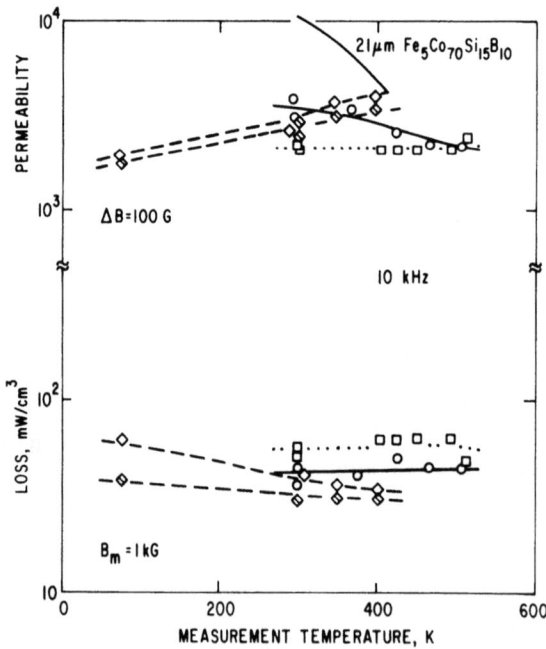

Fig. 11 Loss and permeability at 10 kHz of some amorphous alloys vs measurement temperature.

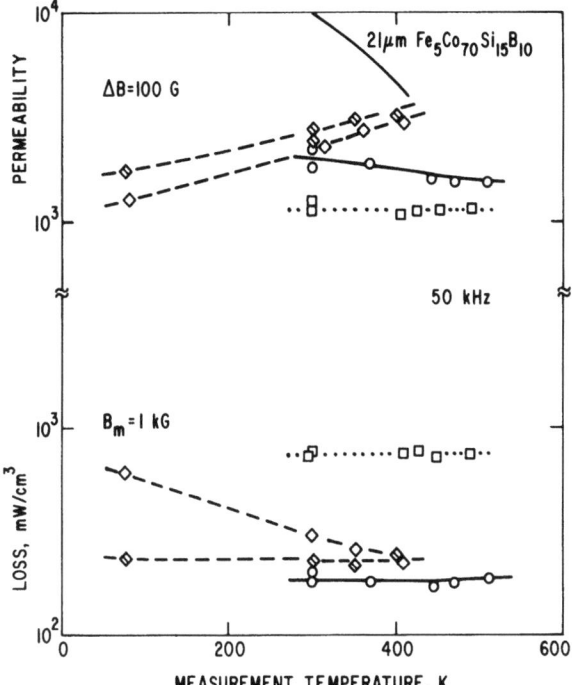

Fig. 12 Loss and permeability at 50 kHz of some amorphous alloys vs measurement temperature.

and from our own work [27] is summarized in Fig. 13 and in Table III. The life times shown here are, as expected, considerably shorter than for crystalline alloys where recrystallization, oxidation, or phase changes limit their life. The data define the maximum fabrication and operating time-temperature exposures. For example, we see that the $Fe_{80}B_{20}$ is the least stable of the alloys discussed as candidates for application. For this alloy, crystallization will start (if we assume a linear Arrhenius extrapolation) at 175 °C after 550 years or at 200 °C after 25 years. This lifetime appears reasonable for all but the most severe application requirements.

Magnetic Stability

Magnetic annealing is expected to occur in most of these amorphous alloys [9, 22]. It has been observed at temperatures and times well before the onset of crystallization and is, therefore, the effect which determines the lifetime of the alloy for some applications. Magnetic annealing changes the magnitude and direction of the magnetically induced anisotropy, K_u. We define a worst-case condition when K_u is perpendicular to the average magnetization, M,

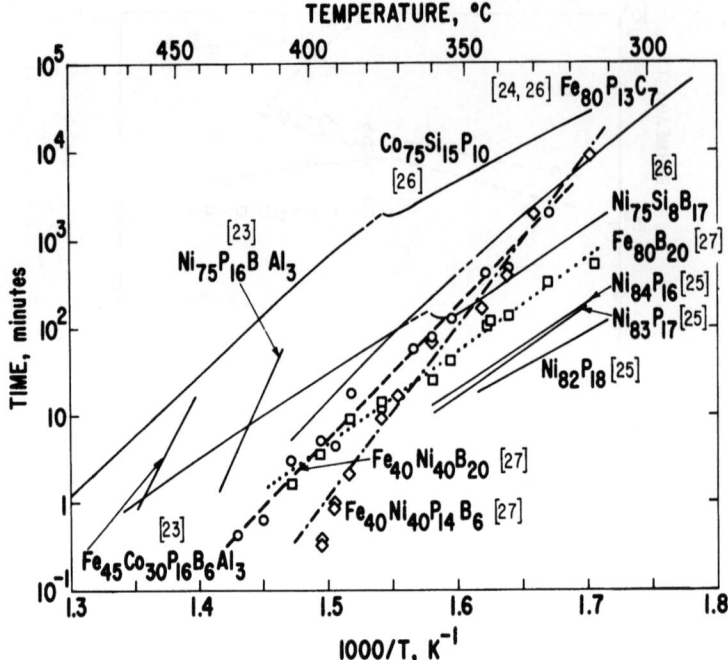

Fig. 13 Time for the start of crystallization as a function of temperature. References are shown in brackets.

during operation. K_u then rotates with time from its original orientation into the direction of M. We have studied this reorientation kinetics for two of the amorphous alloys: $Fe_{40}Ni_{40}P_{14}B_6$ and $Fe_{40}Ni_{40}B_{20}$. In Figure 14 the time constants obtained are shown compared to results reported in the literature [28-34] for amorphous alloys and crystalline Fe-Ni alloys. The $Fe_{40}Ni_{40}P_{14}B_6$ (METGLAS® 2826) time constants are short enough to make the stability of any application of this alloy very suspect. However, the phosphorus-free alloy has a sufficiently long time constant for any foreseeable application. But in most applications the average direction of magnetization will be along the induced anisotropy axis, and thus no change in direction of K_u is expected.

SUMMARY AND CONCLUSIONS

Amorphous alloys have been prepared with a wide range of values of saturation magnetization, coercive force, hysteresis loop squareness, loss, and permeability. Even after complete stress relief the losses still depend on the magnetostriction. This may be the result of residual stresses developed during cooling from the anneal temperature, or the result of a magnetostrictive contribution to the losses. It is suggested that the losses may be further reduced by

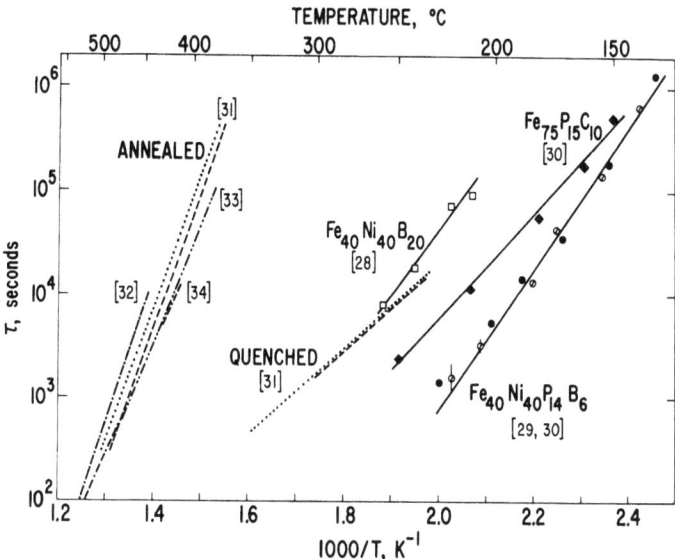

Fig. 14. Time constants for the reorientation of magnetically induced anisotropy in iron alloys. Amorphous alloys shown by solid lines; crystalline alloys in as-quenched and in the annealed state given by nonsolid lines. References are shown in brackets.

surface polishing. The higher resisitivity of the amorphous alloys should result in less deterioration in characteristics with increase in frequency. This has not been observed for many of the alloys—probably again because of poor surface characteristics.

For application in small electronic devices the amorphous alloys have somewhat poorer losses and permeabilities than those of the conventional Fe-Ni alloys but better than those of the Fe-Co and Fe-Si alloys. Where the design optimization requires the lower cost of the amorphous alloys, their higher induction compared to the Fe-Ni alloys or their lower losses compared to the Fe-Co, Fe-Si, and the Fe-Ni at higher frequencies, all will favor the use of the amorphous alloys.

For applications in large power equipment the $Fe_{80}B_{20}$, as thin tape, has about one-fourth the sine B loss of the best quality grain-oriented Fe-3.2 Si. However, because of the differences in thickness and the lower saturation flux of the Fe-B, it is not clear what the final cost/performance trade-off will be for the complete transformer.

The temperature dependencies of all of the characteristics of the amorphous alloys are like those for high Curie temperature alloys. Temperature dependencies are equivalent to temperature dependencies found in crystalline alloys.

Using the maximum life, defined as the onset of crystallization, the $Fe_{80}B_{20}$ amorphous alloy is shown to have the shortest life; extrapolated to 175 °C its lifetime will be 550 years, and at 200 °C its life will be 25 years. The rotation and change in magnitude of the induced anisotropy make up a lower energy process than crystallization and may limit the life of amorphous alloys in those applications where the average direction of magnetization is not along the induced anisotropy axis.

In summary, the application possibilities of amorphous alloys look promising, even based on the limited number of compositions and treatments so far reported. New alloy compositions and methods of treating them will undoubtedly result in further improvements.

ACKNOWLEDGMENTS

I am grateful for the support of many individuals in our laboratory. I am especially indebted to J.L. Walter and W. Rollins for supplying the amorphous ribbons, to B.J. Drummond for the various measurements on them, and to P. Frischmann for making available the equipment for loss at high flux levels and in helping with the measurements. The magnetostriction measurements of P.J. Flanders are also greatly appreciated. I am pleased to acknowledge the partial support of this work by the Office of Naval Research.

REFERENCES

[1] I.S. Jacobs, J. Appl. Phys. 40, 917 (1969).
[2] P.J. Flanders, C.D. Graham, Jr., and T. Egami, IEEE Trans. Magnetics, MAG-11 1323-1325 (1975).
[3] T. Egami, P.J. Flanders, and C.D. Graham, Jr., AIP Conf. Proc. No. 24, Magnetism and Magnetic Materials-1974, 691-701 (1975).
[4] E.M. Gyorgy, H.J. Leamy, R.C. Sherwood, and H.S. Chen, AIP Conf. Proc. No. 29, Magnetism and Magnetic Materials-1975, 198-203 (1976).
[5] G.S. Cargill III, AIP Conf. Proc. No. 24, Magnetism and Magnetic Materials-1974, 138-144 (1975).
[6] M.F. Littman, IEEE Trans. on Magnetics, MAG-7, 48-60 (1971).
[7] S. Taguchi, T. Yamamoto, and A. Sakakura, IEEE Trans. on Magnetics, MAG-10, 123-127 (1974); also, Tech. Report, Orientcore Hi-B, Technical Res. Inst., Yamata Works, Nippon Steel Corp. (Dec. 1972).
[8] H. Lieberman and C.D. Graham, Jr., AIP Conf. Proc. No. 34, Magnetism and Magnetic Materials-1976, Paper 6D1, Joint MMM-INTERMAG Conf.
[9] F. Luborsky, J.J. Becker, and R.O. McCary, IEEE Trans. Magnetics, MAG-11, 1644 (1975).

[10] F.E. Luborsky, R.O. McCary, and J.J. Becker, Proc. of Second International Conf. on Rapidly Quenched Metals, Nov. 1975, eds. N.J. Grant and B.C. Giessen, MIT Press, Cambridge, Mass. (1976).
[11] IEEE Standard 106-1972 or Arnold Catalog 7C-101B Arnold Engineering Co., Marengo, Ill., 60152 (1972).
[12] R. Hasegawa, R.C. O'Handley, and L.I. Mendelsohn, AIP Conference Proc. No. 34 Magnetism and Magnetic Materials-1976, Paper 8B4, Joint MMM-INTERMAG Conf.
[13] J.J. Becker and F.E. Luborsky, paper to be published.
[14] H. Fujimori, M. Kikuchi, Y. Obi, and T. Masumoto, Sci. Repts. A26, 36-47, Research Inst., Tohoku Univ. (1976).
[15] R.C. O'Handley, AIP Conference Proc. No. 29, Magnetism and Magnetic Materials-1975, 206 (1976).
[16] M. Kikuchi, H. Fujimori, Y. Obi, and T. Masumoto, Japan J. Appl. Phys. 14, 1077 (1975).
[17] R.C. O'Handley, L.I. Mendelsohn, R. Hasegawa, R. Ray, and S. Kavesh, J. Appl. Phys. (to appear in Oct. 1976).
[18] D.E. Polk and H.S. Chen, J., Non-Crystalline Solids 15, 165-173 (1974).
[19] L.A. Davis, R. Ray, C.P. Chou, and R.C. O'Handley, Scripta Met. (to appear).
[20] F.E. Luborsky, J.L. Walter, and D. LeGrand, IEEE Trans. Magnetics, MAG-12, XXX (1976), Paper 6D4, Joint MMM-INTERMAG Conf.
[21] K.I. Arai, N. Tsuya, M. Yamada, and T. Masumoto, IEEE Trans. Magnetics, MAG-12, XXX (1976), Paper 6D7, Joint MMM-INTERMAG Conf.
[22] F.E. Luborsky and J.L. Walter, submitted to IEEE Trans. Magnetics.
[23] E. Coleman, Materials Sci. and Eng. 23, 161-167 (1976).
[24] T. Masumoto and R. Maddin, Materials Sci. and Engineering 19, 1 (1975).
[25] W.G. Clements and B. Cantor, Proc. Second-International Conf., Rapidly Quenched Metals, Section I, p. XXX, eds. N.J. Grant and B.C. Giessen, MIT Press, Cambridge, Mass. (1976).
[26] T. Masumoto, Y. Waseda, H. Kumura, and A. Inoue, Sci. Repts. A26, 21-35, Research Insts., Tohoku Univ. (1976).
[27] F.E. Luborsky, to be published.
[28] F.E. Luborsky and J.L. Walter, submitted to Materials Science and Eng.
[29] F.E. Luborsky, AIP Conference Proc. No. 29, Magnetism and Magnetic Materials-1975, 209-210 (1976).
[30] B.S. Berry and W.C. Pritchet, AIP Conference Proc. No. 34, Magnetism and Magnetic Materials-1976, paper 8B3, Joint MMM-INTERMAG Conf.
[31] A. Ferro, G. Griffa and G. Montalenti, IEEE Trans. Magnetics, MAG-2, 764-768 (1966).
[32] E.T. Ferguson, J. Appl. Phys. 29, 252-253 (1958).
[33] O.S. Lutes and R.P. Ulmer, J. Appl. Phys. 38, 1009-1010 (1967).
[34] R.M. Bozorth and J.F. Dillinger, Physics 6, 285-291 (1935).

DISCUSSION

N. D. Heiman: Could you explain what you meant by the statement that the vapor deposited material was not as interesting as the quenched material?

F. E. Luborsky: The reason I said that regarding the vapor deposited material is that their coercivities were much higher than the coercivities of the quenched samples. This had to do with internal structural effects.

N. D. Heiman: So, it is not substrate problem.

F. E. Luborsky: No.

HIGH MAGNETIC PERMEABILITY AMORPHOUS ALLOYS OF THE Fe-Ni-Si-B SYSTEM

T. Masumoto

The Research Institute for Iron, Steel and Other Metals
Tohoku University, Sendai, Japan

K. Watanabe, M. Mitera and S. Ohnuma

The Research Institute of Electric and Magnetic Alloys
Sendai, Japan

ABSTRACT

Low-field magnetic properties of the amorphous $(Fe_{1-x}Ni_x)_{.78}Si_yB_{.22-y}$ alloys ($x=0 \sim 0.7$; $y=0.04 \sim 0.11$) produced by a roller type quenching method were measured in the as-quenched state and the annealed state. Coercivity and remanence were found to be strongly dependent upon the value of both x and y, and heat treatment. The best soft magnetic properties were obtained for the $(Fe_{.80}Ni_{.20})_{.78}Si_{.08}B_{.14}$ alloy annealed at 380°C.

INTRODUCTION

In the last several years transition metal (Fe, Ni and Co) based amorphous alloys made by the continuous rapid quenching methods have been collecting an increasing interest as a favorable soft magnetic materials.[1-4] Most of the works reported so far on the magnetic properties have been concentrated intensively on alloy systems with the limited combinations of metalloids such as P-C, P-B and P-B-Al; e.g., Metglas 2826 with nominal composition $Fe_{.40}Ni_{.40}P_{.14}B_{.06}$ and a zero magnetostrictive $(Fe_{.04}Co_{.96})_{.75}P_{.16}B_{.06}Al_{.03}$ alloy. In order to develop technologically more useful soft magnetic materials, the optimum composition of alloys should be chosen from various combinations of metals and metalloids.

Recently, our research group has demonstrated that a Si-B system is both mechanically very hard and structurally quite stable, furthermore, amorphous alloys can be formed in a wide region of metalloid contents[5]. As a result, we have found a zero magnetostrictive $Fe_{.05}Co_{.70}Si_{.15}B_{.10}$ alloy having the excellent magnetic and mechanical properties[6]. In the present paper the further works on the effects of compositions on the low-field magnetic properties of Fe-Ni-Si-B system are reported.

EXPERIMENTAL

Samples were made by the roller quenching technique in the form of ribbons 0.035 mm thick and about 1.5 mm wide. Quenching rate of this technique was estimated to be about 5 x 10^5 °C/sec by using the simple calculation used by Chen et al.[7] Under this condition amorphous samples were reproducibly obtained in the ranges of $x = 0 \sim 1.0$ and $y = 0.04 \sim 0.11$ for the $(Fe_{1-x}Ni_x)_{.78}Si_yB_{.22-y}$ system. The amorphous nature of samples was confirmed by x-ray diffraction.

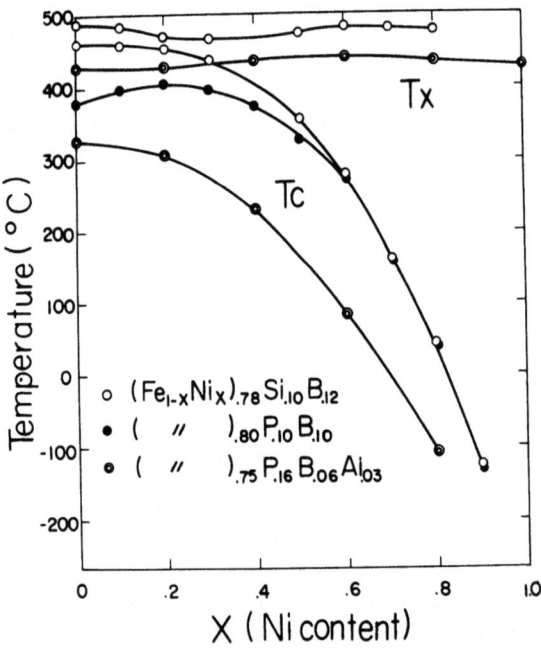

Fig. 1. Composition dependence of the crystallization temperature T_c of amorphous Fe-Ni alloys (o, present data; •, T. Mizoguchi et al.[8]; ⊚, T_x by E. Coleman[9], T_c by R. C. Sherwood et al.[1])

The temperature dependence of magnetization was measured by a pendulum magnetometer between -196° and 900°C. The low-field magnetic measurements were made on straight samples with 30 ~ 40 cm in length placed carefully in a solenoid to avoid a stress.

RESULTS

Fig. 1 illustrates the composition dependence of the crystallization temperature T_x and the Curie temperature T_c in the amorphous Fe-Ni base alloys. The open circles represent our data obtained from the temperature dependence of magnetization and the other symbols the data reported by other authors. As is clear in the figure, T_x of the Si-B system is higher by about 40° to 70°C than that of the P-B-Al system and T_c of this system is also the highest of the three systems. These properties are more desirable in practical use.

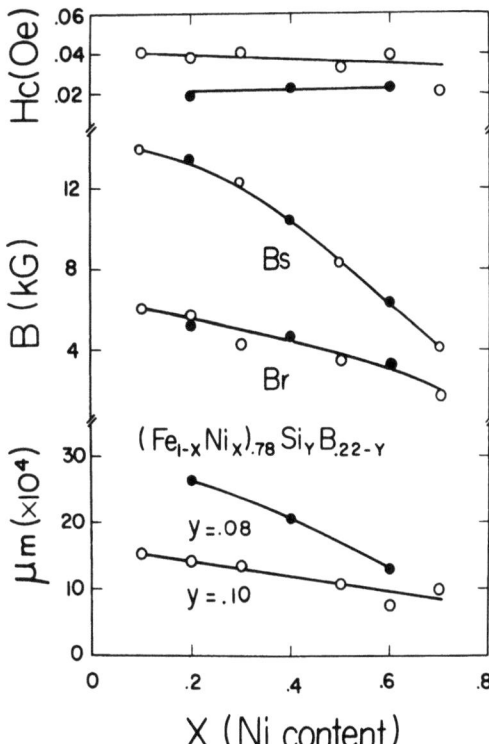

Fig. 2. Composition dependence of the magnetic properties of as-quenched amorphous Fe-Ni-Si-B alloys

Fig. 2 illustrates the composition dependence of the low-field magnetic properties of the as-quenched amorphous $(Fe_{1-x}Ni_x)_{.78}Si_yB_{.22-y}$ alloys. The hysteresis loops of these alloys show large asymmetrical jumps. The coercive force H_c is low, ranging from 0.02 to 0.04 Oe, almost independent of the nickel content. The saturation magnetization B_s decreases gradually with increasing nickel content, in a similar manner to the other report[8]. The remanence B_r decreases with increasing nickel content, corresponding to the decrease in B_s. The maximum permeability μm which is the ratio of remanence to the coercive force shows the same tendency as that of B_r. In this figure, it is also seen that the alloys with y=0.08 have lower coercivity and higher permeability than those for y=0.10.

The low-field magnetic properties of ferromagnetic amorphous alloys are sensitive to heat treatments, and are generally greatly improved due to the relief of internal stress[10]. The present alloys also are not exceptions. Representative data are shown in Figs. 3 and 4. Fig. 3 shows the effect of heat treatment on the low-field magnetic properties for $(Fe_{.60}Ni_{.40})_{.78}Si_{.08}B_{.14}$ alloy.

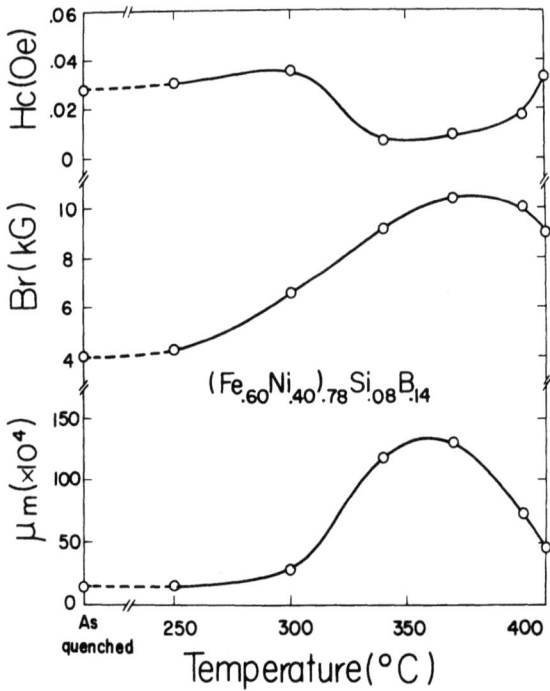

Fig. 3. Effect of the heat treatment on the magnetic properties of $(Fe_{.60}Ni_{.40})_{.78}Si_{.08}B_{.14}$ alloy shown as a typical example for amorphous Fe-Ni alloys (annealed for 1hr)

Each of the samples were held at different temperatures for one hour and then cooled. The magnetic properties are improved by annealing above about 300°C, that is, H_c becomes less than 0.01 Oe and B_r increases, therefore μ_m increases. The best soft magnetic properties are obtained at about 370°C. Fig. 4 shows the relation between the magnetic properties and the nickel content for the two systems with y=0.08 and 0.10 after annealing at the optimum temperature for each composition. The values of H_c in the system of y=0.10 are in the range of 0.02 to 0.03 Oe with minor dependence on nickel content. In the system of y=0.08 except for x=0, on the other hand, H_c becomes as low as 0.006 Oe. B_r decreases linearly with increasing nickel content and therefore μ_m decreases linearly in the system of y=0.10. In the case of y=0.08, on the other hand, the curve of μ_m shows the maximum peak of 2×10^6 at x=0.2.

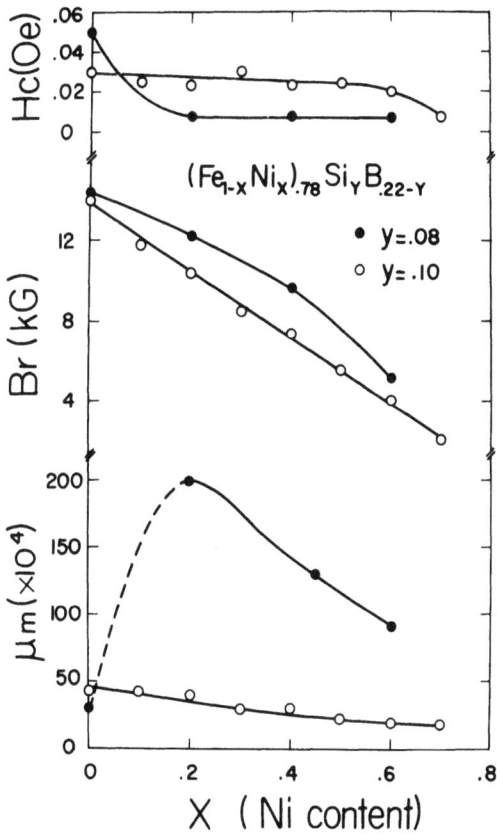

Fig. 4. Composition dependence of the magnetic properties of amorphous Fe-Ni-Si-B alloys annealed for 1 hr at optimum temperature

The above results suggest that the magnetic properties depend strongly on the concentration of metalloid elements. For the purpose of checking this effect, we have examined the effect of y on the magnetic properties in both as-quenched and annealed states. The results obtained are shown in Fig. 5. In the as-quenched state, H_c, B_r and μ_m are only slightly dependent upon the value of y. In the annealed state, however, H_c shows a more pronounced minimum, resulting in a large peak of μ_m around y=0.08. It is of particular interest that the slight change of Si content is remarkably reflected on the improved magnetic properties in the annealed state as seen in Fig. 5.

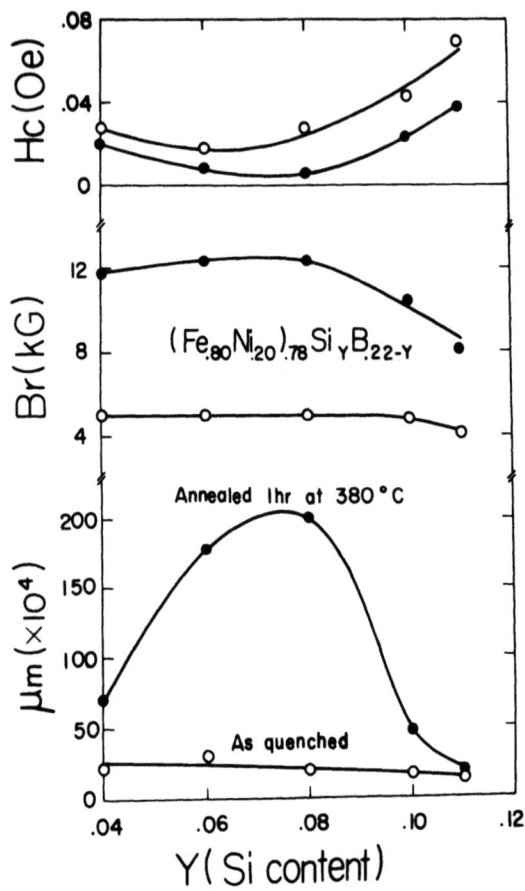

Fig. 5. Composition dependence of the magnetic properties of amorphous $(Fe_{.80}Ni_{.20})_{.78}Si_yB_{.22-y}$ alloys

It was found that the anisotropy K defined by $\int_0^{Ms} H \, dM$[11] exhibits a similar behavior, as shown in Fig. 6. The annealing is most effective in the range of y=0.6 ~ 0.8, and the value of K is decreased by two orders of magnitude by annealing. At the concentrations of y=0.04 or 0.11, the decrease in the value of K is only by the factor of 3 ~ 4. At this moment it is unclear why the heat treatment is effective only in the limited range of composition. A possible explanation is that the alloys with the marginal composition such as y=0.04 or 0.11, may contain small crystalline precipitates[12] or microscopic compositional fluctuations. If that is the case, it may be said that the magnetic measurement is best suited to locate the compositional range within which stable homogeneous amorphous alloys can be formed, while other conventional methods such as bending fracture test or x-ray diffraction define the range in a looser manner.

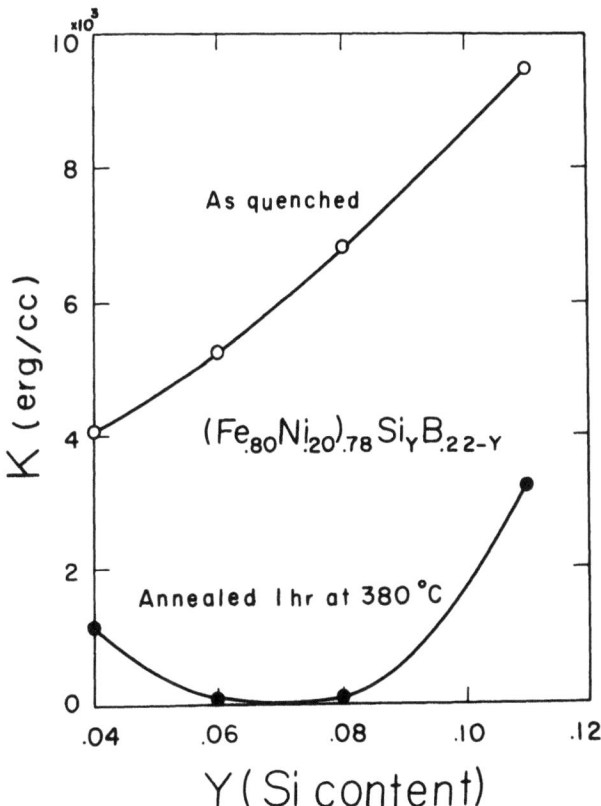

Fig. 6. Effect of the composition and heat treatment on the anisotropy K of amorphous $(Fe_{.80}Ni_{.20})_{.78}Si_y B_{.22-y}$ alloys

Table 1. Magnetic properties of several amorphous $(Fe_{1-x}Ni_x)_{.78}$-$Si_yB_{.22-y}$ alloys after various heat treatments

Property	$(Fe_{80}Ni_{20})_{.78}Si_{.10}B_{.12}$				$(Fe_{80}Ni_{20})_{.78}Si_{.08}B_{.14}$				$(Fe_{60}Ni_{40})_{.78}Si_{.08}B_{.14}$			$(Fe_{40}Ni_{60})_{.78}Si_{.08}B_{.14}$		
	a	b	c	d	a	b	c	d	a	b	d	a	b	d
T_c (°C)	456				460				400			277		
H_c(Oe)	.04	.023	.013	.015	.018	.006	.006	.006	.022	.006	.006	.024	.005	.005
B_r(kG)	5.8	10.8	12.0	12.2	5.4	12.2	12.2	12.4	4.6	9.8	10	2.9	4.5	4.8
B_r/B_s(%)	44	83	92	93	41	93	93	95	44	95	97	46	72	76
$\mu_m(\times 10^4)$	14	47	90	80	30	200	200	200	21	160	170	12	90	95
T_a(°C)	—	(	380	)	—	(	380	)	—	(370	)	—	(350	)

a) As quenched.
b) Annealed for 1 hr at T_a.
c) Annealed for 1 hr at T_a under $2 \sim 3$ kg/mm².
d) Annealed for 1 hr at T_a in the field of 200 Oe.

Table 1 summarizes the low-field magnetic properties of several amorphous Fe-Ni-Si-B alloys after various treatments. In this table, it should be noted that the superior magnetic properties of $(Fe_{.80}Ni_{.20})_{.78}Si_{.08}B_{.14}$ alloy is obtained even by a simple annealing at 380°C without more complicated treatments such as c) and d) in the table.

CONCLUSIONS

The crystallization temperature and the Curie temperature of amorphous Fe-Ni alloys containing Si and B are noticeably higher than alloys containing other metalloid elements. The soft magnetic properties such as coercivity and remanence were found to depend not only on the Fe to Ni ratio, but also on the ratio of Si and B content. Within the compositional range in which a stable amorphous alloy can be formed, coercivity was found to have a minimum at a certain composition. We obtained excellent soft magnetic properties of H_c=0.006 Oe, B_r=12.2 KG and μ_m=2 x 10^6 for $(Fe_{.80}Ni_{.20})_{.78}Si_{.08}B_{.14}$ alloy annealed at 380°C.

ACKNOWLEDGEMENTS

The authors with to express their thanks to Dr. T. Egami of University of Pennsylvania and Dr. H. Fujimori of Tohoku University for their valuable discussions and suggestions. The authers are also indebted to Directer H. Masumoto of the Research Institute of Electric and Magnetic alloys for constant guidance in the course of the work, and also to Mr.Y. Kobayashi of the same Institute for preparing samples.

REFERENCES

1. R. C. Sherwood, E. M. Gyorgy, H. S. Chen, S. D. Ferris, G. Norman and H. J. Leamy, AIP Conf. Proc., 24, 697(1975).
2. T. Egami, P. J. Flanders and C. D. Graham, Jr., Appl. Phys. Letters, 26, 128(1975); AIP Conf. Proc., 24, 607(1975).
3. H. Fujimori, T. Masumoto, Y. Obi and M. Kikuchi, Japan. J. Appl. Phys., 13, 1889(1974).
4. E. M. Gyorgy, H. J. Leamy, R. C. Sherwood and H. S. Chen, AIP Conf. Proc., 29, 198(1976).
5. T. Masumoto, K. Hashimoto and H. Fujimori, Sci. Rep. RITU, A-25, 232(1975), (Review).
6. M. Kikuchi, H. Fujimori, Y. Obi and T. Masumoto, Japan. J. Appl. Phys., 14, 1077(1975); Sci. Rep. RITU, A-26, 36(1976).
7. H. S. Chen and C. E. Miller, Rev. Sci. Instr., 41, 1237(1970).
8. T. Mizoguchi, K. Yamauchi and H. Miyajima, in Amorphous Magnetism, ed. H. O. Hooper and A. M. de Graaf, Plenum, NY pp. 325(1973).
9. E. Coleman, 2nd Intern. Conf. on Rapid Quenched Metals, ed. N. J. Grant and B. C. Giessen, Section II, 161(1975).
10. F. E. Luborsky, J. J. Becker and R. O. McCary, IEEE Trans. Mag., MAG-11, 1644(1975).
11. T. Egami and P. J. Flanders, AIP Conf. Proc., 29, 220(1976).
12. T. Masumoto and R. Maddin, Mat. Sci. Eng., 19, 1(1975).

MAGNETOSTRICTION OF METALLIC GLASSES

R. C. O'Handley

Materials Research Center, Allied Chemical Corporation
Morristown, NJ 07960

ABSTRACT

Linear saturation magnetostrictions λ_s are reported for $(FeNi)_{80}B_{20}$ and $(FeCo)_{80}B_{20}$ glasses. In the former series, λ_s decreases monotonically from its value at $Fe_{80}B_{20}$, approaching zero with the magnetization. In the latter series, λ_s passes through zero at the composition $Co_{75}Fe_5B_{20}$, approximately the same Co/Fe ratio for which $\lambda_s = 0$ in crystalline iron-cobalt alloys, viz. $Co_{92}Fe_8$. Band models which adequately describe the magnetostriction of crystalline Fe-Ni alloys do not appear to apply to Fe-Ni-base or Fe-Co-base glasses. A structure-sensitive, pseudodipolar model accounts well for the sign of λ_s in crystalline Fe and Co. When generalized to consider non-crystalline alloys, the model suggests that the short-range order of cobalt-rich metallic glasses resembles that their crystalline counterparts. Interpretation of the data in terms of a quantum-statistical-mechanical theory suggests that magnetostriction in Fe- and Ni-rich glasses will be dominated by one mechanism for anisotropic spin correlation and that another mechanism is important in Co-rich glasses.

I. INTRODUCTION

Of central importance to the understanding of ferromagnetism in noncrystalline materials is knowledge of any property that depends strongly on both structure and magnetic interactions. Two such properties are magnetic anisotropy and its strain derivative, magnetostriction. While the investigation of these two phenomena should go hand in hand, studies of magnetoelastic effects[1-10] in transition metal-metalloid (TM-M) glasses outnumber those of anisotropy.[1,11-15] This is a consequence, at least in part, of the difficulty in characterizing the generally weak, and often random, in-plane anisotropy of these materials[16] as well as of the difficulty in obtaining samples of sufficient size and dimensional tolerance to perform sensitive torque measurements. On the other hand, the ribbon form of many TM-M glasses, such as the samples used in the present study, is well suited to magnetostriction measurements using semiconductor, or metal-foil, strain gauges.

Early theoretical work on anisotropy and magnetostriction focused on true dipole[17] and pseudodipolar interactions based on spin-orbit coupling.[18] The close relation between anisotropy and magnetostriction was put clearly by Kittel[19]: "There will be no linear magnetostriction if the anisotropy energy is independent of the state of strain of the crystal. ... a crystal will deform spontaneously if to do so will lower the (anisotropy) energy." With the incorporation of this relation into a quantum-statistical-mechanical theory by Callen and Callen,[20] the transformation of magnetostriction from an important engineering property to one that is also of fundamental physical significance was complete. Nevertheless, our present understanding of magnetoelastic effects, particularly in transition metals, is incomplete. Noncrystalline ferromagnets represent a novel medium in which to observe these effects, offering an opportunity to study them as functions of composition and temperature without the complicating interference of structural phase transformations.

II. EXPERIMENTAL

The metallic glass samples were prepared by continuous, rapid quenching ($\sim 10^6$ K sec^{-1}) from the melt, using high purity elements. The resulting ribbons, about 0.04 x 2 mm^2 in cross section, were shown to be noncrystalline by x-ray diffractometry. Magnetostriction measurements were made using semiconductor strain gauges. Short strips of glassy ribbon about 1 cm in length were bonded to each side of the gauge which formed one arm of an ac (~ 28 Hz) null bridge. Departure from null upon application of parallel ($H||\delta\ell$) or perpendicular ($H\perp\delta\ell$) in-plane fields was detected using a differential amplifier and an r.m.s. voltmeter. From the expression for strain in an isotropic material[19]

$$\delta\ell/\ell = 3/2 \; \lambda \; [\cos^2\theta - 1/3] \tag{1}$$

where θ measures the angle between the magnetization and the direction in which $\delta\ell/\ell$ is measured, we obtain for the magnetostriction

$$\lambda = \frac{2}{3} \left[\left(\frac{\delta\ell}{\ell}\right)_{||} - \left(\frac{\delta\ell}{\ell}\right)_\perp \right]. \tag{2}$$

Saturation magnetostriction is obtained by extrapolation of the $\lambda(H)$ data [H = 1 - 10 kOe] to H = 0.

Spontaneous Hall effect measurements were made using copper wire pressure contacts to the glassy ribbon. Since accessible magnetic fields (applied normal to the ribbon plane) were limited to 10 kOe, the data were extrapolated (when $4\pi M_S > 10{,}000$ gauss) to $H = 4\pi M_S$ in order to estimate the spontaneous Hall coefficient, R_S:

$$\rho_H = \frac{E_y^H}{j_x} = R_o B_z + R_s 4\pi M_s. \qquad (3)$$

III. RESULTS & DISCUSSION

Magnetostriction of $TM_{80}B_{20}$ Glasses

Figure 1 shows the linear saturation magnetostrictions (295 K) for the new[10] METGLAS® alloys (®Trademark, Allied Chemical Corporation) in the series $Fe_{80-x}Co_xB_{20}$ ($0 \lesssim x \lesssim 80$ at.%) and $Fe_{80-x}Ni_xB_{20}$ ($0 \lesssim x \lesssim 60$). These are the first such data for single-metalloid, transition-metal glasses, as well as the first for a complete series of Fe-Ni-based glasses.

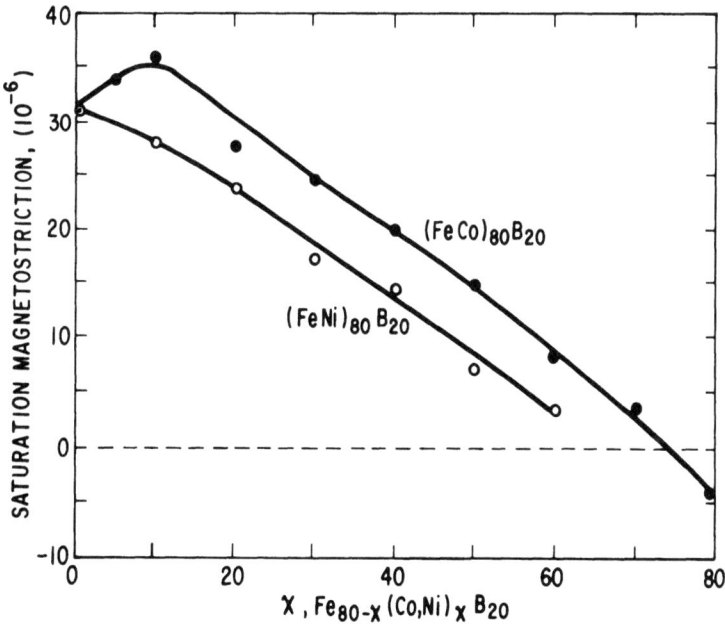

Fig. 1. Linear, saturation magnetostriction (295 K) for $(FeCo)_{80}B_{20}$ (●) and $(FeNi)_{80}B_{20}$ (o) glasses.

In the $(FeNi)_{80}B_{20}$ glasses, λ_S decreases monotonically from its value at $Fe_{80}B_{20}$, approaching zero approximately as the square of the room temperature saturation magnetization.[10] Similar behavior is observed in the magnetostriction of iron-nickel-based glasses containing a mixture of metalloids rich in phosphorous.[2,7] The data for $(FeNi)_{80}B_{20}$ glasses are noteworthy for the absence of a zero in the magnetostriction at or near the TM content of the well known zero-magnetostriction "permalloys" ($\sim$80% Ni in Fe) (Fig. 2a).

The magnetostriction of the $(FeCo)_{80}B_{20}$ glasses peaks at low cobalt concentration, then decreases, passing through zero at $Co_{75}Fe_5B_{20}$. The zero-magnetostriction composition here occurs at a Co/Fe ratio ($\sim$0.93) close to that for which $\lambda_S = 0$ in polycrystalline Fe-Co alloys[9] (Fig. 2b). The behavior of magnetostriction for the rest of the Fe-Co glassy alloys does not parallel that of the polycrystalline alloys. Magnetostriction of iron-cobalt-based glasses of different metalloid content[4,6,8] also shows these features; i.e. $\lambda_S = 0$ near Co/Fe = 0.93 and no correlation with the crystalline data over the rest of the series (Fig. 3).

The occurrence of $\lambda_S = 0$ in cobalt-rich Fe-Co-base glasses but not in glasses with TM content comparable to that of other crystalline alloys where $\lambda_S = 0$ (Fe-rich with Ni or Co and Ni rich Fe-Ni, Fig. 2) is now discussed from the point of view of various models of magnetostriction.

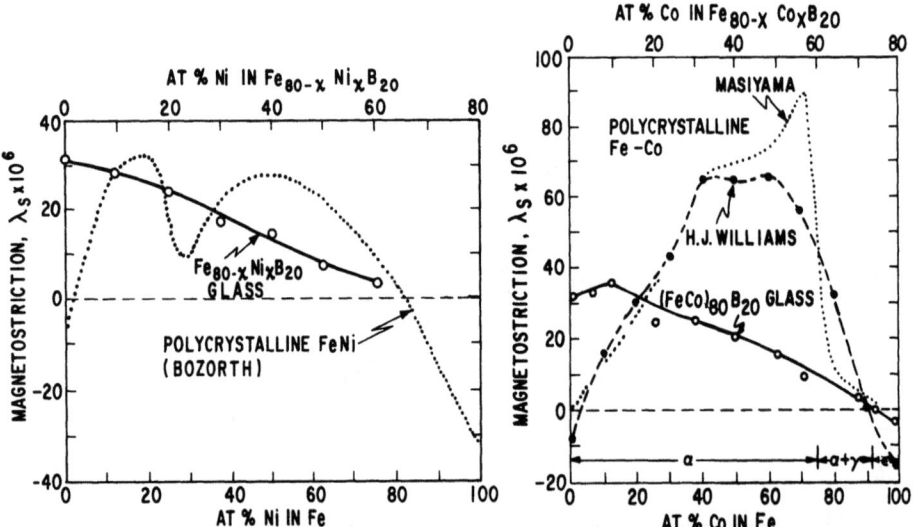

Fig. 2. Magnetostrictions of glasses in Fig. 1 compared with those of comparable polycrystalline TM alloys: (a) Fe-Ni-base alloys; (b) Fe-Co-base alloys. Polycrystalline data from Ref. 29, p. 669 and p. 664 respectively.

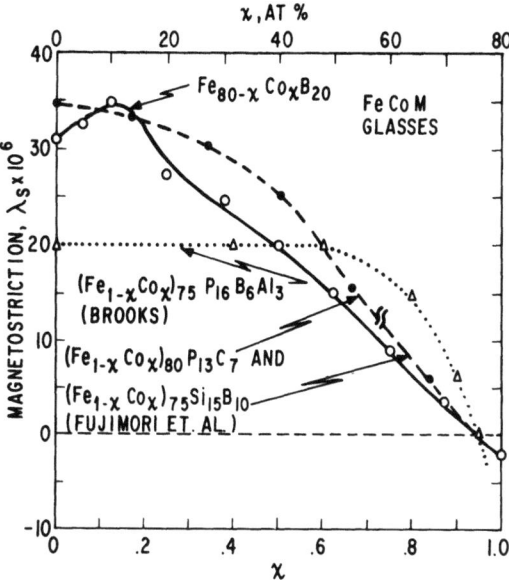

Fig. 3. Magnetostrictions of Fe-Co-base glasses of different metalloid content. Mixed-metalloid data from Refs. 6 and 8.

Band Models

In polycrystalline FeNi and CoNi alloys, singularities are observed in the magnetoresistance (a maximum) the extraordinary Hall coefficient (a zero) and the magnetostriction (a zero) at an electron concentration of 27.7 per atom.[21,22] These three properties are generally attributed to spin-orbit interactions. Berger[21] has interpreted these anomalies to result from a 3d-band degeneracy which is partially lifted by spin-orbit interaction. The magnetostriction (or the extraordinary Hall coefficient) is shown to be proportional to the spin-orbit energy shift of the d-bands at the Fermi energy and hence changes sign as the Fermi level moves through the lifted degeneracy.

In another band model Campbell[23] relates the equilibrium strain to a distortion of the Fermi surface and a resulting change in Fermi wave vectors parallel and perpendicular to the direction of magnetization. The effect has opposite signs for electrons and holes and hence magnetostriction vanishes when the Fermi surface has compensating electron and hole character.

Both of these models predict a zero-magnetostriction composition that is primarily sensitive to the location of the Fermi level in the d-band. Existing data suggests that the metalloids exhibit effective (with respect to TM d-bands) valences in TM-M glasses[24,25] ranging from about 1.6 e/Boron-atom through 2.4 e/Phosphorus-atom and higher for aluminum (the population of the transition-metal d-states is increased by metalloid presence).[12] This is most clearly seen by comparing saturation moments for a given TM content,[25] e.g., $n_B(Co_{80}B_{20})$ = 1.27 μ_B/TM atom, n_B $(Fe_{40}Ni_{40}P_{14}B_6)$ = 1.14 μ_B/TM atom, and $n_B(Co_{75}P_{16}B_6Al_3) \simeq 1$ μ_B/TM-atom.[4] If a band model were to apply to metallic glasses, one would expect λ_S = 0 to occur on the Fe-rich side of the "perm-alloy" ratio, Fe/Ni $\simeq$ 0.25 (Fig. 2a). No such zero is observed; instead, λ_S approaches zero approximately as $[\sigma(295\ K)]^2$.[10] Magnetostrictions of Fe-Ni-base glasses are, therefore, in conflict with these band models.

While the band models have not previously been applied to Fe-Co alloys, it is worth noting that their λ_S = 0 composition (Fig. 2b) is unaffected by the nature, or presence, of a metalloid which shifts the Fermi level. Furthermore, preliminary data on the Hall effect in $(FeCo)_{80}B_{20}$ glasses show no sign change in R_S near the λ_S = 0 composition. This is seen in Fig. 4 where the Hall resistivity, measured to 10 kOe (and extrapolated to H = $4\pi M_S$ where $4\pi M_S$ > 10 kG) is displayed for five of these glasses. The value of R_S for $Fe_{80}B_{20}$ glass compares well with that measured by Lin[26] on $Fe_{80}P_{13}C_7$ glass, 6 x 10^{-10} V-cm/A-G. The spontaneous Hall coefficient shows no sign change in the $(FeCo)_{80}B_{20}$ system. It is, therefore, unlikely that the zero in magnetostriction of Fe-Co-base glasses is a band structure effect or particularly related to a spin-orbit-lifted degeneracy in the TM d-bands as appears to be the case in the nickel-rich crystalline alloys where λ_S = 0.

Pseudodipolar Model

One of the simplest and earliest approaches to the problem of magnetostriction in crystalline materials considers the strain-induced change in the energy of the dipole-dipole interaction between neighboring spins.[17] While the true dipolar interaction is too weak to explain the magnetostriction in most metals, the spin-orbit interaction is of the proper magnitude and reducible to dipolar form.[18] Although the symmetry of a dipolar interaction between parallel spins causes it to vanish in a cubic crystal, it appears in second order perturbation when the spins are non-parallel (T > 0), and in first order when the cubic lattice is strained. No strain or spin misalignment is necessary for dipolar effects to appear in uniaxial structures. Quadrupolar interactions, which

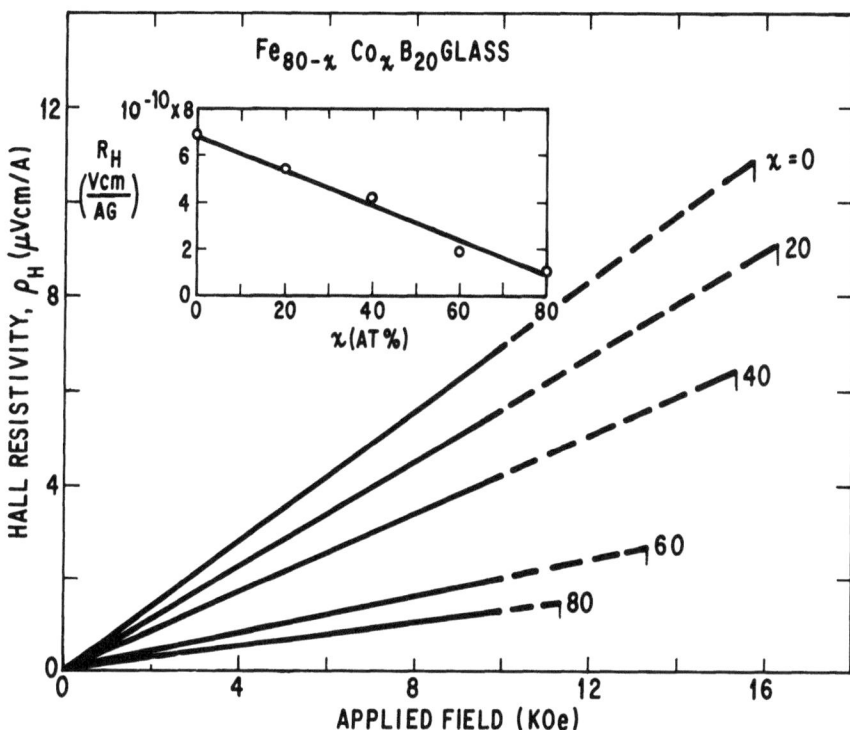

Fig. 4. Hall resistivity as a function of applied field for (FeCo)$_{80}$B$_{20}$ glasses. Data extrapolated to $4\pi M_s$ where necessary. Inset: spontaneous Hall coefficient from extrapolations of ρ_H.

are compatible with cubic symmetry, are important only for $S > 1/2$.[18]

We might, therefore, expect a pseudodipolar model to give a reasonable first approximation to the magnetostriction in some ferromagnetic crystals and in noncrystalline (uniaxial) materials, particularly where $S \lesssim 1/2$. Because little is published about this model,[17,18,27] its general form is outlined in an appendix, it is believed, for the first time. The surprising extent to which it describes magnetostriction in crystalline iron and cobalt will be shown there.

This model was applied to computer-generated, random structures of TM-M alloys.[28] A cluster of 45 atoms (9 of them metalloids) at the center of the structure was selected because it is in this region that the computed RDF and pair correlation function are in best agreement with experiment.[28] Figure 5 shows: (a) the location of the first four TM atoms (the third and fifth atoms are metalloids), and (b) their dominant effect on the strain derivative of the pseudodipolar interaction energy $\Sigma U_D'$, for three directions of strain. Similar calculations have been carried out for the ten independent permutations of these directions. The results show that it is possible to have non-zero magnetostriction of either sign ($\Sigma U_D' > 0 \Rightarrow \lambda < 0$ and vice versa) in this model provided the short-range order is not randomly oriented throughout the material, i.e. provided the pair correlation function is anisotropic.

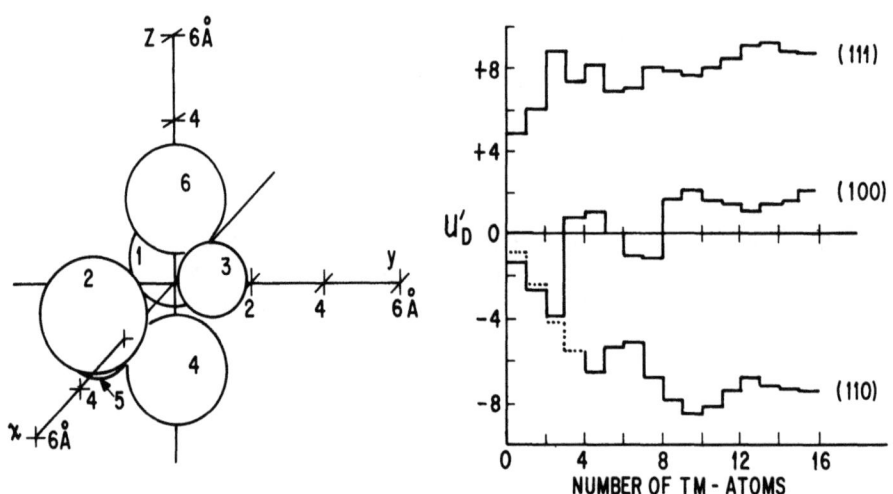

Fig. 5. (a) First six atoms in a computer-generated, random structure of a TM-M (large circles-small circles, respectively) glass; (b) U_D' summed over first 16 TM atoms for three directions of constant-volume strain.

It is suggested, therefore, that the change in sign of magnetostriction in cobalt-rich Fe-Co-base glasses shown in Fig. 1, (regardless of metalloid content, Fig. 3) is related to the existence of a short range structural order in the noncrystalline alloys similar to that in the crystalline alloys (both are close packed with approximately 12-fold coordination), and that this short range order has a nonuniform distribution of orientations throughout the material. This is difficult to verify directly. Nevertheless, directional ordering is consistent with the observed macroscopic anisotropy of ferromagnetic glasses[16] as well as with their susceptibility to field annealing.[1,16]

The lack of correlation between the magnetostriction of crystalline and noncrystalline iron-rich alloys (Fig. 2) may then be related to the fact that the former have bcc structure with a coordination of 8 while the latter are best described by random dense packed models with coordinations of about 12.[29]

Quantum-Statistical-Mechanical Theory

Callen and Callen have developed a comprehensive and rigorous quantum-statistical-mechanical theory for anisotropy and magnetoelastic effects.[20] Their theory shows these phenomena to arise from three mechanisms: two-ion isotropic exchange (giving rise to K_0 terms and volume magnetostriction), and two-ion anisotorpic exchange and single-ion anisotropy (both giving rise to linear magnetostriction and various components of anisotropy). The data of Fig. 1, when plotted as a function of the saturation magnetization of each alloy,[10,25] show completely different behavior for Fe- and Ni-base glasses on the one hand and Co-containing glasses on the other. It is, therefore, suggested that one of the anisotropic mechanisms dominates in Fe-Ni-based glasses while another mechanism is also important in Co-base glasses. Temperature dependence of magnetostriction is being measured to determine the operating spin interaction(s).[10]

IV. SUMMARY

The magnetostriction of TM-M glasses shows a dependence on TM content unlike that of comparable polycrystalline materials except in the case of cobalt-rich alloys. Observed magnetostrictions of iron-nickel-base glasses are incompatible with existing band models for iron-nickel crystalline alloys. It is also unlikely that the zero in the magnetostriction of Fe-Co-base glasses is a band structure effect. A localized spin model which sums the strain derivative of pseudodipolar interaction energies over nearest neighbors adequately describes the magnetostrictions of crystalline Fe and Co. This model suggests short range order in cobalt-rich Fe-Co-base glasses similar to that in the corresponding crystalline alloys. Consideration of the theory of Callen and Callen suggests that the magnetostriction of iron- and nickel-based glasses is dominated by one anisotropic exchange mechanism whereas another is important in Co-rich glasses.

APPENDIX

The pseudodipolar model is expressed in general terms and applied to the magnetostrictions of crystalline Fe and Co.

The strain derivative (assuming no change in volume) of $U_D = C\mu^2(1-3\cos^2\alpha)/r^3$, the pseudodipolar interaction energy, is given by:

$$U_D' = \frac{\partial U_D}{\partial \alpha} d\alpha + \frac{\partial U_D}{\partial r} dr = 3C(\mu^2/r^3)[\sin 2\alpha \, d\alpha - (1-3\cos^2\alpha)\frac{dr}{r}], \quad (A1)$$

where α and r locate the second ion relative to an axis of symmetry (the direction of $\bar{M}$) which passes through the first. The constant C is the ratio of spin-orbit to true dipolar interaction strength and μ is the magnetic moment at each site. Using the relations

$$\frac{dr}{r} = \lambda(\cos^2\alpha - \frac{1}{2}\sin^2\alpha), \text{ and } d\alpha = -\frac{3}{4}\lambda\sin 2\alpha, \quad (A2)$$

Eq. A1 can be reduced to

$$U_D' = 3C\mu^2\lambda \, (1+3\cos^2\alpha - 15/4\sin^2 2\alpha)/2r^3. \quad (A3)$$

For a state of strain to be stable, $U_D' < 0$ when summed over near neighbors. Fig. A1 shows the dependence of U_D' on α. The solid (dashed) line shows the results for a fractional volume change of zero (equal to λ). Clearly, in this model, for linear magnetostriction to be positive along a given crystallographic direction, the distribution of nearest neighbors should be concentrated on cones about the direction of $\bar{M}$ with semi-apex angles in the range 35°-75°.

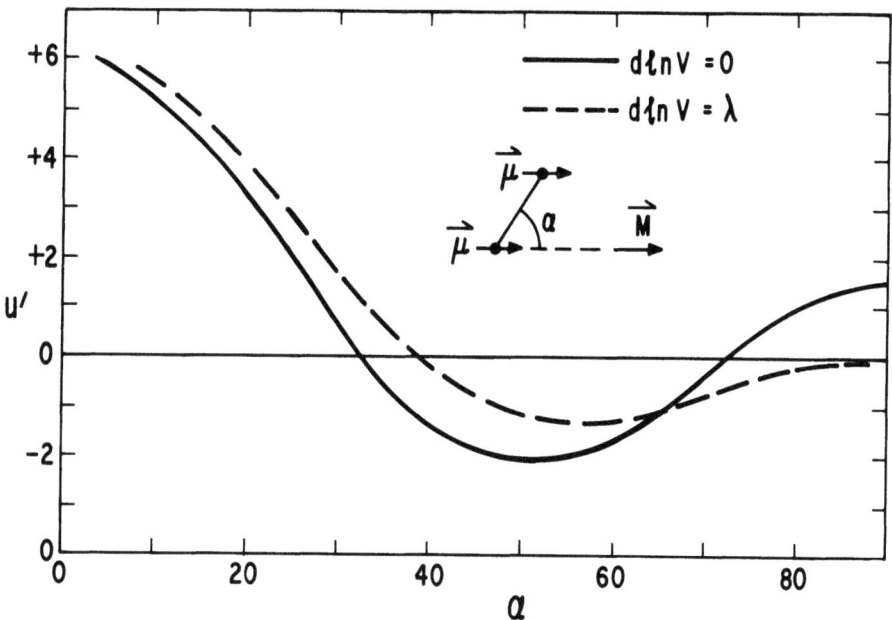

Fig. A1. Dependence of strain derivative of dipole-dipole interaction energy (units of $C\mu^2\lambda/r^3$) on azimuthal angle, α. See Eq. A3 in text.

TABLE A1

Constant-volume strain derivative of dipolar interaction energy (Eq. A3) summed over nearest, next nearest, and next-next-nearest, neighbors (units of $C\mu^2\lambda/a^3$, a = respective lattice constant), and magnetostrictions for single crystals of Fe and Co.

STRAIN DIRECTION	SC U'_D	BCC(α) U'_D	λ(Fe)[a]	FCC(γ) U'_D	HCP(ε) U'_D	λ(Co)[b]
100	+11.7	−9.8	+21	−10.6	+4.0	−50 (1000)
110	−2.9	+2.5	−10	+2.7	+3.5	−46 (1100)
111	−7.8	+11.6	−21	+7.1	+10.5	−90 (0001)

[a] Ref. 23, p. 653.

[b] R. M. Bozorth, Phys. Rev. 96, 311 (1954).

single crystal and polycrystalline Fe and Co are included for comparison. Where $U_D' < 0$, experiment shows $\lambda_s > 0$, and vice versa, indicating that the magnetostriction of these metals is well described by this model; it does not describe λ_s in Ni.

Using arguments similar to those outlined here, Goldman and Smoluchowski[27] showed that the effects of the order-disorder transformation on magnetostriction in 50% Fe-Co could be properly accounted for, disorder lowering the saturation magnetostriction.

Generalization of this model to polycrystalline or noncrystalline materials is tenuous because one might expect any direction of magnetization to have a nearly uniform distribution (in α) of near neighbors. Hence the model would predict zero magnetostriction because

$$\int_0^{\pi/2} U_D' \sin\alpha \, d\alpha = 0. \tag{A4}$$

Nevertheless, the model does correctly identify the sign change of λ_s in cobalt-rich polycrystalline Fe-Co with the $\gamma \to \varepsilon$ phase transformation ($\sim$92% Co,[30] see Fig. 2b and Table A1). Consistent with this, the negative magnetostriction of polycrystalline cobalt (ε phase) becomes positive upon heating through the $\varepsilon \to \gamma$ transformation[30,31] at about 400°C.

REFERENCES

1. B. S. Berry and W. C. Pritchet, Phys. Rev. Lett. <u>34</u>, 1022-1025 (1975); B. S. Berry, Joint 3M/INTERMAG Conf., June 15-18, 1976, Pitt., Pa.
2. T. Egami, P. J. Flanders and C. D. Graham, Jr., Appl. Phys. Lett. <u>26</u>, 128-130 (1975); T. Egami, P. J. Flanders and C. D. Graham, Jr., AIP Conf. Proc. No. 24, p. 697-701 (1975).
3. A. W. Simpson and W. G. Clements, IEEE Trans. Mag. <u>11</u>, 1338-1340 (1975).
4. R. C. Sherwood, E. M. Gyorgy, H. S. Chen, S. K. Ferris, G. Norman and H. J. Leamy, AIP Conf. Proc. No. 24, p. 745-746 (1975).
5. N. Tsuya, K. I. Arai, Y. Shiraga and T. Masumoto, Phys. Lett. <u>51A</u>, 121-122 (1975); N. Tsuya, K. I. Arai, Y. Shiraga, M. Yamada and T. Masumoto, phys. stat. sol. (a) <u>31</u>, 557-561 (1975).
6. H. Fujimori, K. I. Arai, H. Shirae, H. Sato, T. Masumoto and N. Tsuya, Japan. J. Appl. Phys. <u>15</u>, 705-706 (1976); M. Kikuchi, H. Fujimori, Y. Obi and T. Masumoto, Japan. J. Appl. Phys. <u>14</u>, 1077-1078 (1975).
7. R. C. O'Handley, AIP Conf. Proc. No. 29, 206-208 (1976).

8. H. A. Brooks, J. Appl. Phys. 47, 334-335 (1976).
9. R. C. O'Handley, L. I. Mendelsohn and E. A. Nesbitt, Joint 3M/INTERMAG Conf., June 15-18, 1976, Pitt., Pa.
10. R. C. O'Handley, submitted for publication.
11. R. Hasegawa, AIP Conf. Proc. No. 29, 216-217 (1976).
12. R. Hasegawa and C.-L. Chien, Sol. St. Comm. 18, 913-916 (1976).
13. C.-L. Chien and R. Hasegawa, AIP Conf. Proc. No. 29, 214-216 (1976); J. Appl. Phys. 47, 2234-2236 (1976); Joint 3M/INTERMAG Conf., June 15-18, 1976, Pitt., Pa.
14. R. Hasegawa, R. C. O'Handley, L. E. Tanner, R. Ray and S. Kavesh, Appl. Phys. Lett. 29, 219-221 (1976); R. Hasegawa, R. C. O'Handley and L. I. Mendelsohn, Joint 3M/INTERMAG Conf., June 15-18, 1976, Pitt., Pa.
15. T. Egami and P. J. Flanders, AIP Conf. Proc. No. 29, 220-221 (1976); F. E. Luborsky, AIP Conf. Proc. No. 29, 209-210 (1976); R. C. Sherwood, E. M. Gyorgy, H. J. Leamy and H. S. Chen, Joint 3M/INTERMAG Conf., June 15-18, 1976, Pitt., Pa.
16. Domain observations on several metallic glasses (H. J. Leamy, S. D. Ferris, G. Norman, D. C. Joy, R. C. Sherwood, E. M. Gyorgy and H. S. Chen, Appl. Phys. Lett. 26, 259-260 (1975); and Ref. 14, present work) suggest a local uniaxial anisotropy whose direction is determined largely by flow patterns during quenching. FMR (ref. 11) and Mössbauer spectroscopy (ref. 12, 13) indicate that the macroscopic anisotropy is uniaxial with the easy axis generally in-plane.
17. N. S. Akulov, Zeits. fur Physik 52, 389 (1928); 59, 254 (1930); 69, 78 (1931); R. Becker, Zeits. fur Physik, 62, 253-269 (1930).
18. J. H. VanVleck, Phys. Rev. 52, 1178-1198 (1937).
19. C. Kittel, Rev. Mod. Phys. 21, 541-583 (1949).
20. E. R. Callen and H. B. Callen, Phys. Rev. 129, 578-593 (1963); Phys. Rev. 139A, 455-471 (1965); and E. Callen, J. Appl. Phys. 39, 519-527 (1968).
21. L. Berger, Phys. Rev. 138A, 1083-1087 (1965); H. Ashworth, D. Sengupta, G. Schnakenberg, L. Shapiro and L. Berger, Phys. Rev. 185, 792-797 (1969).
22. J. Smit, Physica 17, 612 (1951); L. Berger, Physica 30, 1141-1159 (1964).
23. I. A. Campbell, Sol. St. Comm. 10, 953-955 (1972).
24. R. C. O'Handley and D. S. Boudreaux, AIP Conf. Proc. No. 29, p. 161; K. Yamauchi and T. Mizoguchi, J. Phys. Soc. Japan 39, 541-542 (1975).
25. R. C. O'Handley, R. Hasegawa, R. Ray and C.-P. Chou, Appl. Phys. Lett. 29, 330-332 (1976).
26. S. C. H. Lin, J. Appl. Phys. 40, 2175-2177 (1969).
27. J. E. Goldman and R. Smoluchowski, Phys. Rev. 75, 140-147 (1949).
28. D. S. Boudreaux and J. M. Gregor, Second Int. Conf. on Rapidly Quenched Metals, Nov. 17-19, 1975 (MIT Press); D. S. Boudreaux, this conference.

29. G. S. Cargill, III, Sol. St. Phys. 30, Eds., H. Ehrenreich, F. Seitz and D. Turnbull (Academic Press, NY, 1975) p. 227-320.
30. M. Hanson, Constitution of Binary Alloys (McGraw Hill, NY, 1958) p. 472.
31. R. M. Bozorth, Ferromagnetism (Van Nostrand, NY, 1955) p. 659.

DISCUSSION

E. Callen: As a way of comment, I would like to mention that one must be careful at looking at a room temperature magnetostriction measurement and interpreting it as the intrinsic magnetostriction, because the magnetostriction has a temperature dependence which is based upon $M(T)/M(0)$. So that even though you may not change the intrinsic magnetostriction, by simply alloying one changes the Curie temperature and as a result changes the reduced magnetization. This makes it easy to see some changes in the room temperature magnetostriction which could falsely be attributed to changes in the intrinsic magnetostriction. The other comment I have is that in a metal it is very difficult to analyze what the intrinsic magnetostriction is because it comes about from a whole bunch of complicated mechanisms such as the spin-orbit coupling in the crystal field. The anisotropy is relatively easy to do but one envisions a fixed crystal field that rotates the magnetization around so that one is measuring the difference in energies in different directions. When one distorts the crystal the electrons redistribute, and that redistribution changes the screening of the crystal field and that in turn may affect the sign of the predominant term and that is very hard to estimate theoretically.

R. C. O'Handley: As a comment not directly to that but since you did refer to the anisotropy certainly a study of anisotropy and magnetostriction should go hand in hand but the nature of the materials currently on hand makes such a study difficult. Ideally one needs a disk-shaped sample with well defined dimensions so that one can do torque measurments on it. This may be one reason why studies of magnetostriction have gone ahead of anisotropy studies in these materials.

ON THE MAGNETICALLY INDUCED ANISOTROPY IN AMORPHOUS FERROMAGNETIC ALLOYS

H. Fujimori, H. Morita, Y. Obi and S. Ohta

The Research Institute for Iron, Steel and Other Metals

Tohoku University, Sendai, Japan

ABSTRACT

Compositional dependence of the magnetically induced magnetic anisotropy in the Fe and Co based amorphous alloys has been studied experimentally. The induced anisotropy constant is 30-60 erg/g for $Fe_{78}Si_{10}B_{12}$ and for $Co_{78}Si_{10}B_{12}$ but is 400-500 erg/g for $(Fe_{1-x}Co_x)_{78}Si_{10}B_{12}$ in $0.25<x<0.8$. The addition of elements such as Ni, Pd and Cr into $Co_{78}Si_{10}B_{12}$ results in the increase in induced anisotropy constant. The directional order of metal-metal pairs is most likely to be responsible for the induced anisotropy.

INTRODUCTION

Recently, there has been a fair amount of interest in the high permeability amorphous alloys as possible magnetic materials[1,2,3,4]. It has been shown experimentally that the magnetic properties of the amorphous alloys can be altered not only by constitutional alloy composition, heat-treatment or fabrication condition but also by magnetic field cooling.

Berry and Pritchet[5] have first examined the field cooling effect on the $Fe_{75}P_{15}C_{10}$ amorphous alloy. The effect on other amorphous alloys have then been studied by many workers[4,6,7,8]. The mechanism concerning the magnetic anisotropy induced by field cooling, however, has not yet been made clear.

We studied the compositional dependence of the field induced magnetic anisotropy in the amorphous ferromagnetic alloys of nominal compositions $(Fe_{1-x}Co_x)_{78}Si_{10}B_{12}$ and $(Co_{1-x}Y_x)_{78}Si_{10}B_{12}$, where Y is Pd, Ni and Cr, for the purpose of clarifying the mechanism of the field induced anisotropy.

EXPERIMENTAL

The specimens in the form of ribbon (2mm x 30μm in cross-section) were prepared by using a standard rapid quenching technique. Some of the as-quenched specimens and magnetically annealed specimens were confirmed to be in the amorphous state by the X-ray diffraction method. The electrical resistance of all the as-quenched alloys showed a sharp drop as the temperature was increased. This critical temperature has been found to correspond to the crystallization temperature which has been determined by the differential calorimetric experiment for the same compositional amorphous alloys[9]. The saturation magnetization per gram was also measured at various temperatures by the conventional magnetic balance. The data of the crystallization temperature (T_{cry}), the Curie temperature (T_c) and the saturation magnetization (M_s,RT) are summarized in Fig. 4-a.

The magnetization vs. magnetic field curves at room temperature were measured along the ribbon axis of 8cm long specimen by means of the conventional Cioffi recording fluxmeter. The energy value of the field induced magnetic anisotropy was determined from the area between the two magnetization curves each obtained after the heat-treatments in a field applied parallel to the ribbon axis and in a field applied perpendicular to the axis. The heat-treatment was carried out in an evacuated quartz tube.

RESULTS AND DISCUSSIONS

It is known that the as-quenched ribbons show a magnetic anisotropy as is evident from the magnetization (M) vs. magnetic field (H) curve measured along the ribbon axis. The energy value of the anisotropy (K) is often defined as $\int_0^{M_s} HdM$, where M_s is the saturation magnetization. The annealing at temperatures below the crystallization temperature decreases the anisotropy. Such a magnetic anisotropy existing in the as-quenched specimen has been explained in terms of the internal stress-magnetostriction[8,10,11]*.

---

*Takahashi et al. have pointed out that the as-quenched ribbons may involve a very small amount of micro-crystallites and its effect on the anisotropy is important[12].

The magnitude of the anisotropy defined above changes differently when heat-treated in magnetic fields. Fig. 1-a shows the example that the value of K depends on the annealing temperature (T_a) as well as the direction of the field applied during the annealing. The annealing was carried out for 15 min. at each temperature from about 100°C up to near T_{cry}. As can be seen in the figure, the value of K obtained after the annealing in a field applied along the ribbon axis is always smaller than that obtained after the annealing in a field applied across the ribbon width. This is obviously due to the field induced magnetic anisotropy having an easy magnetization direction parallel to the applied field direction. The magnitude of the induced anisotropy (K_{ind}) is given by the difference in the value of K between these two cases. Fig. 2 shows the dependence of K_{ind} on T_a for some of the amorphous alloys examined. K_{ind} vs. T_a curve always takes a maximum value around 200°C–300°C and K_{ind} disappears near the Curie temperature of each alloy. These curves suggest that the alloys respond to the magnetic annealing at temperatures below the Curie temperature but high enough for diffusion to occur. The maximum values of K_{ind}

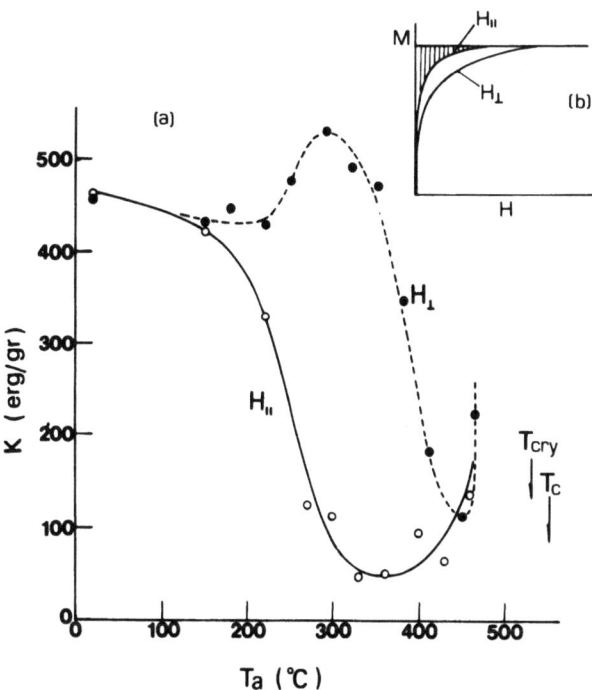

Fig. 1. Magnetic anisotropy constant (K) vs. annealing temperature (T_a) curve for the $Fe_{58}Co_{20}Si_{10}B_{12}$ amorphous alloy.
$H_{\parallel}$: annealed in a field applied along the ribbon axis.
$H_{\perp}$: annealed in a field applied across the width.
K corresponds to the hatched area to the left of the magnetization curve shown in the schematic picture of (b).

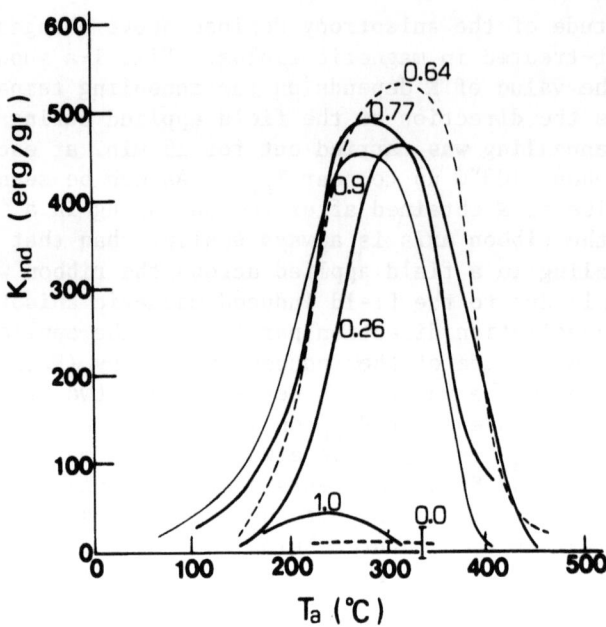

Fig. 2. Induced magnetic anisotropy constant (K_{ind}) vs. annealing temperature (T_a) curves for the $(Fe_{1-x}Co_x)_{78}Si_{10}B_{12}$ amorphous alloys. Values in the figure indicate the concentration x.

thus obtained are plotted as a function of $x = Co/(Fe + Co)$ in Fig. 4.

In an another annealing sequence, the sample was first heated up to 380°C below the crystallization temperature and then cooled in a field applied along the ribbon axis to the room temperature with a constant cooling rate (1.6°C/min.). The magnetization of this sample saturates easily as shown in Fig. 3 (curve-$H_\parallel$). Next, the sample was re-heated up to 380°C and then cooled in a field applied across the width of the ribbon. The magnetization in this case reaches its saturation at a higher field as shown in Fig. 3 (curve-$H_\perp$). The area between the two magnetization curves for each cooling cases gives the magnitude of the magnetic anisotropy induced during the field cooling (K_{ind}). The changes in magnetization curve due to the induced anisotropy were reversible during the repeated cooling sequence. The magnetic torque measurements revealed that the induced anisotropy can also be reversibly changed by switching the field direction at the annealing temperature, with an activation energy of about 0.6 eV[13]. The data obtained from the field cooling are also plotted in Fig. 4. As can be seen in the figure, the values of K_{ind} obtained by the two different annealing processes

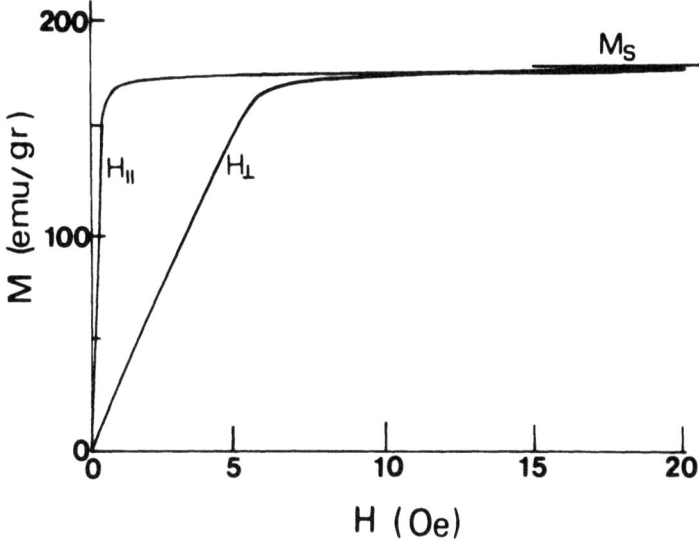

Fig. 3. Magnetization (M) vs. magnetic field (H) curves measured along the ribbon axis for the $Fe_{58}Co_{20}Si_{10}B_{12}$ amorphous alloy subjected to the field cooling from 380°C. $H_{\parallel}$: cooled in a field applied along the ribbon axis. $H_{\perp}$: cooled in a field applied across the width. M_s is the saturation magnetization.

described above are in good agreement, indicating that the obtained K_{ind} represents the saturated value of the induced magnetic anisotropy.

An important feature of these results is that the magnitude of the induced anisotropy depends strongly on the concentration of x. The values of K_{ind} for $Fe_{78}Si_{10}B_{12}$ are as small as 30-60 erg/g, while K_{ind} increases sharply with increasing x as Fe in $Fe_{78}Si_{10}B_{12}$ is substituted with Co in the range of 0<x<0.25. With further increasing x, K_{ind} takes the value of 400-500 erg/g in the range of 0.25<x<0.8 showing a plateau. In the range of x>0.8, K_{ind} decreases sharply down to the small value for $Co_{78}Si_{10}B_{12}$. As far as the alloys in the Fe side are concerned, the values of K_{ind} of the present amorphous alloys are in the same order of magnitude with those of the bulk crystalline Fe-Co alloys subjected to a field cooling[14,15] but are smaller than those of the Fe-Co thin films produced by the evaporation in a magnetic field[15]. The value of 30-60 erg/g for K_{ind} of $Fe_{78}Si_{10}B_{12}$ is significantly smaller than 117 erg/g (900 erg/cc) of K_{ind} for a single crystal 3.25 at% Si-Fe alloy[16].

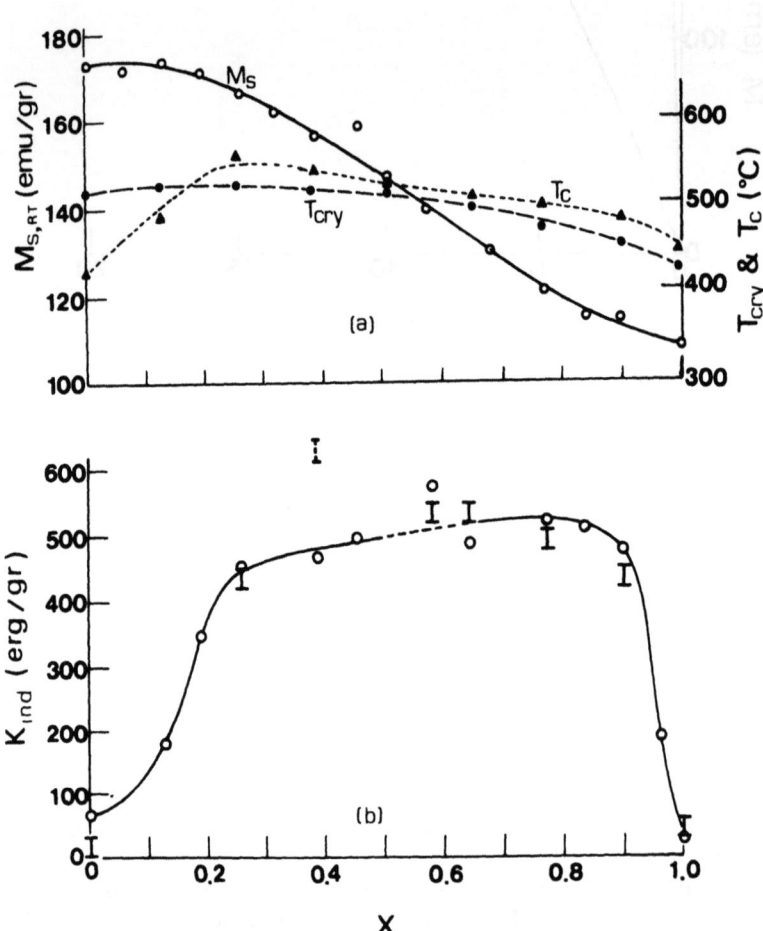

Fig. 4. Saturation magnetization ($M_{s,RT}$), crystallization temperature (T_{cry}), Curie temperature (T_c) and induced magnetic anisotropy (K_{ind}) for the $(Fe_{1-x}Co_x)_{78}Si_{10}B_{12}$ amorphous alloys as a function of x. In the figure (b), I : obtained from the first field annealing sequence (see text), maximum values of K_{ind} in Fig. 2 are shown. o : obtained from the second field annealing sequence.

The above result suggests that the co-existence of the different kinds of metals of Fe and Co is essentially important in obtaining the large values of K_{ind}. It is then expected that the large anisotropy can also be induced in other pseudo-binary alloys. For the purpose of checking this view, we have examined the induced magnetic anisotropies for the amorphous alloy systems of $(Co_{1-x}Y_x)_{78}Si_{10}B_{12}$, where Y is Ni, Pd and Cr. The annealing was done for 15 min. at each temperature from about 100°C up to near T_{cry} of each alloy. The results obtained are summarized in Fig. 5-a together with the data of $(Co_{1-x}Fe_x)_{78}Si_{10}B_{12}$ in the Co rich region. It can be clearly seen in the figure that the addition of either ferromagnetic or nonferromagnetic element into the $Co_{78}Si_{10}B_{12}$ results in the increase in K_{ind}, excepting for the cases of higher concentration of Cr. The small values of K_{ind} in the case of the Cr alloys may be due to their extremely small magnetization at the annealing temperature. Fig. 5-b shows the saturation magnetizations at room temperature as a function of concentration of solute atom.

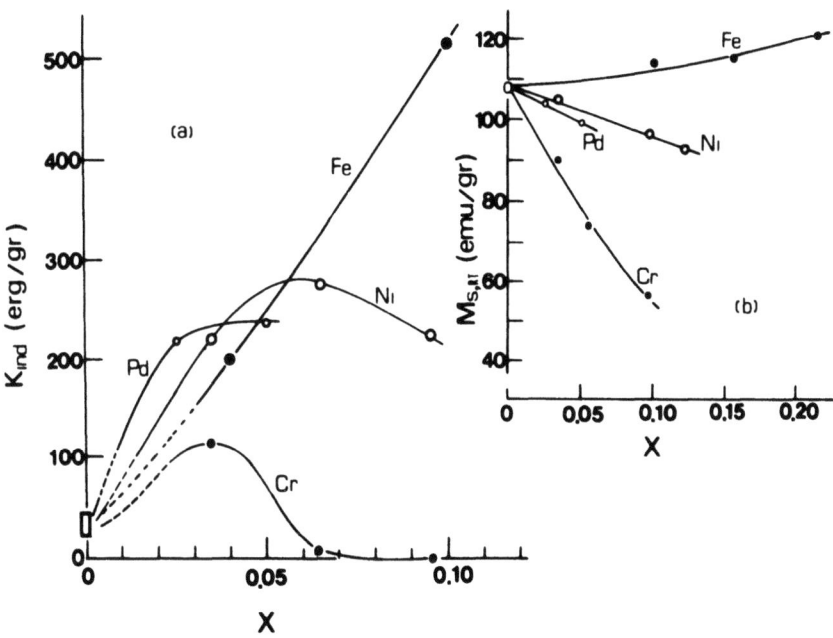

Fig. 5. Induced magnetic anisotropy (K_{ind}) and saturation magnetization ($M_{s,RT}$) for the $(Co_{1-x}Y_x)_{78}Si_{10}B_{12}$ amorphous alloys as a function of x, where Y is Fe, Ni, Pd and Cr.

Many theories have so far been proposed in order to interpret the behaviour of the field induced magnetic anisotropy[17]. The magnetostriction of the present amorphous alloys takes the largest value in $Fe_{78}Si_{10}B_{12}$ and decreases monotonically with increasing x in the similar manner to the case of the $(Fe_{1-x}Co_x)_{80}P_{13}C_7$ alloys[18]. Therefore, the compositional dependence of K_{ind} shown in Fig. 4-b can not be explained in terms of the stress-magnetostriction. From the structural feature of the amorphous alloys, it is considered that the precipitation theory on induced magnetic anisotropy must also be excluded as possible interpretation. The directional-short range order of atom pairs has first been considered by Berry and Pritchet[5] as a possible cause of the induced magnetic anisotropy for the amorphous alloy. Present results appear to be best explained from that point of view.

It should be noted that, in the amorphous structure, the metals involved are nearly equivalent to each other in their relative posisions because of their similar atomic size, while the metalloids Si and B are not equivalent to the metals. In this situation, it is likely that the directionality in distribution of atom pairs between the metals is easy to occur without destruction of amorphous structure. In contrast to this, the inducement of the directionality in distribution of atom pairs between the metals and the metalloids requires a relatively higher energy and is unlikely to happen. This simple speculation is quite consistent to the fact that the observed values of K_{ind} are very small in $Fe_{78}Si_{10}B_{12}$ and $Co_{78}Si_{10}B_{12}$ but are large in the pseudo-binary $(Fe_{1-x}Co_x)_{78}Si_{10}B_{12}$ and $(Co_{1-x}Y_x)_{78}Si_{10}B_{12}$ alloys.

The compositional dependence of K_{ind} for $(Fe_{1-x}Co_x)_{78}Si_{10}B_{12}$ further supports the view that the field induced anisotropy involves the directional ordering of the metal atoms. According to Taniguchi's and Neel's theories[19] concerning the field cooling effect of the disordered alloy, the induced uniaxial magnetic anisotropy constant (K_u) is proportional to $-(Nl_ol_o')/kT$, where N is the parameter representing the probability of finding numbers of unbalanced pairs, l_o and l_o' the mean coefficients of the dipole interactions at measuring temperature and annealing temperature respectively, k the Boltzman constant and T the annealing temperature. In the disordered binary alloy, N can be assumed to be proportional to $c^2(1-c)^2$, where c is the alloy concentration. l_o and l_o' are proportional to the value of M_s^2 at each temperature. Under the assumption that the above relations can be applied to the case of the $(Fe_{1-x}Co_x)_{78}Si_{10}B_{12}$ amorphous alloy, we have tried to evaluate the values of $K_i/(M_{RT}/M_o)^2(M_{300}/M_o)^2$ as a function of x, where M_{RT}, M_{300} and M_o are the saturation magnetizations at room temperature and 300°C and 0°K, respectively. The result is shown in Fig. 6*.

*Such a treatment was shown to be important by Luborsky for the amorphous Fe-Ni base alloy[20].

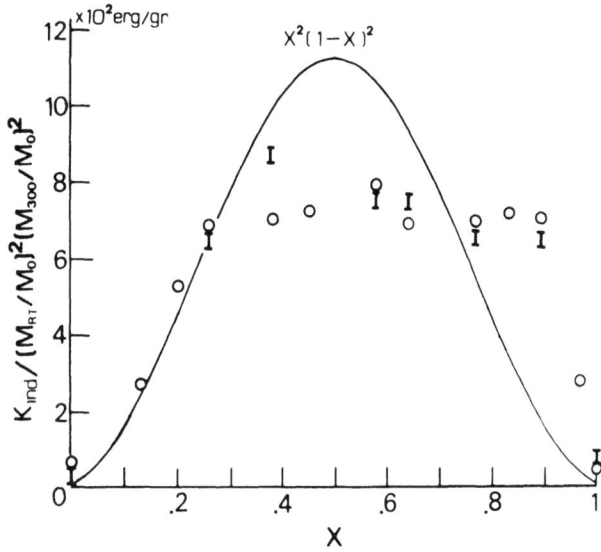

Fig. 6. Comparison between the compositional dependence of $K_i/(M_{RT}/M_o)^2(M_{300}/M_o)^2$ and the curve of $x^2(1-x)^2$ vs x for the $(Fe_{1-x}Co_x)_{78}Si_{10}B_{12}$ amorphous alloys.

In the same figure, the curve of $x^2(1-x)^2$ vs x is also shown for the comparison. As can be seen in the figure, the experimental curve of $K_i/(M_{RT}/M_o)^2(M_{300}/M_o)^2$ vs x agrees fairly well with the curve of $x^2(1-x)^2$ vs x but in the limited range of $0 < x < 0.3$. This agreement implies that the induced magnetic anisotropy in this range is most likely to be due to the directional order of atom pairs.

However, as can also be seen in Fig. 6, the discrepancy betweeen the experimental curve and the theoretical curve in the range of 0.3 x 1 is quite large. This discrepancy suggests that another mechanism in addition to the directional order must be introduced for a complete explanation of the field induced anisotropy in the Co rich amorphous alloys.

ACKNOWLEDGMENT

The authors wish to express their thanks to Prof. T. Masumoto of Tohoku University for his help and encouragement throughout the present work. The authors are also indebted to Prof. Y. Nakagawa, Prof. M. Takahashi of Tohoku University and Prof. T. Egami of University of Pennsylvania for their valuable discussions.

REFERENCES

1. T. Egami, P.J. Flanders and C.D. Graham, Jr., AIP Conf. Proc. 24(1975) 697.
2. R.C. Sherwood, E.M. Gyorgy, H.S. Chen, S.D. Ferris, G. Norman & H.J. Leamy, AIP Conf. Proc. 24(1975) 745. E.M. Gyorgy, H.J. Leamy, R.C. Sherwood and H.S. Chen, AIP Conf. Proc. 29(1976) 198.
3. M. Kikuchi, H. Fujimori, Y. Obi and T. Masumoto, Japan. J. Appl. Phys. 14(1975) 1077. H. Fujimori, M. Kikuchi, Y. Obi and T. Masumoto, Sci. Rept. RITU 26(1976) 36.
4. F.E. Luborsky, J.J. Becker and R.O. McCary, IEEE Trans. Mag. 11(1975) 1644.
5. B.S. Berry and W.C. Pritchet, Phys. Rev. Letters 34(1975) 1022.
6. H. Fujimori and T. Masumoto, Trans. JIM 17(1976) 175.
7. H.S. Chen, S.D. Ferris, E.M. Gyorgy, H.J. Leamy and R.C. Sherwood, Appl. Phys. Letters, 26(1976) 405.
8. T. Egami and P.J. Flanders, AIP Conf. Proc. 29(1976) 220.
9. A. Inoue, T. Minemura, M. Kikuchi and T. Masumoto, 78th. Annual Meeting of the Japan Institute of Metals, April(1976).
10. H. Fujimori, Y. Obi, T. Masumoto and H. Saito, Mat. Sci. and Eng. 23(1976) 281.
11. R.C. O'Handley, AIP Conf. Proc. 29(1976) 206.
12. M. Takahashi, F. Ono and K. Takakura, Japan.J. Appl. Phys. 15(1976) 183. M. Takahashi, T. Suzuki and K. Takakura, Japan. J. Appl. Phys. 15(1976) 711.
13. H. Morita et al. to be published.
14. M.J. Marechal, J. Phys. Rad. 16(1955) 122S.
15. M. Takahashi, private communication.
16. H.C. Fiedler and R.H. Pry, J. Appl. Phys. 30(1959) 109S.
17. S. Chikazumi and S.H. Charap, Physics of Magnetism (Wiley, New York, 1964). p. 359. C.D. Graham, Jr., in Magnetic Properties of Metals and Alloys (American Society for Metals, Cleveland, Ohio, 1959), p.288.
18. H. Fujimori, H. Shirae, K.I. Arai, T. Masumoto and N. Tsuya, Japan. J. Appl. Phys. 15(1976) 709.
19. L. Neel, Comp. Rend. 237(1953) 1613. S. Taniguchi, Sci. Rept. RITU A7(1955) 269.
20. F.E. Luborsky, private communication.

APPLICATION OF DOMAIN WALL PINNING THEORY TO AMORPHOUS FERROMAGNETIC MATERIALS

David I. Paul

Columbia University

New York, N.Y. 10027

INTRODUCTION

Much work has been done on the coercive force of amorphous ferromagnetic materials.[1-4] Although qualitative and some quantitative aspects of this property are beginning to be understood, further quantitative explanations are still desirable. Without these, it is difficult to arrive at any basic understanding of the interplay of the various characteristics of the medium on the coercive force or to set tolerances on the fabrication techniques for these materials.

In this paper, we shall attempt to apply the theory of Friedberg and Paul[5] on domain wall pinning to soft (low coercive force) amorphous materials. This theory, derived for bulk materials, is concerned with those aspects of pinning caused by "reflection and tunneling" of the bound magnetic moment disturbance (that represents the domain wall) in the presence of changes in the medium. The mechanisms of domain wall pinning considered in this theory do not include those based on magnetostatic energy changes in the vicinity of the defects - originally calculated by Neel[6] for the low coercive force materials. (For the high coercive force materials, such magnetostatic energy considerations are insufficient to account for the large experimentally observed values.) By the definition of the coercive force as the field necessary to demagnetize the substance, that mechanism which proves to be the largest of the obstacles to wall motion becomes the measure of the resistant force - assuming that conditions prevail within the material which allow that mechanism to operate. Two different types of magnetostatic effects were considered by Neel: that caused by non-magnetic inclusions is illustrated in Fig. 1. In amorphous materials where inclusions are

Fig. 1. Illustration of coercive force mechanism due to non-magnetic inclusions.

more realistically represented as small crystalline defects which remain ferromagnetic with a saturation magnetization roughly comparable to that of the amorphous material, this mechanism would be inoperative. The second type is that caused by random strain fluctuations. The assumption is made that the direction of magnetization within a domain is partially randomized by the effective anisotropy resulting from regions of strain irregularly distributed within the material. These result in the formation of magnetic poles -the energy from which may be reduced by appropriate positioning of domain walls. We shall discuss this further in the appendix where we consider its application to amorphous $Fe_{80}P_{13}C_7$.

The theory of Friedberg and Paul represents a general solution to the non-linear differential equations governing the mechanism of 180° domain wall motion in the presence of a planar barrier. The geometry considered is shown in Fig. 2 where regions 1 and 3 are identical and represent the homogeneous material while region 2 represents the energy barrier characterized by an abrupt change in the properties of the magnetic material. The total energy is

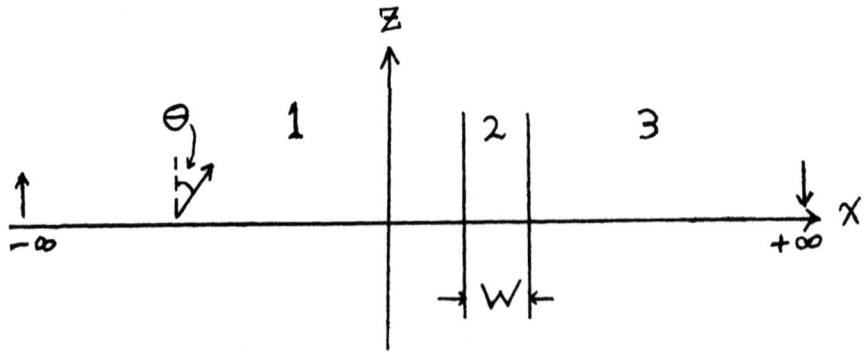

Fig. 2 Geometry of medium for theory of Friedberg and Paul.

$$\int [A\theta'^2 + K\sin^2\theta - H_e M\cos\theta] dv. \qquad (1)$$

The resultant solution is given by the formula for the coercivity,

$$H_c = \pm 0.38 \frac{K_1 W}{M_1 \delta_1}\left[\frac{A_1}{A_2} - \frac{K_2}{K_1}\right], \qquad (2)$$

where θ is the angle between the magnetic moment and the z axis, H_e represents the external magnetic field, K is the magnetic anisotropy, M is the saturation magnetization, W is the width of the defect region, A is the exchange energy constant, and δ is the domain wall width parameter given by $(A/K)^{1/2}$. The subscripts 1 and 2 refer to the regions shown in Fig. 2 and the sign is chosen such that $H_c > 0$.

The quantity $K_1/M_1\delta_1$ represents the "intrinsic resistance" of the material to the coercive force. The ratio W/δ_1 indicates the importance of the defect width to the domain wall width. The factor $W[(A_1/A_2) - (K_2/K_1)]$ characterizes the planar defect or barrier and is a measure of its pinning strength. Any changes in the medium which affect the original pulse of domain wall shape will act as obstacles to its propagation. From the above, bounds can be placed on the spatial extent or permissible changes of a defect region before they seriously affect the coercive force.

We shall apply Eq. (2) as it stands to the iron rich amorphous ferromagnets presently being produced in the form of ribbons and to the amorphous $GdCo_4$ material used in bubble domain work – both of which are low coercive force materials. We recognize that, in real materials, defects are usually not planar nor necessarily parallel to domain walls. The increased complexity for more general geometries is such that neither equivalent closed form algebraic solutions nor equivalent computer type calculations have as yet been attained for the coercive force. Therefore, we shall assume that, for almost all geometries, there is a "break-away" region where part of the domain wall is parallel to part of the defect region and obeys Eq. (2). Further, we recognize that, even when domain walls have been experimentally observed in a material, the dominant cause of the coercive force may not be due to domain wall pinning but to domain wall nucleation. Thus, Eq. (2) tells us the resistance to wall motion – given that the wall preexists or is easily formed. Bearing the above in mind, we attempt in the next two sections to demonstrate some quantitative features of the effects of the fabrication process for amorphous material on the coercive force.

AMORPHOUS $Fe_{80}P_{13}C_7$

We take as our first example the amorphous iron alloy $Fe_{80}P_{13}C_7$, made by rapid quenching from about 1200°C onto a rotating copper drum by Prof. M. Takahashi and his group at Tohoku University, inasmuch as its characteristics have been extensively measured.[7-9] The sample was prepared in the form of a flat plate with thickness of about 0.035 mm from which circular disc specimens 2.15mm in diameter were made. Values obtained for this material are tabulated in Table I. Using these numbers, the intrinsic resistance of the material, K_1/M_1S_1, may be calculated. For A_1, we use the approximate formula, $A = 3k_BT_c/za$ where we take z, the number of nearest iron neighbors to be ~10, and the "lattice distance", a, to be the interatomic distance - measured by C. Tsuei[10] - of 2.6 Å. We get for the exchange constant, A, the value 1.0×10^{-6} ergs/cm. The resultant expression for H_c of amorphous $Fe_{80}P_{13}C_7$ is from Eq. (2).

$$H_c = \pm 1.2 \times 10^6 W[(A_1/A_2) - (K_2/K_1)] \text{ Oersteds.} \quad (3)$$

We are now ready to consider the effect of spatial variations in the sample. In particular, we note the reaction upon crystallization of the material,

TABLE I. Experimental Values of the magnetic parameters. Refs.7-9,20.

$Fe_{80}P_{13}C_7$	M - μ_B per iron atom.	$T_c(°K)$	$K_{eff} \times 10^{-4}$ ergs/cm^3	$\lambda \times 10^{-6}$	H_c (Oe.)
	2.1 (1062 Gauss)	587	2.32	31	0.12 unannealed 0.06 annealed
Fe_3P	1.8	716			
Fe_3C	1.9	483			

GdCo$_4$	M (G)	$A \times 10^{-7}$ ergs/cm	$K_u \times 10^5$ ergs/cm^3	$h \times 10^{-4}$ cm	$\ell \times 10^{-4}$ cm	$D \times 10^{-4}$ cm	H_c (Oe)
unannealed	66	6.15	1.8	0.83	0.25	1.9	7.9
annealed	75.5	6.15	1.7	0.83	0.185	1.4	2.5
annealed w/H_e	64	6.15	1.1	0.83	0.20	1.45	0.8

$$Fe_{80}P_{13}C_7 \rightarrow 13Fe_3P + 7Fe_3C + 20Fe. \tag{4}$$

(The compound Fe_3C is a metastable state and its formation is dependent on the metallurgical preparation. R. Hasegawa and R. O'Handley as well as C. Chien and R. Hasegawa[10,11] have examined the crystallization of $Fe_{80}P_{16}C_3B_1$ (Metglass 2615); C. Tsuei et al[12] have looked at $Fe_{80}P_{12.5}C_{7.5}$ and P. Rastogi and P. Duwez[13] at $Fe_{75}P_{15}C_{10}$ both prepared by rapid quenching from the liquid state using "piston and anvil" techniques; and N. Kazama et al[14] have looked at $Fe_{80}P_{13}C_7$ prepared by roller techniques as described above. All report substantial amounts of Fe_3C although there does exist a dependence on method of treatment.) We consider the possibility of small crystallite regions being present in the sample. Using the Curie temperature values[15] tabulated in Table I, Eq. (4) gives a weighted average crystalline Curie temperature of 716°K. Approximating A_1/A_2 by T_{c1}/T_{c2}, we get the value 0.80 for this ratio. The change in magnetic anisotropy as the amorphous $Fe_{80}P_{13}C_7$ is crystallized has been measured by Takahashi[7]. The crystalline value is $K_{cryst.} = 2.45 \times 10^4 ergs/cm^3$. Therefore, the factor K_2/K_1 is given by 1.06. Substituting these two values into Eq. (3) yields

$$H_c = 3.1 \times 10^5 W \text{ Oersteds.} \tag{5}$$

If we take for W the value 15Å (about the limit of high resolution instruments for observing correlations due to crystallinity), then H_c = 46mOe. which is within the observed coercive force value of 60mOe. observed after stress relief annealing. For crystalline defects of the order of 300Å one approaches coercive forces of one Oersted. (Note that a cluster of pure iron defects would have a much larger effect wherein $H_c = 2.1 \times 10^7 W$. Thus, a 15Å pure iron crystalline defect would pin domain walls sufficiently to withstand magnetic fields up to 3.1 Oersteds - i.e., the compounds Fe_3P and Fe_3C play an important role in the coercive force by lowering the effective crystalline anisotropy and Curie temperature of the defect region.)

Finally, we look at the effects of stress variation in this material. Such variations in stress can be written as a contribution to the effective anisotropy. We shall only consider 180° walls though we recognize that additional contributions come from 90° walls[16]. When the magnetostriction is isotropic, the magnetoelastic energy can be written as

$$E_{me} = (3/2) \lambda \sigma \sin^2 \Theta, \tag{6}$$

where Θ is the angle between M and σ. If the stress is along the z axis, then from Eq. (1) we obtain the familiar result that the effective anisotropy is $K_{me} = (3/2) \lambda \sigma$ where λ is the saturation magnetostriction and σ represents the internal stresses. Placing

$A_1 = A_2$ in Eq. (3), we get

$$H_c = 1.2 \times 10^6 W (\Delta K)_{me}/K_1 \quad \text{Oersteds.} \tag{7}$$

Except for the additional numerical factor of 0.6, this is equivalent to the formula of Kersten[17] for wall pinning by stresses.

Two different types of internal stress measurements have been done on iron rich amorphous material. One is by Berry and Pritchet[18] on splat cooled amorphous $Fe_{75}P_{15}C_{10}$ prepared by the "piston and anvil" technique. They used internal friction measurements - obtaining a value of 4.3×10^8 dynes/cm^2 for σ in the unannealed state and a linear decrease with annealing temperature to a value of 1.9×10^8 dynes/cm^2 at an annealing temperature of 300°C. The second measurement was done by O'Handley[19] on $Fe_{80}P_{16}C_3B_1$ by measuring the effective anisotropy in the unannealed state and after a 225°C two hour anneal. His results show a 10% drop in anisotropy. If we assume that the linear decrease as a function of annealing temperature obtained by Berry and Pritchet on $Fe_{75}P_{15}C_{10}$ also applied here, then Metglass 2615 exhibits approximately a 15% drop on annealing at 300°C compared to a 43% drop in $Fe_{75}P_{15}C_{10}$. Also, the change in the coercivity of Metglass 2615 upon annealing is of the order of 25% compared to a 50% change for $Fe_{80}P_{13}C_7$ - indicating less internal stresses. The smaller stresses in Metglass 2615 may be explained by 1) the addition of boron which appears to reduce the internal strains[3] and 2) the slower quenching rate used in the preparation of Metglass. From the above, we suspect that $Fe_{80}P_{13}C_7$ has values between Metglass 2615 and Prof. Duwez's material ($Fe_{75}P_{15}C_{10}$), but considerably closer to $Fe_{75}P_{15}C_{10}$. Assuming that the variations in stress are of the same order of magnitude as the stress itself and using the values got by Berry and Pritchet, Eq. (7) yields for $Fe_{80}P_{13}C_7$

$$H_c = \begin{cases} 1.1 \times 10^6 W & - \text{ unannealed} \\ 4.5 \times 10^5 W & - \text{ 300°C anneal.} \end{cases} \tag{8}$$

The extent of the variation in stress in an amorphous material is at present unknown. To be effective in pinning domain walls, it must be less than the domain width. Otherwise, it appears as a constant stress to the domain wall. If we use 10^{-4}cm as a nominal figure for W, we see that our coercive forces are way too large. On the other hand, if we assume that the extent of the individual stress variations are confined to the effective correlation length and use as previously the value of 15Å as an upper bound for this length, then H_c(unannealed) = 160mOe. and H_c(300°C anneal) = 67mOe. in very good agreement with the experimental values. It is interesting to note the "tradeoffs" necessary for a low coercive force - i.e., a fast quench rate implies a minimum in microcrystallite size but larger internal stresses while an annealing process reduces the stress but too heavy an anneal may increase microcrystallite size.

AMORPHOUS GdCo$_4$

The treatment of the coercive force due to domain wall pinning of strip domains in amorphous GdCo$_4$ material made by sputtering contains some interesting aspects. Accepted values of the coercive force for this material range from one to ten Oersteds. Table I gives representative values as used by R. Hasagawa[20] for the magnetic parameters describing this material.

We first examine the effect on the coercivity of the magnetic moment perpendicular to the plane of the film and the associated magnetostatic energy at the surface. The fact that the perpendicular magnetic anisotropy energy in a "commercially good" material is sufficiently strong to force the magnetic moment to orient in this direction (i.e., a high Q material) implies that the magnetostatic energy plays a minor role in determining the "intrinsic resistance", $K_1/M_1\delta_1$, but not necessarily so in determining the "defect pinning strength", $W[(A_1/A_2) - (K_2/K_1)]$, in Eq. (2). To examine this more quantitatively, we assume a periodic magnetic moment distribution in the yz plane with the periodicity 2D in the x direction. The magnetic poles appear in the z direction only and are given by $M\cos\Theta$ where the angle Θ contains the periodicity 2D. We note that there also exists a y component of the magnetization with poles at $y = \pm\infty$. Thus, from symmetry, the magnetic fields are calculated from M_z only and the magnetic energy density is given by $(1/2)M_zH_z$. The solution for H requires a knowledge of the magnetic moment distribution function $\Theta(x)$ which is in turn dependent on H as well as on the quantities given in Eq. (1). If we assume for the purposes of computing H_z that the angle Θ in a given domain is relatively constant except over a small region, then this quantity may be written as[21]

$$H_z(x,z) = \pm 8M_z \sum_{n=0}^{\infty}(2n+1)^{-1}\sin[(2n+1)\pi x/D]\exp[\mp(2n+1)\pi z/D]. \quad (9)$$

The magnetic field energy per unit surface area allowing for both sides of the material is

$$1.63M_z^2 D \sum_{n=0}^{\infty}(2n+1)^{-3}[1 - (1 + [2n+1]\pi h/D)\exp(-[2n+1]\pi h/D)], \quad (10)$$

where the factor under the summation sign takes into account the finite thickness of the material h. For $D \sim h$, this factor may be replaced by the quantity,

$$f(h/D) = 1.05 - [1+ \pi(h/D)]\exp(-\pi h/D), \quad (11)$$

where $f(h/D)$ is of order unity. Using this in $K\sin^2\theta$ of Eq. (1) yields for the effective anisotropy caused by magnetostatic poles at the surface,

$$K_{ms} \simeq -1.63 DM^2 h^{-1} f(h/D). \tag{12}$$

The domain width, D, is got from the intrinsic length, ℓ, and the thickness h, using the method of Fowlis and Copeland[22] and values are given in Table I. Substituting into Eq. (12), we find the magnetostatic energy contribution K_{ms} is of the order of 1×10^4 ergs/cm^3 or more than an order of magnitude smaller than the perpendicular anisotropy K_u. Thus, the effect of this magnetostatic term on the "intrinsic resistance" of the material in Eq. (2) may be ignored. However, as we shall see, K_{ms} plays an important role in determining the "pinning strength" due to variations within the material. Substituting numerical values for the intrinsic resistance, we get,

$$H_c = \pm B10^8 W[(A_1/A_2) - (K_2/K_1)] \text{ Oersteds,} \tag{13}$$

where B equals 5.7 in the unannealed state, 4.5 when the material is annealed at 200°C without an applied field, and 2.7 when annealed at 180°C in the presence of an applied field of 3kOe. We note that the changes in the anisotropy and magnetization due to annealing with and without an external field are insufficient to account for the experimentally observed change in coercivity (shown in Table I), and that in agreement with the conclusions of Hasegawa[20], one must look at changes in the spatial variations within a given sample to explain drops in the coercivity.

Spatial Variations

Unlike the amorphous $Fe_{80}P_{13}C_7$ foils, it does not appear reasonable to consider the probability of small crystalline regions in $GdCo_4$. We note that this material crystallizes to $GdCo_5$ plus Gd_2Co_7 by the formula[23]

$$3GdCo_4 \rightarrow Gd_2Co_7 + GdCo_5. \tag{14}$$

$GdCo_5$ possesses one of the highest magnetic anisotropies known - of the order of 10^8 ergs/cm^3, and, although the magnetic anisotropy of Gd_2Co_7 has not as yet been measured, it is suspected of also having a very high value. From Eq. (2) and from experience with crystalline rare earth-cobalt compounds, crystallite regions of these materials would produce large in plane magnetic anisotropy, extremely strong wall pinning, and unusually high coercive forces - not observed in amorphous $GdCo_4$. Experimentally,[24,25] within the resolution limit of x-ray flourescence analysis, α-backscattering, and microprobe analysis, the ratio of gadolinium to cobalt was found to be independent of position while from transmission electron microcopy, the coherent scattering regions were determined by the resolution limit of the microscope as being less than 15Å. Presumably, for a uniform "frozen-in" atomic distribution, the energy of activation for diffusion and nucleation to form two separate $GdCo_5$ and Gd_2Co_7 microcrystalline defects is quite high.

We now consider spatial variations in the effective magnetostatic anisotropy, K_{ms}. In particular, spatial variations in the magnetization (caused by small variations in composition) can be represented by taking variations, ΔM, in Equ. (12) - recognizing that the domain width D is also a function of M, (i.e., to first order $D \propto M^{-1}$ and to $L^{1/2}$). We get from Eq. (13),

$$H_c = B 10^8 W (K_{ms}/K_1)(\Delta M/M) \text{ Oersteds,} \qquad (15)$$

where B is defined in Eq. (13) and we have placed the ratio of the exchange constants, A_1/A_2, equal to unity. Although fluctuations in the exchange constant are known to exist due to variations in the interatomic distance in an amorphous material, the width W of such fluctuations should also be of the order of the interatomic distance and therefore make a negligible contribution to Eq. (15).

Substituting the values from Table I into Eq. (12), we get for K_{ms} the results 1.3×10^4, 1.4×10^4, and 1.0×10^4 ergs/cm^3 for the unannealed, annealed, and annealed with a 3kOe. applied field samples respectively. Thus, from Eq. (15), we get

$$H_c = B' 10^7 W (\Delta M/M) \text{ Oersteds,} \qquad (16)$$

where B' equals 4.2 in the unannealed state, 3.7 when the material is annealed, and 2.6 when annealed in an external field.

Measurements of variation in domain width in a given sample indicate an average magnetization fluctuation of ~2% though individual domain width measurements have not been made.[22,26] These changes in magnetization result from small fluctuations in the chemical composition - the relation being given by the magnetization vs composition curve in Ref.24. The extent of the spatial variations W appears to be too small for measurement within the resolution of the equipment.

For a coercive force of 8 Oersteds, with $\Delta M/M$ equal to 2%, one must take W to be of the order of 10^{-5}cm. This coercivity corresponds to the observed value for the unannealed sample. To reduce H_c by a factor of 3 corresponding to the observed decrease upon annealing and a further factor of 3 when annealed in the presence of a 3 kOe. field would require corresponding reductions in the variation of $\Delta M/M$ and possibly of W. As shown in Ref. 24, M is very sharply dependent on composition. Thus, a compositional smoothing equivalent to a change in local composition of 1 part in 3000 would yield the required reduction in the coercive force of a factor of 3, i.e., from a local composition of 80.04% cobalt (equivalent to a 2% magnetization variation) to 80.013% cobalt. Thus, this appears to be a feasible mechanism.

A similar formula arises for local thickness - i.e., within a domain width rather than a general long range increase in thickness. Such local variations are essentially neglible when materials are prepared by the sputtering process due to an inherent feedback in the system and measurements having resolutions of 100Å have failed to observe any such variations.[26] Thus, $\Delta\bar{h}/h$ is negligible and makes very little contribution to the coercive force.

We summarize this section by suggesting variations in magnetization resulting from small spatial variations in composition in a given sample as a possible major contribution to the coercive force of $GdCo_4$ and further suggesting that 1) thermal annealing tends to smooth out these variations in composition while 2) thermal annealing in the presence of an external magnetic field tends to enhance that compositional smoothing correlated with the perpendicular magnetization.

APPENDIX

The effective pinning of domain walls by strain irregularities distributed within the material thru magnetostatic interactions have been calculated by Neel[6] as discussed in the Introduction. He obtains the formulas

$$H_c = \begin{cases} \dfrac{0.19 v \lambda^2 \sigma^2}{KM}\left(1.4 + \dfrac{1}{2}\ln\dfrac{2\pi M^2}{K}\right), & \dfrac{3}{2}\lambda\sigma \ll K, \quad (17a) \\[2ex] \dfrac{0.46 v \lambda\sigma}{M}\left(1.4 + \dfrac{1}{2}\ln\dfrac{4.5 M^2}{\lambda\sigma}\right), & \dfrac{3}{2}\lambda\sigma \gg K, \quad (17b) \end{cases}$$

where v is the fractional volume of strains. Utilizing the data given in Table I for $Fe_{80}P_{13}C_7$, we find that for the unannealed material, $(3/2)\lambda\sigma$ equals $2.0 \times 10^4 ergs/cm^3$ and is of the same order of magnitude as K while for the material annealed at 300°C, the quantity $(3/2)\lambda\sigma$ equals $8.6 \times 10^3 ergs/cm^3$ and is less than K. If Eq. (17a) is used, H_c(unannealed) = 2.6v Oersteds and H_c(300°C anneal) equals 0.49v while Eq. (17b) yields H_c equals 24v Oersteds for the unannealed case. The usual procedure is[6] to take v = 1. However, as we see, this gives unrealistic values for the coercive force of this material. Also, this theory predicts a change in the coercive force upon annealing at 300°C of anywhere from a factor of 5 to a factor of 50 while the experimental results show at most only a factor of two change in agreement with that given by Eq. (7). Thus, we must conclude that the formulas of Eqs. (17) do not apply to the amorphous iron rich strips.

REFERENCES

*This research was supported in part by NSF Grant DMR72-03118-A02.
1. H. S. Chen, et al, Appl. Phys. Ltr. 26 405, (1975)
2. T. Egam, P. Flanders, C. Graham Jr., A.I.P.Conf.Proc. 24, 697 (1975).
3. R. Hasagawa, R. O'Handley, and L. Mendelsohn, A.I.P.Conf.Proc. MMM Pittsburgh Conf - to be published.
4. P. Chaudhari, J. Cuomo, and R. Gambino, IBM J. Res.Dev.17, 66 (1973).
5. R. Friedberg and D. Paul, Phys. Rev. Ltrs 34, 1234 (1975), and and D. I. Paul, A.I.P.Conf. Proc. 29, 545 (1975).
6. L. Neel, Physica 15, 225 (1949).
7. M. Takahasi, F. Ono, and K. Takakura, Jpn Jrl Appl. Phys. 15, 183 (1976).
8. H. Fujimori, et al, Jpn Jrnl Appl Phys 13, 1889 (1974)
9. T. Masumoto and R. Maddin, Int'l Conf on Metastable Metallic Alloys, Yugoslavia, (1970).
10. R. Hasegawa and R. O'Handley, 2nd Int'l Conf on Rapidly Quenched Metals, Boston (1975) to be published.
11. C. Chien and R. Hasegawa, MMM Conf Pittsburgh (1976) to be published.
12. C. C. Tsuei, et al, Phys Rev. 170, 603 (1968).
13. P. Rastogi and P. Duwez, Jrl. of Non-Cryst Sol. 5, 1 (1970).
14. N. Kazama et al, MMM Conf. Pittsburgh (1976) to be published.
15. N. Kazama et al, Jrnl. Phys Soc Jpn 37, 1171 (1974).
16. D. Craik and R. Tebble - Ferromagnetism and Ferromagnetic Domains, John Wiley Inc., New York (1965).
17. M. Kersten, Z. Angew Phys 7, 397 (1955)
18. B. Berry and W. Pritchet, Jrl Appl. Phys 47, 3295 (1976).
19. R. C. O'Handley- AIP Conf Proc. 29, 206 (1975).
20. R. Hasegawa, Jrnl Appl. Phys. 45, 4036 (1974)
21. I. Privorotskii, Thermodynamic Theory of Domain Structures, John Wiley and Sons Inc. New York (1976).
22. D. Fowlis and J. Copeland, AIP Conf Proc. 5, 240 (1971).
23. R. Lemaire, Cobalt Monograph Series 33, 207 (1966).
24. P. Chaudhari et al, IBM Jrl. Res. andDev. 17, 66 (1973).
25. S. Herd and P. Chaudhari Phys.Stat.Col.(a) 18, 603 (1973).
26. Private Communication with P. Chaudhari and R. Kobliska, IBM.

MAGNETIC CHARACTERIZATION OF SEMI-AMORPHOUS NICKEL ON ALUMINA

DISPERSIONS: CORRELATIONS WITH THEIR METHANATION AND CHEMISORPTION

ACTIVITIES

L.N. Mulay[†], R.C. Everson[*†], O.P. Mahajan, P.L. Walker, Jr.

Material Sciences Dept. and Materials Research Laboratory

The Pennsylvania State Univ., University Park, Pa. 16802

ABSTRACT

Commercial samples of semiamorphous nickel supported on alumina, typically used as methanation catalysts have been characterized by using magnetization (σ) versus field (up to 20 KOe) and temperature (77-600 K) type curves and by electron microscopy. Hysteresis curves yielded coercive force (H_c) remanence (I_r) and saturation magnetizations (σ_s), following heat treatment of the catalysts between 400-700°C. These parameters are interpreted in terms of the following properties of constituent particles: (a) superparamagnetic (b) single-domain anisotropic and (c) multi-domain. Needlelike "b" and "c" type particles are formed when a sample A with 43 wt % of Ni is heated to about 600 and above 650°C respectively. Another sample B with a higher loading of Ni (67 wt %) consisted of a larger fraction of "a" type particles which on heat treatment increased H_c and σ (sat) as expected, but was more thermally resistant to the formation of "c" type particles. Unlike sample A, B did not show a significant decrease in the methanation activity upon the formation of "c" type particles. The methanation reaction $CO + 3H_2 \rightarrow CH_4 + H_2O$ was followed using standard gas chromatographic techniques. The distribution of the superparamagnetic particles in the presence of multi-domain particles in various heat treated samples has been calculated and shown to correlate well with their hydrogen chemisorption activities.

[†]Inquiries should be addressed to these authors.
[*]Present address: Dept. of Chem. Eng. Univ. of Natal, Durban, S. Africa.

INTRODUCTION

Magnetic properties of amorphous materials, including metal dispersions on oxides are fascinating indeed and have attracted considerable attention during recent years. One of us (LNM) previously reported on the magnetic properties of dispersion of α-Fe_2O_3 in zeolites[1] and of Fe^{3+} ions in silicates[2,3] and glassy carbons.[4] Among dispersed magnetic systems the very fine dispersions of 3d metals such as Fe, Co, and Ni on various (diamagnetic) substrates (e.g. SiO_2, Al_2O_3) are technologically most significant because of their widespread use as industrial catalysts in tonnage quantities. Significant fractions of such dispersions display superparamagnetic properties and are considered to be in part amorphous.

Correlations between the (superpara)magnetic properties of the 3d transition metal dispersions and chemisorption have been well recognized and widely investigated over the past three decades.[5-11] Most of the work has been on materials containing very low amounts (usually below 5%) of the metal which was dispersed on substrates to yield, so to speak ideal,"single phase" superparamagnetic systems.[12,13] While such studies have undoubtedly enhanced our understanding of the mechanism of chemisorption of electron donor and acceptor type molecules, very few attempts have been made to thoroughly characterize commercial catalysts containing large amounts of metal (up to 70%) and to correlate their magnetic properties with catalytic and chemisorptive activities.

In this paper we present typical results on two commercial grade nickel on alumina dispersions containing 43 and 67% Ni with special reference to (i) the delineation of the properties of their magnetic components (superparamagnetic,[12,13] single-domain anisotropic[14] and multi-domain ferromagnetic), (ii) the change in magnetic properties upon heat treatment to elevated temperatures and (iii) correlation of such changes with corresponding changes in catalytic activity for the reaction: $CO + 3H_2 \rightarrow CH_4 + 2H_2O$ as well as correlations with their hydrogen chemisorptive properties.

EXPERIMENTAL

Two nickel on Al_2O_3 dispersions A and B with the physical characteristics shown in Table I were investigated.

Magnetization (σ) measurements were performed on a vibrating sample magnetometer (made by Princeton Applied Research Labs., Princeton, N.J.) as a function of the field (up to 20 k Oe) and over a range of temperature (77-600 K) to yield especially values for saturation magnetization per gram (σ sat) of nickel in the catalyst, the coercive force (H_c) and the remanence (I_r). Special quartz sample holders were designed to accommodate in vacuum the as received

Table I: Characteristics of the As Received Ni Samples

	A	B
Nickel, wt %	43	67
Pellet size, in	1/8	3/16
BET surface area, m^2/g	51	117
Average crystallite size (Ni), Å	185	74

samples, and those after reaction and/or heat treatment as described below.

Heat treatment of the catalyst was conducted by (a) exposure to H_2 for 1 hour at selected heat treatment temperatures (HTT) in the range 400-700°C followed by (b) exposure to the synthesis H_2/CO mixture for an additional 1 hour at HTT.

Catalytic activities were measured in terms of the methanation reaction which was carried out by passing a mixture of 17 mole percent of carbon monoxide (CO) in hydrogen (H_2) at atmospheric pressure through a vertical reactor packed with a known weight of catalyst pellets. These pellets were crushed to give fine powders (100-170 mesh) and low residence times (0.003-0.019 secs) were used so that activities could be measured unaffected by diffusion effects.[15] Methane content in the product stream was monitored using a Hewlett Packard chromatograph with a column of 100-120 mesh fraction of Carbosieve B, obtained from Supelco, Inc. The column was operated at 50°C. Methane production at 300°C, after reduction of the catalyst for 1 hour in H_2 and reaction for 1 hour, was used as the basis for comparison of activities of catalysts taken to different HTT.

RESULTS AND DISCUSSION

Typical plots of the coercive force (H_c) measured at three temperatures (T_m = 77°, 298°, 423°K) as a function of the HTT for catalysts A and B are shown in Fig. 1. Maxima in H_c are observed around HTT's of 600°C and 650°C respectively for catalysts A and B. A comparison of these curves with Kneller and Luborsky's results[13] suggests that the maxima are probably due to two competing processes: first, conversion of "a" superparamagnetic to "b" single-domain anisotropic particles producing an increase in H_c and then a conversion of such "b" particles to "c" type multi-domain particles with zero H_c. These two conversions are believed to be characteristic of what occurs

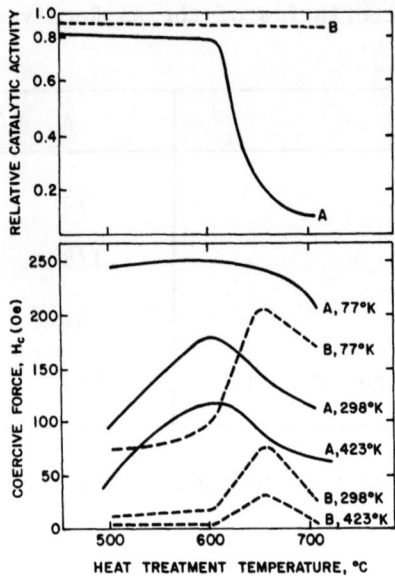

FIG. 1 Rel. activity (top) and H_c (bottom) vs HTT °C. Magnetic measuring temperature in °K.

during the sintering of nickel supported on alumina. However, a partial conversion of "a" to "c" due to formation of closed multi-domain configurations is not ruled out during heat treatment.

Curves for catalyst A indicate the formation of large particles in the multi-domain region corresponding to a HTT range, 600-700°C. The relative activity for "A" (Fig. 1) decreases sharply in this region. The temperature at which the activity of A commences to decrease sharply coincides with the HTT, at which the maxima in H_c occurs. By contrast the activity of B does not drop at its corresponding maxima in H_c. This aspect is being investigated. It should be noted here that B shows no loss in activity even up to a HTT of 700°C. We have estimated in Table II the fraction of superparamagnetic particles in various heat treated samples and their mean diameters, using essentially the approaches and approximations described in the appendix.[10] Table II shows that for catalyst B (despite its high concentration of nickel), the size of the superparamagnetic particles remains nearly constant and the fraction of such particles decreases less over the range of HTT's as compared to A.

Thus the magnetic technique provides a better characterization of the particles in terms of the critical sizes of superparamagnetic particles as well as their abundance than the gross estimates of crystallite sizes often reported in the catalysis literature. The small fraction (0.12) of superparamagnetic particles of A at HTT of

Table II. Weight Fraction of Superparamagnetic Particles in Heat Treated Samples A and B

Heat Treatment Temperature (HTT)°C	Wt. Fraction of Superparamagnetic Particles		Mean Particle Size, Å (diam) of Superparamagnetic Fraction	
	A	B	A	B
As received	0.36	0.58	25	24
500	0.27	0.48	29	25
600	0.21	0.40	54	28
700	0.12	0.26	37	26

700°C indeed represents the least active catalyst. Since the chemisorptive properties of CO and H_2 have been shown to be dependent on the d-band characteristics of such particles[5,6] the magnetic technique may be said to provide additional characterization of the particles at the microscopic level. This aspect is discussed later.

The shape anisotropy of the particles is believed to be responsible for the increase in H_c up to the transition temperature (T_t), and their disappearance is expected to cause the subsequent decrease in H_c. This conclusion is supported by evidence from electron microscopy, which clearly shows the formation of needles or chainlike clusters up to T_t and their conversion to more regularly shaped particles, resembling those with multi-domain properties, above T_t. The average crystallite size increased from 185 to 256Å for catalyst A, and from 74 to 160Å for B over the range of HTT.

Curves for the remanence (I_r) measured at the same temperatures as H_c and as a function of HTT resemble those of H_c and are not shown here. Their unique feature is that the peaks appeared sharper and at somewhat lower values of HTT corresponding to smaller particles than in the corresponding H_c curves as one would expect from the theoretical treatment.[13] Thus, the parameters H_c and I_r are indeed useful in characterizing the presence of single-domain anisotropic particles, which are formed at least partially during the transition from the essentially superparamagnetic to the essentially multi-domain regions. Furthermore, the HTT at which the single-domain anisotropic particles are formed appears to be characteristic of the "dispersive" state of the metal on the catalyst support.

Another distinction between the two samples was observed in terms of their hydrogen chemisorption, which was studied by standard techniques.[5] These techniques yield the number of hydrogen atoms chemisorbed on the surface of the catalyst, which in turn corresponds to the number of surface nickel atoms.

From these results the percent degree of dispersion (f) is calculated simply as the fraction f = (Ni atoms at the surface ÷ total number of nickel atoms in the bulk per gram of catalyst) x 100. Relevant information is given in Table III, which again shows that the catalyst B has an overall better degree of dispersion (and hence better catalytic activity) than catalyst A over the entire range of HTT.

The degree of dispersion is plotted as a function of the weight fraction of superparamagnetic particles in Fig. 2, for various heat treated samples. The average particle sizes from Table II are also shown next to the data points for the two catalysts. This figure clearly indicates that the dispersion for B is indeed relatively large (7.6 to 10%), as we would expect, for smaller particles (24 to 28Å) with a better distribution (26 to 58%), thus this type of correlation between the results of a strictly chemisorptive technique and the magnetic technique is helpful in quantitatively characterizing the dispersions of metals such as Ni on diamagnetic supports.

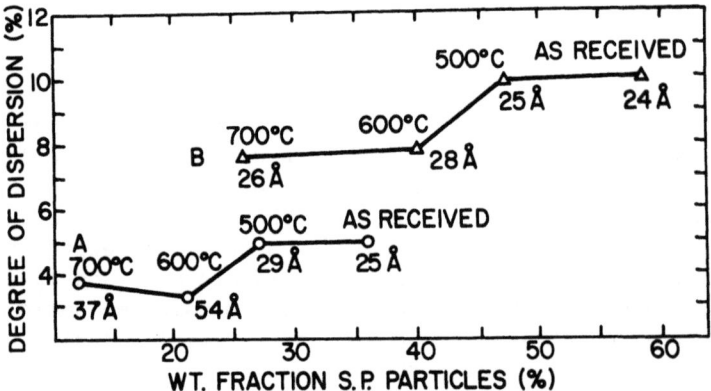

FIG. 2 Plots of the degree of dispersion (f) vs weight fraction of superparamagnetic particles for various heat treated samples.

Table III. Results of Hydrogen Chemisorption

HTT °C	Sample A		Sample B	
	No. of Ni atoms per g (x 10^{20})	Dispersion f (%)	No. of Ni atoms per g (x 10^{20})	Dispersion f (%)
As received	2.26	5.12	6.93	10.01
500	2.45	5.54	6.72	9.78
600	1.48	3.36	5.38	7.82
700	1.67	3.78	5.24	7.62

APPENDIX

The weight fractions x_i of the superparamagnetic ("a" type) and hence of the multi-domain ("c" type) particles in the various heat treated samples were estimated from the observed curves of the relative magnetization, σ/σ_∞ versus H/T (at 573°K). Typical magnetization curves for catalyst B only are shown in Fig. 3.

Romanowski's equation[10] for a two component system was assumed for deriving the weight fractions of the superparamagnetic ("b" type) and ferromagnetic ("c" type) particles:

$$\frac{\sigma}{\sigma_\infty} = \left(1 - \sum_i x_i\right) + \sum_i x_i L\left(\frac{I_{sp} v_i H}{kT}\right)$$

The expressions under the summation signs refer to superparamagnetic particles; L denotes the Langevin function and v_i the mean volumes of the particles within the x_i fraction. This expression is applicable at high fields such that saturation magnetization of multi-domain particles show no change and superparamagnetic particles obey the well known Langevin function. The average moment μ_i of the "a" type particle is defined by, $\mu_i = I_{sp} \cdot V_i$ (a magnetization curve for superparamagnetic particles of 30Å radius, and at 573°K, assuming the Langevin function is also shown in Fig. 3 for comparison with the observed curves).

By assuming an overall mean volume v for all superparamagnetic particles, as in Selwood[5], a nonlinear regression procedure was developed for estimating the weight fraction (x) of the "a" type

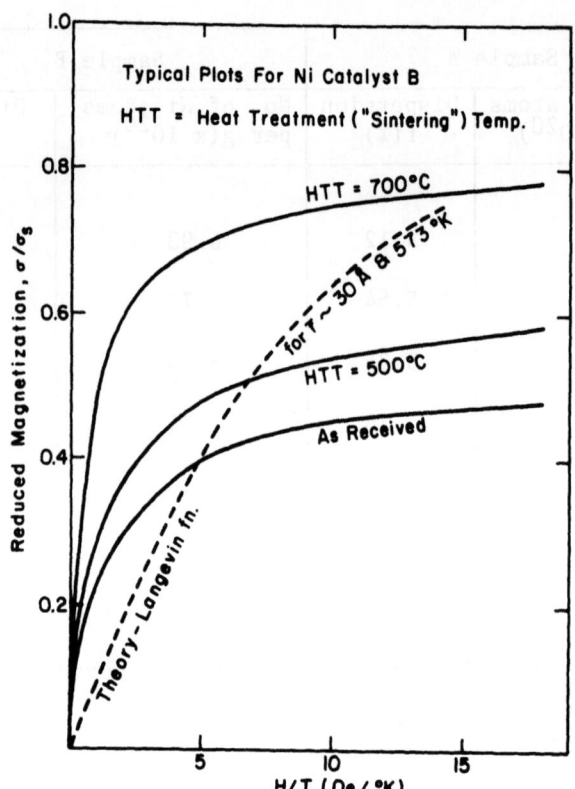

FIG. 3 Typical magnetization (σ/σ_s) vs (H/T) curves for catalyst B for the as received and heat treated samples. A magnetization curve for superparamagnetic particles (30Å radius at 573 K is shown for comparison (see Appendix).

particles, and the mean particle diameter of this fraction based on spherically shaped particles. The regression program used accomplished convergence very easily with a conventional hill climbing subroutine.

REFERENCES

Only key references to selected reviews and papers are given.

1. L.N. Mulay and D.W. Collins, "Amorphous Magnetism," M. Hooper and A.M. deGraff, Eds. Plenum Press, N.Y. (1973).

2. D.W. Collins and L.N. Mulay, J.Am. Ceram. Soc. 53, 74 (1970); 54, 52, 69 (1971).
3. D.W. Collins and L.N. Mulay (Proc. Intermag. Conf. 1968), IEEE Trans. Magnetics 4, 470 (1968).
4. L.N. Mulay, A. Thompson, P.L. Walker, Jr. in "Amorphous Magnetism" (See Ref. 1).
5. P.W. Selwood, "Adsorption and Collective Paramagnetism," Academic Press, N.Y. (1962). See also "Chemisorption and Magnetization," Academic Press, N.Y. (1975).
6. J.H. Sinfelt, in "Annual Revs. in Materials Science," Part 2, R.H. Bube and R.W. Roberts, Eds. Annual Reviews, Inc. Palo Alto, Calif. (1972).
7. T.E. Whyte, Jr., Catal. Revs. 8, 117 (1973).
8. J.I. McNab and R.B. Anderson, J. Catal. 29, 328 (1973).
9. J.L. Carter and J.H. Sinfelt, J. Catal. 10, 134 (1968); J. Phys. Chem. 70, 3003 (1966).
10. W. Romanowski, Zeits. anorg. allgem. Chem. 351, 180 (1967); V.W. Romanowski, H. Dryer and D. Nehring, ibid. 310, 286 (1961).
11. S.D. Robertson, S.C. Kloet, W.M.H. Sachtler, J. Catal. 39, 234 (1975).
12. I.S. Jacobs and C.P. Bean in "Magnetism" Vol. III, Academic Press, N.Y. (1963).
13. E.F. Kneller and F.E. Luborsky, J. Appl. Phys., 34(3), 656 (1963
14. C. Kittel and J.K. Galt in "Solid State Physics," F. Seitz and D. Turnbull, Vol. 3, 508 (1956).
15. R.A. Dalla-Betta, A.G. Piken and M. Shelef, J. Catal. 35, 54 (1974).

DISCUSSION

E. Siegel: Can you distinguish between the particles that are ferromagnetic and those that are superparamagnetic?

L. N. Mulay: I would say that is is probably the metallic nickel which is superparamagnetic and not the nickel aluminate because from a chemical point of view it is certainly easier for the nickel metal atoms to form a cluster.

R. Hasegawa: In the title of your talk you have the work semi-amorphous. I did not see any indication that you are dealing with an amorphous material here.

L. N. Mulay: By definition, if one supposes the entire system to be superparamagnetic, that system would lack order and could be considered amorphous. Here we have a system that is partially ferromagnetic and I associated that part with the crystalline part of the matrix, so that the superparamagnetic particles would contribute to that semi-amorphous reference I used.

SURFACE EFFECTS IN AMORPHOUS FERROMAGNETS WITH RANDOM ANISOTROPY

R. Micnas[a], A. R. Ferchmin[b], S. Krompiewski[b] and

B. Szczepaniak[a]

[a]Institute of Physics, A. Mickiewicz University, Poznan

[b]Polish Academy of Sciences, Institute of Molecular

Physics, Ferromagnetics Laboratory, Poznan, Poland

INTRODUCTION

Considerable attention is being given at present to the surface problem of magnetics. In this paper we would like to discuss how surfaces influence the properties of amorphous magnetics [1,2]. To this aim, we have selected, from the various available models of amorphous magnetics [1,2,3,4], two interrelated models [3,4] stressing the role of random local magnetic anisotropy in these materials. Accordingly, we adopt a Hamiltonian consisting of an exchange term and an anisotropic term:

$$H = - \sum_i V_i - \sum_i \sum_j K_{ij} J_i J_j \qquad (1)$$

Specifically, the direction of the anisotropy varies randomly in the Harris-Plischke-Zuchermann (HPZ) model [3]. In the related model [4] proposed by Taggart, Tahir-Kheli and Shiles (TTS), it is the value of the anisotropy D_i which is subject to random variation.

MODEL FOR AN AMORPHOUS THIN FILM

To deal with the problem of surfaces, we propose an approach analogous to that known from the theory of crystalline thin films, i.e., applying difference equations with appropriate boundary conditions. There arises a problem of the number of nearest interacting neighbors of a spin situated at the surface as contrasted to one inside the sample. It is reasonable to admit [7] that a surface atom feels a lack of about three neighbors from the total of about twelve nearest neighbors (the structure of amorphous metals and alloys is most probably a topologically disordered, closely packed one [6]). Six more of its neighbors sit approximately at the surface, and the remaining three inside the sample. We similarly classify Z nearest neighbors of an inner atom into sets of 3-6-3 (or more generally a-b-a, conforming to the rule Z = 2a+b) atoms according to their respective distances from the surface. We have checked that our results do not depend qualitatively on the particular choice of the numbers a and b. This classification of neighbors allows us to divide mentally the material into very thin sheets of a thickness equal to the mean interatomic spacing. We assign the label i = 0,1, ..., L-1 to any atom with its center within the sheet i; therefore, despite the irregular positions of the individual atoms in the amorphous structure, we have labeled them discretely. A more detailed classification of neighbors would not improve the results appreciably, but would enormously complicate our calculations. In the spirit of HPZ and TTS models we stress the importance of anisotropy rather than the isotropic exchange integral K. The local anisotropy operator

$$V_i = -D_i(\vec{J}_i \cdot \vec{z}_i)^2 \qquad (2)$$

is presumed to be sensitive to the changes in symmetry introduced by the presence of a surface. We assume $D_i = D_0$ for the surface sheets i = 0 and L - 1, and $D_i = D$ otherwise. Moreover, for the TTS model, D is subject to random fluctuations with the very strongly simplified rectangular distribution

$$P(y) = \frac{ZK}{D} \text{ for } \frac{D}{ZK} < y < \frac{D + \Delta D}{ZK} \qquad (3)$$

and zero otherwise. For the HPZ model, we introduce [3,8-11] isotropic probability distribution for the directions z_i with respect to the overall magnetization direction z

$$P(\mu) = 1 \text{ for } 0 \leq \mu \leq 1, \qquad (4)$$

where $\mu = \vec{z} \cdot \vec{z}_i = \cos \theta_i$. Equation (4) implies that $P(\mu_i)$ is assumed identical for all sheets i within the film.

MOLECULAR FIELD EQUATIONS

By a variational procedure (see Appendix) we obtain the difference equation

$$m_{i-1} - xm_i + m_{i+1} = 0, \qquad (5)$$

with the boundary conditions

$$(x - A)m_0 - m_1 = 0,$$

$$m_{L-2} - (x - A)m_{L-1} = 0. \qquad (6)$$

We have introduced the following notations:

SURFACE EFFECTS IN AMORPHOUS FERROMAGNETS

$$m_i = \langle J_i^z \rangle , \tag{7}$$

where $\langle \ \rangle$ is a double average: a thermal average, as in standard mean-field theory, and an average over the random distribution of the local anisotropies, as specified in the Appendix Eq. (A4). Further notations are:

$$x = \frac{1 - bKF}{aKF} , \tag{8}$$

$$A = \frac{1 - bKF}{aKF} - \frac{1 - bKF_o}{aKF_o} , \tag{9}$$

$$F = 2\int dy \int d\mu U P(y) P(\mu) , \tag{10}$$

where

$$U = \frac{\beta\mu^2 \exp(\beta ZKy) + \frac{1 - \mu^2}{ZKy} [\exp(\beta ZKy) - 1]}{1 + 2\exp(\beta ZKy)} ,$$

and F_o arises from F by putting $D = D_o$ and/or by letting $P(\mu)$ have a form appropriate for the surface. For the HPZ model $\Delta D = 0$. For the TTS model and the surface of either model we put $P(\mu) = \delta(1-\mu)$ and $P(y)$ is given by Eq. (3); for the surface of a sample described by the HPZ model we have also checked the case $P(\mu) = \delta(\mu)$. This corresponds to our conviction that the statistical distribution of the anisotropies at the surface can exhibit a rather sharp peak about the normal or, alternatively, in the surface plane. We obtain

$$F = \frac{2}{3} \frac{\beta \exp(\beta D) + 2D^{-1}[\exp(\beta D) - 1]}{1 + 2\exp(\beta D)} , \tag{11}$$

and

$$F_o = \frac{2\beta \exp(\beta D_o)}{1 + 2\exp(\beta D_o)} \tag{12}$$

on assuming that both the surface anisotropy and the average magnetization tend simultaneously to be normal or parallel to the surface of the sample for the HPZ model. If the surface anisotropy tends to be normal to the surface but the overall magnetization tends to be parallel thereto, then for the HPZ model

$$F_o(T) = \frac{2D_o^{-1}[\exp(\beta D_o) - 1]}{\exp(\beta D_o) + 2} . \tag{13}$$

Similarly, applying the angular distribution $P(\mu) = \delta(1 - \mu)$, we get for the TTS model

$$F(T) = \frac{1}{\Delta D} \ln \frac{1 + 2\exp[\beta(D + \Delta D)]}{1 + 2\exp(\beta D)} \tag{14}$$

and an analogous expression for F_o. However, for either model, one can also discuss the simplest case of "free boundaries", i.e., with the same value of F everywhere, including the surfaces.

CURIE TEMPERATURE

Our Eqs. (5) and (6) can be easily solved by methods developed in crystalline thin films theory [12], leading us to a spatial distribution of magnetization as a function of the distance from the surface and an implicit equation for the Curie temperature T_c. We shall concentrate here on the last parameter. We obtain the following transcendental equation for the transition temperature T_c:

$$x(T_c) = 2 \cos k(T_c) . \qquad (15)$$

The allowed values of k follow from the condition

$$\tan Lk = \frac{r \sin k}{\cos k - p} \qquad (16)$$

where $r = (A^2 - 1)/(A^2 + 1)$, $p = 2A/(A^2 + 1)$. The above equations admit of L solutions, we choose from them the one giving the highest value of the transition temperature. The value of T_c of our film follows from Eq. (15) provided $T_c \leq T_{cB}$ (Curie temperature of the bulk material). One obtains the value of T_{cB} from the condition

$$x(T_{cB}) = 2 . \qquad (17)$$

This value is readily seen to be identical with either the HPZ or TTS expression if we keep in mind that our 2a+b is the coordination number. The surface type solutions correspond to complex values of k = it in Eq. (15) which should be re-written as

$$x(T_c) = 2 \cosh t(T_c) \qquad (18)$$

and Eq. (16) as

$$\tanh Lt = \frac{r \sinh t}{\cosh t - p} \qquad (19)$$

The semi-infinite medium can be described, as usual, by taking the limit $L \to \infty$. In particular, the surface Curie temperature corresponding to the transition to surface magnetic ordering fulfills the equation

$$\frac{F_o(T_c) - F(T_c)}{F(T_c)} = \frac{x(T_c) + [x^2(T_c) - 4]^{\frac{1}{2}}}{\frac{2b}{a} + x(T_c) - [x^2(T_c) - 4]^{\frac{1}{2}}} . \qquad (20)$$

RESULTS AND DISCUSSION

The preceding formulas show that films presenting a higher degree of disorder and/or reduced thickness undergo phase transitions at less elevated temperatures. This is illustrated in Fig. 1, which shows graphs of the Curie temperature T_c versus the film thickness L for the simplest free boundary conditions at the surfaces. Boundary conditions different from the free boundary type cause no significant changes in this behavior provided they do not generate the surface magnetism phenomenon discussed below.

In the following we shall concentrate on the problem of existence of surface magnetism. As mentioned above, formally the surface ordering temperature T_c follows from Eqs. (20) and (18). As

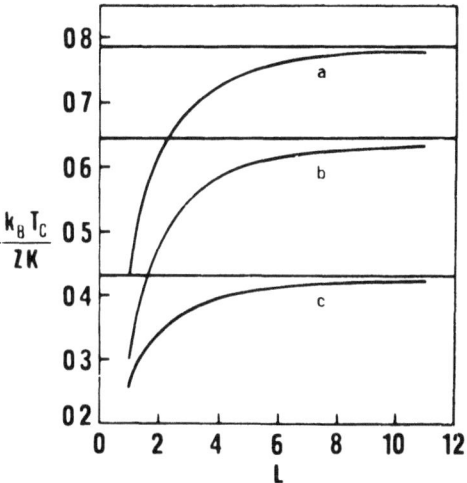

Fig. 1. The reduced Curie temperature versus the film thickness L for D/ZK = 0.5. Free boundary conditions, a/b = 0.5. a - crystalline film, b - amorphous HPZ film, c - amorphous TTS film, ΔD = 0.01 D. For comparison, the asymptotic bulk T_c - values are plotted.

is well known [13,14], in crystals the magnetic surface ordering may be due to surface induced changes in exchange or crystalline field level structure.

Let the value of K at the surface be K_o with crystal field parameters and single-ion susceptibilities unchanged at the surface. Then for either model discussed here, in the molecular field approximation, the condition for existence of the surface magnetism is identical with that for the Heisenberg (Ising) model, i.e., $(K_o-K)/K \geq b/a$. In the following we put $K_o = K$.

We consider the following possible situations for the HPZ model:

I. $D < 0$, singlet-doublet, within the bulk and $D_o < 0$ at the surface.
II. $D < 0$ and $D_o > 0$.
III. $D > 0$, doublet-singlet and $D_o > 0$.
IV. $D > 0$ and $D_o < 0$.

For the TTS model we consider only the $D > 0$ and $D_o > 0$ case.

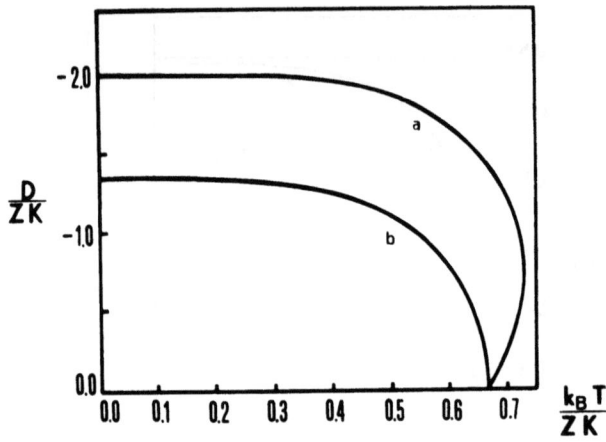

Fig. 2. Phase boundaries for crystalline and amorphous singlet-doublet systems bulk material. a - crystalline, b - amorphous HPZ.

The analysis of the bulk HPZ model with $D < 0$ singlet-doublet has not been available in the literature. Let us discuss this case in some detail before passing to the surface magnetism problem. It is well known that $D < 0$ causes the induced magnetism. This means that there exists a critical value of $\lambda = 2ZK/D$ and for $\lambda \geq 1$ the material orders magnetically even at $T = 0$. We find the critical value of λ for the HPZ model larger than for the crystal. For finite temperatures the corresponding phase diagram is given in Fig. 2.

Our numerical analysis concerning the surface magnetism $T_c \geq T_{cB}$ leads to the following conclusions. For the HPZ model:

I. Surface magnetic solutions exist for a wider range of parameters with respect to the crystal. We present the corresponding phase diagram for $\lambda < 1$ and $\lambda > 1$ in Fig. 3.

II. and III. Surface magnetism exists (Fig. 4). It is of interest that in the case of III the corresponding crystal does <u>not</u> exhibit surface magnetism. The existence of the surface magnetism in our case is due to the change in single-ion susceptibility F caused by ampophization.

In the remaining cases (IV for HPZ model and the TTS model) we have not found any surface solutions. The largest influence of D_o/D ratio on the surface ordering temperature is noted in the case I. for the HPZ model when $\lambda \leq 1$. For growing values of $\lambda > 1$ the influence of D_o/D on T_c decreases systematically. This enlarges the possibility of finding magnetic surfaces on intrinsically paramagnetic substances first proposed by Peschel and Fulde for crystals [14].

Our results show that in every case considered amorphization favors surface magnetism enabling it for a wider range of physical parameters and sometimes even if it is not allowed in the crystal. Amorphous materials containing rare earth ions seem therefore to be good candidates for experiments aimed at search for this phenomenon. The experimental methods suggested up to now for this purpose include LEED, photoemission or field emission of electrons.

We took all possible pains to extract information independent of the classification of neighbors, i.e., independent of the a/b ratio. Nonetheless, this classification remains somewhat arbitrary. One way to circumvent this inconvenience would be to re-formulate the problem using a continuous spin density model and differential rather than difference equations. This will hopefully by the subject matter of a separate paper. However, we thought the present simpler approach might prove useful in a first attack on the problem.

We would like to thank Dr. G. Kamieniarz for helpful discussions.

APPENDIX

In order to derive the molecular field equations for our film we apply a variational procedure [5]. Our starting point is the following general inequality for the Gibbs free energy G

$$G \leq G_o = G_t + \langle H - H_t \rangle , \qquad (A1)$$

where

$$G_t = - \beta \ln \text{Tr} \exp(- \beta H_t) , \qquad (A2)$$

$\beta = (k_B T)^{-1}$, and $\langle A \rangle$ denotes double averaging of A:

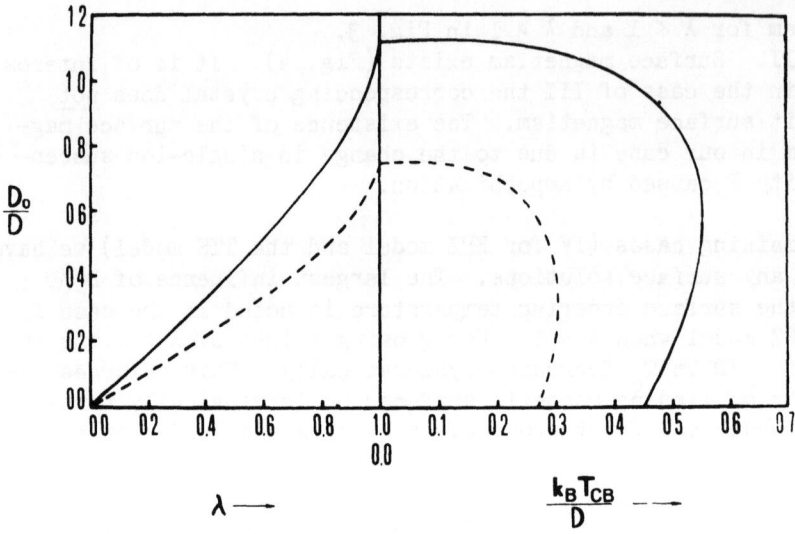

Fig. 3. Phase diagram for the surface magnetism for a singlet-doublet system. In the region below the different curves surface magnetism occurs.
---- crystalline, $\lambda = 2ZK/D$
——— amorphous (HPZ), $\lambda = 4ZK/3D$

$$\langle A \rangle = \langle\langle A \rangle_t \rangle_r \,,$$

$$\langle A \rangle_t = \frac{\mathrm{Tr}\, A \exp(-\beta H_t)}{\mathrm{Tr} \exp(-\beta H_t)} \,, \tag{A3}$$

$$\langle\langle A \rangle_t \rangle_r = \int dy \int d\mu \langle A \rangle_t P(\mu) P(y) \,, \tag{A4}$$

H_t denotes a trial Hamiltonian, which we specify to be a strictly diagonalizable molecular field Hamiltonian:

$$H = -\sum_i (h_i + \eta_i) J_i^z - \sum_i D_i (\vec{J}_i \cdot \vec{z}_i)^2 \,. \tag{A5}$$

This Hamiltonian depends on the variational parameters η_i, the values of which can be found from the conditions for the minimum of G_o:

$$\frac{\partial G_o}{\partial \eta_i} = 0 \,. \tag{A6}$$

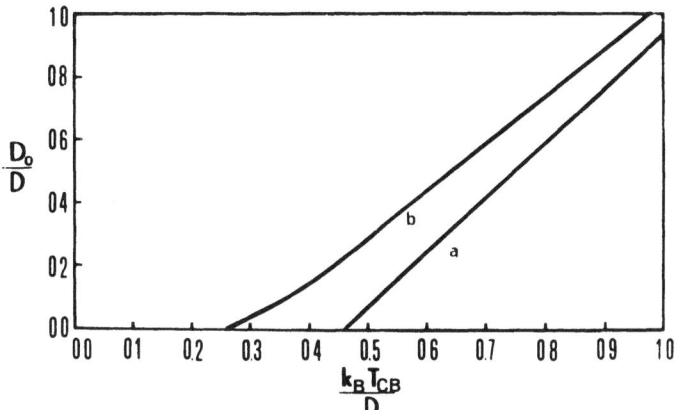

Fig. 4. The same as Fig. 3. a - amorphous bulk singlet-doublet (HPZ) system with a surface of the doublet-singlet type. b - amorphous bulk doublet-singlet (HPZ) system with a surface of the doublet-singlet type.

The local anisotropy axis and the overall magnetization direction form an angle θ_i. We have denoted $\mu_i = \cos\theta_i$. For the sake of simplicity, we restrict ourselves to the case of spin $S = 1$. The eigenvalues of H_t fulfill the cubic equation

$$E_i^3 + 2D_i E_i^2 - (\eta_i^2 - D_i^2)E_i - (1 - \mu_i^2)D_i \eta_i^2 = 0 . \tag{A7}$$

From Eq. (A6), we obtain

$$\eta_i = 2\sum_i K_{ij} m_j , \tag{A8}$$

where $m_j = \langle J^z \rangle$ can be obtained by differentiation of G_o over the local fictitious field h_j to give

$$m_j = \left\langle \frac{-\sum_{\ell=1}^{3} \dfrac{\partial E_j^{(1)}}{\partial \eta_j} \exp(-\beta E_j^{(1)})}{\sum_{\ell=1}^{3} \exp(-\beta E_j^{(1)})} \right\rangle_r , \tag{A9}$$

with $E^{(1)}$ - eigenvalues of H_t. The averaged free energy can be expressed as follows:

$$G_o = -\beta^{-1} \sum_j \ln \sum_{\ell=1}^{3} \exp(-\beta E_j^{(1)}) + \tfrac{1}{2} \sum_{ij} K_{ij} m_i m_j >_r \quad (A10)$$

Assuming a continuous transition at T_c, we expand for small m's in the vicinity of T_c and arrive at the difference equation given in the main text (Eqs. (5) and (6)).

Work supported by the Institute for Low Temperature and Structure Research, Polish Academy of Sciences, Wroclaw, and the Institute of Physics, Polish Academy of Sciences, Warsaw, Poland.

REFERENCES
[1] H. O. Hooper and A. M. de Graaf (editors) "Amorphous Magnetism", Plenum Press, New York, 1973.
[2] K. Handrich "Amorphe Magnetika", in: "Magnetische Eigenschaften von Festkörpern" (in Germany), VEB Deutscher Verlag für Grundstoffindustrie, Leipzig 1975, p. 169.
[3] R. Harris, M. Plischke and M. J. Zuckermann, Phys. Rev. Lett. 31 (1973) 160.
[4] G. B. Taggart, R. A. Tahir-Kheli and E. Shiles, Physica 75 (1974) 234.
[5] S. V. Tyablikov "Methods in the Quantum Theory of Magnetism" (in Russian), Nauka, Moscow 1965 (English translation: Plenum Press, New York 1967).
[6] D. E. Polk, Acta Met. 20 (1972) 485 and references therein.
[7] Mu Shik Jhon and H. Eyring in: "Physical Chemistry - an Advanced Treatise" Vol. VIIIA, H. Eyring, D. Henderson and W. Jost (Eds.), Academic Press, New York 1971, p. 366.
[8] R. W. Cochrane, R. Harris and M. Plischke, J. Non-cryst. Solids 15 (1974) 239.
[9] R. W. Cochrane, R. Harris, M. Plischke, D. Zobin and M. J. Zuckermann, Phys. Rev. B 12 (1975) 1969.
[10] R. W. Cochrane, R. Harris, M. Plischke, D. Zobin and M. J. Zuckermann, J. Phys. F 5 (1975) 763.
[11] P. M. Richards, Phys. Lett. 55A (1975) 121.
[12] H. Puszkarski, Surf. Science 34 (1973) 125.
[13] D. L. Mills, Phys. Rev. B 3 (1971) 3887.
[14] I. Peschel and P. Fulde, Z. Physik 259 (1973) 145.

SPIN WAVE EXCITATION AND PROPAGATION IN AMORPHOUS BUBBLE FILMS

R. F. Soohoo

University of California

Davis, California 95616

ABSTRACT

The excitation of long wavelength spin wave resonance modes in amorphous ferromagnetic bubble films is examined both theoretically and experimentally in some detail. Whereas resonance of long wavelength spin waves with wavelengths in the order of the film thickness ($\sim 1\mu$) can occur, spatial fluctuations on the atomic scale in lattice constant, coordination number, etc. in the amorphous state lead only to a broadening of the resonance line. Experimental determination of gyromagnetic ratio γ, magnetization $4\pi M$, uniaxial anisotropy constant K_1, saturation field H_s, line width ΔH and exchange constant A for GdCoMo amorphous films are reported and compared with our theoretical calculations.

INTRODUCTION

Although many spin wave resonance experiments have been performed on thin crystalline magnetic films since they were first observed in permalloy by Seavy and Tannewald,[1] very few such experiments on amorphous magnetic films have been reported.[2-6] The discovery that $Gd_{1-x}Co_x$ with $0.7 < x < 0.9$ prepared by r-f sputtering has a uniaxial anisotropy perpendicular to the plane of the film[7,8] has excited considerable interest in the magnetic properties of amorphous films. Such films can support sub-micron size cylindrical domains with their axes perpendicular to the film and are therefore potentially very useful as bubble memory materials. Since then, amorphous film of composition RFe_2 (R = rare earth)[9] and of ternary elements[10] have been prepared by sputtering. Amorphous films of Ni-Fe[11] and a wide range of binary alloys[12] have also been prepared by vacuum deposition upon a cooled substrate.

Before the preparation of amorphous GdCo films, single crystal garnet films grown on GGG substrates using LPE (liquid phase epitaxy) method were used exclusively as bubble memory materials. The bubbles observed in amorphous films can be submicron in size rather than micron in size as in crystalline films. Thus, potentially at least, amorphous films can give rise to a considerably higher bubble storage density. Also, deposition or sputtering method of amorphous film preparation are quite compatible with circuit fabrication technology. Whereas garnet SWR spectra have been studied by many workers, little has been reported for amorphous films. Even a more fundamental reason for our specific interest is the fact that amorphous magnetism is nowhere as well understood as magnetism in crystalline solids.

Spin wave resonance in thin films can yield important information on their magnetic properties. In particular, such fundamental quantities as the gyromagnetic ratio γ, magnetization $4\pi M_s$, uniaxial anisotropy contact K_1, anisotropy dispersion, saturation field H_s, and exchange constant A can all be measured using SWR techniques. Referring to Fig. 1, if the applied field H_o is perpendicular to the film, the resonance condition is given by:[13]

$$\frac{\omega}{|\gamma|} = H_{o\perp} + H_k - 4\pi M_s + \frac{2A}{M_s}\left(\frac{n\pi}{d}\right)^2 \qquad (1)$$

where ω is the frequency, A the exchange constant, d the thickness of the film, and $n = 1, 3, 5 \ldots$. The derivation of Eq. (1) assumes that the spins are pinned at the surfaces of the film. If the spins are not completely pinned at the surface, the expression will change somewhat.[14] For bubble memory materials, the uniaxial anisotropy field H_k must be larger than the saturation magnetization $4\pi M_s$ in order that the remanent magnetization will lie along either the +y or -y directions. Cylindrical domains with M in the +y directions, say, imbedded in the surrounding material magnetized in the -y direction can then be nucleated with the application of a -y-directed magnetic field of appropriate magnitude. A lattice of such bubbles, of micron size, can be used as a high density memory.

On the other hand, if the field H_o is applied parallel to the film plane, the resonance expression is given by:

$$\frac{\omega}{|\gamma|} = \left[(H_{o\parallel} - H_k)(H_{o\parallel} - H_k + 4\pi M_s)\right]^{1/2} \qquad (2)$$

If stress is present in the film, $4\pi M_s$ should be replaced by $(4\pi M_s + H_s)$ where H_s is an equivalent stress field.[15]

By observing the resonance spectra parallel and perpendicular to the uniaxial anisotropy axis (or perpendicular and parallel to the film respectively), we can determine, $|\gamma|$, H_k, $4\pi M_s$, and A with the aid of Eqs. (1) and (2).

EQUATION OF MOTION

The equation of motion of the magnetization $\vec{M}$ is governed by the Landau-Lifshitz equation.[16]

$$\frac{d\vec{M}}{dt} = \gamma(\vec{\tau} + \vec{M} \times \frac{2A}{M^2} \nabla^2 \vec{M}) + \frac{\alpha}{M} \vec{M} \times \frac{d\vec{M}}{dt} \quad (3)$$

where γ is the gyromagnetic ratio, $\vec{\tau}$, the torque (other than exchange and damping) acting on $\vec{M}$, A the exchange constant, and α, the Gilbert damping parameter. If the ferromagnetic sample is a good conductor, Eq. (3) should in principle be solved in conjunction with Maxwell's equations subjected to the appropriate electromagnetic and exchange boundary conditions.[17] However, the main effect of the presence of conductivity is to broaden the resonance line as a consequence of eddy current damping, without a significant shift in the resonance field. Since α is but a composit experimentally-determined damping parameter, the presence of conductivity merely increases the value of α. Thus, inasmuch as the incorporation of Maxwell's equations greatly complicates the mathematics involved without the benefit of much additional insight, we shall solve Eq. (3) by itself.

The torque $\vec{\tau}$ of Eq. (3) may be found from a generalized force $\vec{F} = -\nabla E$ using the relation $\vec{\tau} = \vec{r} \times \vec{F} = -\vec{r} \times \nabla E$ where $\vec{r}$ is a radius vector and E is the free energy of the system. In general, E is given by:

$$E = 2\pi M^2 \sin^2\theta \sin^2\phi - M H_o \sin\theta \cos(\phi_H - \phi) + E_k(\theta,\phi) \quad (4)$$

θ and ϕ are the angles that $\vec{M}$ makes with the z and x axis respectively with the film lying parallel to the x-z plane. ϕ_H is the angle that $\vec{H}_o$ makes with the x-axis with $\vec{H}_o$ lying in the x-y plane as shown in Fig. 1. The first term in Eq. (4) represents the demagnetizing energy of the film while the second term is the Zeeman energy of interaction between $\vec{M}$ and $\vec{H}_o$. Finally the function $E_k(\theta,\phi)$ depends on the type of anisotropy energy under consideration.

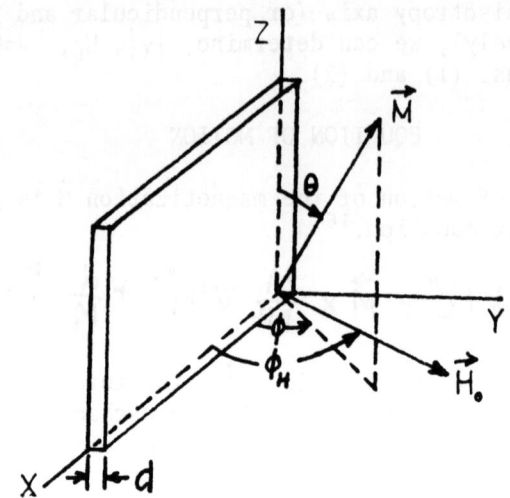

Fig. 1 Amorphous bubble film in a static field H_o of arbitrary orientation.

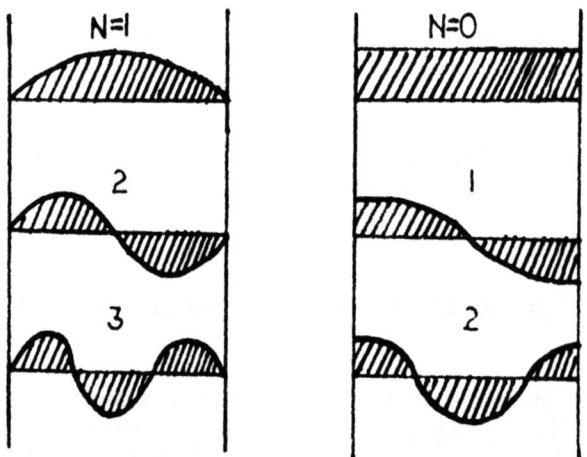

Fig. 2 Resonance modes of y-directed spin waves. Left: pinned and right: unpinned.

For small excursions of $\hat{M}$ from equilibrium, we may let $\theta = \pi/2 + \delta_\theta$, $\phi = \phi_{eq} + \delta_\phi$, $\vec{M} = \hat{f} M_s + \hat{\theta} m_\theta + \hat{\phi} m_\phi$, $m_\theta = M \delta_\theta$ etc., and neglect second order terms in m_θ and m_ϕ. In a bubble film, the anisotropy is uniaxial with the easy axis perpendicular to the surface of the film or $E_k(\theta,\phi) = E_k(1-\sin^2\theta\sin^2\phi)$. Combining Eqs. (3) and (4), we find for $\alpha = 0$:

$$\frac{\partial m_\theta}{\partial t} = -\gamma \frac{2A}{M_s} \frac{\partial^2 m_\phi}{\partial y^2} + \gamma(4\pi M_s - H_K)\cos 2\phi_{eq} m_\phi$$
$$+ \gamma H_0 \cos(\phi_H - \phi_{eq}) m_\phi \qquad (5)$$
$$+ \gamma M_s [(4\pi M_s - H_K)\sin 2\phi_{eq} - H_0 \sin(\phi_H - \phi_{eq})]$$

and

$$\frac{\partial m_\phi}{\partial t} = \gamma \frac{2A}{M_s} \frac{\partial^2 m_\theta}{\partial y^2} + \gamma(4\pi M_s - H_K)\sin^2 \phi_{eq} m_\theta$$
$$- \gamma H_0 \cos(\phi_H - \phi_{eq}) m_\theta \qquad (6)$$

where $H_k = 2K_1/M$ is the anisotropy field and ϕ_{eq} is the equilibrium angle determined by the relation:

$$H_0 \cos\phi_H \sin\phi_{eq} = [H_0 \sin\phi_H - (4\pi M_s - H_K)\sin\phi_{eq}]\cos\phi_{eq} \qquad (7)$$

BOUNDARY CONDITIONS

Close to the air-film interface, there could be a composition gradient due, for example, to oxidation. This would lead to not only a different M but a different K_1 near that surface as well. For this surface region, the component equations of motion would still be of the same form as those given by Eqs. (5)-(7) above except that $4\pi M_s$ and H_k would correspond to those of the surface region. If the easy axis of the surface region should be parallel rather than perpendicular to the film, $E_k(\theta,\phi)$ of Eq. (4) should be replaced by $K_1\sin^2\theta\sin^2\phi$ and the sign of the H_k terms in Eqs. (5)-(7) should also be changed. The presence of such a surface region would give rise to a distinct surface resonance mode as we shall see later in connection with experimental results on GdCoMo amorphous films. However, due to the presence of exchange forces between spins at the surface-"bulk" boundary, the two regions are not entirely decoupled.

The situation discussed above is further complicated by the fact that the exchange boundary condition at the air-film and substrate-film boundaries for the solution of Eq. (3) is dependent upon the relative effects of lower order symmetry at the surface and surface anisotropy energy density K_s.[18] In general, the surface r.f. magnetization m and its derivative are both finite. In the limiting cases where $m = 0$ (for $K_s \to \infty$) and $\partial m/\partial y = 0$ ($K_s \to 0$), the surface spins are said to be pinned and unpinned respectively.

EXCITATION AMPLITUDE AND LINEWIDTH

Consider the simple case where there is no surface region and $\vec{H} = \hat{y} H_o$. The spin wave wave vector $\vec{k}$ can be assumed to be directed in the + and -y directions since the film thickness d is much smaller than the other dimensions of the film. In this case, $\phi_H = \pi/2$ and according to Eqs. (5)-(7), $\phi_{eq} = \pi/2$ and the resonance field H_o is given by (compare with Eq. (1)):

$$H_o = \frac{\omega}{|\gamma|} + 4\pi M_s - H_K - \frac{2A}{M_s} K^2 \qquad (8)$$

For either the $m = 0$ (pinned) and $\partial m/\partial y = 0$ (unpinned) boundary condition, reference to Fig. 2 shows that d must be equal to an integral multiple of $\lambda/2$ where λ is the spin wave wave length for resonance to occur.

Thus, since $k = 2\pi/\lambda$, we have:

$$K = \frac{n\pi}{d} \qquad (9)$$

where n is an integer including zero (uniform mode) for the unpinned case.

According to Fig. 2, the spin wave resonant amplitude for the pinned case is appreciable only for n odd. This is so because if a homogeneous film is excited by a uniform filed, the excitation amplitude is directly proportional to the net shaded area under the curve. For the unpinned case, correspondingly, only the uniform mode would be excited. If the film is not homogeneous, however, all spin wave modes can have finite amplitude.

In a perfect single crystal, the linewidth ΔH can be related to a relaxation time τ via the expression $\Delta H/2 = 1/(\gamma_e)\tau$.[19] Here τ is a composite relaxation time due to spin-spin and spin-lattice relaxations. In an amorphous ferromagnetic film, such as GdCoMo, other sources of contribution to the linewidth will occur. To begin with, we have already mentioned earlier that eddy current damping broadens the resonance line and thus lower the resonance amplitude of the spin wave modes. For example, spatial fluctuations on the atomic scale in lattice constant, coordination number in the amorphous state will further broaden the resonance lines. Although such fluctuations will lead to random variations in local $\vec{M}$, K_1 and A, k can not adjust its local value rapidly enough to maintain a sharp resonance peak. To do so would have required the excitation of spin waves with wavelength in the order of a lattice constant resulting in very large exchange energy. Consequently, the atomic fluctuation in $\vec{M}$, K_1 and A result in variations instead in local field and line broadening (see Eq. (8)). For gradual variations in M, K_1 and A ($\lambda \sim 1\mu$), on the other hand, exchange energy is relatively small. Consequently k can adjust itself locally to yield sharp resonances.

EXPERIMENTAL RESULTS

Ferromagnetic resonance spectra at T = 20° and 300°K of two samples of GdCoMo films sputtered in argon are shown in Figs. 3 to 6. The composition of these films is in the general range of $Gd_{.13}Co_{.72}Mo_{.13}Ar_{.02}$. For sample 1 (Figs. 3 and 4), d = 2.05μ, stripe width w = 1μ, H_{sat} = 1,076 oe and for sample 2 (Figs. 5 and 6), d = 2.09μ, w = 1.25μ, H_{sat} = 803 oe. It is noted that the spectra at 20°K are considerably more complicated than those at 300°K and f = 9.2 GHz.

To determine γ and H_k, we solve Eqs. (1) and (2) simultaneously; note that Eq. (1) is identical to Eq. (8) while Eq. (2) can be obtained from Eqs. (5)-(7). The result is:

$$H_K = \frac{H_{o\parallel}^2 - H_{o\perp}^2 + 4\pi M_s (H_{o\parallel} + 2H_{o\perp} - 4\pi M_s)}{2(H_{o\perp} + H_{o\parallel}) - 4\pi M_s} \quad (10)$$

and

$$|\gamma| = \frac{\omega}{H_{o\perp} - 4\pi M_s + H_K} \quad (11)$$

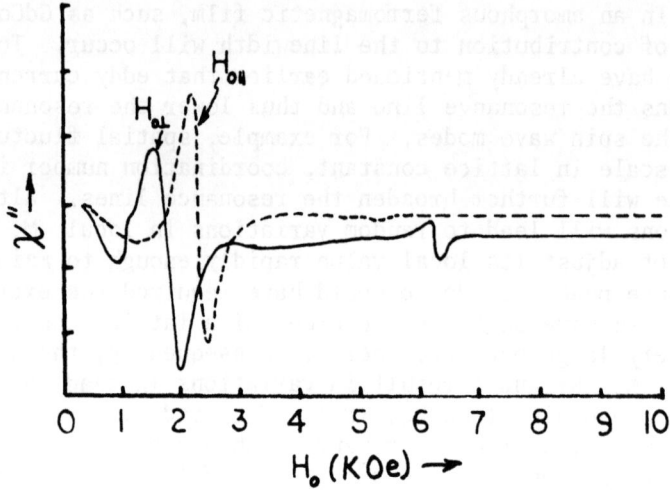

Fig. 3 Susceptibility spectra with H_o parallel and perpendicular to GdCoMo amorphous film 1 at 300°K

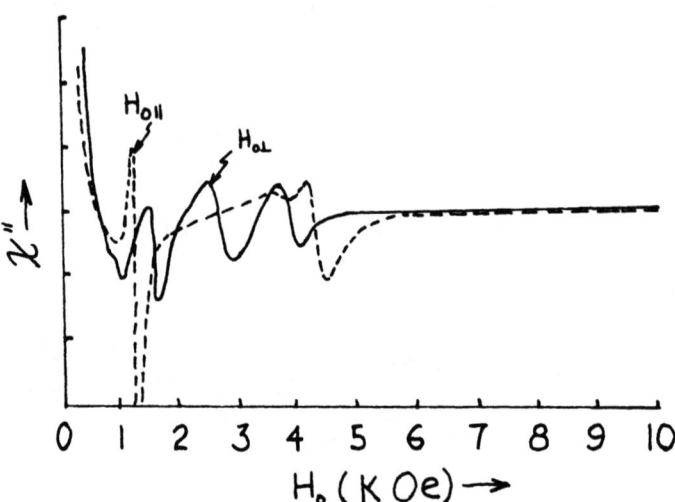

Fig. 4 Susceptibility spectra of GdCoMo film 1 at 20°K

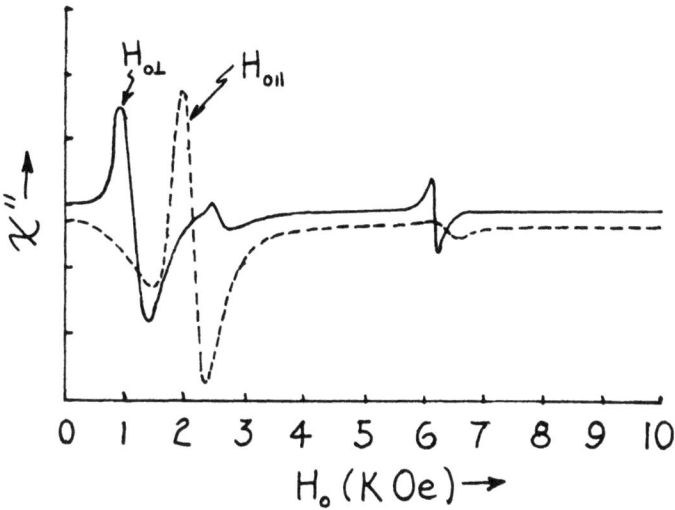

Fig. 5 Susceptibility spectra of GdCoMo film 2 at 300°K

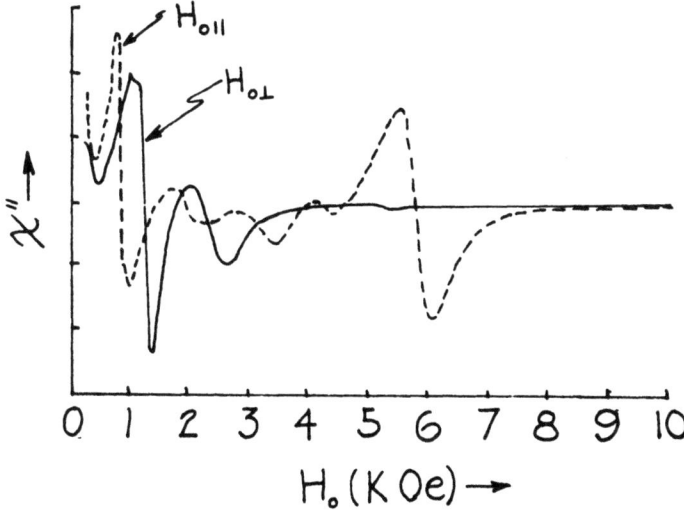

Fig. 6 Susceptibility spectra of GdCoMo film 2 at 20°K

Using these formulas we find for sample 1, at 300°K (Fig. 3) H_k = 1374 oe, $|\gamma|$ = 6.8x10^6 oe^{-1} sec^{-1} while for sample 2 (Fig. 5), H_k = 1434, $|\gamma|$ = 7.4x10^6 oe^{-1}sec^{-1}. In this calculation, we used the ratio w/d to determine $H_s/4\pi M_s$ from Figs. 3 and 7 of Cape and Lehman.[20] In this way, $4\pi M_s$ was found to be 1,793 gauss and 1,338 gauss for samples 1 and 2 respectively. It is interesting to note that the $Q(H_k/4\pi M_s)$ of No. 1 sample is less than 1. However, stripes were nevertheless observed indicating that vertical domains rather than a single domain magnetized in the plane of the film exist; the multi-domain state is evidently one of lower energy.

Comparing Fig. 3 with 4 and 5 with 6, we see that whereas $H_{o\perp}$ is less than $H_{o\parallel}$ at 300°K, the reverse is true at 20°K. This is due to a rise in M as well as a change in K_1 at low temperatures. Notice also that there is a high-field resonance mode at both temperatures for both samples. This mode is due to a surface region with different M and K_1 mentioned earlier.

In addition to the main and surface modes, we see in Figs. 3 to 6 also some spin wave modes. For example, a spin wave mode exists at about 1,100 oe in sample 1 (Fig. 3) and several such modes are evident at low temperatures (Figs. 4 and 6). An attempt was made to calculate the exchange constant A for these films but the result is considered not sufficiently accurate due to the overlapping of these modes as a result of rather broad lines. Furthermore, it is likely that there is inhomogeneity in H_k even in the "bulk" region which affects the spacing between modes. Finally, in principle, H_{sat} can be determined from the shape of the curves at low fields. However, resonance begins to occur before the sample is saturated rendering such determination also inaccurate.

REFERENCES

1. M. H. Seavy and T. E. Tannenwald, Phys. Rev. Letters 1, 168 (1958); J. Appl. Phys. 30, 227 S (1959).
2. B. Elschner and H. Gartner, Z. Angew, Phys. 29, 342 (1966).
3. Y. Ajiro, K. Tamura, and H. Endo, Phys. Letters 35A, 275 (1971).
4. D. C. Cronemeyer, AIP Conf. Proc., No. 18, pt. 1, Magnetism and Magnetic Materials, 1973 (19th Annual Conference, Boston), p. 85.
5. H. Nosé, "Spin Wave Resonance in Amorphous Thin Films," presented at the 6th Int'l. Conf. on Magnetic Thin Films, Minsk, USSR (Aug., 1973).
6. J. R. Mc Coll, D. Murphy, G. S. Cargill III and T. Mitzoguchi, 20th Ann. Conf. Magnetism Magnetic Materials, Am. Inst. Phys. Conf. Proc. No. 29, ed. C. D. Graham, Jr., G. H. Lander and J. J. Rhyne (1976), p. 172.
7. P. Chaudhari, J. J. Cuomo, and J. Gambino, IBM J. Res. & Develop. 17, 66 (1973); Appl. Phys. Letters 22, 337 (1973).
8. S. Herd and P. Chaudhari, Phys. Stat. Sol. (a) 18, 603 (1973).

9. J. J. Rhyne, S. J. Pickard, and H. A. Alperin, Phys. Rev. Letters 29, 1562 (1972).
10. P. Chaudhari, J. J. Cuomo, R. J. Gambino, S. Kirkpatrick and L. J. Tao, AIP Conf. Proc. No. 24 (1975), p. 562.
11. S. Fujime, Japanese J. Appl. Phys. 5, 59, 643, 739, 764, 778 and 1029 (1966) and 6, 270, 305 (1967).
12. N. Heiman and K. Lee, AIP Conf. Proc. No. 24 (1975), p. 108.
13. R. F. Soohoo, Thin Magnetic Films, Harper & Row Publ., New York, N. Y. (1965), p. 235.
14. R. F. Soohoo, J. Appl. Phys. 34, 1149 (1963); Phys. Rev. 131, 594 (1963).
15. H. Nosé, J. Phys. Soc. Japan 23, 937 (1967).
16. L. Landau and E. Lifshitz, Physik Zeitschrift Sowjetunion 8, 153 (1935).
17. R. F. Soohoo, p. 233, Ref. 13 above.
18. R. F. Soohoo, p. 230, Ref. 13 above.
19. R. F. Soohoo, Theory and Application of Ferrites, Prentice-Hall, Inc., Englewood Cliffs, N. J. (1960), p. 74.
20. J. A. Cape and G. W. Lehman, J. Appl. Phys. 42, 5732 (1971).

DISCUSSION

L. M. Roth: Can you get a dispersion relation from this treatment?

R. F. Soohoo: Yes, I have showed one for the perpendicular case. In my paper I will derive the general expression including the electromagnetic and exchange boundary conditions.

G. S. Cargill: The experimental results you have look like they are insufficient to determine what the dispersion is for these materials.

R. F. Soohoo: I agree, it is something difficult to do mainly with the film being so thick and the line being broad. However we made an attempt and it looks like it is in the order of 10^{-7} erg/cm depending on what kind of assumptions you make.

TEMPERATURE DEPENDENCE OF THE EXTRAORDINARY HALL EFFECT IN

AMORPHOUS Co-Gd-Mo THIN FILMS

R. J. Kobliska and A. Gangulee

IBM Thomas J. Watson Research Center

Yorktown Heights, New York 10598

ABSTRACT

The temperature dependence of the extraordinary Hall resistivity and the ordinary resistivity of amorphous Co-Gd-Mo-Ar thin films have been investigated from 4.2°K to the ferrimagnetic Curie temperature; the nominal composition of these films was $Co_{65}Gd_{10}Mo_{15}Ar_{10}$. The extraordinary Hall resistivity reverses sign at the compensation temperature and reflects the paramagnetic susceptibility above the Curie temperature. The temperature dependence of the Hall resistivity can be adequately expressed as $R_1(Co)M_{Co} + R_1(Gd)M_{Gd}$, where M_{Co} and M_{Gd} are the subnetwork magnetizations determined from a mean field fitting procedure. This fit emphasizes the inappropriateness of spin fluctuation scattering models which are relevant to the crystalline phases. Such models predict a minimum in Hall resistivity at low temperatures which is not observed experimentally.

INTRODUCTION

Amorphous transition metal-rare earth alloy thin films have generated considerable interest because of their potential applications in bubble domain devices. These TM-RE alloys often exhibit a large extraordinary Hall signal, which is conveniently used for characterizing these materials. However, in contrast to the crystalline materials, only rudimentary understanding exists about the origin of the extraordinary Hall effect in amorphous materials. The Hall effect in amorphous Co-Gd binary alloys thin films have been most extensively studied. The magnetic properties of these films are reasonably well characterized,[1] and this alloy system is understood to be a ferrimagnetic one in which the Co and Gd spins couple antiferromagnetically. Okamoto et al.[2] have shown that the Hall voltage in this system changes sign at the compensation temperature (T_{comp}). They interpreted their results in terms of scattering from each subnetwork and attributed the signal primarily to the Gd subnetwork. Ogawa et al.[3] have studied the binary alloys Co_3Gd, Co_7Gd_2 and Co_5Gd, in their amorphous and crystalline states; in each case they found a sign reversal of the Hall voltage at T_{comp}. They interpreted their crystalline state data by assuming a thermal spin fluctuation model in which the extraordinary Hall resistivity varies as $< (M - <M>)^2 >$. As the temperature is lowered, thermal spin fluctuations decrease and the Hall resistivity approaches zero.

In order to obtain a better understanding of the extraordinary Hall effect in amorphous materials, a careful analysis of the temperature dependence of the extraordinary Hall resistivity, in the temperature range from 0°K to the ferrimagnetic Curie temperature (T_c), was thought to be desirable. The amorphous Co-Gd binary alloys are unsuitable for this purpose because of the annealing and crystallization effects that take place near their Curie temperatures. Amorphous Co-Gd-Mo thin films, on the other hand, can be fabricated with Curie temperatures below 425°K and show negligible annealing effects at temperatures up to 500°K.[4] An investigation of the electrical resistivity and the Hall resistivity in two such films in the temperature range 4° - 500°K is reported in this paper.

HALL RESISTIVITY

The magnetic units employed in this paper are not the traditional CGS units but the SI units suggested by the National Bureau of Standards.[5] In this system of units $B = \mu_o(H + M)$ and the Hall resistivity for a ferromagnetic material can be expressed as

$$\rho_H = \frac{Vd}{I} = \rho_{HO} + \rho_{HA} = \mu_o R_o H + \mu_o R_1 M \tag{1}$$

where $H(A/m)$, $M(A/m)$, $V(V)$ are the magnetic field, magnetization, and the voltage measured perpendicular to the current $I(A)$ for a sample of thickness $d(m)$ in the standard Hall measurement configuration. The permeability of free space μ_o, is equal to $4\pi \times 10^{-7}$ ($H/m = V \cdot s/A \cdot m$). The ordinary Hall resistivity, ρ_{HO}, is related to the applied field via the ordinary Hall coefficient R_o (m^3/coul = Ω-m/Tesla), whereas the anomalous (or extraordinary) Hall resistivity ρ_{HA} is characterized by the extraordinary Hall coefficient R_1 (m^3/coul).

This phenomenological description of the extraordinary Hall effect is relevant to ferromagnetic materials, but for the ferrimagnetic films discussed here the following expression is more appropriate:

$$\rho_{HA} = \mu_o < \Sigma\, R_{1i}\, \overline{M}_i > \qquad (2)$$

where the symbol < > denotes spatial average and $\overline{M}_i$ is the projection of $\vec{M}_i$ along the field direction, where M_i and R_{1i} are the magnetization and the extraordinary Hall coefficient of the i-th subnetwork, respectively.

The extraordinary Hall resistivity of a ferromagnetic material for temperatures above the Curie temperature is related to the magnetic susceptibility ($\kappa = M/H$) by the expression:

$$\rho_{HA}/\mu_o H = [R_o + \kappa^*(R_1 - NR_o)] \qquad (3)$$

where $\kappa^* = \kappa/(1 + N\kappa)$ is the effective susceptibility of the sample and N is the demagnetizing factor. For a ferrimagnet this expression can be generalized by expressing the net susceptibility in terms of the subnetwork susceptibilities, $\kappa_i = M_i/H$. The sign of κ_i is determined by whether $\vec{M}_i$ and H are parallel or antiparallel, and the total susceptibility is simply $\kappa = \Sigma \kappa_i$.

The Hall resistivity for a ferrimagnet is

$$\rho_H/\mu_o H = R_o(1 - N\kappa^*) + \Sigma\, R_{1i}\kappa_i^* \qquad (4)$$

Since only κ^* is accessible by direct measurement of M and H, the values for κ_i^* can not be obtained directly without additional data.

EXPERIMENT TECHNIQUES AND RESULTS

The films studied were RF bias sputtered Co-Gd-Mo films of nominal composition $Co_{65}Gd_{10}Mo_{15}Ar_{10}$. The details of film fabrication techniques and the dependence of magnetic properties on composition are summarized elsewhere.[6] The room temperature magnetic properties and the exact compositions for films A and B are listed in Table I.

To minimize experimental complications due to spatial compositional non-uniformities, the Hall pattern shown in Figure 1 was etched in each film.

TABLE I: Room temperature properties of the films investigated.

Film Composition (at. %)

#	Co	Gd	Mo	Ar
A	67.0	10.5	14.5	8.0
B	61.6	8.9	14.0	15.5

#	Stripwidth (μm)	$\mu_o M_s$ (Tesla)	K_{u3} (J/m^3)	d (μm)
A	2.0	3.4×10^{-2}	3.5×10^3	1.88
B	0.9	5.0×10^{-2}	2.5×10^3	1.17

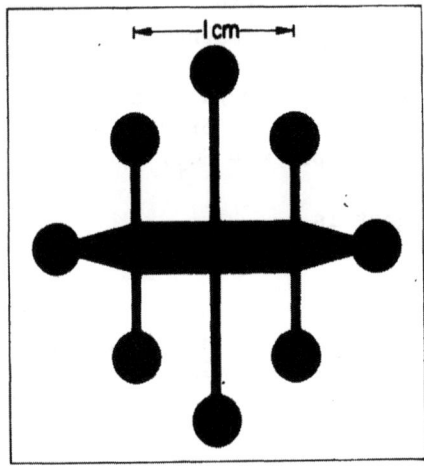

Figure 1: The Hall pattern etched in the amorphous Co-Gd-Mo thin films.

The low temperature measurements were performed in a liquid He immersion dewar, whereas the measurements between liquid nitrogen temperature and the Curie temperature were carried out in a specially designed vacuum dewar/oven. Hall voltages were measured using both AC and DC techniques. The best signal to noise ratio was obtained with a 280 Hz AC current supply and a lock-in detection scheme. Currents of 10^{-2}A were employed in order to minimize sample heating. Thermal EMF and magnetoresistive effects were minimized through current and magnetic field reversal. The magnetization and the magnetic susceptibility in the temperature range 4°-500°K were obtained using a vibrating sample magnetometer.

The ordinary resistivity was measured over the temperature range 4° - 500°K and could be fitted with a function of the form $\alpha + \beta T$ where $\alpha = 1.9$ and 2.3×10^{-6} Ω-m and $\beta = -2.3$ and -3.6×10^{-10} Ω-m/°K for films A and B. No anomalies were observed in the resistivity at either T_{comp} or T_c. Since the resistivity neither decreases at low temperature nor deviates from the $\alpha + \beta T$ dependence for temperatures up to 100°K above T_c, we attribute the resistivity to disorder ($\sim$ impurity) scattering. The small decrease in resistivity with increasing temperature is not readily explained, but one may speculate that the total resistivity is indirectly reduced by the decreased extraordinary Hall scattering at high temperatures.

Hall voltage ($\rho_H \cdot I$) vs. applied field hysteresis loops for film A are shown in Figure 2 for T = 295°K and 417°K ($T_c \sim 420$°K). It should be noted that the field at which the stripe domains collapse, H_{sc}, and the coercivity H_c can be readily measured from the hysteresis loop in Fig. 2(a). In these Co-Gd-Mo films $R_o << R_i$, so that the measured Hall voltages essentially reflect the extraordinary Hall resistivities.

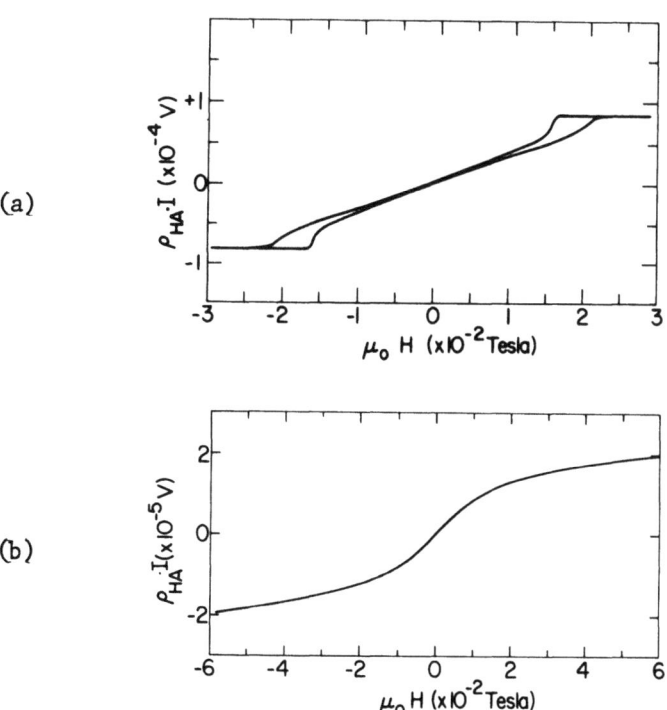

Figure 2: a) Hall voltage hysteresis loop for film A at 295°K.
b) Corresponding hysteresis loop at 417°K illustrating Hall voltage susceptibility.

The temperature dependence of the extraordinary Hall resistivity is shown in Figs. 3(a) and 4(a) for films A and B. We note that Hall resistivity data exhibits an abrupt reversal in sign at the compensation temperature. The width of the sign reversal transition region is $\sim 10°K$ which is much less than the $50° - 100°K$ width reported by Ogawa et al.[3] for binary Co-Gd films. This difference in transition width is attributed to the lesser degree of spatial composition non-uniformity in the films studied here.

The ρ_{HA} data were fitted with Eq (2) using subnetwork magnetizations calculated from a mean field analysis of the saturation magnetization versus temperature data.[7] Implicit in this fitting procedure is the assumption that the R_{1i}'s are temperature independent; the importance of this assumption will be discussed later.

The subnetwork magnetizations and the net magnetization for films A and B are shown in Figs. 3(b) and 4(b). The mean field analysis provides a best fit of the magnetization data with values for the exchange constants (J_{i-j}) and spin values (S_i) which are listed in Table II. It should be noted that the Mo addition has reduced S_{Co} from 0.77, characteristic of pure Co, to 0.35.

TABLE II: Mean field fitting parameters, experimental and calculated values for T_{comp} and T_c, and the fitted values for the extraordinary Hall coefficients.

Mean Field Parameters

#	S_{Co}	g_{Co}	S_{Gd}	g_{Gd}	J_{Co-Co} (J)	J_{Co-Gd} (J)	J_{Gd-Gd} (J)
A	.35	2.22	3.5	2.0	1.96×10^{-21}	-3.29×10^{-22}	2.7×10^{-23}
B	.349	2.22	3.5	2.0	2.10×10^{-21}	-3.24×10^{-22}	2.7×10^{-23}

	Experimental		Mean Field		R_1(Co)	R_1(Gd)
#	T_c	T_{comp}	T_c	T_{comp}	($\times 10^{-8}$ m^3/coul)	
A	430°K	225°K	409°K	232°K	+ 2.44	− 3.17
B	410°K	175°K	393°K	174°K	+ 3.80	− 4.75

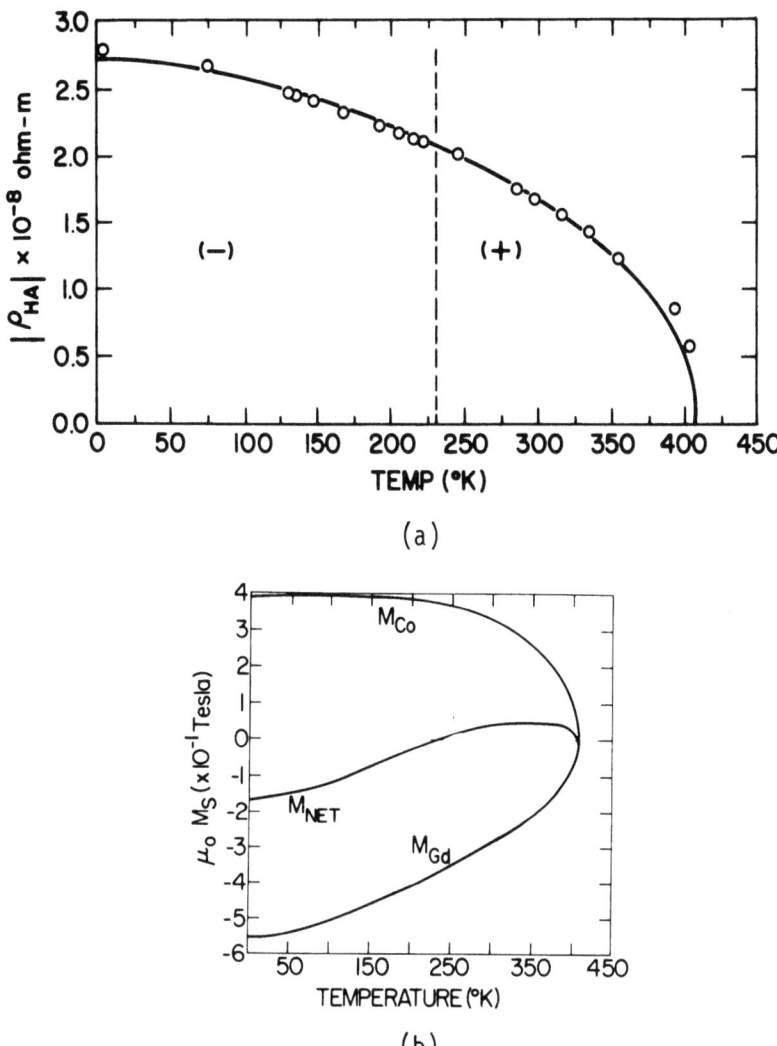

Figure 3: a) The Hall resistivity data for film A(o) and the fitted curve.
b) The subnetwork and net magnetizations for film A (1 Tesla = 10000G)

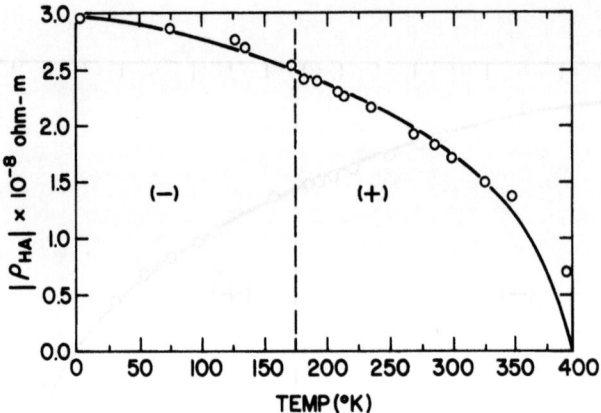

(a)

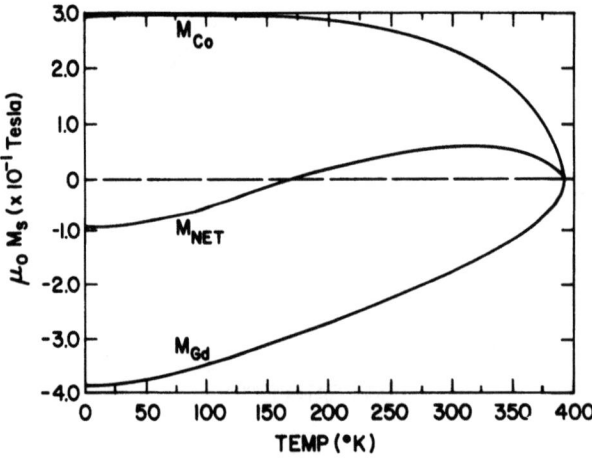

(b)

Figure 4: a) The Hall resistivity data for film B(o) and the fitted curve.
b) The subnetwork and net magnetizations for film B (1 Tesla = 10000G).

The least squares fit for $\rho_{HA} = R_1(Co) M_{Co} + R_2(Gd) M_{Gd}$ is shown as the smooth curve in Figs. 3(a) and 4(a). The only substantial disagreement of the theoretical fit and the experimental data is for temperatures near T_c, because the results of mean field calculations are somewhat unreliable near this temperature due to compositional nonuniformities. This discrepancy is evident in the differences in theoretical and experimental values for T_c. The values derived for $R_1(Co)$ and $R_1(Gd)$ are listed in Table II. For comparison, the largest values reported for ρ_{HA} and R_1 for crystalline Gd are -7×10^{-8} Ω-m and -5×10^{-8} m^3/coul, respectively at 243°K. Similarly, pure Co at $T = 1290°K$ exhibits a ρ_{HA} of 2.7×10^{-8} Ω-m and R_1 of 3.5×10^{-8} m^3/coul.

The difference in the absolute value for $R_1(Co)$ and $R_1(Gd)$ for these two films is attributed to the higher resistivity of the film B due to larger argon incorporation. Irrespective of the absolute values for R_1, an instructive ratio is $R_1(Co)/R_1(Gd)$, which indicates the contribution to the total Hall signal from each subnetwork. Since the number of atoms per unit volume is much higher for Co than for Gd, the scattering efficiency per atom of Co must be much smaller.

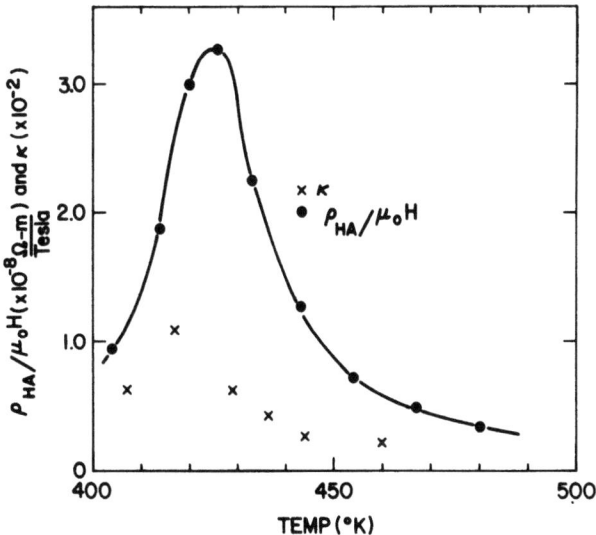

Figure 5: The Hall resistivity susceptibility ($\rho_{HA}/\mu_o H$) and the magnetic susceptibility ($\kappa = M/H$) as a function of temperature for film A.

The signs of $R_1(Co)$ and $R_1(Gd)$ are consistent with the observations of McGuire et al.,[10] which indicate that the signs of R_1 for amorphous Co and Gd are positive and negative, respectively. Since the subnetworks are ferrimagnetically coupled, the resultant Hall signal is the sum indicated in Eq (2), and is negative or positive depending on whether the Gd or the Co subnetwork is dominating the magnetization.

The field dependence of ρ_{HA} above T_c is expected to reflect the subnetwork susceptibilities as stated in Eq (4). The $\rho_{HA}/\mu_o H$ and M/H data is plotted against T in Figure 5. With no independent measurement of the subnetwork susceptibilities, and since the mean field theory is not particularly reliable near or above T_c, we are not able to test the validity of Eq. (4). It may be noted, however, that the susceptibility of the Hall resistivity reflects the net magnetic susceptibility and has a sign characteristic of a Co dominated magnetization.

DISCUSSION

The temperature dependence of the extraordinary Hall effect in amorphous materials differs dramatically from the same phenomena in crystalline materials. The predominant cause of this difference is the scattering mechanism responsible for the ordinary resistivity. The resistivity of these amorphous metal films is dominated by disorder ($\sim$ impurity) scattering and is temperature-insensitive. In crystals the total resistivity, ρ, has contributions from phonon scattering, ρ_{ph}, impurity scattering, ρ_o, as well as scattering from thermal spin flucuations, ρ_M. Theories proposed to explain the extraordinary Hall effect in crystals predict a temperature dependence of R_1 which is proportional to the first or second power of one constituent of the resistivity, the exact functional dependence determined by the scattering mechanism and the temperature range.[11]

The transition metals typically exhibit an R_1 which is roughly proportional to ρ^2. The temperature dependence of R_1 is dominated by the temperature dependence of the total resistivity. Most crystals exhibit a residual resistivity which is constant and contributes a temperature independent contribution to R_1.[12] For Gd, Lee and Legvold[8] report a ρ^2 dependence of R_1 with substantial deviation near T_c, whereas Volkenshtein et al.[13] fitted their data with a $[1 - c(M(T)/M(0))^2]$ dependence for R_1 which corresponds to a ρ_M dependence.

In crystalline materials R_1 can vary by two or more orders of magnitude between 0°K and T_c, while amorphous films have the distinction of possessing a temperature-insensitive resistivity and therefore probably a temperature independent R_1. Whether R_1 is truly temperature independent will require a careful study of an amorphous ferromagnet; reports of a constant R_1 in foils of amorphous Fe-P-C are relevant in this respect.[14] If we assume that the scattering mechanism is equivalent to impurity scattering, then a ρ^2 dependence of R_1 dominates. This functonal form of R_1 on resistivity reconciles the difference in the R_1's obtained for the two films in this investigation. A direct comparison of R_1 for crystalline and amorphous materials is difficult since the dominant scattering mechanisms are different. However, the calculated R_1's for the constituents of these ferrimagnetic films are comparable to their maximum crystalline values.

In summary, the extraordinary Hall resistivity of amorphous TM-RE alloys is determined by the magnitudes of the subnetwork magnetizations and the associated extraordinary Hall coefficients. The temperature dependence of the extraordinary Hall resistivity is characteristic of contributions from both the Co and the Gd constituents. For the compositions investigated here, these contributions are comparable in magnitude and additive since the extraordinary Hall coefficients are of opposite signs and the subnetworks are antiferromagnetically coupled. The electrical resistivity is approximately constant in these Co-Gd-Mo films, and hence it was not possible to determine the functional dependence of R_1 on the resistivity. A careful analysis of amorphous films having substantially different resistivities but consisting of the same elements will allow the determination of this functional dependence, and result in a unified understanding of both the compositional and temperature dependence of the extraordinary Hall resistivity.

ACKNOWLEDGEMENTS

The authors are grateful to Dr. D. E. Cox for fabricating the amorphous specimens. They also wish to thank Messrs. H. R. Lilienthal, V. A. Ranieri and V. C. Richardson for their assistance with the experimental work and Mrs. C. M. Hand for preparing the typescript.

REFERENCES

1. R. C. Taylor and A. Gangulee, J. Appl. Phys. 47 (1976), in press.
2. K. Okamoto, T. Shirakawa, S. Matsushita and Y. Sakurai, IEEE Trans. Magnetics, MAG-10, 799 (1974).
3. A. Ogawa, T. Katayama, M. Hirano and T. Tsushima, Jap. J. Appl. Phys. 15, 87 (1976).
4. R. J. Kobliska and A. Gangulee, AIP Conf. Proc. No. 24, 257 (1975).
5. L. H. Bennett, C. H. Page and L. J. Schwartzendruber, AIP Conf. Proc. No. 29, xix (1976).
6. C. H. Bajorek and R. J. Kobliska, IBM J. Res. Develop., 20, 271 (1976).
7. a) A. Gangulee and R. J. Kobliska, to be published.
   b) R. J. Kobliska, A. Gangulee, D. E. Cox and C. H. Bajorek, to be published.
8. P. S. Lee and S. Legvold, Phys. Rev., 162, 431 (1967).
9. I. A. Tsoukalas, Phys. Stat. Sol. 23, K41 (1974).
10. T. R. McGuire, R. J. Gambino, and R. C. Taylor, to be published.
11. C. M. Hurd, The Hall Effect in Metals and Alloys, (Plenum Press, New York-London, 1972).
12. J. M. Luttinger, Phys. Rev. 112, 739 (1958).
13. N. Y. Volkenstein, I. K. Grigorova, and G. V. Fedorov, Soviet Physics - JETP 23, 1003 ,(1966).
14. S.C.M. Lin, J. Appl. Phys., 40, 2175 (1969).

DISCUSSION

R. Hasegawa: Do you have any information about the magnitude of R_o?

R. J. Kobliska: R_o is probably about three orders of magnitude smaller than R_1 in these materials. You cannot see it.

CURRENT VIEWS ON THE STRUCTURE OF AMORPHOUS METALS II[*]

G. S. Cargill III

IBM T. J. Watson Research Center

Yorktown Heights, N. Y. 10598

Presently available experimental data on structural and chemical order in amorphous metals and alloys are examined in terms of similarities and differences in short range order between the amorphous materials and related crystalline solids. Recent x-ray scattering data on amorphous, electrodeposited Co-P alloys[1] are discussed with regard to composition dependent structural features, microstructural anisotropy, and annealing processes. Effects of tetrahedron perfection constraints in dense random packing models with one and with two sphere sizes are reviewed, including effects on density and porosity of model structures.[2] Results for binary packing models are compared with experimental data for amorphous Gd-Co alloys.[2] Recent results for soft sphere (relaxed) model structures are compared with those for hard sphere structures. Most of the material presented in this review is being published elsewhere.[1-5]

REFERENCES

1. G. C. Chi and G. S. Cargill III, presented at Intern. Conf. on Structure and Excitations of Amorphous Solids, Williamsburg, Va., Mar. 1976, to be published in the conf. proc.; Mat. Sci. Engr. 23, 155 (1976); AIP Conf. Proc. 29, 147 (1976).

2. G. S. Cargill III and S. Kirkpatrick, presented at Intern. Conf. on Structure and Excitations of Amorphous Solids, Williamsburg, Va., Mar. 1976, to be published in the conf. proc.

3. G. S. Cargill III and S. Kirkpatrick, to be published.

[*]Invited paper.

4.   G. S. Cargill III, in <u>Solid State Physics</u>, Vol. 30, H. Ehrenreich, F. Seitz, and D. Turnbull, eds., (Academic Press, N.Y., 1975) p. 227.

5.   G. S. Cargill III, to appear in Proceedings of the Second International Conference on Rapidly Quenched Metals, N. J. Grant and B. C. Giessen, eds. (MIT Press, Cambridge, Mass., 1976).

DISCUSSION

R. C. O'Handley: When you showed the small angle scattering data you suggested that it is indicative of some sort of phase separation that takes place before the onset of crystallization. Is there a well defined way to tell from this type of data, when crystallization does indeed occur?

G. S. Cargill: The reason for saying that crystallization did not occur was based on the failure to see anything of a crystalline nature in the large angle x-ray scattering pattern or the transmission electron diffraction pattern.

S. J. Pickart: What exactly is that energy minimization process you have referred to when discussing the dense random packing model?

G. S. Cargill: The energy minimization process involves starting up with the random dense packing of hard sphere structures which was generated on the computer, then choosing a Lennard-Jones potential between pairs of atoms and using a conjugate gradient method to choose displacements of atoms which reduce the sum of the Lennard-Jones energies. We have carried this out for structures containing one size of spheres and the agreement between the structure radial distribution function and that for example for amorphous cobalt is strikingly improved. Similar efforts of this type have been previously reported by Von Heimendahl and others. We have simply extended this to larger clusters and have looked at the interference function as well as the distribution function.

D. S. Boudreaux: Did you consider trying a Morse potential as Von Heimendahl did in his treatment of the problem?

G. S. Cargill: We have not tried a Morse potential. In fact, we have not optimized the Lennard-Jones potential to be well representative of a particular material. However, our results are qualitatively similar to those of Von Heimendahl. We have simply looked at larger numbers of atoms.

D. I. Paul: In reference to the rod-shaped defects you saw in Co-P can you give me an idea of what size they were?

G. S. Cargill: From the angular dependence of the scattering, one can conclude that the size in the plane of the film is about 100 Å. Normal to the plane of the film, and these films being 30 μm thick, the size is something greater than 4000 Å.

STRUCTURE SIMULATION OF TRANSITION METAL - METALLOID GLASSES, II

D. S. Boudreaux

Allied Chemical Corporation, Materials Research Center

Morristown, NJ 07960

Computer simulated structural models of transition metal (TM)-metalloid (M) glass systems are generated which consist of dense random packings of hard spheres whose radii are chosen to correspond to the elements included in the alloy. The models, as constructed by the "global" scheme of Bennett, are found to be unacceptable. They exhibit strong angular anisotropies in the pair correlations in regions away from the center of the cluster and predict, for sufficiently different TM and M atomic sizes, physically unreasonable M to TM coordination numbers. It is evident now that pair correlation anisotropies are reasonsible for the shortcomings of earlier single size sphere models.[1,4] Acceptable TM-M models will require either an energetic relaxation of the present models or genesis of a new stacking algorithm.

INTRODUCTION

A detailed investigation is under way to determine the applicability of the concept of dense random packing (DRP) of hard spheres to the structure of binary glasses. The work involves computer simulation using the basic algorithm described by Bennett.[1] The details of the present implementation have recently been described[2]; it was shown that the pair correlations resulting from the Bennett scheme are highly anisotropic, that it is difficult to control the composition of the alloy being simulated, and that the shape of the second peak of the partial pair correlation functions (PCF) depend on composition. The sizes of the spheres used were appropriate for modeling FeP or PdSi glasses. The present paper considers sphere sizes chosen to represent FeB glasses. The ratio of the radii of boron spheres to iron spheres was taken to be 0.52, based on data from the crystal structure of $Fe_3B_{0.9}C_{0.1}$.[3] Other parameters are the same as described in Ref. 2.

ANGULAR DEPENDENCE OF PCF'S

Figure 1 presents the TM-TM partial pair correlations for a model of the structure of $TM_{81}M_{19}$. Parts (a) through (c) are based on all of the atoms in the structure considered as centers; part (d) is based only on atoms within the inner third of the model. In addition, part (a) only considers neighbors within a 39° cone opening from the central atom as apex, toward the center of the structure; part (b) considers neighbors within 6° above or below a plane through the central atom and normal to a model radius; part (c) is similar to (a) but the cone opens away from the model center. Clearly the short range order is highly anisotropic; more

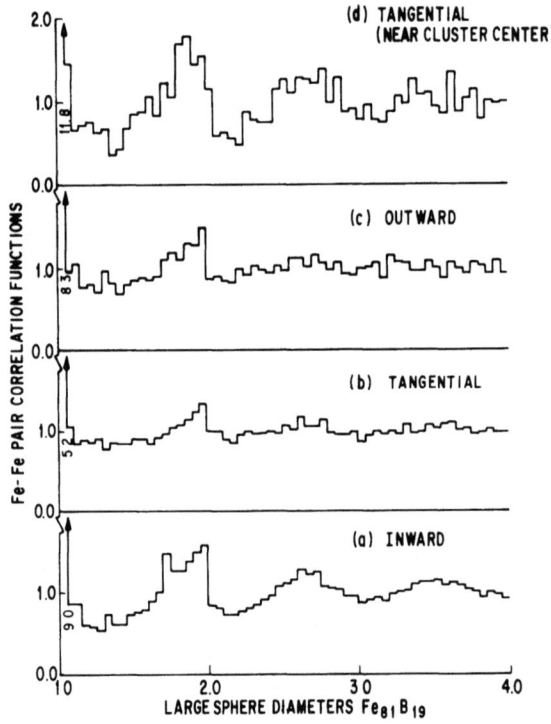

Fig. 1. Fe-Fe partial pair correlation functions for a model of $Fe_{81}B_{19}$. (a), (b) and (c) are calculated using all atoms as center and refers to inward, tangential and outward distributions of neighbors respectively. (d) Considers neighbors in tangential directions but is calculated using central atoms located in the inner third of the whole model.

effective packing is required in the tangential directions. Figure 1(d) is also a picture of tangential pair correlations but using only atoms near the model center to generate the figure; the packing is more effective in the central regions of the cluster.

COORDINATION NUMBERS

A similar anisotropy is exhibited in the coordination numbers. The data for the structures being reported in this paper are given in the table. Bonds between boron atoms are not allowed.

Structure	Fe by Fe	Fe by B	B by Fe
$Fe_{78}B_{22}$	8.4 (7.5)	1.9 (1.2)	4.6 (4.0)
$Fe_{81}B_{19}$	8.7 (7.6)	1.4 (1.0)	4.4 (3.9)
$Fe_{85}B_{15}$	8.8 (8.2)	1.1 (0.7)	4.5 (4.0)

The numbers given to the left of each column are average coordinations for atoms in the central third of the models (i.e., where the anisotropy in PCF's is weakest); the numbers in parentheses are averages over the whole structure. A serious defect in the model is noted in that B atoms have only a little more than 4 Fe neighbors; one expects from intermetallic compound structures to find a number more like 6. A detailed analysis using angular partitions, as for the PCF's, shows that the inner half of the average B atom is nearly properly coordinated (approx. 3) but not so on the outer side. This effect was examined for the structures discussed in ref. 2, and it was found to be less important for the more nearly equal size spheres used there.

The reason for this behavior is clearly connected with the insistence that an ad-atom sit on a triad of surface atoms. The "inner half" coordination is thus necessarily at least 3. But atom packing is insufficient to produce the "proper" number of neighbors above the same ad-atom if it is small, as for B in Fe, and, hence, sits too deeply in a surface pocket.

ALLOY COMPOSITION

As the composition of the simulated alloys varies, one sees changes in the partial pair correlation functions. Figure 2 gives the Fe-Fe PCF's and Figure 3 the Fe-B PCF's for the structure studied. Again, data are taken from the inner third of the models to minimize effects of the anisotropy. In previous work on modeling FeP alloys,[2] a distinct sharpening of the second peak splitting

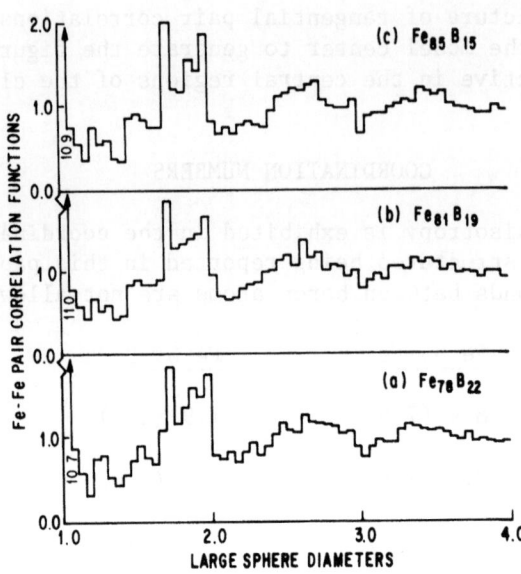

Fig. 2. Partial pair correlation functions for Fe-Fe pairs in (a) $Fe_{78}B_{22}$, (b) $Fe_{81}B_{19}$ and (c) $Fe_{85}B_{15}$. These plots are generated by considering only atoms in the central third of the models.

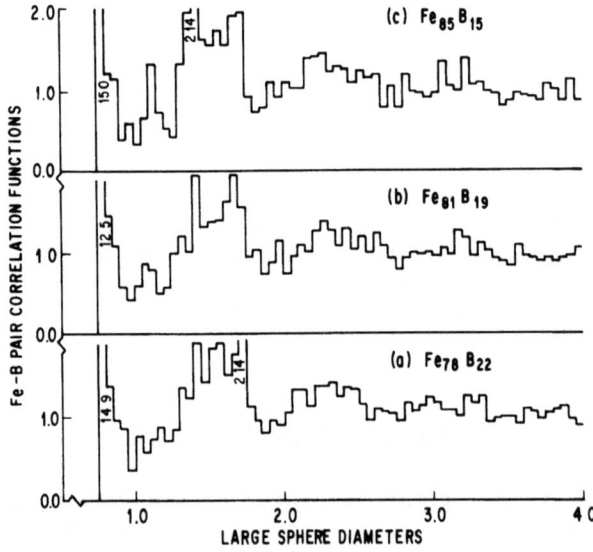

Fig. 3. Partial pair correlation functions: same as Fig. 2 except for the Fe-B pairs.

occurs as the alloy goes through the 80/20 composition region. The same trend seems to exist in the present work, being most evident in Figure 3. As observed experimentally,[4] the first subpeak is more intense than the second. It was previously thought that DRP models could not correctly predict the subpeak intensity ratio; this apparent deficiency is removed, however, when one excludes data from the strongly anisotropic regions of the structural model (c.f. Fig. 2b).

It is believed that the diminished splitting in the second peak away from the 80/20 composition reflects on the structural stability of a glass. Each metal atom is surrounded by a polyhedron whose verticies are formed, primarily, by other metal atoms. The second neighbors sit on the faces of this polyhedron. If the number of faces or the regularity of the polyhedra varies from site to site, one would not expect a sharp distribution of second neighbors. The sharpening of the second peak structure in the PCF's, then, is a measure of the uniformity from site to site and hence, we believe, a reflection of structurally stable packings.

MODEL DENSITIES

The densities and packing fractions measured from the models vary with model size as originally noted by Bennett.[1] This unacceptable feature was suggested[2] to be due to the anisotropies noted above, because the packing deteriorates as one moves away from the cluster center. Radially partitioned data from the present structures show a definite increase in the anisotropies as the structures grow in size.

In·the central regions, the calculated densities are of the order of 6.5 g/cm^3 and the packing fractions approximately 0.65. The densities observed experimentally for FeB glasses are about 13% higher. Elimination of the packing anisotropies observed, should bring the densities to satisfactory values. More troublesome are systematic variations in composition with model size which reflect themselves in the density and packing fraction.

CONCLUSIONS

This paper, as well as an earlier one,[2] have pointed out some serious defects with the conventional computer simulation methods used to generate DRP structures. It has also noted the structural changes which result from alloy composition variations. In particular the second peak of the pair correlation function is found to be more sharply split near the 80/20 compositions; this is believed to be related to glass phase structural stability.

Two things are required for future analysis. The pair correlation anisotropies must be removed; this will require a different stacking algorithm or a subsequent "relaxation" of the structures generated. Secondly, a way to build models with precisely controlled homogeneous alloy composition must be found. These two problems limit the statistical significance of the results by restricting analysis to the central portion of the model. Even in this region, however, the density and TM to M coordination numbers are too low.

REFERENCES

1. C. H. Bennett, J. Appl. Phys. $\underline{43}$, 2727 (1972).
2. D. S. Boudreaux and J. M. Gregor, J. Appl. Phys., to be published (January 1977).
3. R. W. G. Wyckoff, "Crystal Structures," (Interscience, NY, 1964) Vol. II, p. 114.
4. G. S. Cargill, III, Solid State Phys. $\underline{30}$, H. Ehrenreich, F. Seitz and D. Turnbull, Eds. (Academic Press, NY, 1975) p. 227.

DISCUSSION

E. Siegel: In reference to these angular anisotropies, is there any possibility of measuring them?

D. S. Boudreaux: The angular anisotropies in my opinion represent an artifact which is present in all of the computer simulations which have been done using this basic Bennett algorithm and it is the problem that has prevented the agreement between theory and experiment. One of the problems has been the reverse nature of the second peak. In experiments one finds the first subsecond peak to be taller than the second subpeak and in the computer simulations it has always been the other way around. If one looks at the data from the central region of our calculations of the equal size sphere model one finds that subpeak structure to come and register with theory, which suggests the the dense random packing concept is better than people had realized. In doing these calculations they have simply averaged in a lot of data, which has its artificial anisotropy because of the simplicity of the algorithm used.

STRUCTURAL MODELS FOR AMORPHOUS TRANSITION METAL BINARY ALLOYS*

W. Y. Ching

Department of Physics, University of Wisconsin, Madison,
Wisconsin 53706 and Argonne National Laboratory
Argonne, Illinois 60439

Chun C. Lin

Department of Physics, University of Wisconsin
Madison, Wisconsin 53706

ABSTRACT

A dense random packing of 445 hard spheres with two different diameters in a concentration ratio of 3:1 was hand-built to simulate the structure of amorphous transition metal-metalloid alloys. By introducing appropriate pair potentials of the Lennard-Jones type, the structure is dynamically relaxed by minimizing the total energy. The radial distribution functions (RDF) for amorphous $Fe_{0.75}P_{0.25}$, $Ni_{0.75}P_{0.25}$, $Co_{0.75}P_{0.25}$ are obtained and compared with the experimental data. The calculated RDF's are resolved into their partial components. The results indicate that such dynamically constructed models are capable of accounting for some subtle features in the RDF of amorphous transition metal-metalloid alloys.

INTRODUCTION

In order to study the electronic and magnetic properties of amorphous solids, it is necessary to have structural models of atoms which give the correct radial distribution functions (RDF) as determined from experimental measurements. The Dense Random

---

*Supported in part by the National Science Foundation and by the U.S. Energy Research and Development Administration.

Packing of Hard Spheres Model (DRPHS) has been used to represent the structure of amorphous transition metals.[1] For binary alloys involving transition elements, a model of dense random packing of spheres of two different sizes should be a promising one. Using computer calculations, Sadoc et al.[2] have studied the RDF of 300 spheres with various concentrations and diameter ratios. A major drawback of the hard-sphere model is that it only describes the repulsive part of the interaction potential neglecting the attractive part. A more realistic approach is to introduce a suitable pair potential and relax the structure as was done by Heimendahl,[3] and Rahman et al.[4] for the case of single-component atoms. In order to delineate the detailed differences in the structures of amorphous transition-metal-phosphorus alloys with different metallic constituents, we have constructed physical models, each with 445 atoms of which 75% represent the metallic atoms and 25% the phosphorus atoms. By using crystal-data information and experimentally measured RDF and density as guidelines, we select the appropriate parameters of the Lennard-Jones potentials and dynamically relaxed the structures to equilibrium. In this respect, our calculations are different from those of Refs. 3, 4, in which the RDF is studied as a function of the parameters. We obtain the RDF's of amorphous $Fe_{0.75}P_{0.25}$, $Ni_{0.75}P_{0.25}$, $Co_{0.75}, P_{0.25}$ alloys and compared them with the experimental data. It is shown that such dynamically constructed models can explain several subtle features of the RDF's of alloys of different metallic constituents.

MODEL CONSTRUCTION

A round bottom Pyrex flask about 500 cm^3 in volume is cut into two halves near the middle. Steel balls of two different sizes are added to the hemispherical glass bowl one by one with no close contact allowed between the small balls. Slight occasional shaking is necessary for better close packing of the balls. When the bowl is full, the upper half of the flask is joined back to it by adhesive tape and the addition of steel balls is continued until the entire flask is about completely filled up. Hot liquid gelatin (algae-algae) is then poured into the flask solidifying at room temperature. After solidification is complete in about two hours time, the upper half portion of the flask is removed and a frozen structure of spherical dense random packing of two kinds of rigid spheres is obtained. The coordinates of the balls are measured and then removed one by one. The measurements are made with reference to a fixed point using long armed vernier calipers for easy reach of the balls. The recorded coordinates of the balls provide the initial configuration of the structural model for later computer relaxation. Since the structures are to be dynamically relaxed, it is not necessary to measure the coordinates very accurately nor is the choice of ratio of the radii of the spheres critical (1.40 in our case). One can equally well use a computer generated model[2,15] as the initial configuration. In the present work, we find it more expedient to hand-build the initial structure rather than resorting to computers.

MODEL RELAXATION

We used the Lennard-Jones potential of the form

$$V(r) = 4\varepsilon[(\frac{\sigma}{r})^{12} - (\frac{\sigma}{r})^{6}] \tag{1}$$

to describe the interactions between atomic pairs. The parameter σ is related to the equilibrium separation σ' by $\sigma = 0.89089\sigma'$ and ε is the potential depth at the equilibrium distance σ'. The Lennard-Jones potential is best suited for systems of closed-shell atoms and molecules;[6] the validity of its application to the metallic species is less apparent. Our justification here is that we are using it to provide an attractive component to a steep repulsive barrier. This feature indeed can be reasonably well mimicked by a judicious choice of the two parameters in Eq. (1). Generally the value of σ should be chosen so as to reflect the closest distance of approach. Minor adjustments can be made to give better RDF's. As a first step we use the crystalline data as a guide.[7] In the Fe_3P crystal because not all the Fe atoms (or P atoms) are equivalent, the difference in distances between "nearest Fe-Fe neighbors" (or "nearest P-P neighbors") ranges over a few tenths of an Å. Thus we take their average values as σ'_{Fe-Fe} (and σ'_{P-P}). The same criterion can be used to determine σ'_{Fe-P} yielding 2.34 Å. However, if this value of σ'_{Fe-P} is used in conjunction with σ'_{Fe-Fe} = 2.72 Å, the leading structure (at $r \simeq 2.61$ Å) in the calculated RDF would show two sub-peaks as opposed to a single peak at 2.61 Å observed experimentally. Since the first sub-peak corresponds to the Fe-P pairs the second to the Fe-Fe pairs, this discrepancy can be resolved if we slightly increase σ'_{Fe-P} and decrease σ'_{Fe-Fe} so that the density of the model is close to the experimental measured value. The values of σ'_{Fe-P} and σ'_{Fe-Fe} which we adopt are included in Table I. The parameters for the Ni-P alloy are obtained in the same way from the Ni_3P crystal data. For the case of Co, we are not able to find crystalline Co_3P reported in the literature, thus we have to resort to the experimental data of the amorphous Co alloy to guide our selection of the parameters. We notice that the first major peak of the experimental RDF for the amorphous Co alloy occurs at a slightly larger distance than the Ni alloy whereas the density of the Co alloy is appreciably larger than that of the Ni alloy. Accordingly we choose σ_{Co-Co} to be slightly larger than σ_{Ni-Ni} and σ_{Co-P} smaller than σ_{Ni-P}.

TABLE I. Lennard-Jones Potential Parameters for the Amorphous Transition-Metals-Phosphorus Alloys.

	σ'_{M-M} (Å)	σ'_{M-P} (Å)	σ'_{P-P} (Å)	ε_{M-M} (eV)	ε_{M-P} (eV)	ε_{P-P} (eV)
$Fe_{0.75}P_{0.25}$	2.67	2.46	3.53	0.479	0.683	0.138
$Ni_{0.75}P_{0.25}$	2.60	2.34	3.55	0.500	0.723	0.134
$Co_{0.75}P_{0.25}$	2.61	2.37	3.54	0.502	0.695	0.136

The values of σ'_{P-P} deserve some comments. They are much larger than the nearest-neighbor distances in, say, the pure phosphorus crystal, and do not correspond to the onset distance of the repulsive potential between two P atoms. The nearest-neighbor distance between two P atoms in M_3P (M standing for Fe, Co, or Ni) is dictated primarily by the constraint of the 3:1 composition and, to some extent, the M-P interaction, but has little to do with the direct interaction between two P atoms. Since we use a pairwise-interaction description, the composition constraint and the influence of the M atoms on the P-P distances are all folded in σ'_{P-P}, resulting in an effective repulsive barrier which has a much larger onset distance than does the true barrier. For the same reason, the σ'_{M-M} and σ'_{M-P} values do not correspond to the true atom-atom interaction parameters. Thus the usual relation $\sigma'_{AB} \simeq \frac{1}{2}(\sigma'_{AA} + \sigma'_{BB})$ does not hold here.

To find the ε parameters, we start with the <u>pure</u> crystals (FCC structure) of Fe, Co, and Ni. From the nearest-neighbor distance and the cohesive energy,[8] we can derive a "true" Lennard-Jones potential for an M-M pair. The σ'-values for these "true" potentials are found to be slightly lower than the corresponding σ'_{M-M} values exhibited in Table I. We can think of the effect of alloying on the Lennard-Jones potential as a shift of the steep repulsive component to a larger distance while retaining the same attractive behavior. Thus the ε_{M-M} value for an amorphous alloy is taken as the value of the "true" Lennard-Jones potential at the interatomic distance equal to σ'_{M-M} of the amorphous alloy. Likewise we derive a "true" Lennard-Jones potential for a P-P pair and obtain ε_{P-P} for the alloy. As to the ε parameter for M-P interaction, we use the rule of geometrical means between M-M and P-P.

Here the M-M and P-P values are taken from the respective "true" Lennard-Jones potentials at the distance σ'_{M-P} appropriate for the amorphous alloy. The ε parameters for the three amorphous alloys are summarized in Table I. Several trial calculations with slight variations of σ' (and hence ε) were carried out. The parameters listed in Table I give the best overall agreement with the experimentally measured density and RDF.

With the appropriate Lennard-Jones potentials obtained for each atomic pair, the hand-built structure is then relaxed by computer using a process analogous to the molecular dynamics technique. Starting from the outer edge of the cluster, each particle, one at a time, is displaced along a random direction for a certain specified distance. If the new position results in a lowering of the total energy as calculated by summing the pair potentials, the particle is moved to that new position, otherwise it is not moved. The process is repeated in an iterative fashion until there is virtually no change in the calculated RDF. The resulting configuration is taken as the equilibrium structure. To achieve fast convergence, the displacements start from typically 0.05 Å and are gradually decreased to 0.005 Å as the iteration process proceeds. In computing the total energy by summing the pair potentials, it is not sufficient to consider only nearest neighbors. In the present calculation, we consider all the atoms which are less than 6 Å from the one under displacement. Thus the movement of a particle affects the positions of the atoms well beyond the nearest-neighbor distance; in this sense, the relaxation process is truly dynamical.

RESULTS AND DISCUSSION

The experimentally determined RDF for a multicomponent system is a weighted sum of the partial number densities $\rho_{ij}(r)$, i.e.,

$$RDF = 4\pi r^2 \rho(r) = 4\pi r^2 [w_{11}\rho_{11}(r) + 2w_{12}\rho_{12}(r) + w_{22}\rho_{22}(r)]. \quad (2)$$

The weighting factors w_{ij} in general depend on the composition and the scattering factors of the atomic species.[1,11] To a good approximation, the scattering factor for each type of atom can be replaced by its atomic number. In our calculation, we obtained the following weighting factors: $w_{11} = 0.9379$, $w_{12} = 0.5411$, $w_{22} = 0.1041$ for $Fe_{0.75}P_{0.25}$; $w_{11} = 0.9599$, $w_{12} = 0.5142$, $w_{22} = 0.0918$ for $Ni_{0.75}P_{0.25}$; $w_{11} = 0.9492$, $w_{12} = 0.5273$, $w_{22} = .0977$ for $Co_{0.75}P_{0.25}$. The very small weighting factor for the P-P pairs make their contributions to the RDF practically negligible.

In Fig. 1(a)-(c) we display the calculated RDF for the three amorphous alloys $M_{0.75}P_{0.25}$ along with the experimental results for

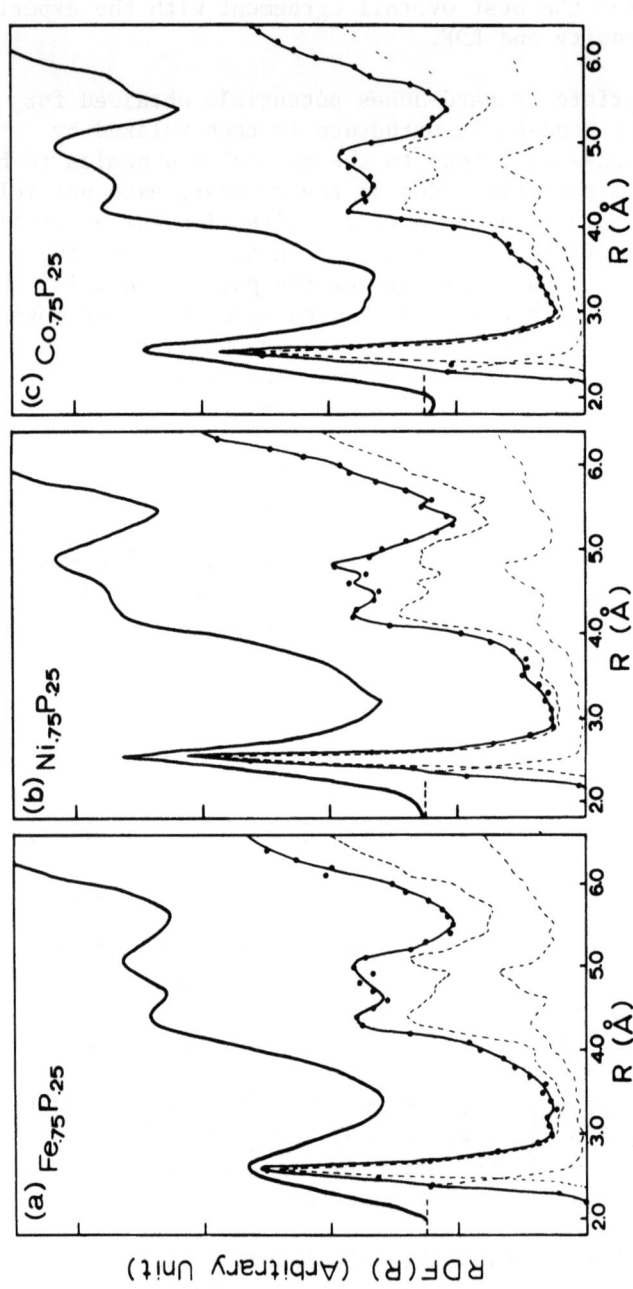

Fig. 1. Radial distribution functions for the amorphous M-P alloys for M = Fe, Ni, and Co in (a), (b), and (c) respectively. The solid curves are the experimental data for $Fe_{0.784}P_{0.216}$, $Ni_{0.77}P_{0.23}$, and $Co_{0.78}P_{0.22}$ (Refs. 9-11). The dots (connected by a line) are the theoretical points for $M_{0.75}P_{0.25}$. The dash and dash-dot curves are the weighted partial components (theoretical) of RDF for M-M and M-P respectively. The weighted contributions from P-P are negligible.

$Fe_{0.784}P_{0.216}$, $Co_{0.78}P_{0.22}$, and $Ni_{0.77}P_{0.23}$.[9,10,11] The calculated RDF's are resolved into their metal-metal, metal-P weighted components. Because of the limited number of atoms in our model, the RDF's are only calculated up to 6.2 Å. It can be seen that all the RDF's calculated agree with their experimental counterparts rather well with regards to the peak position and shape. Of special interest are the following subtle features:

(1) In the experimental curves, the first-nearest-neighbor peak is broadest for the Fe alloy and narrowest for the Ni one; this order is also found in the theoretical RDF's. The calculated peaks are much sharper than the experimental ones because the theoretical calculation does not allow for thermal broadening.

(2) Experimentally $Fe_{0.785}P_{0.216}$ has a more smooth RDF and its double-peak structure is more pronounced whereas the first sub-peak is quite weak for $Ni_{0.77}P_{0.23}$, with the Co alloy having intermediate behaviors. This trend is barely reproduced in the calculated curves although the first calculated sub-peak of the Ni alloy appears to be too strong.

(3) The slopes of the double peak (on both sides) are quite different for different alloys. The calculated curves, to some extent, show similar features.

(4) The position of the second minimum is also well reproduced.

All these interesting features which distinguish the structures of the amorphous transition-metal alloys of different metallic constituents must arise from the slightly different interactions existing between different atoms. The fact that these features are well reproduced by the calculated RDF's adds further justifications for using the Lennard-Jones potential to calculate pair distributions.

From the partial components of RDF, it can be seen that the first major peak is dominated by the contribution from the M-M pairs while the M-P pairs contribute to the asymmetry of the peak. The present results give a larger M-P distance than previously suggested.[1] For the double peak, the first sub-peak is mainly due to the M-M pairs, whereas the second one has considerable contributions from the M-P pairs as well. In constructing these models we assume that there is no contact between two P atoms. In fact the P atoms are kept rather far apart by choosing large σ_{P-P} values. It should be interesting for further study to ascertain whether a closer contact between the P atoms is acceptable. Because of the very small weighting factor for the P-P pairs, the calculated RDF is insensitive to the choice of σ_{P-P}.

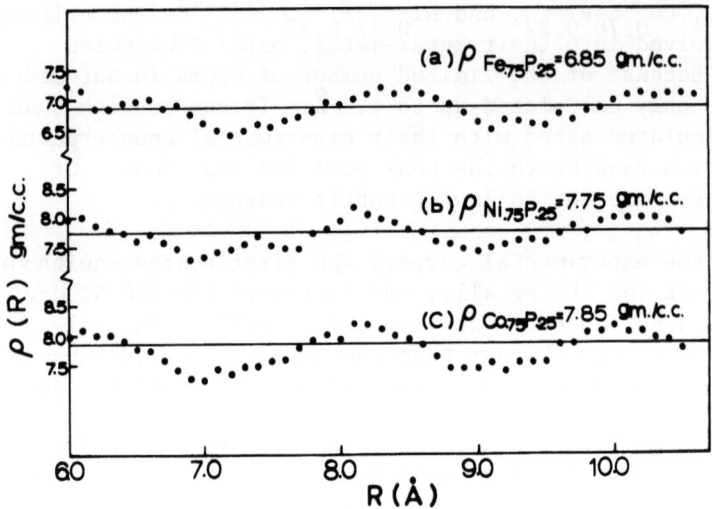

Fig. 2. Density (theoretical) of the amorphous alloys as a function of distance from the centroid of the model for (a) $Fe_{0.75}P_{0.25}$, (b) $Ni_{0.75}P_{0.25}$, and (c) $Co_{0.75}P_{0.25}$. The straight lines are the estimated average densities.

We have also calculated the density of the constructed models as a function of distance from the centroid of the model. The results are shown in Fig. 2. The average densities are in good agreement with the corresponding experimental values.[1,9]

As remarked earlier, the position of the first peak in the RDF is associated with the M-M pair distance. Although the Co alloy shows its first peak at a larger distance than the Ni alloy, yet the former has a higher density than the latter. This observation is suggestive of a substantial difference in the short-range interaction between these two alloys.

In conclusion we have presented a simple dynamical scheme for constructing structural models for amorphous magnetic alloys. Such models are useful for studying the electronic and magnetic properties of these alloys.

ACKNOWLEDGMENT

We wish to acknowledge the invaluable assistance rendered by James Phelps and are benefited by very helpful discussions with Dr. G. S. Cargill.

ACKNOWLEDGMENT

We wish to acknowledge the invaluable assistance rendered by James Phelps and are benefited by very helpful discussions with Dr. G. S. Cargill.

REFERENCES

1. G. S. Cargill III, Solid State Physics 30, 227 (1975).
2. J. F. Sadoc, J. Dixmier and A. Guinier, J. Non-Cryst. Solids 12, 46 (1973).
3. L. v. Heimendahl, J. Phys. F: Metal Phys. L141 (1975).
4. A. Rahman, M. J. Mandell and J. P. McTague, J. Chem. Phys. 64, 1564 (1976).
5. C. H. Bennett, J. Appl. Phys. 43, 2727 (1972).
6. J. O. Hirschfelder, C. F. Curtiss, and R. B. Bird, Molecular Theory of Gases and Liquids, (John Wiley and Sons, Inc., New York, 1954).
7. S. Rundqvist, Ark. Kemi 20, 67 (1962).
8. K. A. Gschneidner, Solid State Phys. 16, 275 (1964).
9. J. Logen, Phys. Stat. Sol. A 32, 361 (1975).
10. G. S. Cargill III, J. Appl. Phys. 41, 12 (1970).
11. G. S. Cargill III and R. W. Cochrane, J. Phys. (Paris) C4-269 (1974).

DISCUSSION

D. S. Boudreaux: I wonder what your thoughts are about using the longer range Lennard-Jones potential as opposed to the shorter range Morse potential?

W. Y. Ching: We have considered using the Morse potential but the mathematical difficulties involved with that treatment are certainly harder to deal with. Moreover we found the Lennard-Jones potential to be quite adequate for our purpose.

SMALL-ANGLE MAGNETIC SCATTERING IN

AMORPHOUS TbFe$_2$*

> S. J. Pickart
> Department of Physics, University of Rhode Island
> Kingston, Rhode Island 02881; Naval Surface Weapons
> Center, White Oak, MD 20910; National Bureau of Standards,
> Washington, D.C. 20234

ABSTRACT

A small-angle scattering measurement of the anomalous magnetic intensity below the Curie point of amorphous TbFe$_2$ was performed with a double perfect crystal arrangement. The intensity at scattering vectors $q \gtrsim .002 \text{A}^{-1}$ follows an exponential curve, from which a radius of gyration of $\sim 800\text{A}$ for the magnetic clusters is derived.

I. INTRODUCTION

The observation of an intense small-angle magnetic scattering below the ferromagnetic ordering temperature has been previously reported[1-3] for three sputtered, amorphous alloy compositions: TbFe$_2$, HoFe$_2$ and, most recently, Tb$_{.018}$Fe$_{.982}$. In all these cases the scattered intensity followed an inverse third power dependence on the scattering vector, intermediate between the asymptotic behavior expected for a random distribution of spherical particles and a random distribution of disk-shaped particles of varying sizes. In order to develop more definite information on the cluster size, the present investigation is an attempt to extend the lowest attainable scattering angle of the observations on TbFe$_2$ to the regime where the exponential approximation is valid.

II. EXPERIMENTAL ARRANGEMENT AND DATA

The earlier measurements were taken with a set of three

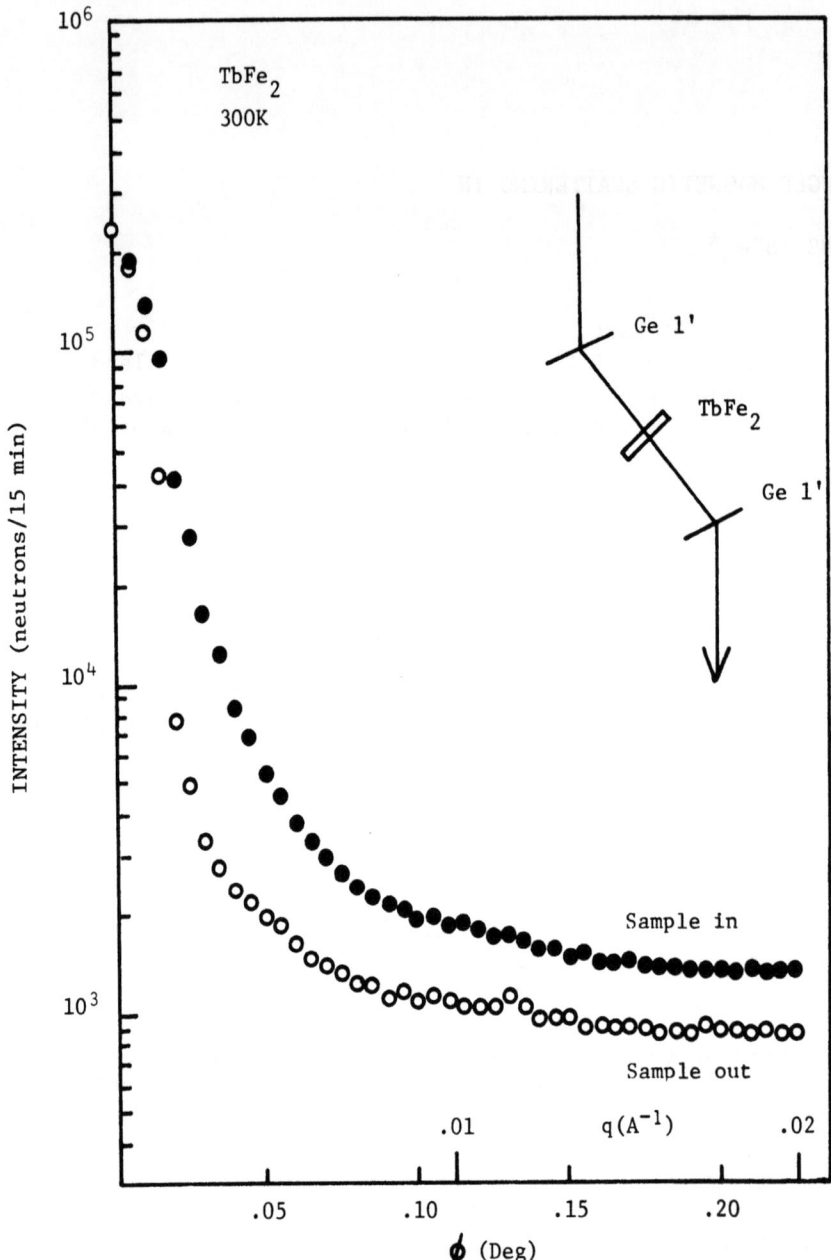

Figure 1. Scattered neutron intensity from TbFe$_2$ at room temperature for $q < .02 A^{-1}$ plotted on a logarithmic scale. "Sample in" data have been corrected for transmission. Inset shows the experimental arrangement.

soller slits of 20', 10' and 10' divergences in combination with a bent graphite monochromator of 24' mosaic width reflecting a wavelength of 2.38 A. Using this arrangement, the lowest attainable scattering vector before the forward scattering became too intense was approximately $.02 A^{-1}$. In the present experiment a set of matched perfect Ge crystals with 1-2' mosaic spread were set up as sketched in the insert of Figure 1, at the same wavelength.

The data were obtained by subtracting the rocking curve of the second crystal with the sample removed from a similar measurement with the sample in place, after correction for transmission. The two sets of measurements made at room temperature are plotted in Figure 1. It is evident from these results that the data are useable down to $\sim 2 \times 10^{-3}$ A^{-1}, where the forward peak begins to rise sharply. Since the shape of this peak is very nearly gaussian, the resolution correction for an exponential scattering law can be obtained from the inverse of the difference of reciprocal widths, according to the convolution theorem of Fourier transforms of exponential functions.

III. ANALYSIS AND DISCUSSION OF RESULTS

Guinier[4,5], in his extensive treatment of the theory of small angle scattering, has particularly emphasized the difficulty of unambiguously deriving the shape of the particles from the form of the scattering. Even in the case where the particles are homogeneous and identical, one can obtain only the <u>radius of gyration</u> of the particles from the small-angle exponential approximation,

$$I(q) \propto \exp(-R^2 q^2/3),$$

where R is defined by

$$R^2 = \int_{particle} r^2 dV/V.$$

In this case, a plot of the log I vs. q^2 shows linear behavior with a slope related to the desired parameter. If one is dealing with a distribution of particles, with a spread of radii, such a plot has a continuously upward curvature as $q \to 0$; but if a small enough angle can be reached, i.e. where $q < \frac{1}{R}$, the curve approaches an asymptotic linear slope from which the radius of gyration of the <u>largest</u> particles may be derived.

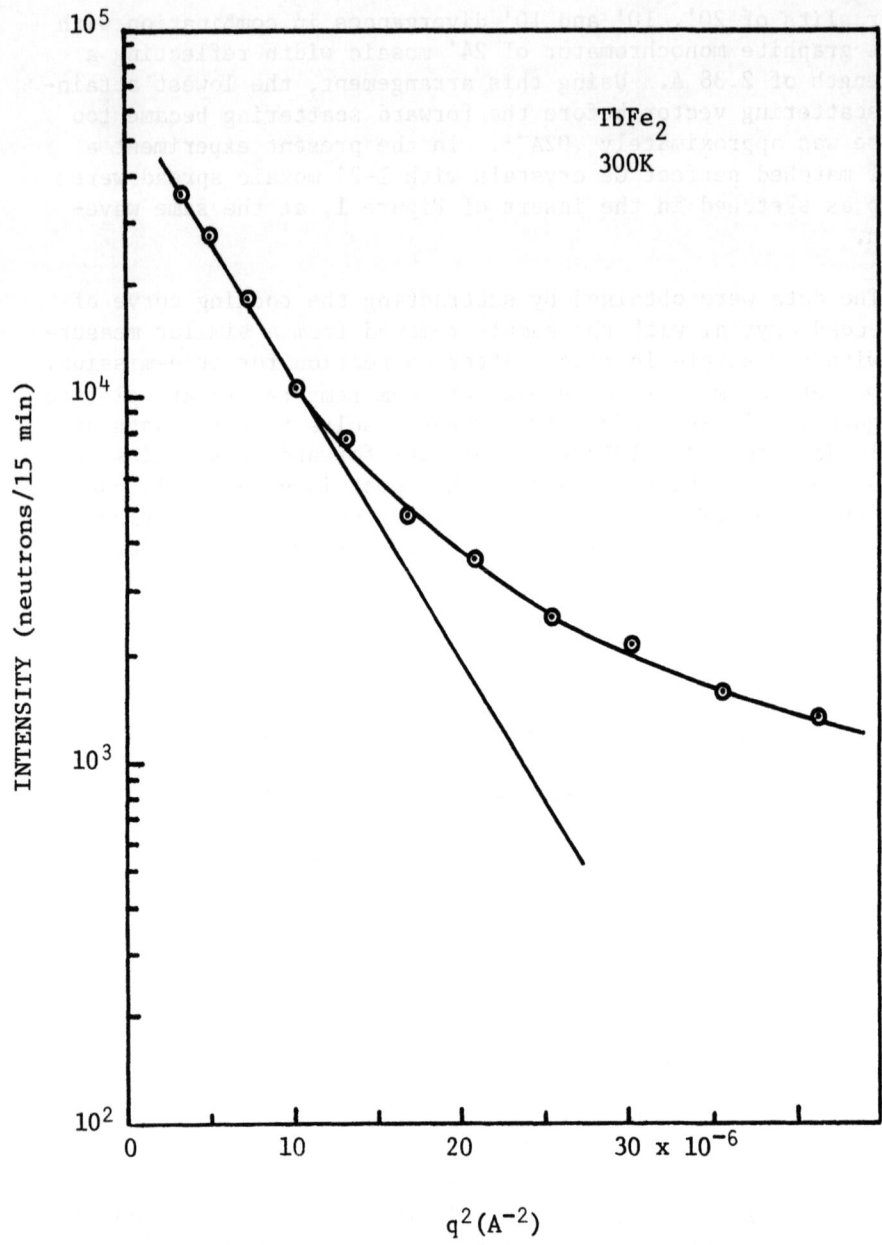

Figure 2. Data for $q < 6.4 \times 10^{-3}$ A^{-1} plotted on a logarithmic scale vs. q^2. The asymptotic slope shown (1.74×10^5 A^2), when corrected for resolution, corresponds to a cluster radius of gyration of 794 A.

In addition to this small angle asymptotic behavior, Porod[6] has pointed out that for the large angle regime where qR >> 1, and for a distribution of radii, the oscillating terms in the expansion of the Bessel function scattering law cancel out, and one is left with a power law for the intensity

$$I(q) \alpha\ m^2 S/q^4,$$

where m is the magnetic moment density (electron density for the x-ray case) and S is the total surface area of the sample. (We chose to scale our neutron measurements to the cross section of vanadium in barns/atom, and so write S as the specific surface area per atom). It is important to note that this relation holds for any particle shape as long as qL >> 1, where L is the smallest dimension of the particles, and for a dense system where diffracted waves from different particles interfere (non-interference has to be <u>assumed</u> for the exponential approximation).

On the other hand, for disk shaped particles of infinitesmal thickness ε and radius R, Kratky and Porod[7] have shown that the asymptotic form is

$$I(q) \alpha\ 2/q^2 R^2$$

as long as qR >> 1 and qε << 1. For a distribution of particle radii this would appear as a weighted average of power law dependences.

In all the previous measurements, which were taken under identical resolution conditions and for widely varying sample thicknesses (but all essentially thin samples), the power law exponent uniformly observed for .02 < q < .14 A^{-1} was −2.4. Assuming multiple scattering corrections are negligible because of the variation in thickness, only the vertical resolution correction is important and would indicate a corrected exponent of −3.0. Since this is intermediate between the above two cases, not much can be said about the particle dimensions, other than the observation of any asymptotic form indicates that R >> 50A.

The data from the present investigation for .002 < q < .007 A^{-1} are plotted in Figure 2 so as to display possible exponential behaviour. It is seen that though the plot continues to show upward curvature for a portion of this range, it does appear to approach a linear slope at its lowest end. The slope shown, after correction for resolution, would indicate a radius of gyration of 794 A. This is not an unreasonable value when compared to the above estimate, and to a straightforward application of the Scherrer formula to the observed line broadening, which would suggest that $\ell \sim$ 2000 A, where ℓ is the mean particle size.

Finally, it is interesting to consider these results in the light of the critical scattering observed[1] above T_c. Isotherms of the inverse intensity plotted vs. q^2 indicated Lorentzian lineshapes down to and even below T_c; however, the resolution-corrected correlation lengths derived from them did not diverge, but approached a finite value of ~ 70 Å at T_c. A consistent though not unique model of the development of the magnetic order may be derived from these results in which the critical fluctuations are "frozen in" near T_c and give rise to inhomogeneous clusters, of undetermined shape, ranging in size from $\sim 10^2$-10^3 Å.

IV. ACKNOWLEDGEMENTS

It is a pleasure to acknowledge numerous and valuable discussions with H. A. Alperin, J. J. Rhyne and J. Cullen.

VI. REFERENCES

1. S. J. Pickart, J. J. Rhyne and H. A. Alperin, Phys. Rev. Lett. 33, 424 (1974).
2. S. J. Pickart, J. J. Rhyne and H. A. Alperin, AIP Conf. Proc. 24, 117 (1975).
3. S. J. Pickart and H. A. Alperin (to be published).
4. A. Guinier and G. Fournet, Small-Angle Scattering of X-Rays, Wiley & Sons, New York, 1955.
5. A. Guinier, X-Ray Diffraction, Freeman, San Francisco, 1963.
6. G. Porod, Kolloid, Z. 124, 83 (1951).
7. O. Kratky and G. Porod, J. Colloid Sci. 4, 35 (1949).

*Supported in part by the NSWC Independent Research Fund.

DISCUSSION

N. D. Heiman: Can you say anything about the differences in cluster size that you see for amorphous $HoFe_2$, $TbFe_2$ and YFe_2?

S. J. Pickart: This is still sort of in the process of being analyzed. But it appears that the size of the cluster increases in going from YFe_2 to $HoFe_2$ to $TbFe_2$.

MAGNETISM AND STRUCTURE OF AMORPHOUS $Fe_{80}P_{13}C_7$ ALLOY[†]

M. Takahashi, M. Koshimura, T. Miyazaki and T. Suzuki

Department of Applied Physics, Tohoku University
Sendai, Japan

INTRODUCTION

Recently much work has been carried out both experimentally[1-7] and theoretically[8-11] as to whether the ferromagnetism in amorphous ferromagnetic materials is the intrinsic nature of the fused state (liquid like structure) or not. In these studies, the amorphous materials are assumed conceptually to be homogeneous in both composition and atomic arrangement. However, the phase diagrams of metals and alloys which can become amorphous are generally complicated and they undergo crystallographical transitions associated with the phase transformation on cooling from the melting points to room temperature. Therefore, it is natural to consider that even if the specimen is quenched rapidly from the liquid, the solidified state may consist of more than 2 phases existing at high temperatures which may be retained in some ways, and moreover, that this state would be highly strained and be unstable due to large quenching effect. On the other hand, it is well known that an amorphous alloy decomposes into the alloys and compounds expected from the phase diagram after crystallization at a certain critical temperature.[12,13] Therefore, it is reasonable to presume that the amorphous state cannot be uniform in composition and in both macroscopic and microscopic structure.

It is interesting to make clear the relation between the magnetic properties and the structure of the amorphous materials,

---

[†] Detailed results and discussions on the temperature dependence of saturation magnetization, susceptibility, magnetic anisotropy, Mössbauer analysis and crystallographic structure will be published elsewhere.

since the magnetic properties are one of the most structure sensitive parameters of solids. Thus, the experimental study on the magnetic properties and the structure of amorphous $Fe_{80}P_{13}C_7$ alloy has been carried out.

EXPERIMENTAL METHOD

An amorphous $Fe_{80}P_{13}C_7$ alloy was made in a ribbon form by rapid freezing from the liquid kept at about 1200°C onto a rotating Cu drum (10 cm in diameter, 4000 rpm), with using a centrifugal solidification technique (pressure at a moment of jet: 3 Kg/cm^2, nozzle diameter: 0.3~0.5 mmϕ). To detect the quenching effect, the rod specimens were also prepared by sucking the molten alloys kept at about 1200°C into the quartz tube (quasi-rapid quenching). The saturation magnetization and the susceptibility at high temperatures were measured by using a vibrating sample magnetometer and a magnetic balance, respectively. The magnetic anisotropy was also measured with a torque magnetometer. The structural analysis was carried out by X-ray (Fe-Kα) and electron microscopy (accelerating voltages of 200 and 500 KV).

EXPERIMENTAL RESULTS AND DISCUSSIONS

1. Temperature Dependence of the Saturation Magnetization

Before proceeding the measurement of the temperature dependence of the saturation magnetization, Ms, it was confirmed that a field of about 10 KOe was enough to saturate the magnetization of all the samples. Therefore, the magnetization at a constant field of 11.5 KOe was measured as a function of temperature. The heating and cooling rate were 100°C/hr.

a) $Fe_{80}P_{13}C_7$ alloy. i) Structure. The X-ray diffraction pattern taken for rapidly quenched specimens shows a diffused halo, indicating the samples are amorphous. On the other hand, the diffraction patterns for the sucked specimens do not show any halos, but sharp rings which correspond to Fe_3C, Fe_3P and $\alpha Fe(P)$.

ii) Ms vs. T. The heating and cooling curves of Ms for the rapidly quenched and sucked samples are shown in Fig. 1. Also, the temperature dependence of electrical resistance measured by a 4-probe method is shown in the same figure. As seen in this figure, Ms for a rapidly quenched (amorphous) sample decreases rapidly up to about 340°C (Curie point of the amorphous state) and then gradually decreases up to about 420°C (crystallization temperature) showing a slight kink at about 200°C. Ms increases abruptly at this crystallization temperature with a hump and decreases

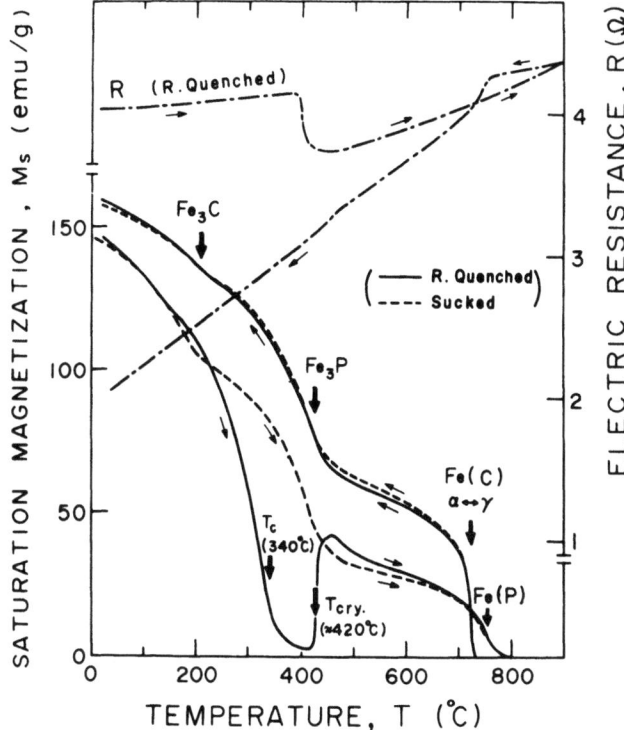

Fig. 1. Temperature dependence of the saturation magnetization for a rapidly quenched and also sucked $Fe_{80}P_{13}C_7$ alloys. Also, the change in electrical resistance with temperature is shown.

gradually and becomes zero at about 800°C, having a tail over temperatures from 700∿800°C.

On the other hand, the Ms vs T curve for the sucked sample exhibits a clear kink at about 200°C for 420°C, and then Ms changes in the similar way to that for the rapidly quenched samples above 420°C.

On the cooling process from 900°C, the magnetization increases abruptly at 720°C (the temperature of α↔γ Fe(C) phase transition) and then it increases with decreasing temperature, exhibiting clearly two kinks near 430°C (Tc for Fe_3P) and 210°C (Tc for Fe_3C).[14] The cooling curves for the rapidly quenched and sucked samples coincide with each other.

The noticeable points in this figure are: (i) There is a large hysteresis between the heating and cooling curves for both rapidly quenched and sucked specimens. (ii) The most remarkable characteristic behavior of Ms(T) depending on the quenching speed is seen in the temperature range between about 200°C and 400°C; namely when the quenching speed is increased, Ms changes remarkably over this temperature range. (iii) The Ms for both the rapidly quenched and sucked samples become zero gradually in the same temperature range from 700°C to 800°C ($\approx$Tc of αFe(P)).[14] The Ms above the crystallization temperature (420°C) results from αFe(P), and its amount is nearly the same for both samples. It should be noticed that the Ms of αFe(C) appears on cooling process only when the samples were heated up to 900°C.

b) Fe-P alloy. i) Structure. The X-ray diffraction patterns for all the samples indicate that they are crystalline and the sharp diffraction lines correspond to Fe_3P and αFe(P).

ii) Ms vs. T. The temperature dependence of Ms for the rapidly quenched and sucked specimens is given in Fig. 2. As seen in this figure, Ms decreases with increasing temperature, exhibiting two kinks at the Curie points of Fe_3P (420°C) and αFe(P) (755°C).[14] It is observed that the amount of Fe_3P increases with the content of P, while the amount of αFe(P) decreases, becoming nearly zero for 26.6%P.

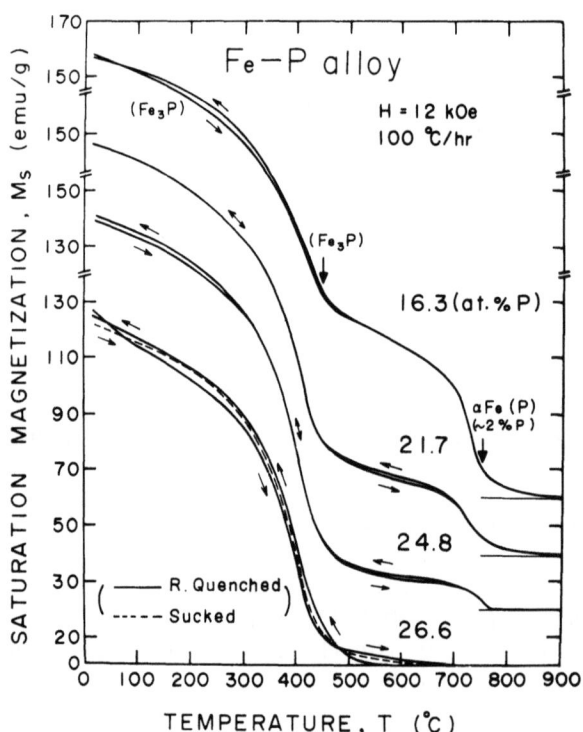

Fig. 2. Saturation magnetization as a function of temperature for both rapidly quenched and sucked Fe-P alloys.

The important points in this figure are: (i) The curves of Ms-T for the rapidly quenched samples are quite the same as those for the sucked specimens, and the quenching effect is not observed in Fe-P alloys. (ii) There is no hysteresis in the Ms-T curves for both of the samples. (iii) The magnetization of αFe(P) decreases gradually, becoming zero at 700°C∿900°C, in a similar way to that of amorphous $Fe_{80}P_{13}C_7$ alloys.

c) <u>Fe-C alloy</u>. i) <u>Structure</u>. The X-ray diffraction patterns show sharp lines for the Fe-C alloys containing 9.3 at.%C or less, whereas the diffraction lines become more smeared out with increasing C. These diffraction lines correspond to Fe_3C, αFe(C) and γFe(C).

ii) <u>Ms vs. T</u>. The change of Ms with temperature for the rapidly quenched and sucked specimens is given in Fig. 3. The

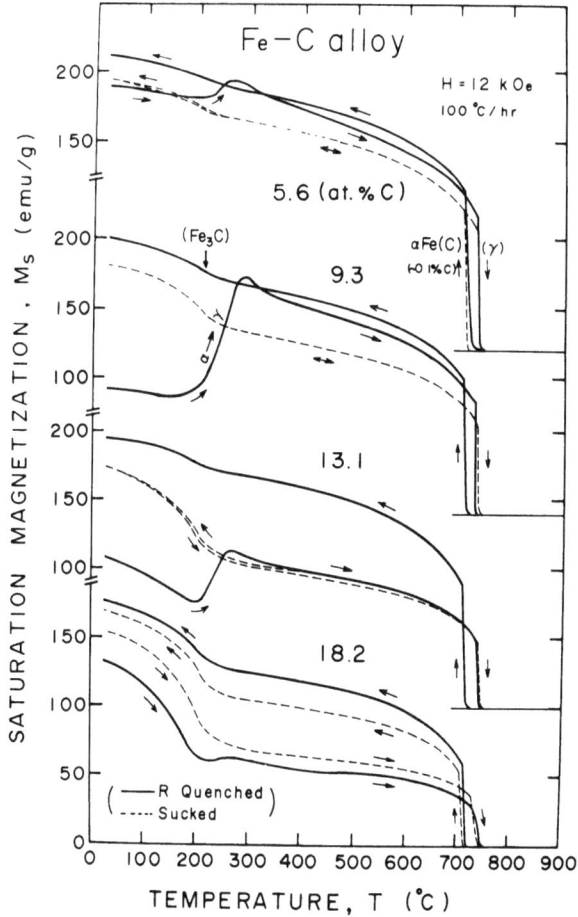

Fig. 3. Saturation magnetization as a function of temperature for both rapidly quenched and sucked Fe-C alloys.

Table 1. The summary of the characteristic points observed on the Ms vs. T curves for Fe$_{80}$P$_{13}$C$_7$, Fe-P and Fe-C alloys.

Alloy		Ms(emu/g) [R.T]		Heating				Cooling			
		A.Q.	A.C.	knick point			[degree]	knick point			[degree]
FeP$_{13}$C$_7$ (8.1-1.7)	R.Q.*	148	159	200°	340°~420° [Tc ~ Cry.]		755°~800°(t) [αFe(P)]	~210°[Fe₃C]	430°[Fe₃P]		725°[αFe(C)](S)
	S	142	156	~210°[Fe₃C]	430°[Fe₃P]	755°~800°[αFe(P)](t)		~210°[Fe₃C]	430°[Fe₃P]		725°[αFe(C)](S)
FeC$_{56}$ (1.3)	R.Q.	189	212	210°~270°		730°[αFe(C)](s)		215°[Fe₃C]			710°[αFe(C)](S)
	S	193	192	210°[Fe₃C]		730°["](s)		210°["]			710°["](S)
FeC$_{9.3}$ (2.2)	R.Q.	91	199	180°~280°		730°["](s)		215°["]			710°["](S)
	S	181	182	210°["]		730°["](s)		210°["]			710°["](S)
FeC$_{131}$ (3.1)	R.Q.	106	195	200°~275°		730°["](s)		220°["]			715°["](S)
	S	173	173	210°["]		730°["](s)		215°["]			715°["](S)
FeC$_{182}$ (4.6)	R.Q.	132	176	210°~270°		755°["](s)		215°["]			715°["](S)
	S	154	168	215°["]		730°["](s)		215°["]			715°["](S)
FeP$_{163}$ (9.8)	R.Q.	156	156		435°[Fe₃P]	755°~800°(t) [αFe(P)]			435°[Fe₃P]		755°~800°(t) [αFe(P)]
FeP$_{217}$ (13.3)	R.Q.	145	145		430°["]	755°~800°(t) [αFe(P)]			430°["]		755°~800°(t) [αFe(P)]
FeP$_{248}$ (15.5)	R.Q.	138	138		430°["]	760°~800°(t) [αFe(P)]			430°["]		755°~770°(t) [αFe(P)]
FeP$_{266}$ (16.7)	R.Q.	123	123		425°[Fe₃P]~700°(t)				430°[Fe₃P]~700°(t)		

( )wt% R.Q.: Rapid Quenched A.Q.: As Quenched (S): Sharp (t) long tail
S: Sucked A.C.: After Cool
*: Amorphous

heating curves for both samples have a clear kink at around the Curie point of Fe_3C (210°C),[14] and Ms disappears abruptly near the phase transformation temperature of $\alpha \rightarrow \gamma$ Fe(C), (730°C).

However, it should be noticed on the heating curve that the Ms for a rapidly quenched sample gradually increases with a hump at about 200~300°C, and then decreases gradually. The appearance of the hump at arount 300°C corresponds to that observed on the heating curve of amorphous $Fe_{80}P_{13}C_7$ alloy as described before (cf. Fig. 1). The gradual increment before the hump is due to the phase transformation of $\gamma Fe(C) \rightarrow \alpha Fe(C)$ (210~220°C).

It should be noted that the sucked specimens have no hysteresis except for the 18.2%C-Fe alloy, but all of the rapidly quenched samples have a large hysteresis over the whole temperature range covered.

In Table 1, the knick,* and the Curie points** obtained from the Ms vs. T curves in Figs. 1, 2 and 3 are summarized. As seen in this table, the knick points appeared on the heating curve of amorphous $Fe_{80}P_{13}C_7$ alloy at about 200°C and the crystallization temperature of 420°C correspond closely to the Curie points of Fe_3C and Fe_3P, respectively. Also, the disappearance of Ms at about 755~800°C coincides with that of $\alpha Fe(P)$. These points become more noticeable on the cooling curve. The Curie points of 340°C for the amorphous sample has no correspondence. However, the gradual decrease in Ms at around 340~420°C seems to be closely related to the disappearance of Ms of Fe_3P alloys.

2. Magnetic Susceptibility, χ

The magnetic susceptibility χ was measured for the samples (20~30 mg) placed inside a quartz basket by using a magnetic balance. The measurement was made in the argon gas atmosphere over the temperature range from 900°C to 1400°C. Fig. 4 shows the temperature dependences of $1/\chi$ up to 950°C, 1150°C and 1380°C for $Fe_{80}P_{13}C_7$ alloys. As seen in this figure, the $1/\chi$ rises sharply at about 780°C at which Ms of $\alpha Fe(P)$ disappears, and increases linearly with temperature with different slopes in each tempera-

---

* The knick points were obtained by extrapolating the curves over the temperatures below and above near the knick points.

**The Curie points for Fe_3C, Fe_3P, $\alpha Fe(P)$ and amorphous samples were obtained by extrapolating the Ms^2 vs. T curve near the Curie points.

ture range: 950∼1000°C, 1000∼1120°C and above 1120°C. By extrapolating the linear curve in each temperature region, the paramagnetic Curie points (cross point) and the assignment of each phase are obtained as follows.

Temperature range	Phase	Cross Point
950°C<T<∼1000°C	(solidus): $Fe_3P+\gamma Fe(C)+\alpha Fe(P)$	380∼400°C
1000°C<T<∼1120°C	(solidus)+(liquidus): $Fe_{2\sim 3}P+L'$	330°C
1120°C<T	(liquidus): L	∼100°C

L': quasi-liquidus in mixture phases

As seen in Fig. 4, the paramagnetic Curie points of "solidus" and "solidus+L" coincide with the Curie point and the crystallization temperature of amorphous $Fe_{80}P_{13}C_7$ alloy, respectively. On the other hand, the paramagnetic Curie point of "liquidus" is below 100°C. Therefore, it is proposed, that the structure of the amorphous state consists of $Fe_{2\sim 3}P(C)$ and L', where L' means the quasi-liquidus containing nuclei of $\gamma Fe(C)$ and $\alpha Fe(P)$.

3. Magnetic Anisotropy

a) <u>Amorphous $Fe_{80}P_{13}C_7$ alloy</u>. To measure the torque curves, the samples in circular disk-shaped form were prepared carefully from the ribbons (1∼3 mm in width, 40 μm in thickness) made by rapid quenching. The diameter of the samples is about 1∼2 mm. The torque curves were obtained in various fields up to 20 KOe, and were Fourier-analyzed. The magnetic anisotropy was determined by extrapolating the curves of the amplitude of $\sin 2\theta$ component vs. $1/H$ to $1/H=0$.

Usually, the striations running along the ribbon axis are observed on the surfaces of the samples made by rapid quenching, as shown in Fig. 5. Thus, in order to eliminate the effect of these striations on the magnetic anisotropy, the striations were removed carefully by mechanical polishing. Figure 6 shows the torque curve before and after removing the striations. As seen in this figure, the sample (as quenched state, A+B+C) with striations shows a large uniaxial magnetic anisotropy.[4] After polishing out the surface A (see figure), the torque curve changed greatly (B+C). It should be noticed here that there is a remarkable increase of

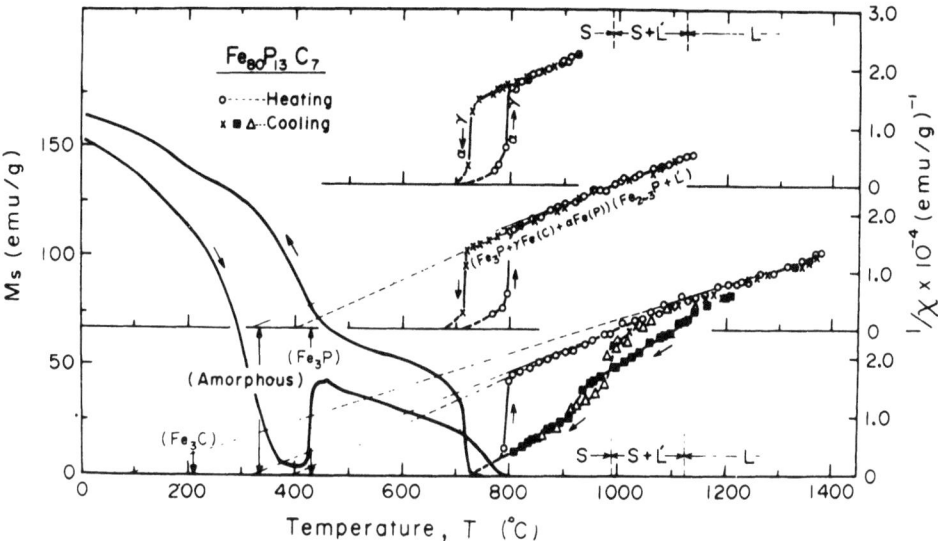

Fig. 4. The inverse magnetic susceptibility $1/\chi$ at high temperatures, together with the temperature dependence of Ms.

the amplitude. This is interpreted as a result of the variation of the direction of the easy magnetization along the thickness. Next, after polishing out the surface C (see figure) the amplitude of the torque curve decreased, but the torque curve of 2-fold symmetry was still observed.

Fig. 7(a) shows the magnetic anisotropy inside the specimen along the thickness[15] and (b) shows the deviation angle of the easy axis δ, from the ribbon axis.[15] As seen in Fig. 7(a), the magnetic anisotropy near the surfaces is larger by an order of magnitude than that in the middle part of the sample. Especially, the anisotropy in the Cu-side which has the deeper and narrower striations as seen in Fig. 5 is larger than that in the nozzle side. It is interesting that even after elimination of the

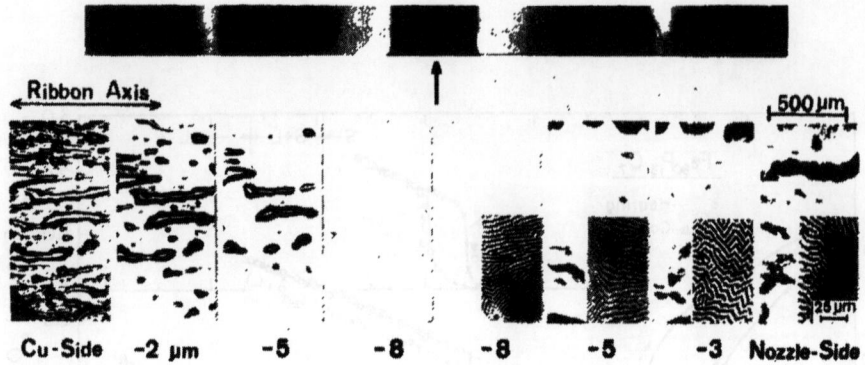

Fig. 5. The surface structure and domain pattern observed by metallurgical microscopy with reducing the sample thickness. The X-ray Debye-Scherrer photograph is also shown.

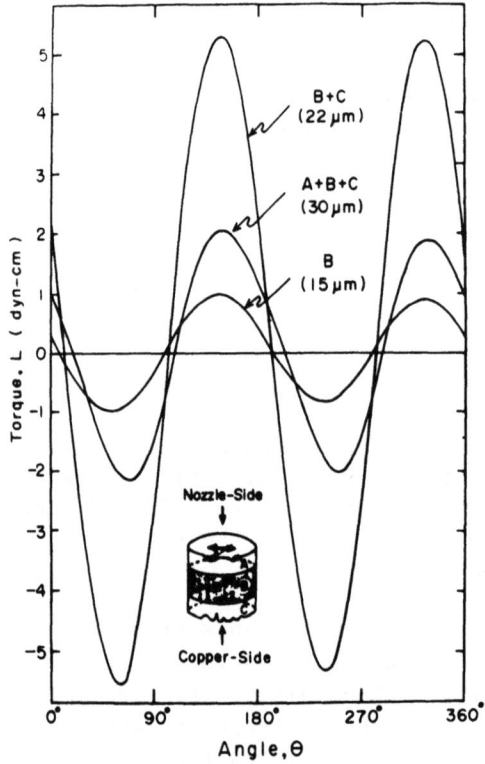

Fig. 6. The torque curves before and after removal of the surface layers, A and C.

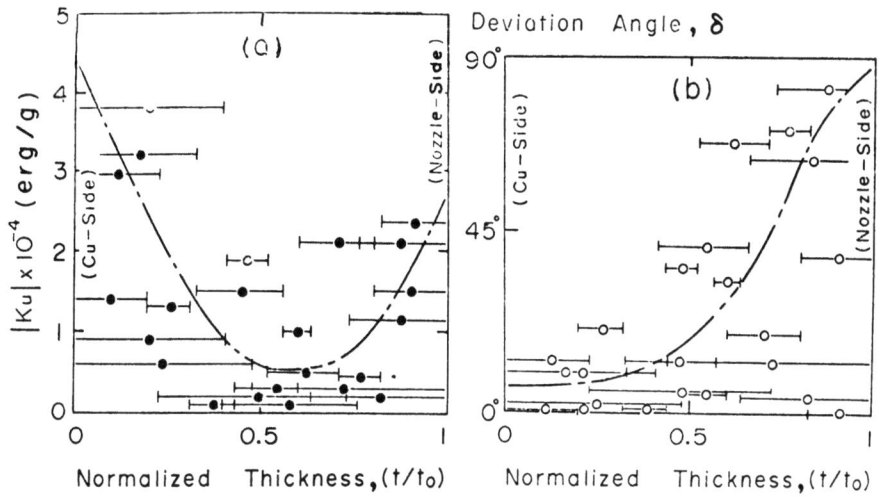

Fig. 7. (a) The variation of the uniaxial magnetic anisotropy constant Ku along the thickness. (b) The deviation angle of the easy magnetization axis δ, from the ribbon axis.

striations, there exists a uniaxial magnetic anisotropy of about 5×10^3 erg/gr (3.5×10^4 erg/cc), which is larger by one order of magnitude than that observed in alloys conventionally cooled in a magnetic field.

As shown in Fig. 7(b), the easy axis is nearly parallel to the ribbon axis on the Cu-side. However, the direction of easy axis changes from the ribbon axis in the Cu-side to the perpendicular direction in the nozzle-side. As seen in this figure, the direction of the easy magnetization around the middle part of the sample is scattered from 0° to 45°.

b) <u>Magnetic anisotropy of Fe-P and Fe-C alloys.</u> Two circular disk-shaped samples were prepared from the rods (3∿6 mm in diameter) which were made by sucking. The samples were made from the disk plates cut parallel (A) and/or perpendicular (B) to the rod-axis (see Figure 8). The diameter and thickness of the disks are 2∿5 mm and 0.05∿0.5 mm, respectively. Even though the samples are polycrystals, the torque curves show 2-fold symmetry, meaning a uniaxial anisotropy.* Fig. 8 shows the compositional dependence of the uniaxial magnetic anisotropy.

---

*The anisotropy is probably due to the existence of fibrous structure developed during sucking. At the present, this is not confirmed yet.

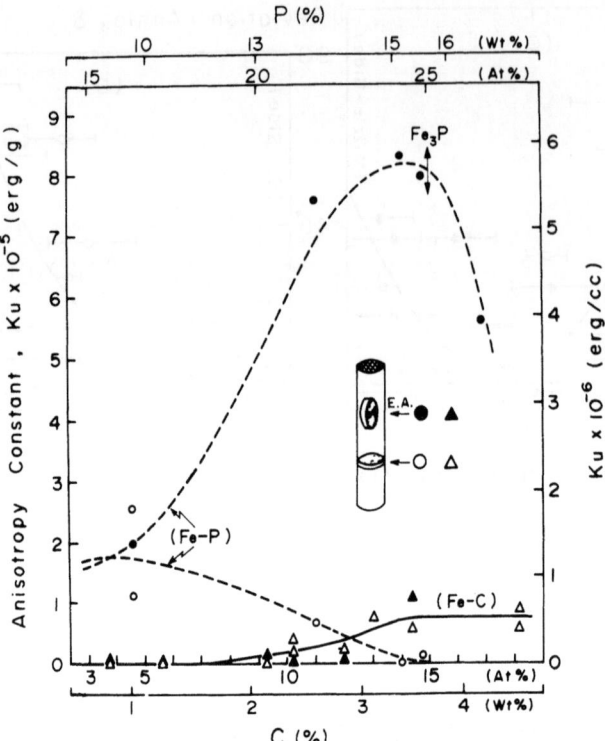

Fig. 8. The uniaxial magnetic anisotropy constant Ku observed for sucked Fe-C and Fe-P polycrystalline alloys against compositions.

In the case of Fe-P alloys, the magnetic anisotropy is nearly the same for both of the samples A and B with the composition less than about 15%P. However, when the concentration of P is increased, there appears a significant difference between A and B. The anisotropy constant of A takes an extremely large value of 6×10^6 erg/cc at around 25%P, while that of B becomes negligibly small.

In the case of the Fe-C alloy, the magnetic anisotropy is not observed for the smaples with concentrations less than 10% C. For the samples with concentrations more than 15%C, the magnetic anisotropy was observed to be a constant value, about one tenth of that for an Fe_3P alloy.

From the results mentioned above, it is easy to consider that if the large magnetic anisotropy of Fe_3P alloys is responsible for the anisotropy observed in the amorphous $Fe_{80}P_{13}C_7$ alloy, the amount of 1% or so of crystallites of Fe_3P, distributed not randomly, is sufficient for explaining the residual magnetic anisotropy (3.5×10^4 erg/cc) in the amorphous state.

4. Structural Analysis

The structural analysis has been made by electron diffraction and transmission microscopy. Fig. 9(a) shows the clear diffraction spots of crystallites and also halo rings. The diffraction spots are identified as those for Fe_2P, Fe_3P and $\varepsilon Fe_{2\sim 3}C$. The transmission micrograph Fig. 9(b) shows small crystallites, the size of which ranges from about 100 Å to 1000 Å. The examination of several samples gives an amount of a couple % in volume of these crystallites.

CONCLUSION

Based on the results of the temperature dependences of Ms and χ, the compositional dependence of Ku, and the structural analysis,

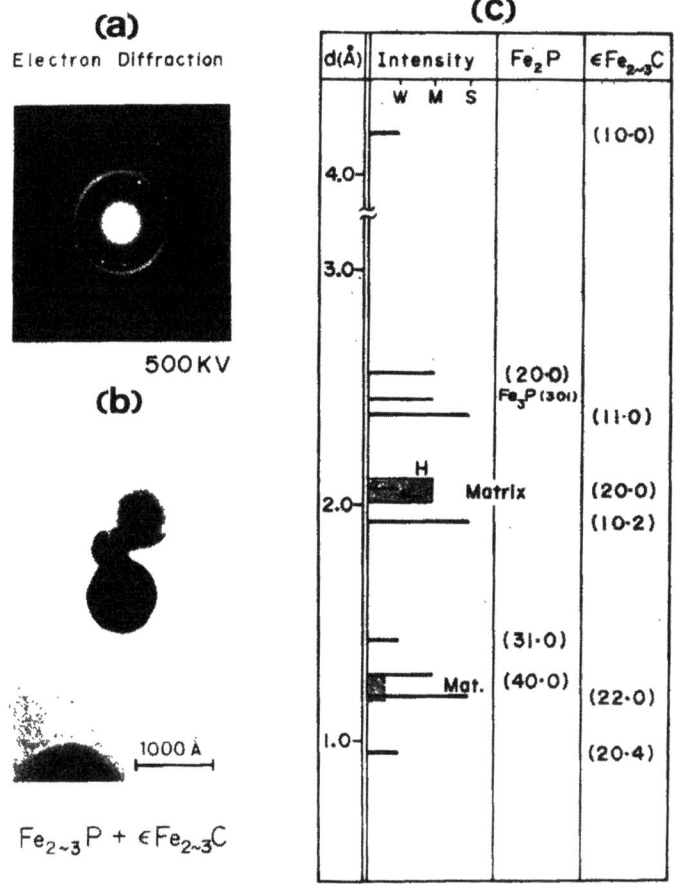

Fig. 9. The electron diffraction pattern (a) and the transmission electron micrograph (b) for a rapidly quenched $Fe_{80}P_{13}C_7$ alloy. Also shown is the result of the analysis of the diffraction pattern (c).

it is concluded that the amorphous $Fe_{80}P_{13}C_7$ alloys are considered to consist of a matrix and fine crystallites of Fe_2P, Fe_3P and $\varepsilon Fe_{2\sim 3}C$. The matrix is supposed to be composed of clusters of several components like $Fe_3P-(C)$, $\alpha Fe(P)$ and $\alpha Fe(C)$.

Amorphous Structure	
Matrix (cluster)	Crystallites
$Fe_3P-(C)$ $\alpha Fe(P)$ $\alpha Fe(C)$	Fe_2P and Fe_3P ($\varepsilon Fe_{2\sim 3}C$)

It is also concluded that the magnetization of the amorphous alloy originates from a highly strained $Fe_3P-(C)$ phase and the magnetic anisotropy is due to that of Fe_3P alloys.

REFERENCES

1. P. Duwez, R. H. Willams and W. Klement, Jr., J. Appl. Phys. 31, 36 (1960).
2. C. C. Tsuei, G. Longworth and S. C. H. Lin, Phys. Rev. 170, 603 (1968).
3. P. Duwez and S. C. H. Lin, J. Appl. Phys. 38, 4096 (1967).
4. Y. Obi, H. Fujimori and H. Saito, Japan. J. Appl. Phys. 15, 611 (1976).
5. M. Takahashi, F. Ono and K. Takakura, Japan. J. Appl. Phys. 15, 183 (1976).
6. M. Takahashi, T. Suzuki and K. Takakura, ibid. 15, 711 (1976).
7. C. C. Tsuei and H. Lilienthal, Phys. Rev. B 13, 4899 (1976).
8. A. I. Gubanov, Fiz. Taverd Tela. 2, 502 (1960).
9. K. Handrich, phys. stat. sol. 32, K55 (1969).
10. J. L. Finney, Proc. Roy. Soc. (London) A319, 479 (1970).
11. G. S. Cargill III(a), J. Appl. Phys. 41, 12 (1970).
12. P. K. Rastogi and P. Duwez, J. Non-Cryst. Sol. 5, 1 (1970).
13. T. Masumoto and H. Kimura, J. Japan. Inst. Metals 39, 273 (1975) (in Japanese).
14. M. Hansen, "Constitution of Binary Alloys," Mc-Graw-Hill, NY (1958).
15. M. Takahashi, N. Ikeda and T. Miyazaki, to be published in Japan. J. Appl. Phys. 15, No. 9 (1976).

THE MAGNETIC AND ELECTRONIC PROPERTIES OF AMORPHOUS NICKEL PHOSPHORUS ALLOYS

P. J. Cote
Rensselaer Polytechnic Institute, Troy, NY and
Benet Weapons Laboratory, Watervliet Arsenal
Watervliet, NY 12189

G. P. Capsimalis
Benet Weapons Laboratory, Watervliet Arsenal
Watervliet, NY 12189

G. L. Salinger
Rensselaer Polytechnic Institute, Troy, NY 12181

INTRODUCTION

The amorphous nickel phosphorus alloy is of interest because of its relative simplicity and the fact that it can be considered a basis system for many of the more complex ternary and quaternary metallic glasses. It is rather well characterized with data available from structure[1], magnetization[2], specific heat[3] and electrical resistivity[4] studies over its 15 to 25 at. % phosphorus composition range.

Some of our measurements of the above properties are discussed in this paper. It is found that the results are consistent with a liquid-like structure for the NiP alloy. In particular, the temperature dependence of the structure factor and the electrical resistivity behavior are seen to parallel that of liquid metals and support the application of liquid transition metal theory[5] to metallic glasses.

EXPERIMENTAL

All samples were prepared by electrodeposition from baths and conditions similar to those used by Cargill[1]. Insoluble platinum anodes were used to avoid possible contamination from soluble anodes. Samples were in the form of flat plates for x-ray diffraction and hollow cylinders for specific heat measurements. A 15 at. % P sample, prepared previously for the resistivity measurements[4], was used for the susceptibility measurements in a Foner type magnetometer.

The x-ray intensity data were obtained using an automated Phillips diffractometer and XRG 5000 x-ray generator. Molybdenum and copper radiations were used with a scintillation counter, a LiF diffracted beam monochromator and a pulse height analyzer to eliminate $\lambda/2$ Compton radiation.

A hot plate sample heater was fabricated for x-ray measurements at various temperatures and mounted directly onto the diffractometer. Sample temperatures were monitored with an iron-constantan thermocouple. Temperature stability during a run was ±2°C.

Sample heat capacities were measured in the 2 to 15K temperature range using a standard adiabatic heat pulse method with a heat switch. Temperatures were measured with a calibrated germanium thermometer. Details of the apparatus are given elsewhere[6]. Heat capacities from 0.15 to 15K were measured in the RPI dilution refrigerator. A calibrated carbon resistor served as the sample thermometer in this range.

MAGNETIC PROPERTIES

Specific heat measurements conducted by Tyan and Toth[3] provide data on the electronic contribution to the specific heat, γT, over the NiP composition range. At 15 at. % P, γ is still near the pure nickel value indicating no significant d-band filling; there is a decrease in γ to two or three times the noble metal values at the highest P compositions. These data correlate with the susceptibility measurements of Pan and Turnbull[2] which show that NiP is still weakly ferromagnetic up to 16-17 at. % P and has a decreasing paramagnetic susceptibility beyond that composition.

The conventional rigid band model predicts filled d-levels and small γ values at approximately 11 at. % P. This is not in agreement with the above NiP data and parallels a similar general result for conventional alloys.[7] Beeby[8] proposed a simple modification of the rigid band model to explain such observations. A main feature in his model is a sinking conduction band which keeps the relative positions of the d-levels and the Fermi level approximately constant. This retains the aspect of the rigid band model which correctly describes the increase in $2k_F$ as a polyvalent solute, say, is added to a monovalent metal[7]. The observed d-level

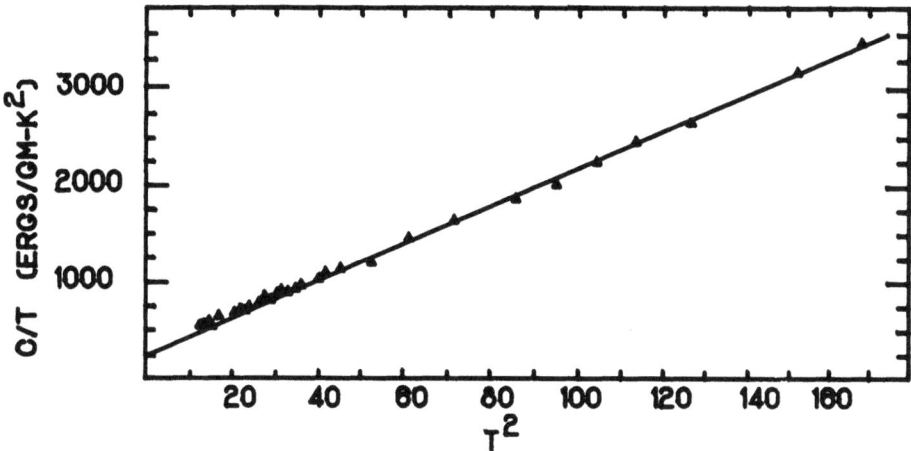

Fig. 1. Specific heat of amorphous 26 at. % P sample between 3 and 15K shown as c/T vs. T^2.

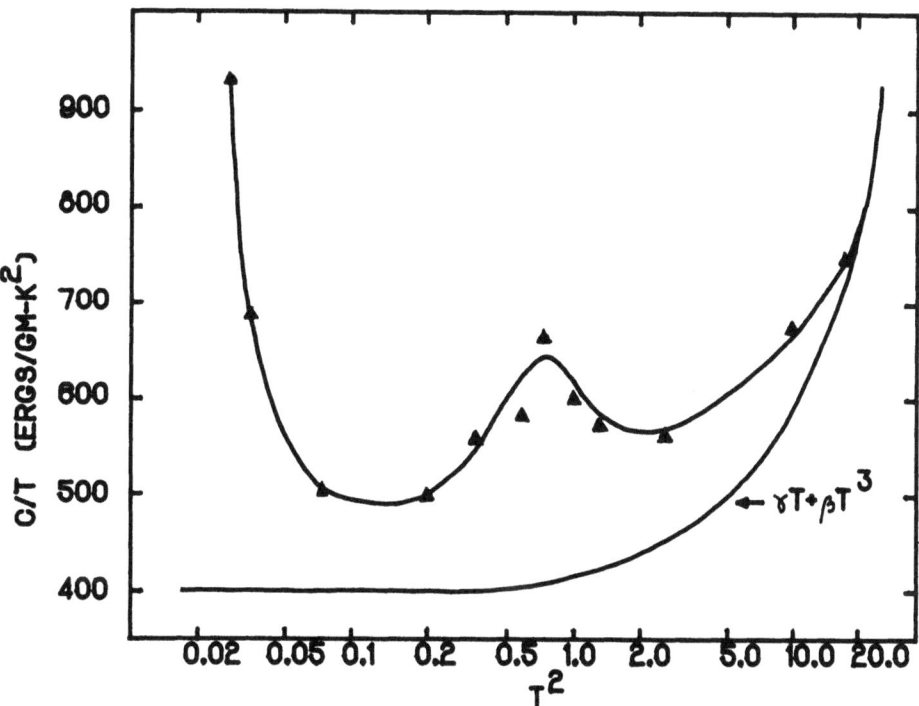

Fig. 2. Specific heat of amorphous 24.5 at. % P sample between 0.15 and 4.2K plotted as c/T vs. T^2. Estimate of normal contribution, $\gamma T + \beta T^3$, is also shown.

filling can be attributed to d-band narrowing upon alloying. Demagnetization occurs by an increased s-d admixture and provides a mechanism, in addition to hole filling, to explain the absence of ferromagnetism. The large γ values for paramagnetic NiP are thus in accord with Beeby's model. This electronic structure is related to the transport properties to be discussed later.

Our specific heat results on a 2.0 gm, at % P sample duplicate Tyan and Toth's[3] measurements near that composition, including an anomalous low temperature upturn on a c/T vs. T^2 plot (Fig. 1). They interpreted this upturn as the high temperature tail of a peak caused by superparamagnetic nickel microcrystals. Our measurements were extended down to 0.15K on a 2.8 gm, 24.5 at. % P sample. The data are given in Fig. 2 on a c/T vs. $\ln T^2$ plot along with an estimate of the normal contribution obtained by using the Debye temperature (267K) from the 26 at. % P sample. The results are consistent with a peak centered at ~2K as suggested by Tyan and Toth. Their model for NiP, which assumes ideal superparamagnetic microcrystallites, is not required however. We suggest an explanation which appears to be in better accord with other properties. This is based on the existence of ferromagnetic clusters in the NiP alloys. Clustering effects are generally observed in crystalline nickel alloys, even for Ni_3Ga and Ni_3Al, studied by de Dood and de Chatel[9], which are considered to be quite homogeneous. The NiP behavior is nearly identical to that for crystalline Ni_3Ga including the upturn at the lowest temperatures ($T^2<0.1$) which they assume is a nuclear contribution. Their analysis, based on Schroeder's model[10] for the specific heat, c, of a cluster of n particles of total moment, μ, forming the cluster with an anistropy field, H_{eff}, gives

$$c = \frac{k}{2} \{C(\Theta_E) - C([2n+1]\Theta_E)\} \quad (1)$$

where $\Theta_E = \mu H_{eff}/k$ and

$$C(\Theta_E) = (\Theta_E/T)^2 \exp(\Theta_E/T)/(\exp(\Theta_E/T)-1)^2. \quad (2)$$

Fitting our results to the above gives approximately 10^{20} clusters per mole with a moment $\sim 5/2\mu_B$ and $H_{eff} = 20kOe$ at $T \lesssim 4K$. This compares with $\sim 10^{20}$ clusters per mole, $\mu \lesssim 5\mu_B$ and $H_{eff} \sim 25$ to $50kOe$ for Ni_3Ga. For NiP, the results are consistent with, for example, a cluster of a dozen or so atoms of total moment $5/2\mu_B$ for every 6000 atoms in the alloy.

Our susceptibility measurements, conducted on a 15 at. % P sample at 293K and 77K, provide support for this cluster interpretation. The results cannot be superimposed on a universal H/T curve as required by the superparamagnetic microcrystalline model. By contrast, the data result in linear Arrott plots (Fig. 3) as do conventional alloys, indicating a largely homogeneous structure.

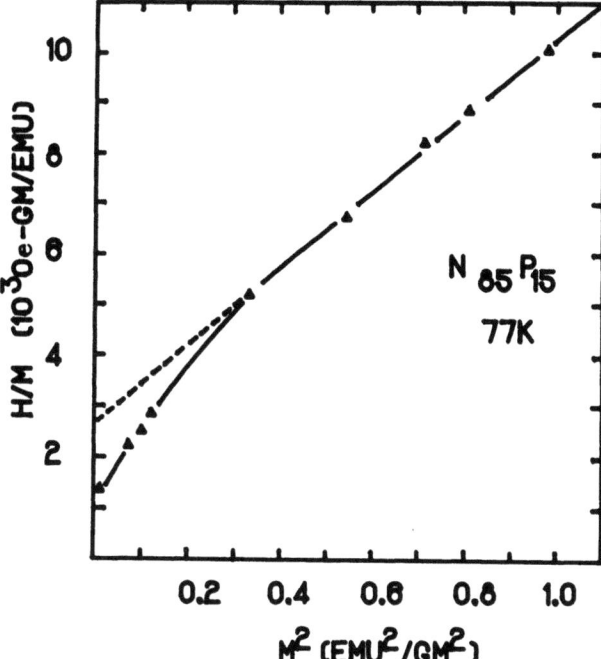

Fig. 3. Arrott plot of susceptibility data for amorphous 15 at. % P sample at 77K.

Pan and Turnbull[2] analyzed their results on this basis. There are large deviations from linearity in the Arrott plots evident in Fig. 3 and in Pan and Turnbull's data[2] at low applied fields. This is the behavior for alloys possessing magnetic inhomogeneities[11] and can be taken as evidence that superparamagnetic clusters are present in the NiP alloys.

The Kondo effect is also known to generate a low temperature specific heat peak and cannot be ruled out completely as the source of (or as a contributor to) the excess specific heat, particularly in view of the low temperature resistivity upturn present in these alloys.[4] The general features, however, indicate a clustering phenomenon. Thus, the overall magnetic properties are consistent with the model of a largely uniform solid solution of phosphorus in nickel, with significant d-band effects present even at these high phosphorus compositions.

ELECTRON TRANSPORT

In view of the non-crystalline nature of the metallic glasses, it can be assumed that liquid metal theory describes their resistivity behavior. Since the metallic glasses are generally transition element based alloys, the extension of the original Ziman theory by Evans et al.[5] and Drierach et al.[12] to liquid transition elements and alloys should be directly applicable. The extended theory is a partial-wave treatment with the large d-level influence reflected in the large contribution to the resistivity, ρ, from the $\ell=2$ phase shifts. In the presence of unfilled d-levels, the $\ell=2$ phase shift, η_2, is the dominant contributor to ρ so that, to a first approximation, the resistivity takes on the simple form

$$\rho = \frac{30\pi^3\hbar^3}{me^2\Omega_o k_F^2 E_F} \sin^2\eta_2(E_F)\, I(2k_F) \tag{3}$$

where $I(2k_F)$ is the structure factor evaluated at $q=2k_F$ and Ω_o is the atomic volume.

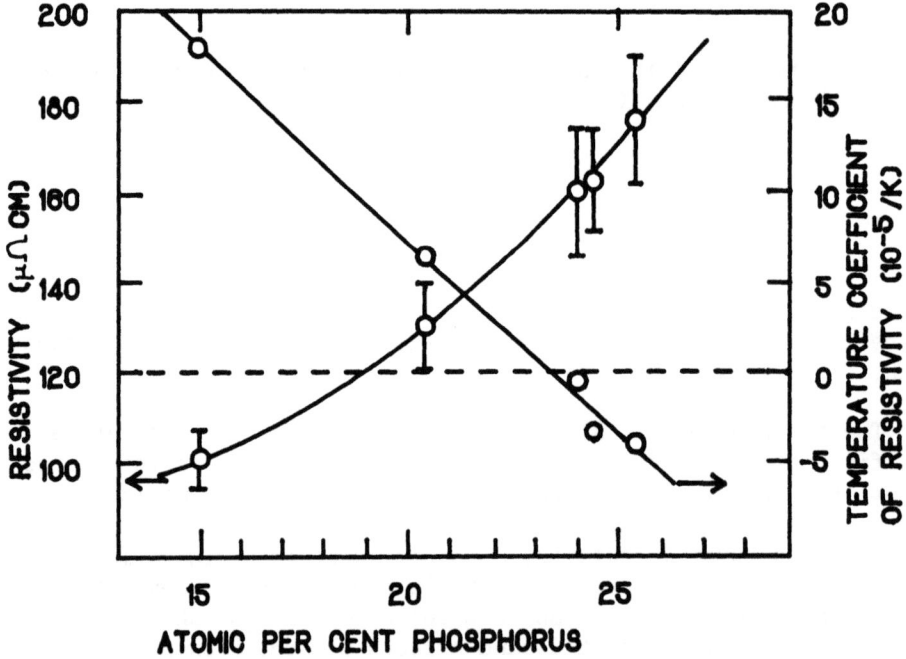

Fig. 4. The resistivity and temperature coefficient of resistivity are shown as a function of phosphorus content. The temperature coefficient becomes negative beyond 23 at. % P.

In an earlier paper[4], we showed that the room temperature NiP resistivity increases from 104μΩcm to 175μΩcm while the temperature coefficient of resistivity (TCR = dρ/ρdT) undergoes a gradual transition from + 1.8x10^{-4}/K x 10^{-4}/K to -0.4 x 10^{-4}/K over the composition range from 15 to 25.4 at. % P. These results are shown in Fig. 4. The transition occurs at approximately 23 at. % P which is remarkably close to a similar transition at 24 at. % P seen by Boucher[15] for the related NiPdP alloy. This nearly identical behavior suggests a common origin for the resistivity of both systems. The similarities occur despite the widely different compositions of magnetic constituents in the two alloys and suggests that magnetic effects such as clustering[16] are not responsible for the negative TCR's seen here. Further support is found in Boucher's demonstration that varying the Ni/(Ni+Pd) ratio from 0.15 to 0.92 produced no change in ρ. The resistivity and magnitude and sign of the TCR are therefore determined by the phosphorus content only. It will be shown, in what follows, that these observations are readily explained in terms of the extended Ziman theory.

Because of the five phosphorus valence electrons, an obvious effect expected from a change in phosphorus composition is a change in the average number of valence electrons per atom. This produces a change in the Fermi sphere diameter, $2k_F$, and is a familiar phenomenon in crystalline alloys.[7] Based on this assumed change in $2k_F$, we have interpreted the changes in ρ and TCR qualitatively in terms of Eq. 3 above[4]. A more complete analysis using the modified Ziman theory is being conducted for NiP by L. V. Meisel at the Watervliet Arsenal and will be published elsewhere. The dominant effects are the large changes in ρ attributed to $2k_F$ moving into the main peak of the structure factor and the negative TCR's which reflect the assumed peak intensity reductions on heating. The height of the main peak in the structure factors for NiP[1] can easily accomodate factors of two or more increase in $I(2k_F)$ upon alloying. However, no data existed on the effects of temperature on $I(q)$ for any metallic glass so that a study was initiated to establish this for a NiP sample.

TEMPERATURE DEPENDENCE OF THE STRUCTURE FACTOR

A 10 mil thick flat 25 at. % P sample was mounted on a hot plate sample heater for x-ray measurements at various temperatures. Because of the small magnitude of the effects to be measured (<10%), a counting time interval was selected which permitted sufficient accuracy at the peaks along with a reasonable total run time (6 to 12 hours). Sample temperatures were varied between 22 and 210°C. This minimized the possibility of crystallization during an experiment while providing measurable changes. The results were reproduced in a dozen or so preliminary runs using different sample orientations and temperatures to minimize the possibilities of experimental error. Molybdenum radiation was used to avoid effects of air heating in the vicinity of the sample heater. Several checks were made using copper radiation which also reproduced the general trends.

Representative results of the measured intensity changes for the first two peaks are shown in Figures 5 and 6 in terms of the scattering angle 2θ. A shift to smaller 2θ values (smaller q values for I(q)) and an intensity reduction on heating are seen in Fig. 5 for the main peak. The second peak exhibits similar behavior as shown in Fig. 6. Using a graphical technique to determine the magnitudes of the peak shifts gives Δ2θ/2θ ~ - 0.29% for the main peak and ~ - 0.28% for the second peak which compares with

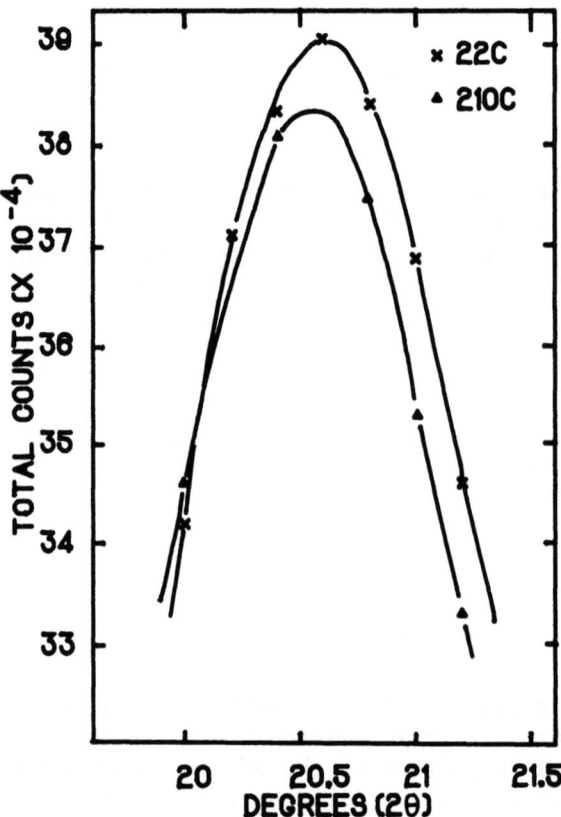

Fig. 5. The effects of temperature on the measured intensities at the main peak in the NiP structure factor.

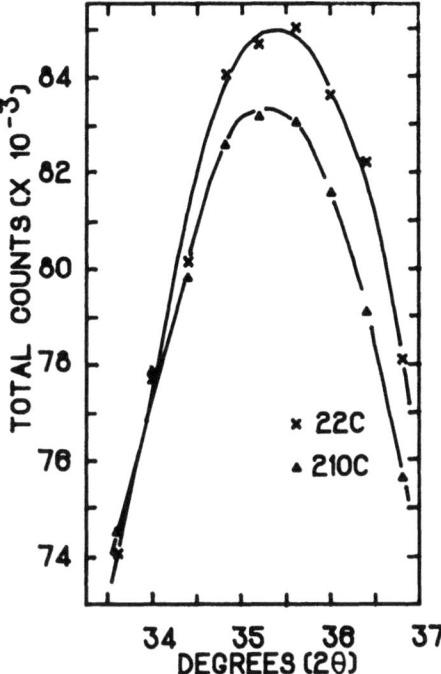

Fig. 6. The effects of temperature on the measured intensities at the second peak in the NiP structure factor.

$\Delta 2\theta/2\theta = -0.24\%$ obtained by assuming a uniform thermal expansion of the alloy and a typical nickel alloy expansion coefficient (1.3×10^{-5}/K) as an approximation for NiP. Although the intensity changes resemble the behavior in liquid alloys, the peak shifts exhibit some differences. For liquids, the main peak may remain constant or shift to smaller q values on heating while the second peak generally shifts to larger q-values.[15,16] These shifts reflect actual structural rearrangements which occur as the liquid is heated. Such structural changes are not apparent in our NiP data, in agreement with what can be expected for a solid.

The relative changes in intensity on heating, $\Delta I/I$, with the thermal expansion shift removed, are shown in Fig. 7. This was obtained from a computer interpolation function curve fit to the data. The peak shifts are not "seen" by the electrons according to the free electron approximation because $2k_F$ shifts to smaller values to exactly compensate for the shifts in $I(q)$ on heating. Also shown

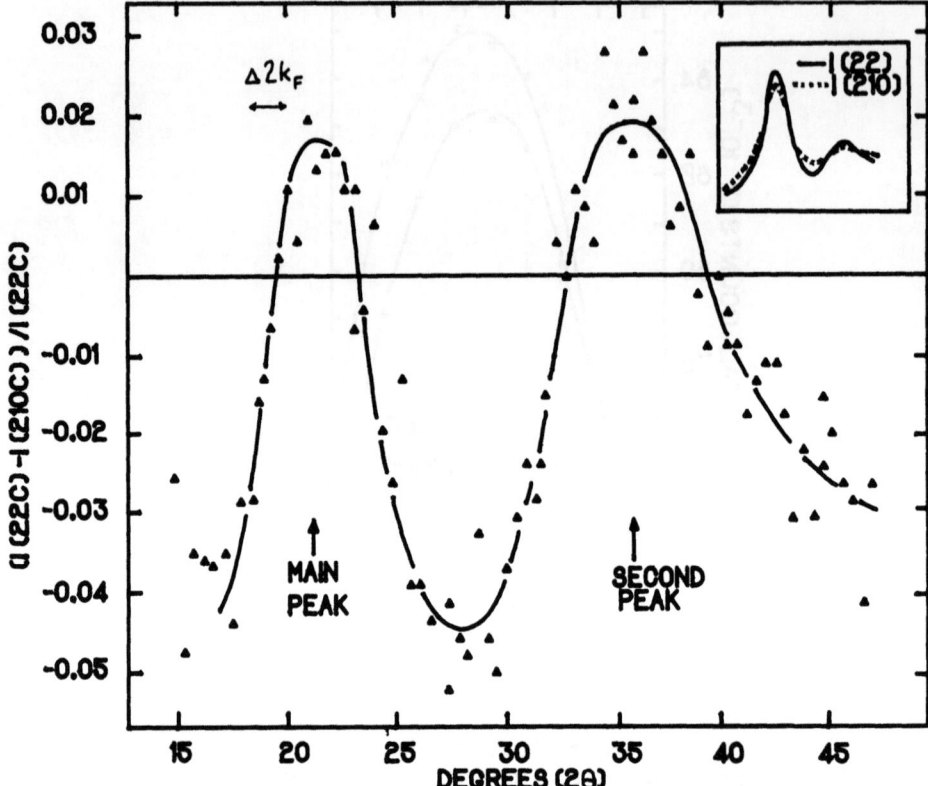

Fig. 7. Relative changes in intensity due to heating. Thermal expansion effects and background were removed. Inset shows schematically the effects of temperature with the thermal expansion correction.

is the range spanned by the Fermi sphere diameter, $\Delta 2k_F$, due to a variation of 1.0 to 1.3 electrons/atom. (The 1.3 electrons/atom is in agreement with the assumptions of nearly filled nickel d-bands at the highest P concentration and five valence electrons for phosphorus.) The transition from positive to negative $\Delta I/I$ over this range of $2k_F$ corresponds to the behavior of the TCR for NiP (Fig. 4). Figure 7 also suggests that the positive TCR's can be factors of two or more larger than the negative TCR's as seen for NiP.

In model calculations, Cargill[17] approximated the effects of thermal disorder on the model structure factor, $I_o(q)$, using

$$I(q) = e^{-2M}I_o(q) + (1-e^{-2M}), \quad (4)$$

where 2M is the Debye-Waller factor. The above represents the Debye-Waller damping of the structure factor and a background increase due to thermal diffuse scattering. Neglecting thermal expansion, Eq. 3 describes qualitatively what we observe; however, the magnitude of the effect gives a Debye temperature of approximately 400K rather than the 267K determined from our specific heat data. One possible reason for this discrepancy is that the independent oscillator approximation for the thermal diffuse scattering (Eq. 3) may be inadequate. In any event, the main observation to be made is that the magnitude of the thermal disorder effects, $\Delta I/I\Delta T$ are of the order of 10^{-4}/K which is the same magnitude as the observed TCR's for NiP[4] and NiPdP[15]. These results can be taken as further experimental confirmation of the validity of the extended Ziman theory for metallic glasses.

Several additional points should be made with regard to this model. At high temperatures, the Debye-Waller factor is linear in temperature and, as a result, the temperature dependent part of the resistivity, using Eq. 4, has the form

$$\rho_T \propto \pm T/\Theta_D^2 \quad (5)$$

with the sign determined by whether I(q) is greater than or less than unity. At low temperatures[18], the Debye-Waller factor approaches a constant value with a T^2 temperature dependence so that

$$\rho_T \propto \pm T^2/\Theta_D^3. \quad (6)$$

This is in accord with the characteristic shape of the ρ vs. T curves for NiP[4] and NiPdP[15]. Hasegawa[19] also noted that a T^2 temperature dependence is commonly seen in metallic galsses at low temperatures. Some deviations from T^2 and T behavior are to be expected in this model due to the possible temperature dependence of Θ_D over these wide temperature ranges. Equations 5 and 6 predict that larger TCR's than those observed above can be expected for alloys with smaller effective Debye temperatures.

In summary, the magnetism and specific heat data for NiP exhibit most of the characteristics of conventional nickel alloys, including clustering and large d-level effects. This and the measured temperature dependence of the structure factor provide support for the use of the extended Ziman theory to explain electron transport in NiP and other metallic glasses.

ACKNOWLEDGEMENTS

We have benefited from helpful discussion with Dr. R. MacCrone and Dr. R. Harper. The assistance provided by Dr. J. Williams and Dr. D. Moon is appreciated. One of us (PJC) thanks Dr. H. J. Guntherodt for many stimulating discussions of liquid metal research. A portion of this work was supported by NASA.

REFERENCES

1. Cargill, G. S., J. Appl. Phys., 41, 12 (1970).
2. Pan, D., and Turnbull, D., AIP Conference Proc., 18, 646 (1973).
3. Tyan, Y. S., and Toth, L. E., J. Elec. Mat., 3, 791 (1974).
4. Cote, P. J., Solid State Comm., 18, 1311 (1976).
5. Evans, R., Greenwood, D. A., and Lloyd, P., Phys. Lett. A, 35, 57 (1971).
6. Cote, P. J., PhD Thesis, RPI (1976).
7. Bennett, L. H., and Willens, R. H., Charge Transfer/Electronic Structure of Alloys, Met. Soc. of AIME, May 1973.
8. Beeby, J. L., Phys. Rev., 141, 781 (1966).
9. de Dood, W., and de Chatel, P. F., J. Phys. F., 3, 1039 (1973).
10. Schroder, K., J. Appl. Phys., 32, 880 (1961).
11. Acker, F., and Huguenin, R., J. Phys. F., 6, L147 (1976).
12. Drierach, O., Evans, R., Guntherodt, H. J., and Kunzi, H. J., J. Phys. F., 2, 709 (1972).
13. Boucher, B. Y., J. Non-Cryst. Sol., 1, 277 (1972).
14. For example, effects attributed to clustering are very sensitive to nickel content in NiCu. Houghton, R. W., and Sarachik, M. P., Phys. Rev. Lett., 25, 238 (1970).
15. Lukens, W. E., and Wagner, C. N. J., J. Appl. Cryst., 9, 159 (1976).
16. Faber, T. E., Introduction to the Theory of Liquid Metals, Cambridge University Press, London and New York (1972).
17. Cargill, G. S., Solid State Physics, 30, 227 (1975).
18. James, R. W., the Optical Principles of the Diffraction of X-Rays, Cornell University Press, NY (1965).
19. Hasegawa, R., Phys. Lett. 36A 425 (1971).

DISCUSSION

C. C. Tsuei: What is your estimate of the Fermi energy for these Ni-P alloys?

P. J. Cote: One can estimate the value of the Fermi energy from the free electron approximation. If one assumes that the d-bands are filled at the highest P concentration we used, one gets approximately 1.3 electrons per atom which in turn gives us a Fermi sphere diameter in this range.

D. I. Paul: Does your resistivity data or your model give any indication of the extent of these magnetic inhomogeneities and do they indicate whether they are due to composition fluctuations or to other factors?

P. J. Cote: The most effective way of detecting these inhomogeneities is through specific heat and susceptibility measurements and not from the resistivity data. The origin of these inhomogeneities is presumably due to compositional variations in the sample.

STRUCTURE AND PHYSICAL PROPERTIES OF AN AMORPHOUS $Cu_{57}Zr_{43}$ ALLOY

T. Mizoguchi[*], S. von Molnar, and G. S. Cargill III

IBM T. J. Watson Research Center
Yorktown Heights, New York 10598

T. Kudo
Department of Physics, Gakushuin University, Mejiro
Tokyo, Japan

N. Shiotani and H. Sekizawa
The Institute of Physical and Chemical Research
Hirosawa, Wako-shi, Saitama, Japan

Some structural, thermal, and electronic properties of amorphous Cu-Zr alloys have been investigated by x-ray scattering, specific heat, positron annihilation, and electrical resistivity measurements. X-ray scattering patterns for sputtered and for rapidly-quenched-from-the-liquid Cu-Zr alloys of ~40 at. % Zr are nearly identical, indicating that the atomic scale structure of these alloys is not grossly affected by preparation method. Rapidly quenched $Cu_{57}Zr_{43}$ has a Debye temperature of ~200°K, which is 13% lower than that found for the crystalline form of this alloy, presumed to be a mixture of Cu_3Zr_2 and $CuZr$, although the electronic specific heat $\gamma = 3 \times 10^{-3}$ J/mole °K^2 is nearly the same for both forms of the alloy and is quite close to that of crystalline Zr. The electron momentum distributions for amorphous and for crystalline $Cu_{57}Zr_{43}$ are very similar, from positron annihilation experiments, and fall between those of polycrystalline Cu and of polycrystalline Zr, lying somewhat closer to the latter. The electrical resistivity of the amorphous binary alloy increases slightly with decreasing temperature, but the resistivity irreversibly <u>increases</u> by about 3% at ~450°C when the alloy undergoes crystallization.

INTRODUCTION

This is a brief review of recent experimental results for amorphous $Cu_{57}Zr_{43}$ alloys. Although many amorphous transition metal-metalloid alloys have been made by rapid-quenching-from-the-liquid, most amorphous metal-metal alloys, e.g. Gd-Co, Tb-Fe, and other rare earth metal-transition metal alloys, have been prepared only by sputtering or by evaporation. The amorphous Cu-Zr alloys described in this paper were made by rapid quenching, except for sputtered films used in an x-ray scattering study of dependence of atomic scale structure on alloy preparation methods. It should be worthwhile to extend measurements of thermal and electronic properties to the amorphous sputtered alloys, to determine the extent to which these properties are intrinsic to the amorphous state of Cu-Zr alloys, i.e. whether they are affected by alloy preparation methods.

X-RAY SCATTERING

Interference functions, i.e. total structure factors, obtained from x-ray scattering measurements for an amorphous $Cu_{57}Zr_{43}$ alloy prepared by rapid quenching and for a $Cu_{60}Zr_{40}$ sputtered film are shown in Fig. 1. These measurements were complicated by Zr-fluorescent radiation, which introduces uncertainties in the vertical scales shown in Fig. 1. Apart from these uncertainties, the two scattering

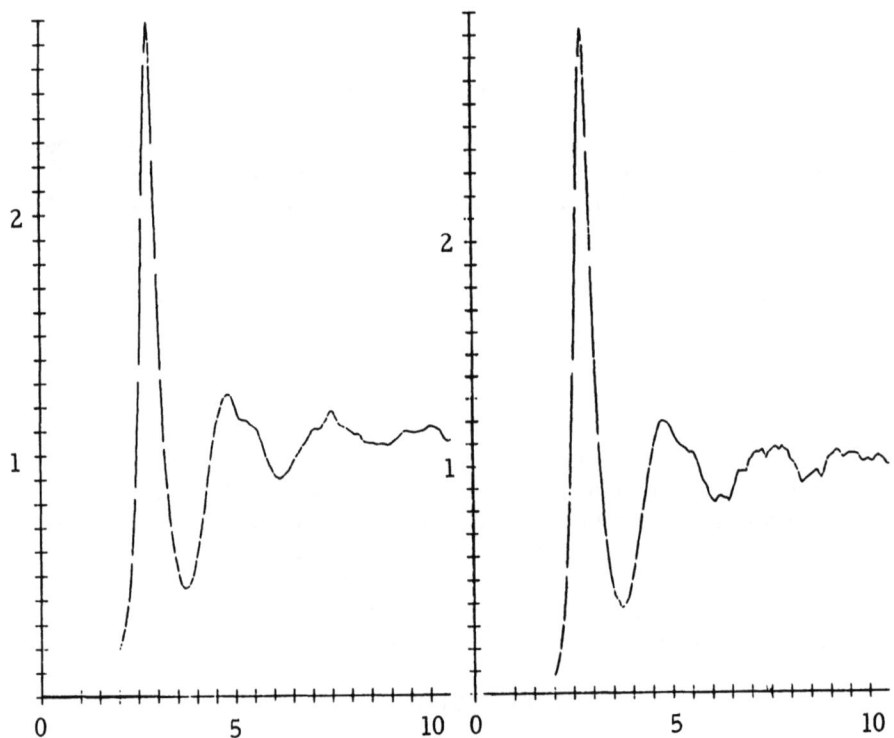

Fig. 1. Interference functions (total structure factors) $S(Q)$ versus $Q = 4\pi \sin\theta/\lambda$ for amorphous $Cu_{57}Zr_{43}$ prepared by rapid quenching (left) and for $Cu_{60}Zr_{40}$ prepared by sputter deposition (right).

patterns are nearly identical; this indicates that the atomic scale structures of the alloys prepared by rapid quenching and by sputtering are very similar. The interference functions of Fig. 1 closely resemble those reported by Waseda and Masumoto[1] for an amorphous $Cu_{57}Zr_{43}$ alloy prepared by rapid quenching. Amorphous alloys prepared by rapid quenching have been used in the experimental studies described in the remainder of this paper.

SPECIFIC HEAT

The specific heat of a $Cu_{57}Zr_{43}$ alloy, in both amorphous and crystalline forms, was measured from 1.7°K to 10°K. The crystalline alloy was obtained by annealing the initially amorphous alloy at 500°C for 3×10^4 mm. The crystallized alloy should be a mixture of Cu_3Zr_2 and $CuZr$, based on the published equilibrium phase diagram. As shown in Fig. 2, the specific heat measurements for both the amorphous and crystalline forms of the alloy can be well

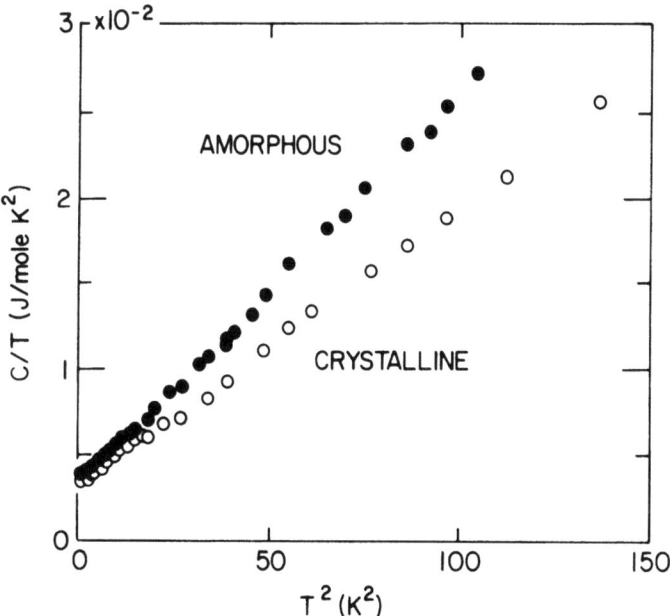

Fig. 2. Specific heat C of amorphous (filled circles) and crystalline (open circles) $Cu_{57}Zr_{43}$ alloys, plotted as C/T versus T^2.

described by the conventional $C(T) = \gamma T + \beta T^3$ expression. The resulting γ-values, for both sets of measurements, fall between 3.0 and 3.4×10^{-3} J/mole °K^2. This is much larger than the value of γ reported for crystalline Cu, 0.696×10^{-3} J/mole °K^2, but is close to that for crystalline Zr, 3.03×10^{-3} J/mole °K^2.[4] The electronic density of states at the Fermi level, $g(E_F)$, for amorphous $Cu_{57}Zr_{43}$, as well as for the corresponding crystalline phase mixture, is apparently similar to that for crystalline Zr. Evaluation of $g(E_F)$ from the experimentally determined γ-value, using the simple electron gas model, $\gamma = \frac{1}{3}\pi^2 k_B^2 g(E_F)$, where k_B is Boltzmann constant, yields ~1.3 states (eV atom)$^{-1}$.

The slope β of C/T vs. T^2 for the amorphous $Cu_{57}Zr_{43}$ alloy, 2.26×10^{-4} J/mole deg^4, is larger than that for the crystallized alloy, 1.55×10^{-4} J/mole deg^4. The Debye temperature for the amorphous alloy, based on this β-value, is roughly 200°K, while the equivalent Debye temperature for the polyphase, crystalline alloy is 13% higher, i.e.

$$\Theta_{am}/\Theta_{cr} = (\beta_{cr}/\beta_{am})^{1/3} \simeq 0.88.$$

Similar relationships between Debye temperatures of amorphous and crystalline metal-metalloid alloys have been reported,[5,6] and the lower Debye temperatures of the amorphous alloys have been shown to result primarily from a reduced shear modulus in the amorphous solids.[6,7]

POSITRON ANNIHILATION

Positron annihilation measurements have been used to probe momentum distributions of electrons in $Cu_{57}Zr_{43}$ alloys, in amorphous, partially crystallized (annealed 10^4 min at 360°C), and totally crystallized forms. Momentum distributions do not differ significantly among the three types of alloy samples; positron lifetimes[8] are 174 ± 1 psec for all of the samples. Insensitivity of positron angular correlation curves to partial or total crystallization has also been reported by Chuang, et al.,[9] for rapidly quenched Pd-Cu-Si alloys. The electron momentum distribution for the $Cu_{57}Zr_{43}$ alloy samples, from the positron annihilation experiments, falls between those of polycrystalline Cu and of polycrystalline Zr,[10] lying somewhat closer to the latter; this suggests that amorphous $Cu_{57}Zr_{43}$, as well as the alloy's equilibrium crystalline phases, are electronically more similar to Zr than to Cu.

The momentum distribution curve for amorphous $Cu_{57}Zr_{43}$ agrees well with the expression, $y = y_0(k_m^2 - k^2)$, with $k_m \simeq 1.58$ Å^{-1}, for $k \lesssim 0.8$ Å^{-1}. Inability to subtract the contribution of core electrons in these positron anihilation data prevent us from extracting an accurate value for k_F, the Fermi momentum. However, k_m provides an upper bound to the actual Fermi momentum.

Nagel and Tauc[11] have proposed a nearly-free electron model for explaining the stability of amorphous metallic alloys. According to their model, the stability of an amorphous alloy for which $2k_F$ falls near the first peak of the interference function is enhanced. As shown in Fig. 1, the first peak for an amorphous

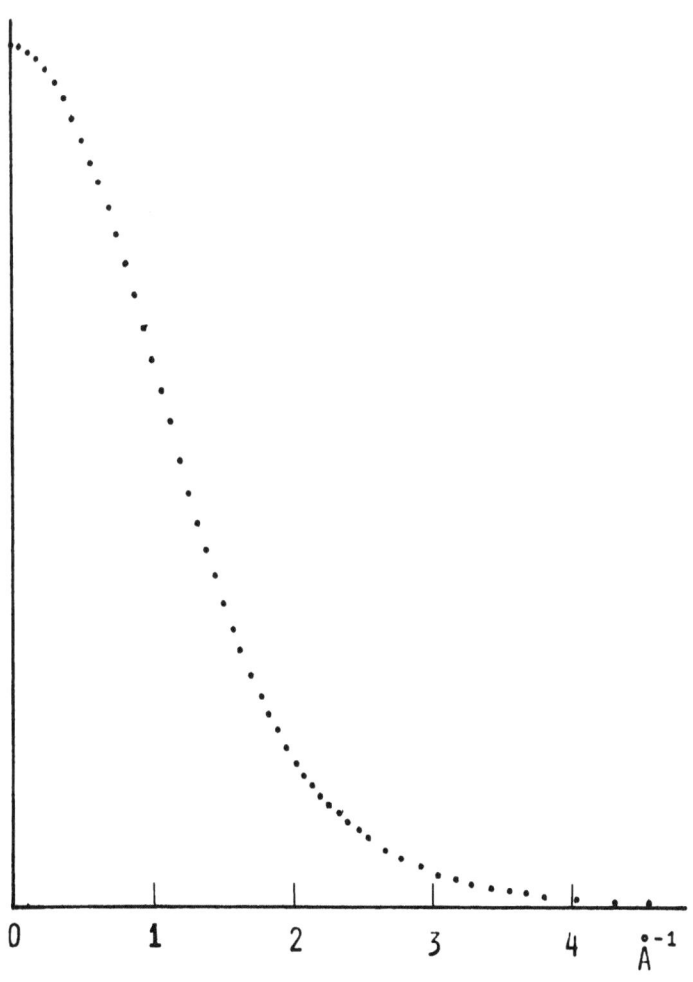

Fig. 3. Momentum distribution of electrons (in arbitrary units) for amorphous $Cu_{57}Zr_{43}$ obtained from positron anihilation experiment plotted against $k_z = p_z/h$.

$Cu_{57}Zr_{43}$ alloy is at $Q_p \simeq 2.7$ Å^{-1}. The momentum distribution in Fig. 3 indicates that $2k_F < 3.16$ Å^{-1}, which is consistant with $2k_F = Q_p$ for this alloy.

MAGNETIC SUSCEPTIBILITY

The magnetic susceptibility of amorphous $Cu_{57}Zr_{43}$ is temperature independent with $\chi = 1.1 \times 10^{-6}$ emu/g.[12] This suggests that the susceptibility arises largely from Pauli spin paramagnetism of the conduction electrons. However, this susceptibility is twice as large as that calculated using the density of states at the Fermi level obtained from specific heat measurements neglecting electron-electron interactions and diamagnetic susceptibilities of conduction electrons and of ion cores.

ELECTRICAL RESISTIVITY

The temperature dependence of the electrical resistivity of amorphous $Cu_{57}Zr_{43}$ below room temperature is shown in Fig. 4. The resistivity increases slightly with decreasing temperature. The same temperature dependence has been observed for an amorphous $Cu_{60}Zr_{40}$ alloy by Szofran, et al.[13] Absolute values of the

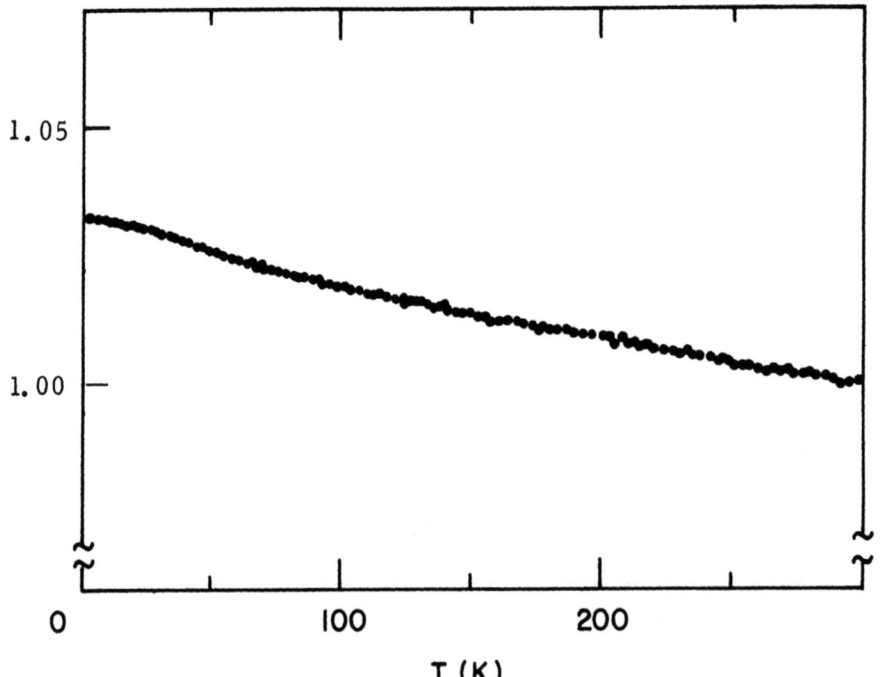

Fig. 4. Normalized temperature dependence of the electrical resistivity, $\rho(T)/\rho(300K)$, of amorphous $Cu_{57}Zr_{43}$ alloy below room temperature.

resistivities varied from sample to sample. The origin of this scatter in absolute resistivities is being investigated. The resistivity above room temperature, shown in Fig. 5, increases abruptly and irreversibly at about 450°C. This resistivity change is associated with crystallization of the alloy. In general the resistivities of amorphous metallic alloys decrease when the alloys crystallize. This is the first observation, to our knowledge, of a resistivity increase on crystallization by a <u>metallic</u> alloy. (The much larger increase in resistivity on transformation of amorphous metallic Bi to its semimetal crystalline form is well known.[14]) The resistivity of the crystallized $Cu_{57}Zr_{43}$ alloy decreases rapidly with decreasing temperature, in contrast to the nearly temperature independent resistivity of the amorphous alloy.

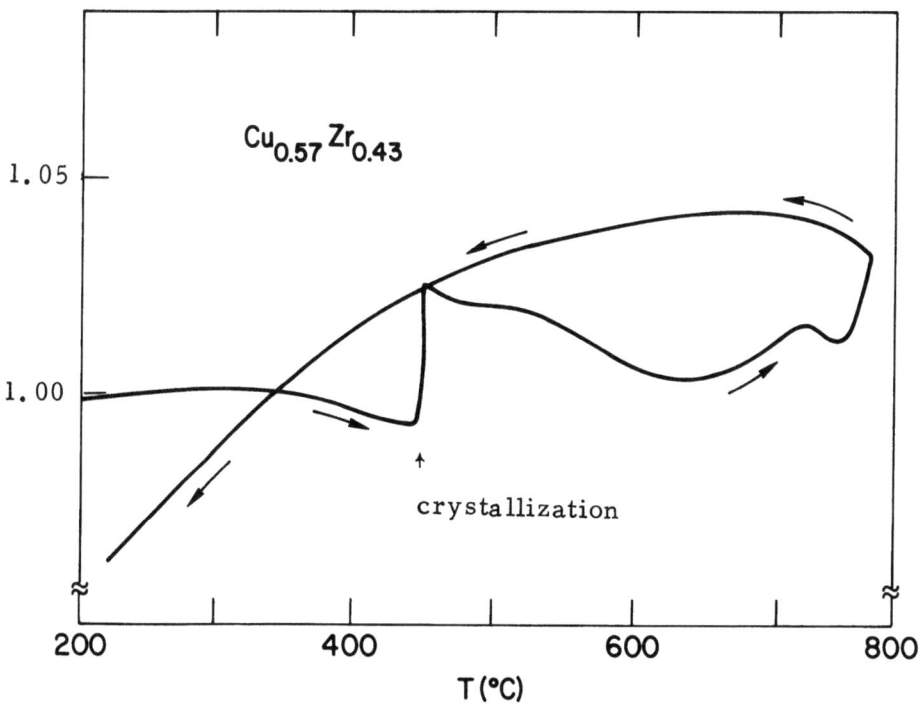

Fig. 5. Irreversible change of the normalized electrical resistivity, $\rho(T)/\rho(300K)$, with increasing temperature. An abrupt increase of resistivity (∼3.3%) was observed at crystallization (∼450°C).

ACKNOWLEDGMENTS

The authors wish to express their sincere thanks to Mr. G. C. Chi for performing x-ray scattering measurements and to Mr. D. S. Yu for preparing sputtered Cu-Zr films. They are also grateful to J. R. Rigotty for assistance with the specific heat measurements.

REFERENCES

*Permanent address: Faculty of Science, Gakushuin University, Mejiro, Tokyo, Japan

1. Y. Waseda and T. Masumoto, Z. Physik B$\underline{21}$, 235 (1975).
2. M. Hansen, Constitution of Binary Alloys, Second Edition (McGraw-Hill, New York, 1958) p. 656.
3. G. Ahlers, Rev. Sci. Inst. $\underline{37}$, 477 (1966).
4. N. M. Wolcott, Phil. Mag. VIII, $\underline{2}$, 1246 (1957).
5. H. S. Chen and W. H. Haemmerle, J. Non-Cryst. Solids $\underline{11}$, 161 (1972).
6. B. Golding, B. G. Bagley and F. S. L. Hsu, Phys. Rev. Letters $\underline{29}$, 68 (1972).
7. D. Weaire, M. F. Ashby, J. Logan and M. J. Weins, Acta Met. $\underline{19}$, 779 (1971).
8. S. Tanigawa, K. Hinode and M. Doyama, private communication.
9. S. Y. Chuang, S. J. Tao, and H. S. Chen, J. Phys. F $\underline{5}$, 1681 (1975).
10. B. Rosenfeld, Acta Physica Polonica $\underline{31}$, 197 (1967).
11. S. R. Nagel and J. Tauc, Phys. Rev. Letters $\underline{35}$, 380 (1975).
12. T. Mizoguchi and T. Kudo, AIP Conf. Proc. $\underline{29}$, 167 (1976).
13. F. R. Szofran, G. R. Gruzalski, J. W. Weymouth, D. J. Sellmyer, and B. C. Giessen, to be published in Phys. Rev. B (1976).
14. R. Hilsch and W. Martienssen, Nuovo cimento Suppl. $\underline{7}$, 480 (1958).

DISCUSSION

J. A. Rayne: In reference to the Debye temperatures of your samples determined from the specific heat data, they look awfully low. Can you comment on that?

T. Mizoguchi: We have had difficulties determining the Debye temperature in these amorphous alloys because we are not dealing here any more with a crystalline state where the number of atoms per unit can be clearly determined.

J. A. Rayne: Did you get values of the Debye temperature for these samples in the crystalline state?

T. Mizoguchi: No, we have not gotten any data for the crystalline state. We have simply concentrated on studying the amorphous state of these alloys. However, since the crystalline state is not a single phase structure, one would also expect problems in determining the Debye temperature for these samples.

A PROPOSED STRUCTURE MODEL FOR AMORPHOUS $Pd_{0.8}Si_{0.2}$ ALLOY

T. Fukunaga, T. Ichikawa and K. Suzuki

The Research Institute for Iron, Steel and Other Metals

Tohoku University, Sendai-980, Japan

ABSTRACT

Computer simulations have been carried out to generate dense random packed hard sphere (DRPHS) models composed of two different-sized spheres. The minimum Si-Si pair distance and the degree of tetrahedral perfection were used as adjustable parameters in this model's construction. A good structural model obtained suggests that Si atoms are not in hard contact with Si atoms in amorphous $Pd_{0.8}Si_{0.2}$.

INTRODUCTION

It has been pointed out that the short-range structure of amorphous transition metal films is not one of instantaneous freezing of the liquid transition metals [1], but rather close to DRPHS structure with a high degree of tetrahedral order [2]. It has also been suggested that the gross structure of amorphous transition metal-metalloid alloys such as amorphous Co-P alloys can be explained by a DRPHS structure containing two differently sized spheres 3 . However, more detailed knowledge about the structure, e.g. metal-metal, metal-metalloid and metalloid-metalloid correlations, is very poor at the present.

The first peak of the radial distribution function (RDF) of amorphous $Pd_{0.8}Si_{0.2}$ alloy determined from time-of-flight pulsed neutron diffraction data in the very wide wave number region of the scattering vector clearly splits into two subpeaks, shown in Fig. 1. They closely correspond to the shortest distances for Pd-Pd and Pd-Si pairs in Pd_3Si crystalline compound [1,4]. If Pd and Si atoms are randomly mixed in amorphous $Pd_{0.8}Si_{0.2}$ alloy, the peak for the Si-Si nearest neighbor pair should appear at a shorter distance than the first peak of the RDF, though it would be small because of the relatively small scattering length and low concentration of Si nuclei in the alloy. The peak for the Si-Si nearest neighbor pair has not been resolved at all in the RDF, as shown in Fig. 1. By examining the nearest neighbor pair distances of the crystalline compound shown in Fig. 1, it can be expected that the Si-Si pairs in amorphous $Pd_{0.8}Si_{0.2}$ alloy are not in hard contact, but rather separated by a distance of 4 Å.

In this work, partial correlations for Pd-Pd, Pd-Si and Si-Si pairs have been derived from computer simulations for the growth of amorphous $Pd_{0.8}Si_{0.2}$ alloy, choosing parameters for the minimum Si-Si pair distance and the degree of tetrahedral perfection so as to reproduce the experimental observations.

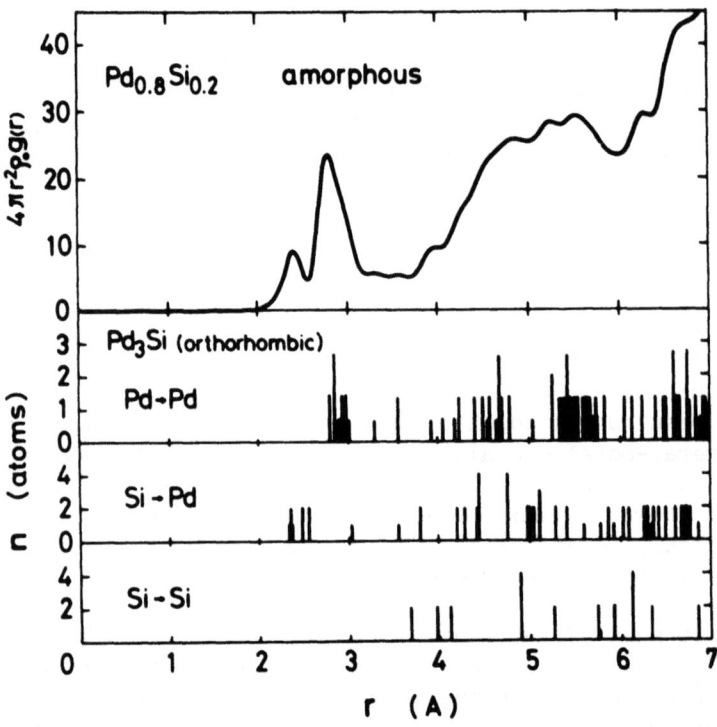

Fig. 1. Comparison between RDF of amorphous $Pd_{0.8}Si_{0.2}$ alloy and Pd-Pd, Pd-Si and Si-Si correlations in Pd_3Si crystalline compound

A PROPOSED STRUCTURE MODEL FOR AMORPHOUS $Pd_{0.8}Si_{0.2}$ ALLOY

A BINARY DRPHS MODEL WITH TWO SPHERE SIZES

A computer program used to generate a DRP model structure with a single-sized sphere by Ichikawa [2] was extended so as to obtain a binary DRPHS model structure containing two differently sized spheres. Essential procedures in generating structural model of amorphous $Pd_{0.8}Si_{0.2}$ alloy are briefly described below.

First, a regular tetrahedron composed of four Pd atoms (diameter $\sigma_{Pd} = 2.81 \text{Å}$) in hard contact with each other was selected as a seed cluster. Next, a Pd or Si ($\sigma_{Si} = 2.03 \text{Å}$) atom was chosen with the probability proportional to their concentration in the alloy. Positions which satisfy the following two conditions were calculated for all three sphere groupings already present on the surface of the cluster; (1) a sphere placed at the position necessarily comes in hard contact with the three spheres, (2) separation, r_{ij}, r_{jk} and r_{ki} between the three spheres are shorter than $k(R_i+R_j)$, $k(R_j+R_k)$ and $k(R_k+R_i)$, respectively, where R_i, R_j and R_k are radii of the spheres i, j and k. These positions are called pockets. The sphere chosen was placed at the position closest to the center of the cluster. In this work, models with k<2.0 were generated. The former model can be said to have short range structure with a much higher degree of order than the latter [2]. The model that best reproduces the experimental observations was obtained by assuming that amorphous $Pd_{0.8}Si_{0.2}$ was composed of a mixture of structures with k<1.2 and k<2.0 at ratio of 3 to 1.

The computer program for this simulation also has parameter that sets a minimum permissible Si-Si pair distance in model structure. Model structures for amorphous $Pd_{0.8}Si_{0.2}$ alloy were produced where Si-Si atoms were considered to be and not to be in hard contact. Finally, the hard sphere potentials were softened by choosing a Gaussian distribution for the sphere sizes. The width of the Gaussian distribution was selected to be equal to the widths of the first Pd-Pd and Pd-Si peaks in the observed RDF [1].

PAIR DISTRIBUTION FUNCTIONS

Various types of model structures for amorphous $Pd_{0.8}Si_{0.2}$ alloy were generated using a total number of 200 spheres. Fig. 2 shows partial pair distribution functions for Pd-Pd, Pd-Si and Si-Si correlation distances for the model structure in which a Si atom must be more than 3.8Å away from the other Si atoms (Model A). The value of 3.8Å is the mean first neighbor distance for Si-Si atom pairs in Pd_3Si crystalline compound [5]. Other partial pair distribution functions are shown in Fig. 3, where hard contact between Si atoms is permitted in the model structure (Model B).

The general shapes of the $g_{PdPd}(r)$ and $g_{PdSi}(r)$ partial pair

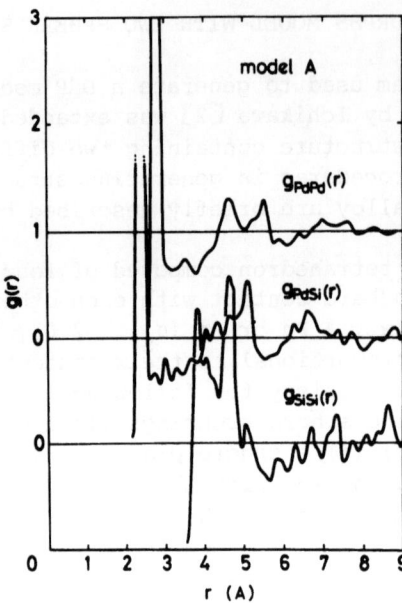

Fig. 2. Partial pair distribution functions of amorphous $Pd_{0.8}Si_{0.2}$ alloy generated by a computer under the constraint that a Si atom must be more that 3.8 A away from the other Si atoms

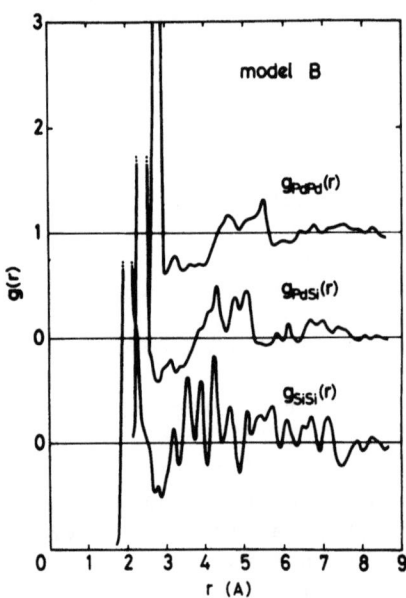

Fig. 3. Partial pair distribtion functions of amorphous $Pd_{0.8}Si_{0.2}$ alloy generated by a computer under the constraint that Si atoms may be in hard contact with each other.

distribution functions are similar in both model A and model B, while in $G_{SiSi}(r)$ the first peak of model A has a broad double peak in contrast to the sharp single first peak of model B. The broad first peak of metalloid-metalloid partial pair distribution function has been found in diffraction experiments for amorphous $Co_{0.8}P_{0.2}$ [6] and $Ni_{0.8}P_{0.2}$ [7] alloys. The $g_{PdPd}(r)$ of model A is closer in overall feature to experimental g(r) for amorphous films of transition metals [8] than the $g_{PdPd}(r)$ of model B.

By summing the three partial pair distribution functions in Fig.2 and 3, taking into account the difference in scattering length and alloy concentration, a total pair distribution function was derived which could be compared with experiment. The results are shown in Fig. 4, together with the experimental g(r) [1,4]. Although the contribution of Si-Si correlation to the total pair distribution function is only 2.2% [4] in neutron diffraction experiment of amorphous $Pd_{0.8}Si_{0.2}$ alloy, the first Si-Si peak can be clearly distinguished from the first Pd-Si and Pd-Pd peaks on the samll distance side of the first Pd-Si peak in g(r) of model B. However, the experimental g(r) never shows such a Si-Si peak, as shown in Fig. 4. Model A better reproduces the overall

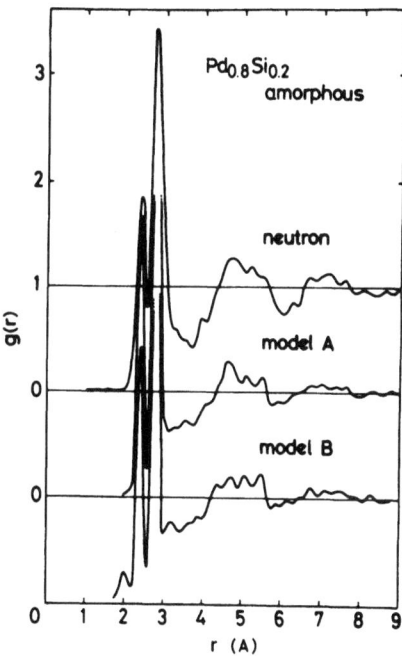

Fig. 4. Comparison of total pair distribution functions of amorphous $Pd_{0.8}Si_{0.2}$ alloy between experimental observation and computer simulations

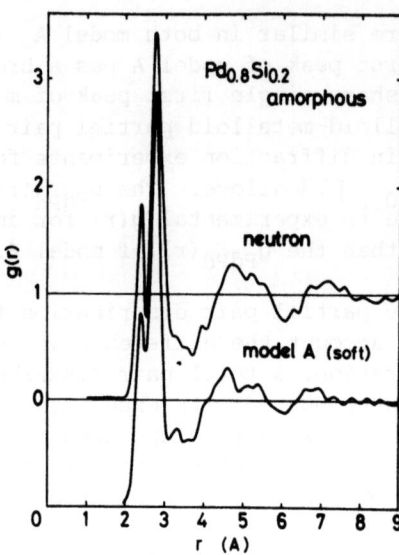

Fig. 5. Comparison of total pair distribution functions of amorphous $Pd_{0.8}Si_{0.2}$ alloy between experimental observation and softened model A

feature of the experimental $g(r)$. Therefore, it was suggested that Si atoms are not in hard contact with adjacent Si atoms in amorphous $Pd_{0.8}Si_{0.2}$ alloy. Fig. 5 shows a modified structure of model A in which the softening of hard sphere potentials is simulated by using a Gaussian distribution for the sphere sizes. Agreement between the experiment and model $g(r)$ is quite good.

INTERFERENCE FUNCTIONS

Partial interference functions shown in Fig. 6 are Fourier transformations of corresponding partial pair distribution functions for softened model A. The second peak in the $S_{PdPd}(Q)$ partial interference function has a shoulder on the high Q side. This shoulder is a common feature in experimental $S(Q)$ for amorphous films of transition metals [8]. Such a shoulder on the second peak has also been observed in experimental metal-metal partial interference functions for amorphous $Co_{0.8}P_{0.2}$ [6] and $Ni_{0.8}P_{0.2}$ [7] alloys. The $S_{PdPd}(Q)$ derived from a combination of neutron and x-ray diffraction data [1,4] assuming that $S_{SiSi}(Q)$ is negligibly small, has an overall feature similar to $S_{PdPd}(Q)$ for model A in Fig. 6.

The partial interference function $S_{PdSi}(Q)$ has a shoulder on the high Q side of the first peak. A similar feature has been

A PROPOSED STRUCTURE MODEL FOR AMORPHOUS $Pd_{0.8}Si_{0.2}$ ALLOY

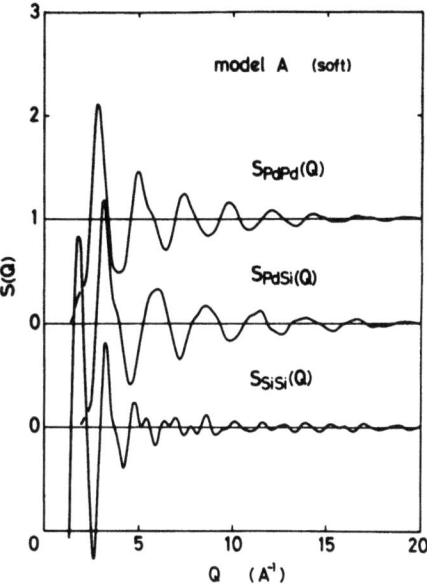

Fig. 6. Partial interference functions of amorphous $Pd_{0.8}Si_{0.2}$ alloy generated for softened model A by computer simulation

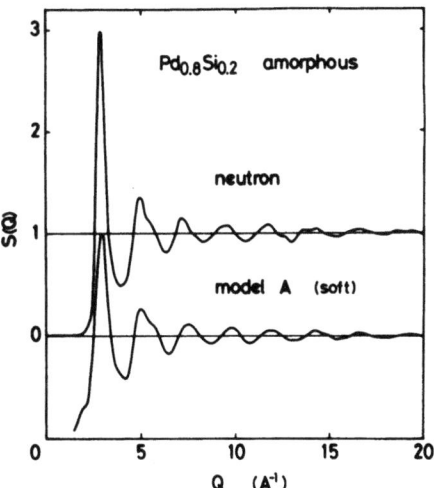

Fig. 7. Comparison of total interference functions between experimental observations and softened model A

found in amorphous $Co_{0.8}P_{0.2}$ [6] and $Ni_{0.8}P_{0.2}$ [7] alloys.

The partial interference function $S_{SiSi}(Q)$ shows many small ripples persisting up to the high Q region. The second peak in $S_{SiSi}(Q)$ is well separated from the first peak and its height is remarkably large compared with those in $S_{PdPd}(Q)$ and $S_{PdSi}(Q)$. This feature has been found in the experimental $S_{PP}(Q)$ for amorphous $Co_{0.8}P_{0.2}$ alloy [6].

The total interference function $S(Q)$ for softened model A is shown in Fig. 7 with $S(Q)$ experimentally determined by neutron diffraction. The general feature of $S(Q)$ for softened model A is in good agreement with that of the experimental $S(Q)$ except in the region of the first peak. Because the height of the first peak depends on the size and density of the cluster rather than the short range structure in a structural model, differences in the first peak height will not have much influence on the short range structure derived by this work.

FINAL REMARK

In spite of the small number of spheres used in this simulation, $g(r)$ and $S(Q)$ calculated were found to be in good agreement with the experimental $g(r)$ and $S(Q)$. One of the troubles claimed in conventional DRPHS models has been a density problem. However, no conclusive statement about the density problem can be made from this work, because the number 200 of spheres used is too small. Further computer simulations are being carried out by using more than 1000 spheres.

REFERENCES

1. K. Suzuki, T. Fukunaga, M. Misawa and T. Masumoto, Sci. Rep. RITU, A26, 1 (1976).
2. T. Ichikawa, phys. stat. sol., (a) 29, 293 (1975).
3. G.S. Cargill III, Solid State Physics (Academic Press, Edited by H. Ehrenreich, F. Seitz and D. Turnbull), 30, 227 (1975).
4. K. Suzuki, T. Fukunaga, M. Misawa and T. Masumoto, Mater. Sci. Eng., 23, 215 (1976).
.5. B. Aronson and Anna Nylund, Acta Chem. Scand., 14, 1011 (1960).
6. J.F. Sadoc and J. Dixmier, Mater. Sci. Eng., 23, 187 (1976).
7. Y. Waseda, H. Okazaki, M. Naka and T. Masumoto, Sci. Rep. RITU, A26, 12 (1976).
8. T. Ichikawa, phys. stat. sol., (a) 19, 707 (1973).

Ground State of an Ising Antiferromagnet with a Dense Random Packing Structure

S. Kobe

Technische Universität Dresden, Sektion Physik

Dresden, GDR

Although the existence of ferromagnetism in amorphous materials is well confirmed both from the theoretical and the experimental point of view the appearance of antiferromagnetism in such substances is dubious. First a simple theory of amorphous antiferromagnetism was given using the molecular field theory.[1,2] The structural disorder was simulated by a sublattice model, i.e., the spins were arranged on lattice points with the antiferromagnetic exchange interactions between them fluctuating randomly. However, in this model all nearest neighbors of a given spin belong to the other sublattice. This assumption is an inadmissable restriction, since in real amorphous systems it is impossible to surround a plus spin by minus spins only or vice versa. Amorphous antiferromagnetism can be disfavored by having the so-called misfit structure, which also occurs in some crystalline antiferromagnets (e.g., fcc lattice with nearest neighbor interaction).[3-5] For an improved description random packing of hard spheres, which succeededs in interpreting the radial distribution functions of amorphous systems, are more suitable structure models. Recently there are first attempts to investigate magnetic properties using these models.[6-8]

Several methods have been given to construct random packings by computer. We follow the procedure proposed by Finney[9] to obtain very dense packings by an algorithm which involves the possibility of local rearrangements of spheres: The coordinates of the centres of the spheres are chosen by a pseudo-random number generator. In the case of overlaps the spheres are moved along the connecting lines between the centres until they just touch, irrespective of other overlaps that may be created. After removing successively all overlaps the sphere diameter is increased by a small amount

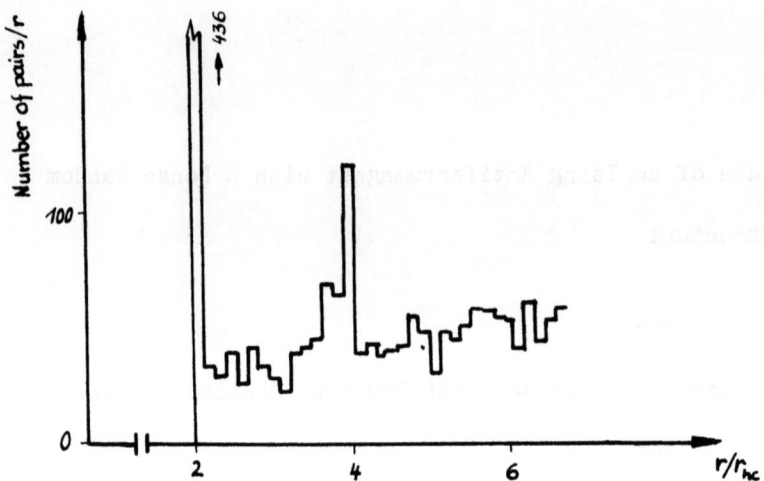

Fig. 1. a. Radial distribution function.

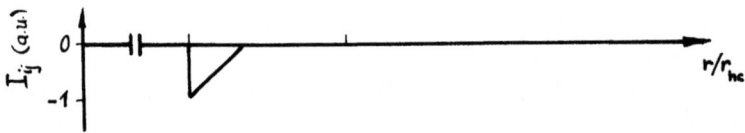

Fig. 1. b. Exchange interaction.

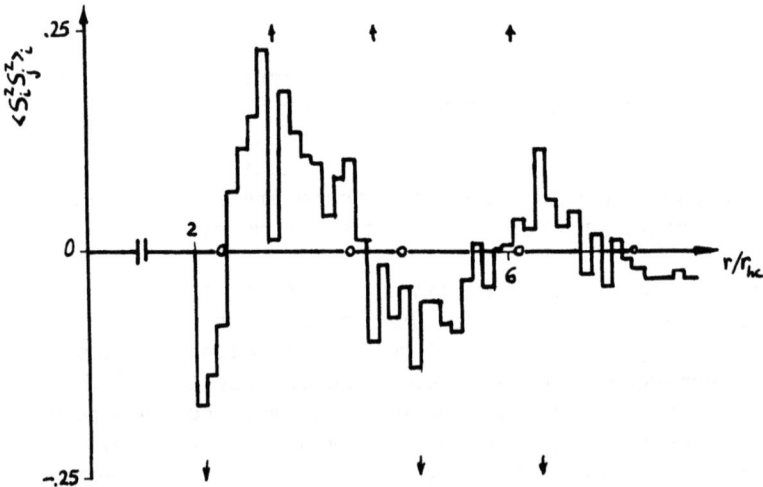

Fig. 1. c. Correlation function $\langle S_i^z S_j^z \rangle_i$; the circles and arrows belong to the triangular and square lattices, respectively, with the same density.

and the procedure is repeated. We have used a two-dimensional version of this algorithm with N = 40 disks in a square box. As opposed to Finney's original work, in which a cluster of 500 spheres with free boundary is considered, we have chosen periodic boundary conditions because of the great portion of surface disks in our small system. In the example of Fig. 2 the density is ρ = 0.68, i.e., 68% of the full square is occupied by disks. The histogram of the radial distribution function is shown in Fig. 1a. For the averaging procedure all disks are successively chosen as centres. An ensemble of 3 systems is constructed with the same density, but with another set of starting coordinates.

Considering the random structure obtained as a fixed one we start with the Ising Hamiltonian in the absence of an external field,

$$H = - \sum_{i<j} I_{ij}(r) S_i^z S_j^z$$

for S = 1/2; and we calculate the exact magnetic ground state, where a simple distance dependent antiferromagnetic exchange interaction $I_{ij}(r)$ with limited range is assumed (Fig. 1b). The result is shown in Fig. 2. If we identify as neighbors all spins within the magnetic interaction range of each other and we find that the average number of neighbors is 4.5. Figure 3 shows the distribution of the effective fields H_{eff} discussed by Simpson[1,5].

A more quantitative representation can be given by plotting the "magnetic radial distribution function". In analogy to the structure description by the radial distribution function we consider the probability of magnetic ordering of a spin having a certain distance from a central spin. Such information is involved in the correlation function $<S_i^z S_j^z>_i$, where $<...>_i$ is the structure average value under the condition that the central spin i has a fixed direction[7,10,11]. A histogram of the distance dependence of this function in Fig. 1c shows a short-range antiferromagnetic order with more and more smeared out extrema at higher distances. (For the average procedure again all spins of the 3 equivalent systems are successively chosen as central spins.) For an amorphous ferrimagnet a corresponding correlation function was obtained by Rhyne et al.[12] with neutron diffraction measurements on $TbFe_2$.

Except for the first one, the extrema in the magnetic and the structural radial distribution functions seem not to be correlated. For a loose 23 spin system the magnetic curve is related to the values for the triangular and the square lattices with the same density[7]. This correspondence is well confirmed in our case.

Since every spin, which is aligned parallel to its neighbor due to misfit, increases the ground state energy E_o compared to

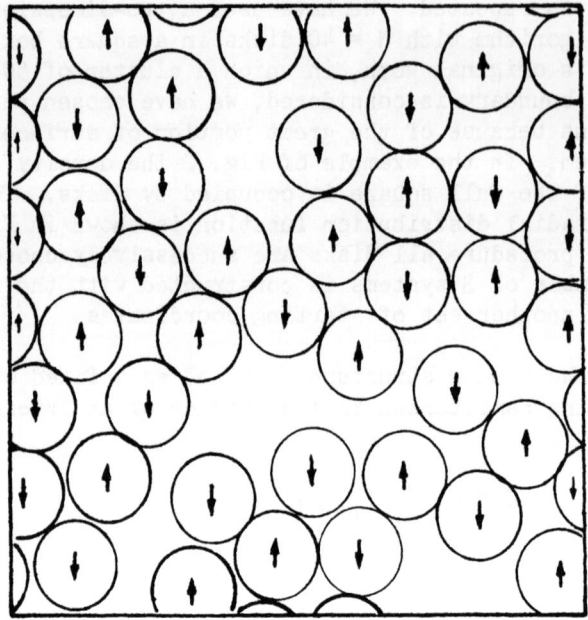

Fig. 2. Ground state of a computer-simulated amorphous antiferromagnetic Ising system with periodical boundary conditions.

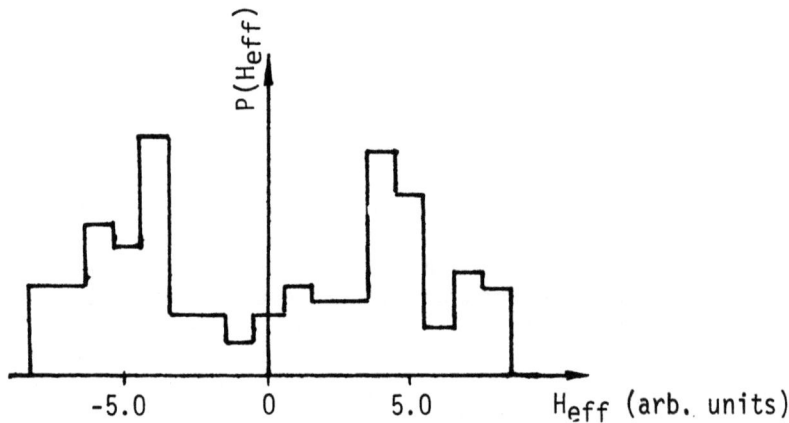

Fig. 3. Distribution of the effective fields H_{eff}.

that of an ideal system without any misfit (E_{id}), a global misfit parameter

$$m = 1 - \frac{E_o}{E_{id}}$$

is proposed[7]. This parameter m is summarized in Table 1 for various crystalline lattices with nearest neighbor antiferromagnetic interaction and computer-simulated amorphous (csa) systems, where

$$\alpha = \frac{\text{range of magnetic interaction}}{\text{hard core diameter}} .$$

Table 1

Dimension	System	m
2	square	0.0
3	body-centered cubic	0.0
2	csa[7], N = 23, ρ = 0.49, α = 1,39	0.31 ± 0.04
2	csa, N = 40, ρ = 0.68, α = 1,36	0.39 ± 0.02
3	csa, N = 30, ρ = 0.58, α = 1,17	0.56
3	face-centered cubic (1st type)	0.67
2	triangular	1.0

Despite the simplifications of our model and the smallness of the investigated computer-simulated random packing clusters, we have found information on amorphous systems with antiferromagnetic interactions and suggested possible descriptions. However, we cannot directly predict the existence of "long-range ordered" amorphous antiferromagnetism and transition temperatures.

The author would like to thank Dr. K. Handrich and Dr. H. Wonn for useful discussions.

References

1. A. W. Simpson, Phys. Stat. Sol. 40, 207 (1970).
2. S. Kobe and K. Handrich, Phys. Stat. Sol. 42, K69 (1970).
3. H. Sato and R. Kikuchi, in: AIP Conf. Proc. No. 18, Part 1, Magnetism and Magnetic Materials, 1973 (C.D. Graham, Jr. and J. J. Rhyne, Eds.) American Institute of Physics, New York 1974 (p. 605).
4. T. Egami, O. A. Sacli, A. W. Simpson, A. L. Terry and F. A. Wedgwood, in: Amorphous Magnetism (H. O. Hooper and A. M. de Graaf, Eds.), Plenum Press, New York/London 1973 (p. 27).
5. A. W. Simpson, Wiss. Z. d. Techn. Univ. Dresden 23, 1029 (1974).

6. R. W. Cochrane, R. Harris and M. Plischke, J. Non-Crystalline Solids 15, 239 (1974).
7. S. Kobe and K. Handrich, Phys. Stat. Sol. (b) 73, K65 (1976).
8. M. C. Chi and R. Alben, Joint Magnetism and Magnetic Materials-Intermag. Conf., 1976, Pittsburgh (to be published in AIP Conf. Proc. and IEEE Trans. Magn.).
9. J. L. Finney, Materials Science and Engineering 23, 199 (1976).
10. S. Kobe and K. Handrich, Phys. Stat. Sol. (b) 54, 663 (1972).
11. J. F. Sadoc, J. Physique 36, C2-75 (1975).
12. J. J. Rhyne, S. J. Pickart, and H. A. Alperin, see Ref. 4 (p. 373).

Magnetic Resonance and Glass Structure[*]

G. E. Peterson

Bell Laboratories, Murray Hill, New Jersey 07974

Detailed glass structure determination by NMR or EPR is by no means easy. It is convenient to think of 3 individual coordinate frames (see Fig. 1). Frame 1 is the glass structure frame in which we consider the atomic arrangements such as bond lengths and angles. Frame 2 contains the Hamiltonian parameters such as g values, A values, quadrupole coupling constants, etc. In Frame 3 one of the axes is the resonance frequency. In amorphous materials, because of the randomness, we expect the parameters in each frame to be described by a joint probability density function. If random vectors $\vec{x}$ and $\vec{y}$ are related by a 1:1 transformation:

$$\vec{y} = \vec{f}(\vec{x}); \quad \vec{x} = \vec{g}(\vec{y}) \tag{1}$$

the density functions are related by [1]:

$$p_{\vec{y}}(\vec{\beta}) = p_{\vec{x}}[\vec{g}(\vec{\beta})] \cdot |J_{\vec{g}}(\vec{\beta})| \tag{2}$$

where $|J_{\vec{g}}(\vec{\beta})|$ is the absolute value of the Jacobian of the transformation $\vec{g}$. Thus we obtain the following very general expression for a magnetic resonance line shape.

$$p_\nu(\gamma) = \int \cdots \int p_{\vec{x}}[\vec{g}(\vec{\beta})] \cdot |J_{\vec{g}}(\vec{\beta})| \, d\beta_1 \cdots d\beta_{K-1} \tag{3}$$

The integration in the above expression removes the unwanted variables that are left after the joint density function is transformed.

[*] Invited paper.

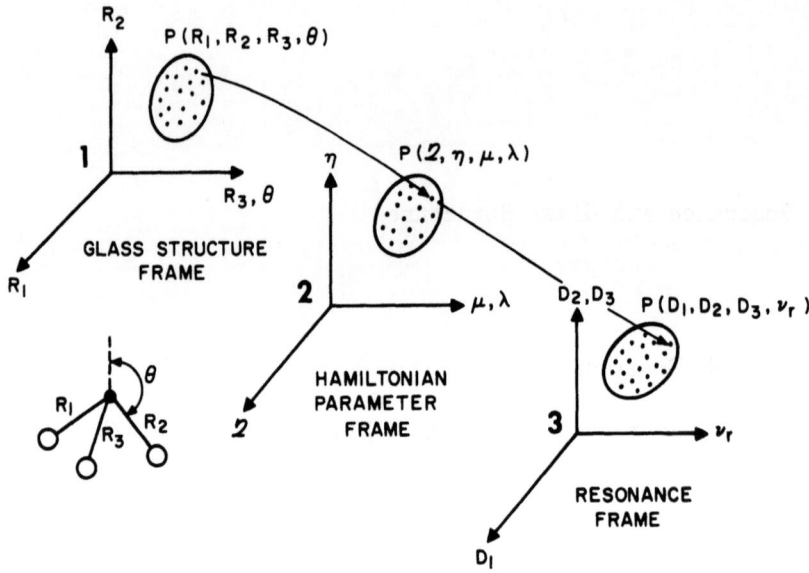

Figure 1

The three reference frames for glass structure analysis using magnetic resonance.

The transformation from Frame 1 to Frame 2 is difficult and simple formulae tend to go astray. A good example is the failure of the Townes and Dailey theory to satisfactorily account for the change in quadrupole coupling constant with molecular distortion [2]. This has forced us to resort to SCF-MO methods whenever possible. The basis functions χ we employ are contracted functions built from Gaussian primitive functions η,

$$\chi_i = \sum_a c_{ia} \eta_a(l, m, n; \alpha) , \qquad (4)$$

where the Gaussian primitive functions η_a are given by:

$$\eta_a(l, m, n; \alpha) = N_a x_a^l y_a^m z_a^n \exp(-\alpha\, r_a^2). \qquad (5)$$

The transformation from Frame 2 to Frame 3 is fairly easy as it only requires a knowledge of the resonance equations. These are usually found in the literature. For example, Fig. (2) shows a family of second order, $I = 3/2$ NMR line shapes and their derivatives for some skewed Gaussian distributions in quadrupole coupling constant. The line shape formula for this case [3] as calculated by Eq. (1) is:

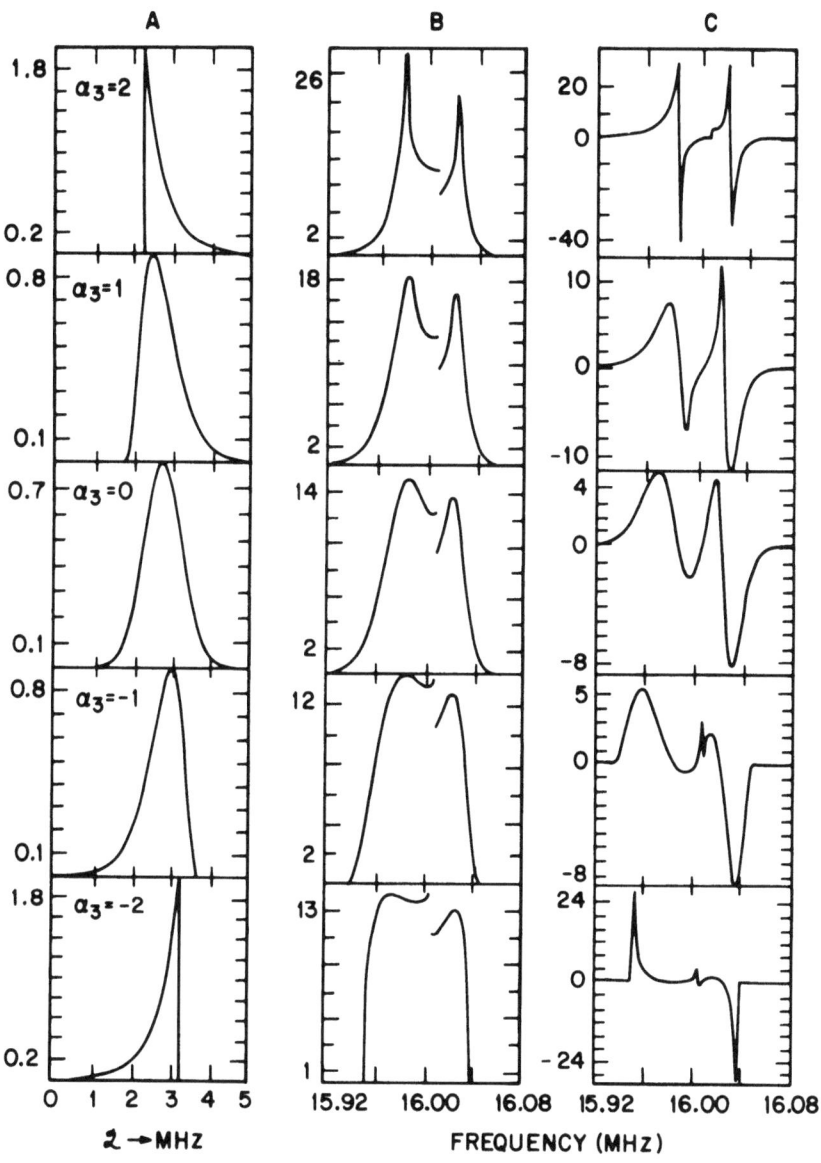

Figure 2

A family of NMR line shapes for skewed Gaussian distributions using Eq. (6). This exact expression incorporates the generalized probability curve Type III of the English biometrician Kark Pearson. A is the skewed density, B is the line shape and C is the derivative of the line shape.

$$p_\nu(\gamma) = \frac{1}{\sigma\sqrt{2\pi}} \int_{\mu=0}^{\mu=1} \frac{(1+\alpha_3 t/2)^{4/\alpha_3^2-1} e^{-2t/\alpha_3}}{\left[\left(\frac{\nu_\ell-\gamma}{\nu_\ell}\right)\left(\frac{3}{64}\right)(1-\mu^2)(9\mu^2-1)\right]^{1/2}} d\mu \quad (6)$$

where:

$$t = \left[\left\{\frac{64\nu_\ell(\nu_\ell-\gamma)}{3(1-\mu^2)(9\mu^2-1)}\right\}^{1/2} - \ell_0\right]\bigg/\sigma$$

$\alpha_3 = 2(\overline{X}-M_o)/\sigma$ = skewness

$\overline{X}$ = arithmetic mean

M_o = mode

It is interesting to note that it is possible to build analogue machines [4] to transform density functions. Suppose we have an electrical noise source whose distribution function (not density function) is $F_1(x)$. Further suppose we wish to convert this into a noise source with a distribution function $F_2(x)$. The required transformation [5] (electrical analogue) is:

$$f(x) = F_2^{-1}[F_1(x)] \quad (7)$$

where F_2^{-1} is the inverse function of F_2. Machines of this sort may have an educational value.

We may also view Eq. (3) as an integral equation for $p_x^{\rightarrow}(\vec{\xi})$. In particular, it is a Fredholm integral equation of the first kind; with all of its well known difficulties [6]. If we algebraize Eq. (3) using some quadrature rule and then employ a stacking operator it becomes a matrix equation. Thus:

$$\vec{\nu} = A\vec{\phi} \quad (8)$$

MAGNETIC RESONANCE AND GLASS STRUCTURE

Here $\vec{\nu}$ is a vector whose components are samples of $p_\nu(\gamma)$ and $\vec{\phi}$ is a vector sampling $p_{\vec{x}}(\vec{\xi})$. We may solve Eq. (8) for $\vec{\phi}$. The result is:

$$\vec{\phi} = \overset{\leftrightarrow}{A}\vec{\nu} \qquad (9)$$

where A^+ is the pseudo inverse of A.

A useful representation of A^+ is in terms of the singular values [7] and vectors of A. If U_i and V_i are the singular vectors associated with the singular value μ_i we have:

$$A = \sum \mu_i U_i V_i^T \qquad (10)$$

and this leads to:

$$A^+ = \sum \frac{1}{\mu_i} V_i U_i^T \qquad (11)$$

The summation in both cases extends up to the pseudo rank [8] of A. The singular values μ_i are all positive and tend monotonically to zero with increasing index i. For this reason the operation specified by A^+ is antismoothing [9]. To obtain a stable solution it is advisable to keep the pseudo rank as low as possible.

To illustrate the above procedure we consider the elementary but important case of B^{11} with $\eta = 0$ and only the distribution in quadrupole coupling constant unknown. We assume that the NMR frequency is 16.0 MHz and that the dipolar broadening is a Gaussian random variable with $\sigma = .004$ MHz. In this case $\vec{\phi}$ represents the distribution in quadrupole coupling constant we wish to solve for. It contains 36 samples covering the range from .05 MHz to 3.55 MHz in .1 MHz intervals. Figure (3) shows the family of singular vectors through index 10 for this case and Fig. (4) the family of functions from which they were calculated. We now apply Eq. (7) to some experimental data for B_2O_3 and obtain the pseudo rank 10 solution shown in Fig. (5). Notice that it peaks as expected at 2.6 MHz and has a roughly Gaussian shape. If we change the pseudo rank we get additional solutions. A rank 11 and rank 16 solution are shown in Fig. (6).

The following two operators [10] are useful in understanding why there are many solutions.

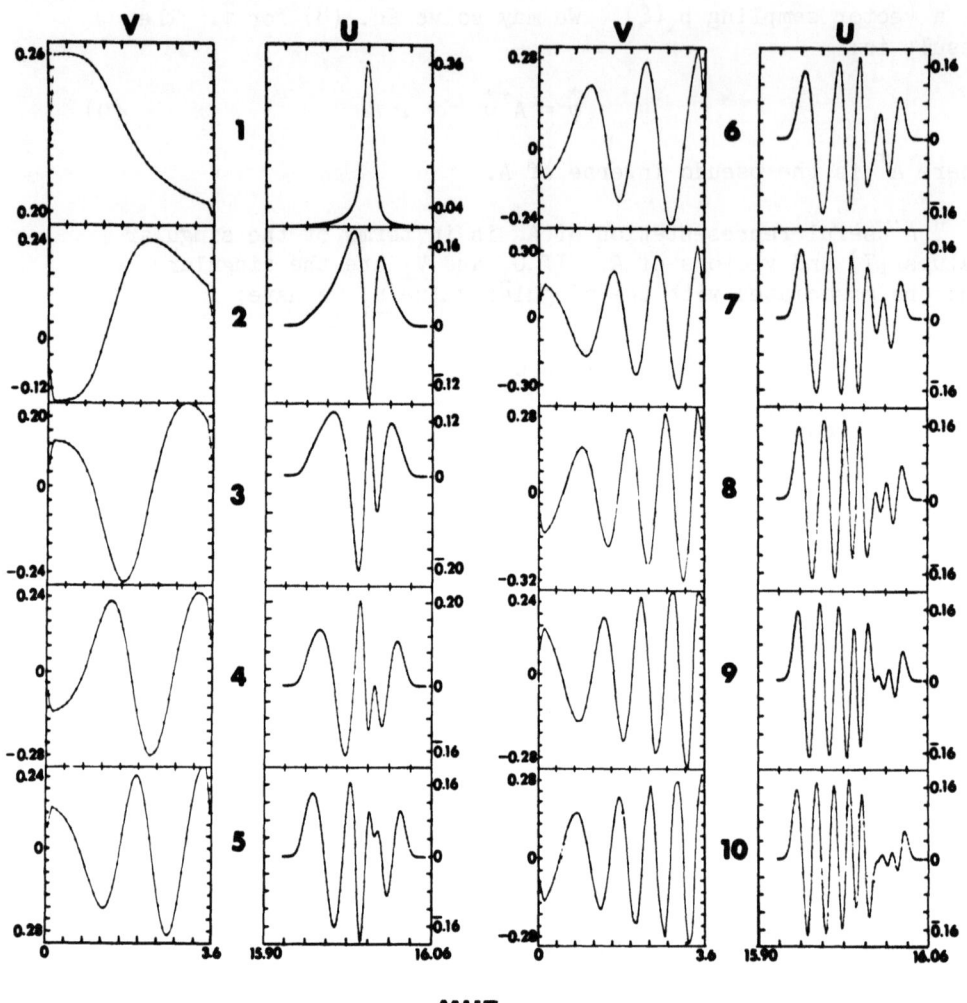

N	SINGULAR VALUE
1	97.6185
2	40.9875
3	23.9284
4	16.4297
5	11.8905
6	8.7599
7	6.4544
8	4.7416
9	3.4526
10	2.4894

Figure 3

Singular values and vectors for the NMR problem discussed in the text. They are the eigenvalues and eigenvectors of the self adjoint operators AA^+ and A^+A.

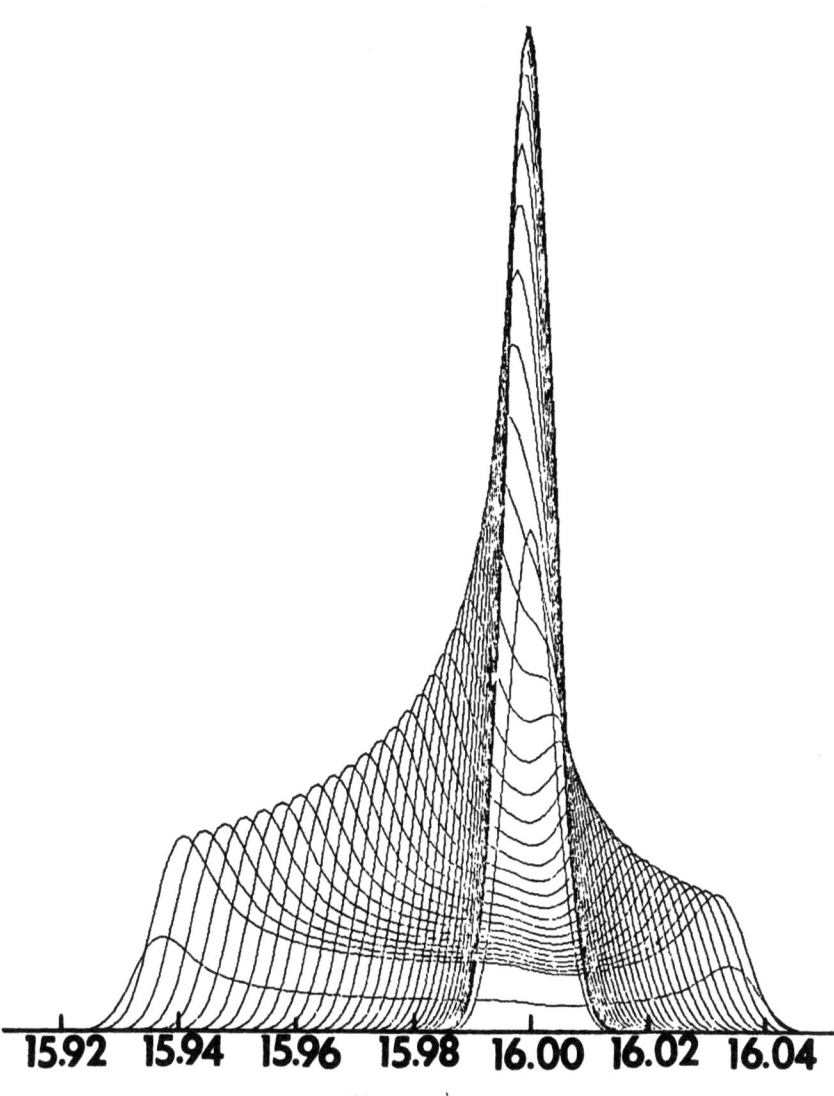

Figure 4

The family of functions stored in the columns of matrix A for the NMR problem discussed in the text. They are the result of applying the trapezoidal quadrature rule to Eq. (3). Each function is sampled at 137 points.

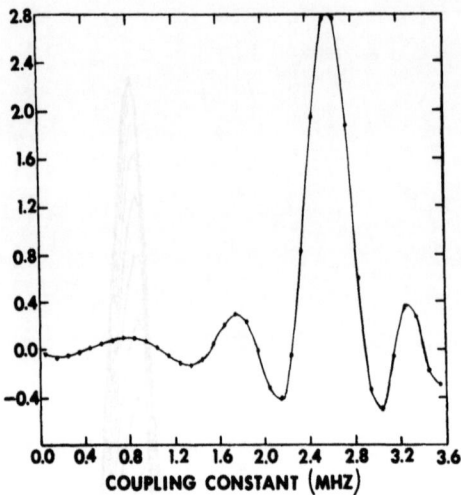

Figure 5

Pseudo rank 10 solution for B_2O_3. The function peaks at 2.6 MHz and has some background rumbles that may or may not be real.

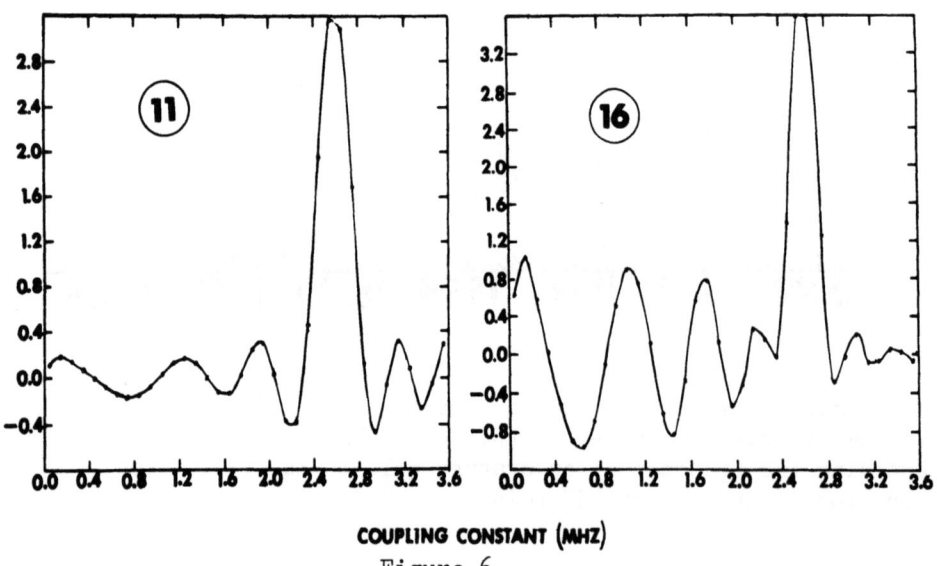

Figure 6

A pseudo rank 11 and a pseudo rank 16 solution.

$$P_R = A^+A \qquad (12)$$

$$P_C = AA^+ \qquad (13)$$

They are the projections into the row and column space of A respectively. In effect they tell us what part of $\vec{\phi}$ or $\vec{v}$ is invertable. Figure (7) shows a $\vec{\phi}$ vector and its pseudo rank 10 projection. Notice in particular the loss of apparent structure. The projection into the null space is given by:

$$P_N = I - A^+A \qquad (14)$$

and this is where the missing structure lies. There is no easy way to uniquely recover the missing structure. Consequently we have a rich family of solutions.

It is useful to estimate the amount of information contained in matrix A. Let each column be considered a vector:

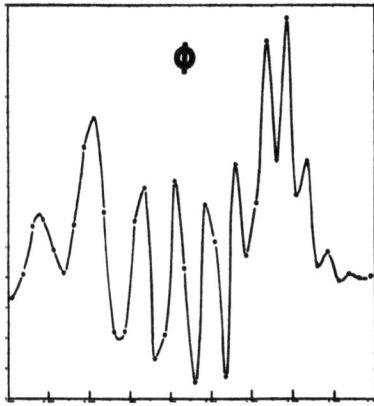 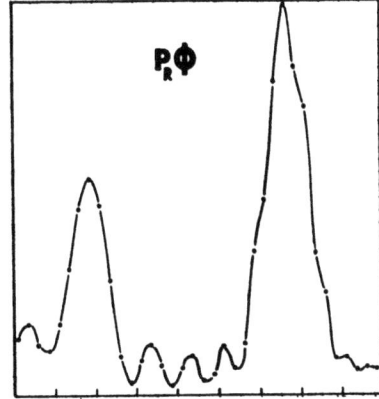

Figure 7

The matrix A splits an arbitrary $\vec{\phi}$ into two parts. One part lies in the row space and the second part in the null space. The part in the null space is transformed to zero. There may be considerable difficulty in uniquely recovering the part in the null space.

$$[c_1 \ldots \ldots c_j]^T \qquad (15)$$

We define the following probability measure [11]:

$$p_j = E[c_j^2]/\sum_j E[c_j^2] \qquad (16)$$

where $E[\ ]$ denotes expectation over all columns. The evenness of the distribution over the coordinates can be expressed by the entropy:

$$S = \sum_j p_j \log p_j \qquad (17)$$

We now seek a new frame in which the information is compressed into as few components as possible. This is the minimum entropy frame and is given by [11]:

$$AA^T W_K = \mu_K^2 W_K \qquad (18)$$

Here $W_1 \ldots W_K$ are the basis vectors for the new frame and the ordering is in terms of the magnitude of μ_K^2. The fraction of the total variance accounted for by using only g of the total j coordinates is:

$$\mu_1^2 + \ldots \mu_g^2 / \mu_1^2 + \ldots \mu_j^2 \qquad (19)$$

In the NMR example just considered $\vec{\phi}$ had 36 components which seemed necessary to give adequate resolution. If we apply Eq. (19) we find that 99% of the total variance is accounted for by 6 components. This suggests that we are asking for too much information and this leads to nonuniqueness.

The method of analysis using a pseudo inverse operator is a departure from the more conventional methods. Usually one assumes some parameters, calculates a line shape and then compares with experiment. We tend to stop when a good fit is obtained and we normally do not methodically search for all possible good fits. It is very natural to start with a set of parameters that seem reasonable. In doing so, however, we bias our solution toward those parameters. This is not necessarily bad. In fact it might be totally justifiable if we have reliable information from other sources.

When we apply the pseudo inverse operator to the experimental data we get many solutions. In fact the richness of solutions tends to be rather overwhelming. It is obviously very hard to be certain which solutions to reject.

When one starts with a set of parameters and shows they fit the experiment there is a feeling of satisfaction. When one has 50 solutions and throws away 49 there is more than a little dissatisfaction. The pseudo inverse techniques have increased our awareness of our limited knowledge of glass structure and have emphasized the necessity of combining many different forms of spectroscopy in the analysis of amorphous material. It has the disadvantage that more accurate calculations, and greater computer speed and storage space are required.

From other studies, for example X-ray or I.R. we may be able to ascertain that solution $\vec{\phi}$ should be close to some known vector $\vec{W}$. We can then modify Eq. (8) to express this preference. We get [12]:

$$\begin{bmatrix} A \\ \lambda I \end{bmatrix} \vec{\phi} = \begin{bmatrix} \vec{V} \\ \lambda \vec{W} \end{bmatrix} \qquad (20)$$

where I is a unit matrix and λ is a scaling parameter indicating the "intensity" of this preference. Sometimes we may have statistical information about $\vec{W}$, for example a covariance matrix. We can derive similar relations incorporating this data into the solution.

We are currently investigating two novel types of EPR measurements [13] aimed at providing additional information about glass structure. They both are electron spin echo studies on divalent copper at 9.3 GHz and 4.2°K. The first is a recording of the envelope of electron spin echoes as a function of time between the echo generating pulses. This yields information on the coupling between copper electron spins and nuclei in the immediate environment. Figure (8) shows some data for $Na_2O \cdot 2B_2O_3:Cu$. The peak near .25 μsec is coupling with Na. There are two additional characteristic patterns, one associated with B^{10} and one with B^{11}. The most striking feature is the persistence of the B^{11} period. This suggests a high degree of symmetry for the boron. The second measurement is based upon the linear electric field effect [13] in EPR. It affords a means of studying the symmetry of the complex formed by the paramagnetic ion and its ligands.

The combination of electric field E and time τ which halves the echo amplitude provides a convenient way of characterizing the shift and of defining the shift parameter σ.

$$\sigma = 1/[6\nu(E\tau)_{1/2}] \qquad (21)$$

The most informative result to date is that the paramagnetic complexes in a variety of copper doped glasses are noncentrosymmetric.

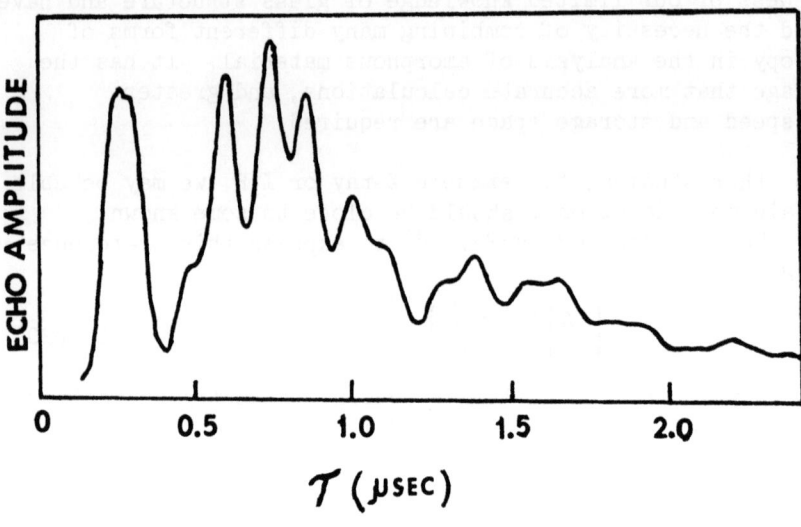

Figure 8

Envelope of electron spin echoes as a function of the time between the echo generating pulses. The sample is $Na_2O \cdot 2B_2O_3$:Cu.

References

1. J. M. Wozencraft and I. M. Jacobs, "Principles of Communication Engineering", (Wiley and Sons, New York, 1967) p. 111; G. E. Peterson, C. R. Kurkjian and A. Carnevale, Physics and Chemistry of Glasses 15, 52 (1974).

2. Lawrence C. Snyder, G. E. Peterson and C. R. Kurkjian, J. Chem. Phys. 64, 1569 (1976).

3. A. Carnevale, unpublished; See also G. E. Peterson, C. R. Kurkjian and A. Carnevale, Physics and Chemistry of Glasses 15, 59 (1974).

4. G. E. Peterson, C. R. Kurkjian and A. Carnevale, Physics and Chemistry of Glasses, August 1976 Issue.

5. Petr Bechmann, "Orthogonal Polynomials for Engineers and Physicists" (Golem Press, Colorado, 1973) p. 179.

6. V. F. Turchin, V. P. Kozlov and M. S. Malkevich, Soviet Phys. Usp. 13, 681 (1971).

7. I. J. Good, Technometrics 4, 823 (1969).

8. G. Golub and W. Kahan, J. SIAM Numer. Anal. 2, 205 (1965). See also Ref. 12, p. 77.

9. L. M. Delves and J. Walsh, "Numberical Solution of Integral Equations" (Clarendon Press, Oxford (1974) p. 179.

10. C. A. Desoer and B. H. Whalen, J. Soc. Indust. Appl. Math 11, 442 (1963).

11. S. Watanabe, "Karhunen-Loeve Expansion and Factor Analysis, Theoretical Remarks and Applications", 4th Prague Conference on Information Theory, p. 635 (1965).

12. Charles L. Lawson and Richard J. Hanson "Solving Least Squares Problems" (Prentice-Hall, 1974) p. 188.

13. W. B. Mims, G. E. Peterson, C. R. Kurkjian, Physics and Chemistry of Glass, to be published; W. B. Mims, Phys. Rev. 133, A835 (1964).

DISCUSSION

N. Y. Rivier: Can you clarify the meaning of a pseudo-inverse?

G. E. Peterson: We are dealing here with a rectangular matrix which should have no inverse. What we are considering is an operator that acts like an inverse but does not give you back a unique solution.

N. Y. Rivier: Is not there the possibility of predicating the missing rows by minimizing the null space in the matrix?

G. E. Peterson: What you are saying is correct; the important part here is to combine other aspects of spectroscopy to reduce this null space somehow, otherwise we do not know what the answer is.

Questioner: What is not the size of the null space related to the size of the number of rows you are throwing out?

G. E. Peterson: There are two types of null spaces. There is the mathematical null space mathematicians talk about where everything is perfectly known in the matrix, and there is, as in our case, the second type of numerical matrix with a fuzzy type of null space; so that one has to make a decision about the size of the null space. The important question here is how many parameters really describe all the parts of the matrix. The matrix we considered had 36 parts which after careful mathematical manipulation boils down to 6 parameters. Since we are asking for 36 we really have a null space of 30 and one gets easily misled if one tries to extract more information than is really there.

A. Bishay: How can Mössbauer spectroscopy be of any specific help in solving the problem?

G. E. Peterson: I am not saying that it will solve the problem, I am simply suggesting that what we want to do is combine various sorts of spectroscopy to extract complementary information which should improve our understanding of the structure of complex glasses.

Magnetic Susceptibility and EPR Studies of

Reduced Titanium Phosphate Glass

C. H. Perry,* D. L. Kinser[Φ] and L. K. Wilson[Φ]

* IBM Corporation, Kingston, New York

[Φ] Vanderbilt University, Nashville, Tennessee

INTRODUCTION

The magnetic properties of Ti^{3+} in glass has not been extensively investigated. There has been little reported research on the magnetic properties of glasses high in Ti^{3+} ions. Most papers report on glasses in which the Ti^{3+} ions are induced by irradation, produced by preparing the glass under a reducing atmosphere, or due to the slight reduction of Ti^{4+} to Ti^{3+} at the melt temperature. Arafa and Bishay [1] were among the first to study the nature of radiation induced Ti^{3+} ions in borate glass using EPR techniques. Kim and Bray [2] also reported on gamma-irradated alkali titanate glasses. EPR studies of Ti^{3+} ions in silicate and phosphate glasses have been reported by Yafaev and Yablokov [3]. Ti^{3+} was induced in these glasses by melting the batch materials in a strongly reducing atmosphere.

In this study, six glasses have been prepared which contain a high mole per cent of Ti_2O_3 and are part of a ternary glass system formed by TiO_2, Ti_2O_3 and P_2O_5. (Other glasses of this system have also been investigated and previously reported.) [4,5] Titanium oxide stoichiometry in the glasses was chosen to be approximately the same as some of the titanium oxide Magneli phases (6). The glasses' magnetic properties were studied for evidence of anomalous behavior similar to that observed in the crystalline titanium oxide Magneli phases (7). Glass composition is given by $XTiO_y - P_2O_5$ where X is the mole ratio of total titanium oxide to P_2O_5, and y gives the titanium oxide stoichiometry.

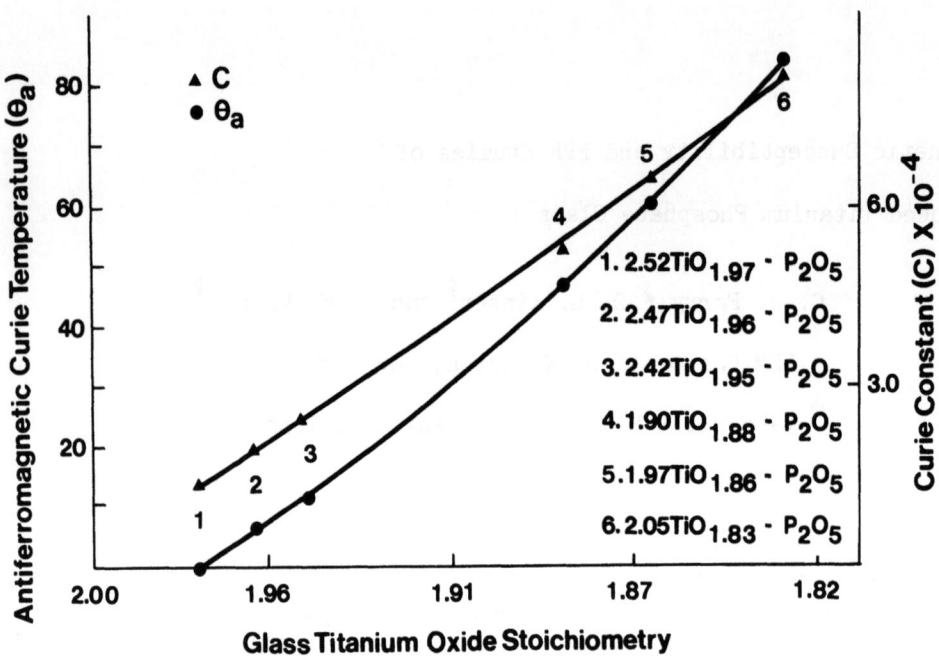

FIGURE 1 Antiferromagnetic Curie temperature and Curie constant as a function of glass titanium oxide stoichiometry.

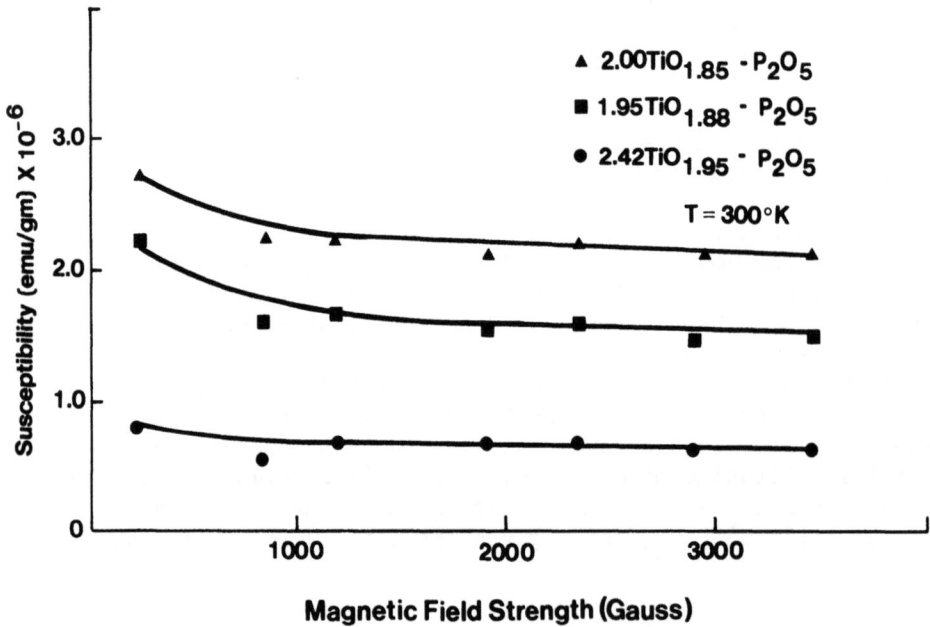

FIGURE 2 Field dependence of magnetic susceptibility at 300°K.

Experimental

The glasses were prepared by combining physical mixtures of reagent grade $(NH_4)H_2PO_4$, TiO_2 and $TiO_{1.56}$ using an alumina mortar and pestle (the $TiO_{1.56}$ oxide was prepared by heating appropriate weights of powdered titanium metal and TiO_2 at $1000°C$ under vacuum for 24 hours). The weight of a batch mix varied from 37 to 40 grams depending on the stoichiometry. All mixtures were heated for 10 minutes at $1350°C$ in covered silica crucibles. After melting, the molten glasses were quenched on water cooled copper blocks. Crucibles, tops and contents were weighed before and after melting to estimate P_2O_5 loss during melting. All "ascast" glasses were analyzed for Ti^{3+} concentration using the oxidation analysis method [8].

Field and temperature dependent magnetic susceptibility was determined using the Faraday method. All glasses generally obeyed a straight-line Curie-Weiss law with negative temperature intercepts, indicating antiferromagnetism. One exception is the least-reduced glass which, within experimental error, gives evidence of paramagnetism. The high temperature (77K to 500K) magnetic susceptibility data is summarized in Figure 1. For all glasses, the antiferromagnetic Curie temperature and Curie constants vary smoothly through the range of titanium oxide stoichiometry. The variation of θ_A and C for the oxidized glasses is similar to the variation of these parameters in the reduced glasses. This implies that any structural differences that may exist between the oxidized and reduced glass do not effect θ_A or C.

Figure 2 shows the variation of the room temperature magnetic field for some of the reduced glasses. As the data shows, χ is sensitive to the magnetic field and generally decreases as H increases. This effect was not observed in the less reduced glasses. It is interesting to note that most of the variation of χ with field occurs within the range of 0 to 1000 gauss. Such variations are not usually observed in antiferromagnetic crystalline compounds.

Low temperature magnetic susceptibility of all glasses indicates behavior typical for amorphous antiferromagnet materials [9]. Figure 3 shows that the inverse susceptibility of all glasses, regardless of composition, departs from Curie-Weiss behavior and bends toward the origin as 0K is approached.

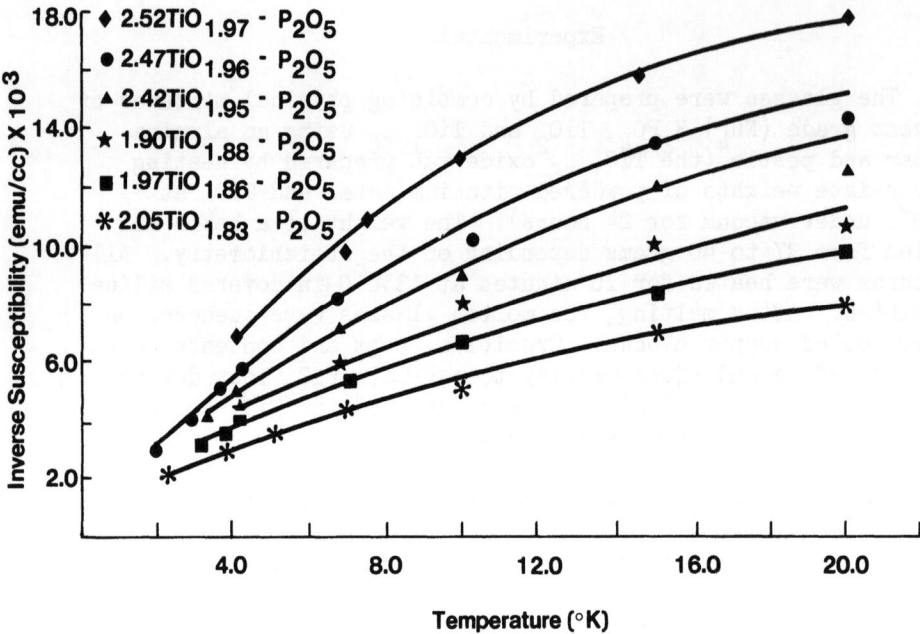

Figure 3 Low temperature inverse magnetic susceptibility as a function of temperature for titanium phosphate glasses.

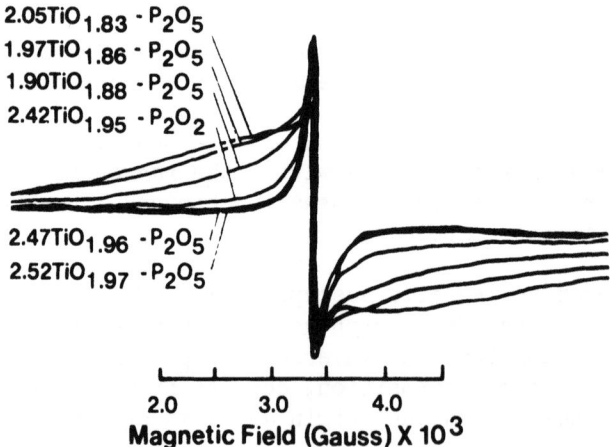

Figure 4 Room temperature electron spin resonance spectra for titanium phosphate glasses (all spectra g = 1.93 ± 0.01).

Temperature dependent ESR measurements were made on the glasses from 77K to 600K on an X-band spectrometer. In general, the spectra were asymmetric singlets with g values of 1.93 ± 0.01. The lineshape and g values of the various glasses do not change significantly with composition or temperature. Figure 4 shows the room temperature spectra of the glasses. The wings of the line broaden with increasing Ti^{3+} concentration, but the peak-to-peak line width is constant with composition. The room temperature spectra of the more reduced glasses appeared to be the sum of two lines. When properly scaled, the spectrum of the least reduced glass, $2.52TiO_{1.97} - P_2O_5$, could be subtracted graphically from each spectrum of the more reduced glasses. This permitted all the spectra of the more reduced glasses to be resolved into two components. Figure 5 shows the resolved room temperature spectra of the most reduced glass. For the narrow components of the ESR spectra in all glasses, the linewidth, ΔH, is 100, 108, and 120 gauss at 600K, 300K and 77K respectively. The broad components of the ESR spectra show an opposite temperature variation as the data in Figure 6 indicates.

DISCUSSION

The resolution of the EPR spectra was accomplished by postulating that all the glasses contained isolated Ti^{3+} ions, and the spectrum of these isolated ions is observed in the least reduced glass. Then, an appropriately scaled spectrum of the least reduced glass was subtracted from the spectra of all the more reduced glasses. This technique assumes that the EPR spectrum of the least reduced glass is due primarily to isolated Ti^{3+} ions. This assumption seems justified from the magnetic susceptibility data. As shown in Figure 1, magnetic susceptibility of the least reduced glass gives evidence for paramagnetic behavior. Therefore, the more reduced glasses have Ti^{3+} ions in two different environments. Isolated Ti^{3+} ions in the glass produce a relatively narrow EPR line that shows a slight broadening as the temperature decreases. Ti^{3+} ions in a different environment produce a broad EPR line that narrows substantially as the temperature decreases.

Garifyanov et al., [10] investigated Ti^{3+} in composite silicate glasses using electron spin resonance. X-band EPR at 300K and 77K produced asymmetric singlets with peak-to-peak linewidths of 71 gauss and g values of 1.946. Yafaev and Yablokov obtained different results for Ti^{3+} ions in composite silicate and phosphate glasses. X-band EPR measurements at room temperature and 77K showed that both the peak-to-peak linewidth and g values for the Ti^{3+} ions in silicate glass changed with temperature.

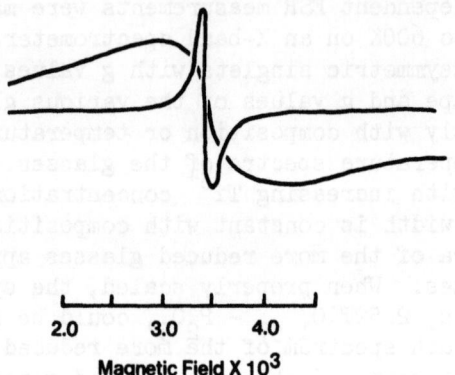

FIGURE 5 Resolved electron spin resonance spectrum for $2.05\text{TiO}_{1.83}\text{-}P_2O_5$ glass. $T = 300°K$, $g = 1.93 \pm 0.01$.

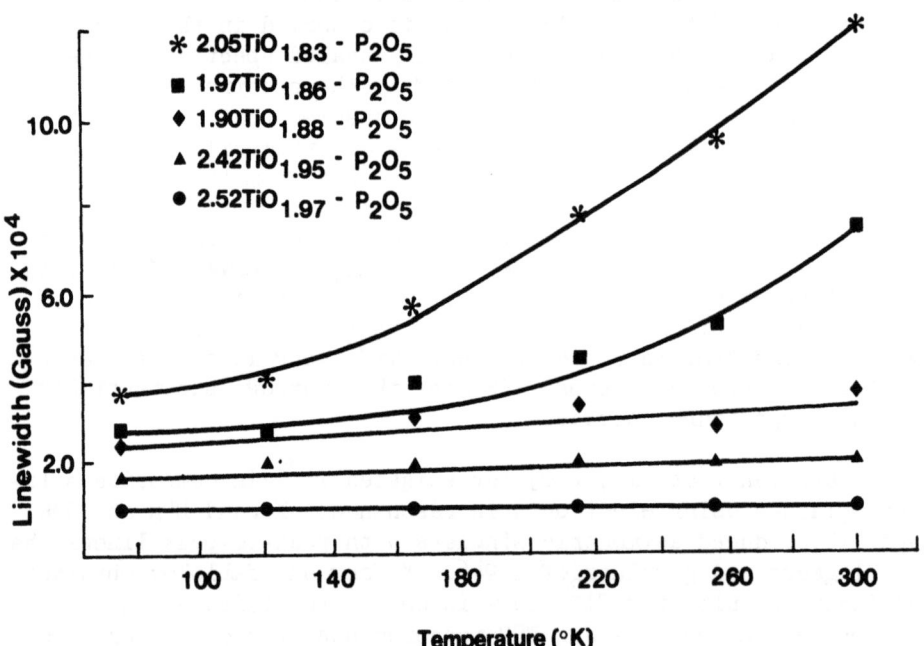

FIGURE 6 Variation of peak-to-peak EPR linewidth with temperature for the broad component of the glass EPR spectra.

The phosphate glasses containing Ti^{3+} ions gave a spectra at 77K, but no spectra could be observed for the phosphate glasses at 300K. The EPR spectra for all the glasses were asymmetric singlets with the asymmetry more pronounced in the phosphate glasses.

These results of electron spin resonance measurements for Ti^{3+} ions in silicate and phosphate glasses give some general properties that can be applied to this research. Ti^{3+} ions in these glasses were induced by preparing the glasses under reducing conditions. The Ti^{3+} ions in the $2.52TiO_{1.97} - P_2O_5$ glass also were produced by reduction of TiO_2 at the melt temperature. It is not always necessary to melt the TiO_2 containing glass in a reducing atmosphere. For some glasses, a fraction of the Ti^{4+} ions are spontaneously reduced to Ti^{3+} at the melt temperature [11].

In the previous studies discussed above, the variation of EPR linewidth with temperature was either constant [10] or increased at lower temperatures [3]. This behavior is generally indicative of spin-spin relaxation as the dominant relaxation mechanism. Spin-spin relaxation is usually independent of temperature. When spin-spin interactions manifest a temperature dependence, the EPR linewidth broadens as the temperature decreases. In order for the spin-spin relaxation mechanism to dominate, the spin-lattice relaxation time must be long [12]. This can happen when the octahedral ligand field surrounding the Ti^{3+} ion is highly distorted. Arafa and Bishay illustrated this in their study of Ti^{3+} produced in borate glass by melting the glass in a reducing atmosphere [1]. By optical absorption measurements, they determined the splitting between the E_g and T_{2g} d orbitals of the Ti^{3+} ion to be on the order of 20,000 cm^{-1}. Optical absorption measurements gave evidence for a high degree of tetragonal distortion. High tetragonal distortion would, therefore, indicate that the splitting of the lower T_{2g} triplet would be large and this would produce a long relaxation time, which is indicated by Arafa and Bishay's data.

The narrow component of the electron spin resonance spectra for the reduced titanium phosphate glasses is probably due to isolated Ti^{3+} ions in a highly distorted octahedral ligand field. Using an equation given by Yafaev and Yablokov and the measured g value, the magnitude of lower T_{2g} orbital triplet splitting is estimated to be 4516 cm^{-1}. A splitting of this magnitude could produce a long spin-lattice relaxation time. As discussed earlier, the variation of EPR linewidth with temperature of the narrow line is almost constant. It does increase slightly, however, as the temperature decreases. This does indicate some evidence of spin-spin dipolar interaction. The width and shape of the narrow EPR line is attributed to anisotropy in the g value since anistropic g in a randomly oriented system can produce a broadened unsymmetric EPR line.

The broad EPR line of the more reduced glasses shows a temperature variation that is quite unusual. The EPR linewidth data shown in Figure 6 are not in accord with previously reported data on Ti^{3+} ions in glass [3, 10]. However, the data are similar to EPR data for the crystalline titanium oxide Magneli phases. EPR measurements on all the titanium oxide Magneli phases show that the peak-to-peak linewidths of the spectra all decrease with decreasing temperature [13, 14, 15]. This behavior is interpreted as an evidence for spin-lattice relaxation being the dominant mechanism in the crystalline titanium oxides.

In order for spin-lattice relaxation to dominate, the average distortion of the octahedral environment surrounding the Ti^{3+} ions must not be too large. This distortion will split the T_{2g} orbital triplets and the magnitude of average splitting will not be large compared to KT. The EPR spectrum of such a system will have a peak-to-peak linewidth variation with temperature similar to that of the crystalline Magneli phases. Depending on the degree of distortion, the spectra will disappear at higher temperatures. In the crystalline titanium oxide Magneli phases all spectra disappear above 190K [13]. In crystals where the octahedral symmetry of the Ti^{3+} centers is high, T_{2g} orbital splitting is small. Thus, the EPR spectra can only be observed at very low temperatures [16]. If the average distortion of the octahedral ligand field is high, T_{2g} splitting will be large compared to KT. This produces long spin-lattice relaxation which is relatively temperature independent, and the spectra can be observed at high temperatures. In the titanium phosphate glasses the narrow EPR component can be observed at 600K but the broad EPR component begins to disappear at 400K.

A similarity does exist between the EPR data for crystalline titanium oxide Magneli phases and the broad component of the titanium phosphate glass spectra. When a sample of the most reduced glass was devitrified and studied by EPR, its peak-to-peak linewidth showed greater variation with temperature. Also, the EPR linewidth at room temperature was twice as wide as the glassy counterpart. Devitrification had reduced some of the octahedral distortion which reduced the average splitting of the T_{2g} orbital triplets. The decreased spin-lattice relaxation time is evident by the increased peak-to-peak EPR linewidth.

The Ti^{3+} ions giving rise to the broad component of the EPR spectra are in a distorted octahedral environment. The average distortion of the octahedral crystal fields is low enough to permit a significant variation of spin-lattice relaxation time with temperature. Yet the distortion is large enough to permit observing the spectra at room temperature. It is probable that structures similar to the titanium oxide Magneli phases are present in the glass. A Ti^{3+} ion located in such a structure would be subject to an octahedral crystal field with higher symmetry than the octahedral field around the isolated Ti^{3+} ions. This gives a reasonable explanation to the unusual EPR peak-to-peak linewidth variation with temperature. From this research, and previous studies on Ti^{3+} induced in glass by reducing conditions, it is apparent that the octahedral environment of the isolated Ti^{3+} ion is highly distorted. This implies that Ti^{3+} in glass produces a local strain field because its requirement for octahedral coordination is not compatible with the tetrahedral coordination of SiO_2 or P_2O_5. When the Ti^{3+} ion can be incorporated into a favorable octahedral symmetry, like a titanium oxide Magneli phase structure, the splitting of the Ti_{2g} triplet is not as great. Then, the EPR spectra shows the peak-to-peak linewidth variation shown in Figure 6.

While the variation of EPR peak-to-peak linewidth with temperature in the reduced glasses can be understood in terms of spin-lattice relaxation, the total width of the broad component of all temperatures needs explanation. The total linewidth of the most reduced glass was on the order of 10,000 gauss. Similar behavior has been observed for exchange coupled chromium ions in phosphate glass [17]. The authors attributed this behavior to variations in single ion axial crystal field interactions and anistropic exchange interactions. It is possible that the width of the broad component of the EPR spectra for the more reduced glasses is due to the same mechanisms. From magnetic susceptibility measurements, it is known that all the more-reduced glasses show evidence of exchange interactions between the Ti^{3+} ions. Since these materials are amorphous, it is not unusual for there to be considerable variation in site symmetry for the Ti^{3+} ions in the titanium oxide Magneli phase structure. It is also possible that the total linewidth of the broad component is partially due to ferromagnetic exchange between some of the Ti^{3+} ions. This is indicated by the data shown in Figure 2.

The narrow component of the EPR spectra corresponds to isolated Ti^{3+} ions in octahedral crystal fields of very low symmetry. To account for the variation of EPR peak-to-peak linewidth with temperature of the broad component, it is necessary to postulate a structure or environment for the Ti^{3+} ions that has higher symmetry. It is, therefore, suggested that the Ti^{3+} ions in the glass, that give rise to the broad component of the EPR spectra, are in a microstructural environment similar to a titanium oxide Magneli phase structure. This would provide octahedral crystal fields around the Ti^{3+} ions of higher symmetry which is evidenced by the EPR data.

Discussion of magnetic data has been devoted primarily to the electron spin resonance measurements on the glasses. A detailed theoretical analysis of the magnetic susceptibility data is in preparation and will be published at a later date.

REFERENCES

1. S. Arafa and A. Bishay, "ESR and Optical Absorption Spectra of Irradiated Borate Glasses Containing Titanium, "Phys. Chem. Glasses 11 (3), 75-82 (1970).

2. Y. M. Kim and P. J. Bray, "Electron Spin Resonance Studies of Gamma-Irradiated Alkali Titanate Glasses, "J. Chem. Phys. 53 (2), 716-723 (1970).

3. N. R. Yafaev and Y. V. Yablokov, "Electron Spin Resonance of Ti^{3+} in Some Silicate and Phosphate Glasses, "Soviet Physics-Solid State 4 (6), 1123-1127 (1962).

4. C. H. Perry, D. L. Kinser and L. K. Wilson, "Magnetic Properties of Titanium Phosphate Glass," Bull. Am. Ceram. Soc., 53 (4), 351 (1974).

5. C. H. Perry, D. L. Kinser and L..K. Wilson, "Magnetic Studies of a Titanium Oxide Phosphate Glass, "Bull Am. Phys. Soc., 18, 258 (1973).

6. S. Anderson, B. Collen, U. Kuylenstierna, and A. Magneli, "Phase Analysis Studies on the Titanium-Oxygen System, "Acta Chem. Scan. 11, 1641-1652 (1957).

7. L. K. Keys and L. N. Mulay, "Magnetic Susceptibility Measurements of Rutile and the Magneli Phases of the Ti-O System, "Phys. Rev. 154 (2), 453-456 (1967).

8. C. H. Perry, D. L. Kinser and L. K. Wilson, "A Quantitative Technique for Oxidation State Determination in Binary Transition Metal Oxide Glasses," J. Am. Ceram. Soc., 57, 227 (1974).

9. L. K. Wilson, E. J. Friebele, and D. L. Kinser, "Antiferromagnetism in the Vanadium, Manganese, and Iron Phosphate Glass Systems," in *Amorphous Magnetism*, H. O. Hooper and A. M. deGraaf, Ed., New York: Plenum Press, 1973, pp. 65-66.

10. N. S. Garifyanov, M. I. Rubstov, and Yu. M. Ryzhmanov, "Electron Paramagnetic Resonance (EPR) in Silicate Glasses Containing Trivalent Titanium," *Glass and Ceramics 20* (3), 11-12 (1963).

11. C. R. Kurkjian and G. E. Peterson, "An EPR Study of Ti^{3+} - Ti^{4+} in TiO_2 - SiO_2 Glasses, "*Phys. Chem. Glasses 15* (1), 12-17 (1974).

12. A. Carrington and A. D. McLachlan, *Introduction to Magnetic Resonance*, New York: Harper and Row, 1967, p. 194.

13. J. F. Houlihan and L. N. Mulay, "Characterization of and Electronic Structural Studies of the Oxides of Titanium: EPR linewidths," *Mat. Res. Bull. 6*, 737-742 (1971).

14. J. F. Houlihan and L. N. Mulay, "Electronic Properties and Defect Structure of Ti_4O_7: Correlation of Magnetic Susceptibility, Electrical Conductivity, and Structural Parameters via EPR Spectroscopy," *Phys. Stat. Sol. (b) 61*, 647-657 (1974).

15. J. F. Houlihan and L. N. Mulay, "Correlation of Magnetic Susceptibility, Electrical Conductivity, and Structural Parameters of Ti_3O_5 via EPR Spectroscopy," *Phys. Stat. Sol. (b) 65*, 513-519 (1974).

16. B. R. McGarvey, "Electron Spin Resonance of Transition-Metal Complexes," in *Transition Metal Chemistry Vol. 3*, R. L. Carlin, Ed., New York: Marcel Dekker, 1969, pp. 89-201.

17. J. T. Fournier, R. J. Landry, and R. H. Bartram, "ESR of Exchanged Coupled Cr^{3+} Ions in Phosphate Glass," *J. Chem. Phys. 55* (5), 2522-2526 (1971).

DISCUSSION

R. A. Levy: How do you explain the change occurring in the wings of your spectra as the composition of the glass is altered?

C. H. Perry: We attributed this line shape to a distribution of exchange parameters arising upon quenching from the locking in of antiferromagnetic interactions in the glass that reduce the intensity of the resonant line. As far as the total width of the line is concerned, we attributed this to an anisotropic exchange mechanism that has also been observed in exchange-coupled chromium ions in phosphate glasses.

D. J. Sellmyer: I did not quite understand one of your slides where you had the inverse susceptibility plotted against temperature and you said it curved down which indicated an antiferromagnetism behavior in these glasses.

C. H. Perry: That plot represented the low temperature data which is characteristic of amorphous antiferromagnetic systems such as what is often observed in iron-bearing glasses. The high temperature plot obeys the Curie-Weiss behavior.

D. J. Sellmyer: I am referring to the low temperature data. In any antiferromagnetic system that I know about the inverse susceptibility curves upward.

C. H. Perry: That is true, but not in amorphous antiferromagnetic systems in which one observes the characteristic downward curvature.

D. J. Sellmyer: But this is not observed in spin glasses though. There seems to be a fundamental difference between what is observed in metallic glasses and insulating glasses.

C. H. Perry: Quite right. There are several theories that claim to explain why the downward curvature occurs in these systems.

R. A. Levy: Did you attempt to examine the change in the oxidation state of titanium as you varied the composition of your glasses?

C. H. Perry: Yes, we established the effective oxidation of titanium by analyzing the glasses. By effective oxidation state we mean the mixture of the Ti^{3+} and Ti^{4+} ions.

L. N. Mulay: How did you distinguish from your analysis the two oxidation states?

C. H. Perry: By using the so-called oxidation analysis method which simply involves oxidizing the glass completely in a system where I can measure the oxygen uptake.

CHARACTERIZATION OF FERROMAGNETIC PRECIPITATES IN GLASSES BY
FERROMAGNETIC RESONANCE

E.J. Friebele and D.L. Griscom

Naval Research Laboratory, Washington, D.C. 20375

C.E. Patton

Colorado State University, Ft. Collins, Co. 80523

ABSTRACT

Ferromagnetic phases precipitated from iron-containing silicate glasses have been identified and characterized by studying the integrated intensity of the ferromagnetic resonance (FMR) as a function of sample temperature. The data show a completely different character for metallic iron precipitates and ferric iron spinel precipitates -- a result which is important because these phases may be indistinguishable in conventional magnetic experiments such as FMR lineshape, Mössbauer, static magnetic analysis, etc. The different behaviors make it possible to determine the relative amounts of the two phases if they are co-present in one sample. The data also show that the temperature variation of the microwave skin depth provides a way to estimate particle sizes for the precipitates.

INTRODUCTION

The ferromagnetic phases which are precipitated from iron-containing silicate glasses may be divided into 3 catagories -- fine particle metallic Fe, coarse particle Fe, and ferric iron spinels. The latter include both stoichiometric and nonstoichiometric Fe_3O_4 and "polluted" magnetite-like phases with other transition metal ions substituting for the iron ions in the spinel lattice.

In concept, it should be possible to discriminate between these three ferromagnetic phases by several methods. In practice, however, complications arise. Magnetite is stable below 600 C under even extreme reducing conditions,[1] so that it is impossible to determine a priori what phases have precipitated. The identification of the phases is further complicated because bcc metallic iron is often indistinguishable from ferric iron spinel phases by static thermomagnetic analysis.[2] The ferromagnetic resonance lineshape is also of little use since FMR spectra characterized by a positive magnetocrystalline anisotropy constant, which is the normal case for bcc metallic iron, have also been observed in magnetite-like phases and nonstoichiometric Fe_3O_4.[3,4] Finally, up to 0.1 wt % large grained ferric iron spinel precipitates can escape detection by Mössbauer techniques,[5] while superparamagnetic particles of both metallic iron and magnetite give rise to an excess area near zero velocity in the γ ray resonance absorption spectrum.[6] Thus, the present study was undertaken to find a simple and reliable way to detect fine-grained metallic iron and ferric iron spinel phases precipitated from iron-containing glassy matrices and to discriminate between them.

EXPERIMENTAL

FMR spectra were obtained using a Varian E-9 electron spin resonance spectrometer operating at X-band (9 GHz). Sample temperatures were varied between 90-573 K by means of a Varian variable temperature accessory and between 4.2-140 K by means of an Air Products LTD-3 Helitran apparatus.

A complete description of the samples used in this study and the preparation techniques involved has appeared elsewhere.[7] With the exception of a sample of crushed mineral Fe_3O_4 (OKA-B) dispersed in powdered SiO_2, all ferromagnetic phases were produced by nucleation and precipitation from glasses which were initially homogeneous. Sample 2.1-600-16 was derived from a calcium aluminoborosilicate glass which as-quenched contained both Fe^{2+} and Fe^{3+} ions. X-ray and electron diffraction studies and transmission electron microscopy of this sample revealed the presence of spherical precipitates ~ 100Å characterized by a lattice constant close to that of γ-Fe_2O_3.[4] On the other hand, glass CG-820-74-810-36 was shown[8] to contain metallic iron precipitates ~ 150Å.

The method used in the present study is critically dependent on obtaining a reliable determination of the integrated FMR intensity. Because the first derivative of the FMR absorption is experimentally observed, it is necessary to perform a double integration of the spectra to obtain the intensity. When the resonance is narrow with respect to the mean resonance field, the method is straightforward, but in the present case when the intensity extends to zero field, reproducible total intensities can be obtained only when sufficient care is taken to establish the zero derivative level.[7] The integrations can be performed numerically either by hand[7] or by a computer,[9] but in any case require that a zero derivative level be established (by running an empty sample tube spectrum, for example).

Partially because of the difficulty in establishing a reliable zero level, it is of value to define a narrow integral as that portion of the area under the absorption curve which lies above a straight line intersecting the curve near zero field and tangent to the curve near 5-6 kG (i.e., the dashed line in Fig. 1). This narrow integral is obtained by making a guess of the zero derivative

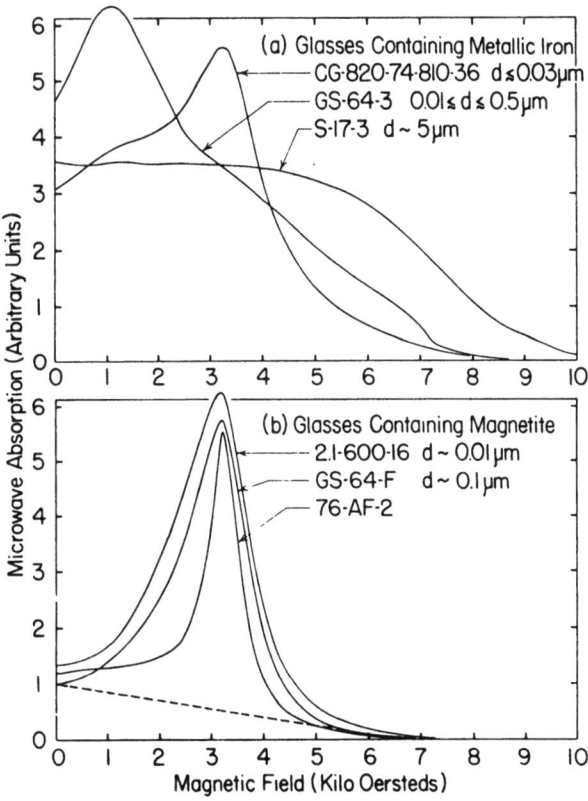

Fig. 1. X-band FMR absorption curves for (a) metallic iron and (b) ferric iron spinel precipitates in glass.

level and then treating the low field part of the FMR absorption as baseline drift in the method outlined by Ayscough.[10] Experience has shown that this method is operationally quite simple and highly reproducible provided the guessed zero derivative level is reasonably good.[7] As discussed below, similar results are obtained with both the total and narrow integrals.

RESULTS AND DISCUSSION

It is possible to qualitatively associate the broad and narrow FMR spectra with different configurations of ferromagnetic precipitates. In principle, multidomain configurations give rise to broad backgrounds which peak at zero field due to the Polder-Smit low field loss which occurs whenever the condition $\omega \gg \omega_m$ is not satisfied, where ω is the microwave frequency and $\omega_m = 4\pi |\gamma| M_s$, where γ is the gyromagnetic ratio and M_s is the saturation magnetization.[11]

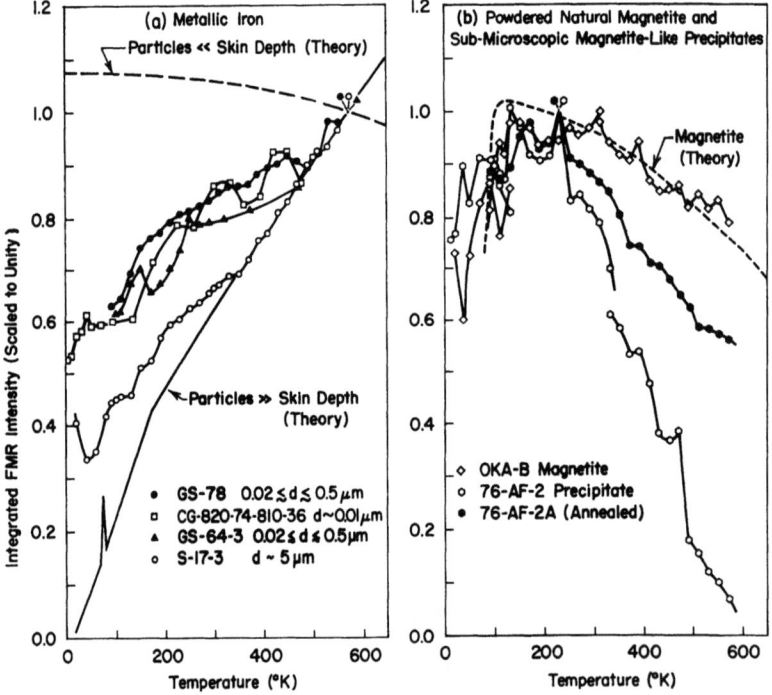

Fig. 2. Total FMR intensity vs. temperature for (a) metallic iron and (b) magnetite and ferric iron spinel precipitates in glass.

For iron and magnetite at room temperature $\omega_m/2\pi \approx 61$ GHz and ≈ 17 GHz respectively,[11] as compared to $\omega/2\pi \approx 9$ GHz at X band. Thus, Polder-Smit losses are expected for both materials, with a larger effect predicted in the case of iron. These domain effects, together with the shape anisotropy of non-spherical particles, are most probably responsible for the zero-field absorption observed for a wide variety of ferromagnetic precipitates in glasses.[12-14] In contrast, single-domain equant particles give rise to narrow resonances.[15,16] Examples of these resonances are shown in Fig. 1 where it can be seen that the FMR absorption of samples containing either iron or magnetite may consist of both a narrow and broad part. This is further evidence that metallic iron and magnetite-like phases cannot be identified on the basis of the FMR lineshape alone.

The FMR intensity can be determined by integration of the loss component of the susceptibility χ_e'' assuming insulator-type materials and spherical single domain particles:

$$\chi_e'' = \alpha \omega \omega_m / [(\omega_{res} - \omega)^2 + \alpha^2 \omega^2)] \tag{1}$$

where $\omega_{res} = |\gamma| H_{app}$, H_{app} is the applied magnetic field, and α is the Landau-Lifshitz damping factor. The FMR intensity is then

$$I = \int_0^\infty \chi_e'' \, dH_{app} = 4\pi M_s [\pi/2 + \tan^{-1}(1/\alpha)] \quad . \tag{2}$$

Thus, in the usual case when $\alpha \sim 0.01$, Eq. (2) reduces to

$$I = 4\pi^2 M_s = \pi \omega_m / |\gamma| \tag{3}$$

and the integrated intensity depends linearly upon the saturation magnetization. If the particles are sufficiently smaller than the skin depth of the microwave radiation so that total particle excitation occurs, the temperature dependence of the FMR intensity will be the same as the temperature dependence of M_s. This is shown in Fig. 2a as the theoretical curve for "Particles $\ll$ Skin Depth".[17] On the other hand, for metallic particles larger than the microwave skin depth $\delta = 1/\sqrt{\pi \sigma \mu \nu}$, where σ is the conductivity, μ is the permeability, and ν is the frequency (mks units), the FMR intensity would also be proportional to the fractional particle volume being penetrated. For δ much smaller than the particle size, one obtains

$$I \propto M_s \delta \propto M_s / \sqrt{\sigma} \quad . \tag{4}$$

Equation (4) should provide a first approximation for the temperature dependence of the integrated intensity for magnetic particles in this category. A normalized plot of $M_s/\sqrt{\sigma}$ versus T for metallic iron is shown in Fig. 2a as the curve "Particles $\gg$ Skin Depth."

As shown in Fig. 2a, there is good agreement between the theory for "size $\gg \delta$" and the data of S-17-3, which is consistent with the belief that $\delta < 5\mu m$ for metallic iron. For smaller particles whose size is intermediate between the two limiting cases, the FMR intensity behavior falls between the two theoretical boundaries (Fig. 2a). However, the total FMR intensity of all metallic iron-containing samples is seen to be an increasing function of temperature, even for particles as small as $0.015\mu m$ (CG-820-74-810-36). This result is evidence that metallic effects are important in the FMR of fine-grained iron precipitates.

Thus, the lower theoretical curve in Fig. 2a represents a semiquantitative lower bound for the integrated FMR intensity versus temperature for metallic iron. Several rather extensive modifications would be necessary to perform more quantitative calculations. In the first place, the μ parameter in the skin depth expression is not a simple scalar number for ferromagnetic metals. For such materials, μ has a tensor form and depends on particle orientation relative to the applied field, as well as other factors. In the second place, the classical skin depth δ as defined by Eq. (3) is only a crude approximation for the actual skin depth of the modes participating in metallic FMR. The effective penetration depth of these modes can vary substantially as the applied magnetic field passes through the region of FMR. The boundary conditions at the surface of the particles are also extremely important and determine the mixture of normal modes in the FMR response. Detailed consideration of these effects can be done with ease only for a planar geometry.[18] However, the general results should apply: the effective skin depth at resonance can be much smaller than δ calculated from Eq. (3) with the static (dc) values of σ and hence the condition "size $\ll \delta$" is not really adequate to justify the assumption of total particle excitation.

Because magnetite has a conductivity substantially less than that of iron, the microwave skin depth calculated from Eq. (3) is relatively large -- $\sim 200 \mu m$ for $T < 120$ K and $\sim 5\mu m$ at higher temperatures. Thus, the effect of the skin depth on the FMR intensity can be neglected for ferric iron spinel particles $< 5\mu m$, and the intensity will be expected to vary as M_s. Furthermore, it is expected that there will be a drop of $\sim 20\%$ in the FMR intensity upon cooling the sample below 120K since the induced magnetization of magnetite measured in static fields ~ 3 kG is known to drop $\sim 20\%$ as the sample is cooled through the Verwey temperature (~ 120 K).[19] These expectations are incorporated into the theory curve shown in Fig. 2b.[19,20]

The agreement between the theory and experimental results for OKA-B magnetite is quite good. The FMR intensity of the ferric iron spinel precipitates decreases more rapidly than that of the Fe_3O_4

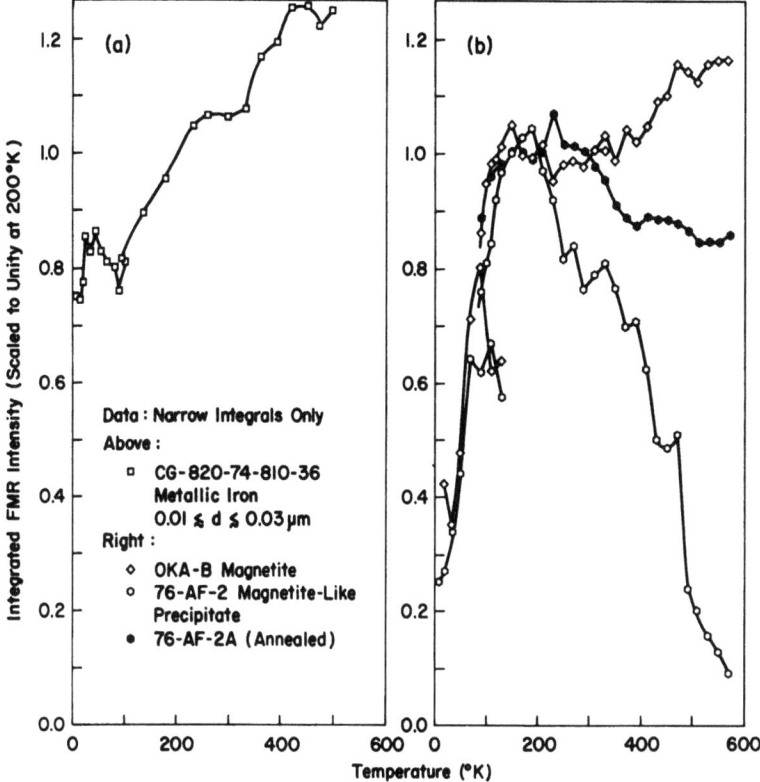

Fig. 3. Narrow integral intensity for (a) metallic iron and (b) ferric iron spinel precipitates.

sample with increasing temperature, which is probably a result of a lower Curie temperature in the "polluted" or nonstoichiometric magnetite precipitates than in pure magnetite. However, all samples display the characteristic drop in intensity near the Verwey temperature.

The data shown in Fig. 2 are derived from the **total** areas under the FMR absorption curves and provide a clear means for distinguishing between fine-grained iron and fine-grained magnetite. However, as discussed above, it is often easier to obtain the intensity of only the narrow part of a resonance. When the narrow integral can be obtained for an iron-containing sample (e.g. for single domain Fe precipitates), it can be seen in Fig. 3a that the temperature dependence is similar to that of the total integral. Likewise, the narrow integral of samples containing ferric iron spinel precipitates in Fig. 3b display the characteristic drop in intensity below the Verwey temperature.[7] As shown in Fig. 4, the broad absorption (the triangular wedge below the dashed line in Fig. 1) of the latter samples actually gains intensity upon cooling.

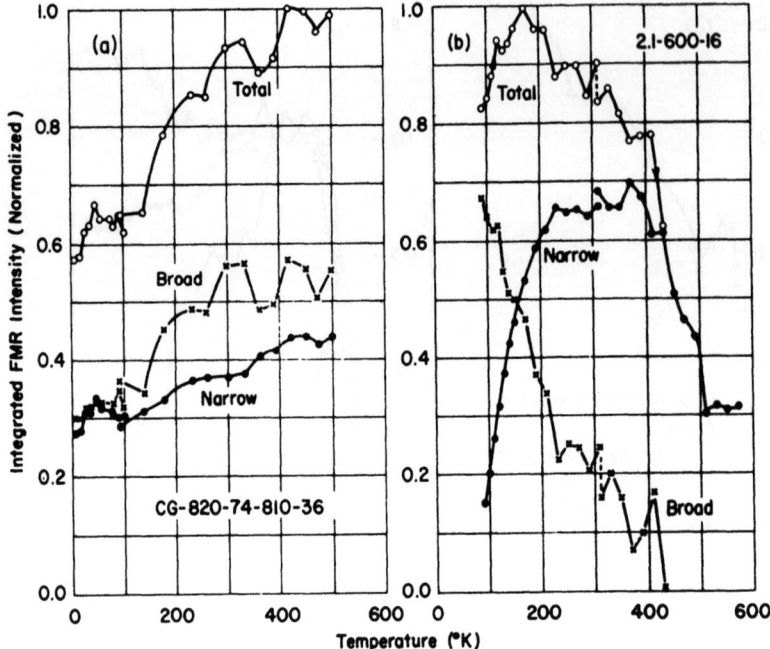

Fig. 4. FMR intensity vs. temperature for (a) metallic iron and (b) magnetite precipitates in glass.

This effect is also observed in Fe_3O_4 where the FMR absorption near g=2 (narrow part) shifts to lower fields (broad part) due to a cubic to uniaxial symmetry change upon cooling through the Verwey temperature.[21] Thus, the drop in the narrow integral of the spinel-containing sample of Figs. 3b and 4b is greater than that of the total integral. The narrow and broad parts of the resonance for metallic iron precipitates (Figs. 3a and 4a) monotonically lose intensity upon cooling, in contrast to the behavior of the spinel precipitates. It should be pointed out that at the present time there is no quantitative theory for the temperature dependence of the narrow integrals; the usefulness of this simpler technique rests on the empirical evidence that it behaves in a manner which is similar to the total integral method.[7]

By using the techniques outlined above, it is possible to distinguish between precipitated iron and/or ferric iron spinels (or magnetite) in glass. A sample which displays an inverted-U-shape FMR intensity vs. temperature behavior in the range 4.2 - 600K can be inferred to contain ferric iron spinel precipitates whether or not metallic iron is present. As little as ∼0.1 wt % spinel phase has been detected in the presence of ∼ 1.0 wt % metallic iron, detected by conventional spectroscopic techniques.[7]

In conclusion, the technique of studying the temperature dependence of the FMR intensity has been shown to be a powerful tool for detecting and identifying fine-grained ferromagnetic precipitates in glass, even when they may escape detection by conventional means. In addition, it is possible to detect ferric iron spinel phases in the presence of as much as 10 times more metallic iron.

REFERENCES

1. R.J. Williams, and E.K. Gibson, Earth Planet. Sci. Lett. 17, 84 (1972).
2. D.L. Griscom, E.J. Friebele, C.L. Marquardt, N. Sugiura and T. Nagata, EOS, Trans. Am. Geophys. Union 55, 329 (1974).
3. D.L. Griscom, Geochim. Cosmochim. Acta 38, 1509 (1974).
4. M.P. O'Horo, Bull. Am. Ceram. Soc. 53, 324 and 356 (1974), and these proceedings.
5. D.W. Forester, Geochim. Cosmochim. Acta, Suppl. 4, Vol. 3, 2697 (1973).
6. R.M. Housley, R.W. Grant, and M. Abdel-Gawad, Geochim. Cosmochim. Acta, Suppl. 4, Vol. 1, 1065 (1972).
7. D.L. Griscom, C.L. Marquardt and E.J. Friebele, J. Geophys. Res. 80, 2935 (1975).
8. G.W. Pearce, R.J. Williams and D.S. McKay, Earth Planet. Sci. 17, 95 (1972).
9. E.H. Cirlin, I.B. Goldberg, R.M. Housley, R.A. Weeks and R. Perhac, Geochim. Cosmochim. Acta, Suppl. 6, Vol. 3 (1975).
10. P.B. Ayscough, Electron Spin Resonance in Chemistry, Methuen (London) 1967, p. 442.
11. J. Smit and H.P.J. Wijn, Ferrites, John Wiley (New York), 1959, pp 22, 82 ff, 157.
12. D.L. Griscom, C.L. Marquardt, E.J. Friebele, and D.J. Dunlop, Earth Planet. Sci. Lett. 24, 78 (1974).
13. R.A. Weeks, J.L. Kolopus, D. Kline and A. Chatelain, Geochim. Cosmochim. Acta, Suppl. 3, Vol. 1, 797 (1970).
14. E.J. Friebele, D.L. Griscom, C.L. Marquardt, R.A. Weeks and D. Prestel, Geochim. Cosmochim. Acta, Suppl. 5, Vol. 3, 2729 (1974).
15. F.D. Tsay, S.I. Chan and S.L. Manatt, Geochim. Cosmochim. Acta 35, 865 (1971).
16. D.L. Griscom, E.J. Friebele, and C.L. Marquardt, Geochim. Cosmochim. Acta, Suppl. 4, Vol. 3, 2709 (1973).
17. A.H. Morrish, The Physical Principles of Magnetism, John Wiley (New York), 1965.
18. For example, see C. Vittoria, R.C. Barker, and H. Yelon, J. Appl. Phys. 40, 1561 (1969).
19. C.A. Domenicali, Phys. Rev. 78, 458 (1950).
20. Handbook of Chemistry and Physics (1956), C.D. Hodgman, ed. Chemical Rubber Publ. Co., p. 2362.
21. L.R. Bickford, Jr., Phys. Rev. 78, 449 (1950).

LOW TEMPERATURE THERMAL CONDUCTIVITY OF $MnO \cdot Al_2O_3 \cdot SiO_2$ GLASS*

A. K. Raychaudhuri and R. O. Pohl

Laboratory of Atomic and Solid State Physics, Cornell

Ithaca, New York 14853

De Graaf and co-workers have shown that cobalt and manganese alumino-silicate glasses have remarkable magnetic properties. They appear to be chemically inhomogeneous on the scale of the order of 10Å, and much of their magnetic behavior can be explained on the basis of this inhomogeneity[1]. Their magnetic excitations dominate the low temperature specific heat in these glasses, as has been shown by Wenger and Keesom[2] and by Kline et al[3]. In Fig. 1 the specific heat of a manganese glass (containing 13.3at%Mn) is compared with the Debye-T^3 specific heat based on speed of sound measurements [3,4], the experimental specific heat of a typical amorphous solid, SiO_2, containing no magnetic impurities, and that of a piece of alumino-borosilicate glass (pyrex), containing 100 ppm by weight of iron[5]. Below 1°K, the specific heat of the manganese-containing glass exceeds that of the SiO_2 glass by three orders of magnitude.

There presently exists good evidence that the anomalous specific heat varying approximately linearly with temperature at low temperatures, as shown in Fig. 1 for SiO_2, is characteristic for the amorphous state[6], and it is generallly believed that the states responsible for this anomaly also act as phonon scatterers to account for the characteristic thermal conductivity observed in all amorphous solids. Two examples of the latter are shown in Fig. 2. All efforts to alter the thermal conductivity of glasses by chemical means have so far been without success. The only ways to influence it have been (a) partial crystallization of the glass, which has lowered the conductivity, probably through phonon scattering by internal surfaces as reviewed by

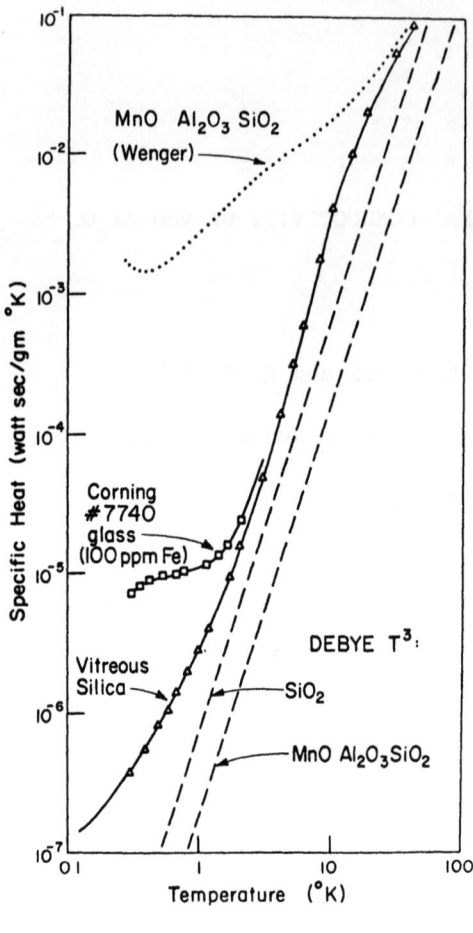

Figure 1

Figure 1 - Specific heat of vitreous silica with no detectable magnetic impurities (ref. 5), of a silica based glass with a small unintentional amount of iron (ref. 5), and of the manganese aluminosilicate glass used in this study. The dashed lines were computed from the measured speeds of sound using the Debye model.

Ashcroft et al[7]; (b) internal or external boundaries (Casimir effect) in glasses containing voids or in thin fibers[8,9]; and (c) high energy neutron irradiation[10,11]. In the latter case, the conductivity was found to increase with irradiation. The cause for this is not understood.

THERMAL CONDUCTIVITY OF MnO·Al$_2$O$_3$·SiO$_2$ GLASS

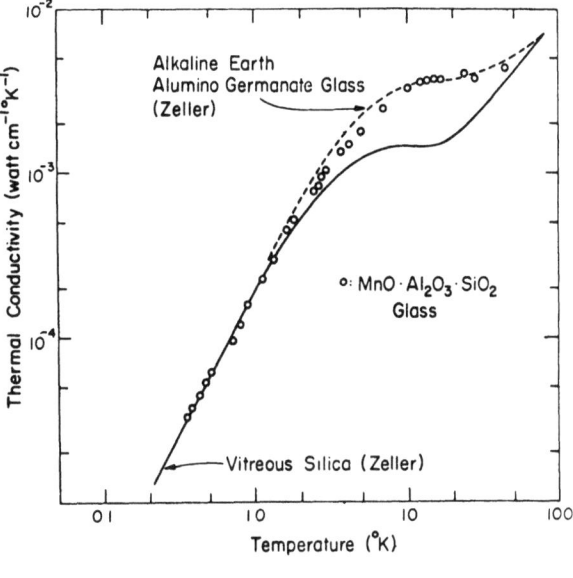

Figure 2

Figure 2 - The solid and the dashed lines are the conductivities of amorphous SiO$_2$ and of a germanate based glass respectively (ref. 5). The data points were obtained in the present investigation.

To date, any efforts to influence the thermal conductivity by magnetic impurities have failed. 100 ppm iron in silica based glasses (Fig. 1) had no effect[5], and neither had the addition of 2000 atomic ppm (0.5 weight%) of iron oxide to a soda-lime glass[12].

A study of thermal conductivity of the manganese or cobalt alumino-silicate glasses was of interest for three reasons. Firstly, these glasses contain considerably higher magnetic ion concentrations than the glasses studied previously. Secondly, the chemical inhomogeneity of these glasses might cause an additional phonon scattering. Thirdly, any crystalline inclusions resulting from the chemical inhomogeneity, which might have been missed in the x-ray diffraction studies, might be observable in such an experiment.

A sample of MnO·Al$_2$O$_3$·SiO$_2$ (13.3 at.%) was provided by Dr. de Graaf. It was identical to the one whose specific heat had been measured previously[3,4]. The data in Fig. 2 shows that even in this glass the phonon scattering is determined entirely by those phonon scattering mechanisms active also in glasses containing no magnetic ions. Another way of representing these results is used in Fig. 3. From the gas-kinetic analogue, the thermal conductivity $\Lambda(T)$ is written as the product of the Debye

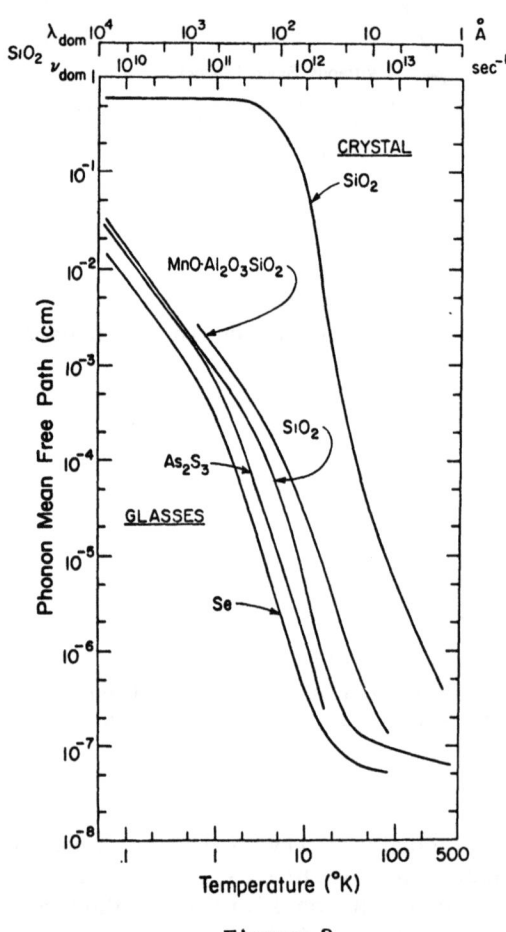

Figure 3

Figure 3 - The phonon mean free path, $\bar{l}$, compared for four different glasses, together with $\bar{l}$ for α-quartz. See text.

specific heat C_D, the Debye speed of sound v_D, and an average phonon mean free path $\bar{\ell}$: $\Lambda = (1/3)C_D v_D \bar{\ell}$. The mean free paths are compared for four different glasses, and show all the same characteristic behavior. In view of the very small variation of the thermal conductivity of all silica based glasses without magnetic ions[5], it appears unlikely that a hypothetical glass identical to the manganese glass in all respects, but without spins, would have a conductivity differing from the one we measured by more than a factor of two. On the basis of this assumption, and from the fact that the specific heat of the manganese glass exceeds that of the iron-free silica by a factor of $\sim 10^3$, we may conclude that the states associated with the magnetic properties of the manganese glass scatter the phonons at least one thousand times less strongly than the average states responsible for the specific heat anomaly in the vitreous silica. In addition, neither the chemical nor any possible physical inhomogeneity in the manganese glass appears to have an influence on its thermal conductivity.

To summarize: As far as the thermal conductivity, i.e. the scattering of thermal phonons is concerned, even this magnetically and structurally extraordinary glass does not behave differently from any other amorphous solid. This observation makes the origin of the generality of the thermal conductivity of all amorphous solids an even more puzzling question.

We would like to thank Dr. de Graaf for suggesting this material to us, for supplying the sample, and for illuminating discussions. We also thank Dr. Wenger for sending us the specific heat data.

References

*Supported by the Energy Research and Development Administration, Contract No. E(11-1) 3151, technical report no. COO-3151-70.

1. R.A. Verhelst, R.W. Kline, A.M. de Graaf, and H.D. Hooper, "Magnetic Properties of Cobalt and Manganese Aluminosilicate Glasses," Phys. Rev. B $\underline{11}$, 4427 (1975).

2. L.E. Wenger and P.H. Keesom, "A Calorimetric Investigation of a $CoO \cdot Al_2O_3 \cdot SiO_2$ Glass", following paper.

3. R.W. Kline, A.M. de Graaf, L.E. Wenger, and P.H. Keesom, "A Calorimetric Study of $MnO \cdot Al_2O_3 \cdot SiO_2$ Glasses", Proceedings of 21st Annual Conference on Magnetism and Magnetic Materials, Philadelphia 1975, American Institute of Physics, New York 1976, page 169.

4. L.E. Wenger, Ph.D. Thesis, Purdue University, 1976.

5. R.C. Zeller and R.O. Pohl, "Thermal Conductivity and Specific Heat of Non-Crystalline Solids", Phys. Rev. B, 4, 2029 (1971).

6. R.B. Stephens, "Intrinsic Low-Temperature Thermal Properties of Glasses", Phys. Rev. B, 13, 852 (1976).

7. N.W. Ashcroft, R.B. Stephens, and R.O. Pohl, "Lattice Vibrations of Non-Crystalline Solids", Technical Report of the Institute of Solid State Physics (ISSP), University of Tokyo, Japan, Ser. B, No. 15, March 1973, page 144.

8. M.P. Zaitlin and A.C. Anderson, "On the Phonon Thermal Transport in Non-Crystalline Materials," Phys. Rev. B 12, 4475 (1975).

9. R.O. Pohl, W.F. Love, and R.B. Stephens, "Lattice Vibrations in Non-Crystalline Solids," in *Amorphous* and *Liquid Semiconductors*, J. Stuke and W. Brenig, eds. Taylor and Francis, 1974, page 1121.

10. A.F. Cohen, "Low Temperature Thermal Conductivity in Neutron Irradiated Vitreous Silica," J. Appl. Phys. 29, 591 (1958).

11. A. Assfalg, A. Raychaudhuri, and R.O. Pohl, to be published.

12. L.H. Challis and C.N. Hooker, "The Effect of Electric and Magnetic Fields on the Thermal Conductivity of Glass at Low Temperatures," J. Phys. C 5, 1153 (1972).

DISCUSSION

E. Siegel: Does the plot of thermal conductivity versus temperature generally fit the behavior of all non-metallic glasses?

R. O. Pohl: Yes. We have been unable to find any deviation from that behavior down to 1 K in all considered glasses.

D. L. Huber: Have you looked at other glasses containing Fe?

R. O. Pohl: Yes. I did leave out in my discussion one study done by Challis and Hooker at Nottingham which studied a borosilicate glass containing 2000 atomic ppm of Fe and they also found no deviation in the behavior of the thermal conductivity. They went a step further in applying a strong magnetic field, of the order of 50 kOe, and also noticed no change in the temperature-dependent behavior of the thermal conductivity.

A CALORIMETRIC INVESTIGATION OF A $CoO \cdot A\ell_2O_3 \cdot SiO_2$ GLASS[†]

L. E. Wenger[*] and P. H. Keesom

Department of Physics
Purdue University
West Lafayette, Indiana 47907

ABSTRACT

In a continuing effort to further the understanding of the magnetic properties of concentrated cobalt aluminosilicate glasses, the specific heat of a sample[1] doped with 14.3 at .% cobalt was measured in the temperature range 0.4 - 40 K. The low-field susceptibility for the same sample exhibits a sharp peak at 6.4 K, seeming to indicate the onset of magnetic ordering. However, the specific heat does not show a cooperative-type peak, but instead a rather broad, smooth maximum centered at 8.2 K. The application of a 25 kOe magnetic field decreases this maximum by approximately 6% and shifts its center to 10.3 K. Previous susceptibility[2] and sound velocity[3] results have been described by a domain model[2] and subsequent application of Néel's theory[4] on small magnetic particles. These calorimetric results will be discussed in terms of this model.

[†]Work supported by NSF Grant No. DMR 72-01821A01 and NSF-MRL Grant No. DMR 72-03018A04.

[*]Present address: Department of Physics, Wayne State University, Detroit, Michigan 48202

INTRODUCTION

Recently, the cobalt and manganese aluminosilicate glass systems have drawn considerable attention owing to their remarkable magnetic properties. In particular, these insulating glasses exhibited a relatively sharp peak in the low-field ac magnetic susceptibility,[2] very similar to the susceptibility peaks observed in the spin-glass alloys.[5] In addition, a broad dip in the sound velocity[3] was found at low temperatures. A prior calorimetric investigation[6] on the manganese aluminosilicate glass system revealed broad maxima in the specific heat centered at temperatures significantly higher than those associated with the susceptibility peaks. More recently, a Mössbauer study[7] of a cobalt aluminosilicate glass showed a dramatic spectral broadening at low temperatures.

All four of these observations have been described by a domain model and subsequent application of Néel's theory of small magnetic particles. In this model, the sample is divided into small regions (domains). Each domain is assumed to have acquired internal antiferromagnetic or ferrimagnetic order at temperatures above the temperature of the susceptibility peak. In addition, the domains are treated as independent, randomly-positioned superparamagnetic particles embedded in a non-magnetic host. Each domain has an anisotropy energy KV which caused the moment of each domain to freeze in its anisotropy direction at low temperatures. This results in a peak in the susceptibility.

The magnetic specific heat associated with superparamagnetic particles has been previously calculated by Livingston and Bean.[8] The Hamiltonian for an anisotropic superparamagnetic particle with finite spin S is:

$$H = - a \left[S_z^2 - S(S + 1)/3 \right] \tag{1}$$

where S_z is an operator with eigenvalues $-S, -S + 1, \ldots, S$. The constant $\underline{a}$ is a measure of the anisotropy energy KV, where KV is defined as the energy difference between the lowest and highest level. For large spin values, the partition function may be treated classically and represented by an integral. The domain specific heat for this case is

$$C_d/k_B = 1/2 - (e^b/4I)(1 - 2b + e^b/I) \tag{2}$$

where

$$b = KV/k_B T \quad \text{and} \quad I = \int_0^1 \exp(bx^2)\,dx . \tag{3}$$

The domain specific heats for both finite and infinite spin values will be compared to the experimental results of the cobalt

A CALORIMETRIC INVESTIGATION OF A CoO·Al$_2$O$_3$·SiO$_2$ GLASS

aluminosilicate glass sample as well as to the prior results on the manganese doped glasses.

EXPERIMENTAL

The glass samples which were provided by Wayne State University were prepared by mixing an appropriate amount of aluminum oxide, pure silica sand, and cobalt carbonate to form the cobalt glasses, and manganese carbonate to form the manganese glasses.[1,2] X-ray diffraction and electron microscopy studies had previously indicated the absence of crystallinity.

All heat capacity measurements were made by using a standard heat-pulse technique throughout the temperature range 0.4 - 40 K. The addenda corrections owing to the thermometer, glyptal, etc., have been previously measured and along with calibration techniques have been described elsewhere.[9] The magnetic susceptibility of the cobalt aluminosilicate glass was measured using a low-field ($\sim$ 1 Oe) ac mutual-inductance bridge at a frequency of 17 Hz. The temperature was monitored in this susceptibility measurement with a calibrated Ge thermometer in good thermal contact with the sample by filling the experimental chamber with ^{4}He exchange gas.

RESULTS AND DISCUSSIONS

The specific heats of the 14.3 at .% cobalt aluminosilicate glass in 0 and 25 kOe fields are displayed in Fig. 1 as a plot of C/T vs $T^{1/2}$, while its low-field susceptibility is shown in the inset. This type of specific heat plot was finally chosen to rep- the data for several reasons. (1) The previous calorimetric results on manganese aluminosilicate glasses were described by the relation, $C = \gamma T + \beta T^{3/2} + \alpha T^3 + \nu C_d$, where γ, β, α and ν are constants and C_d is the domain specific heat as calculated from Eqn. (1). The phonon contribution (αT^3) was extremely small such that it is negligible to first approximation. Thus a C/T vs $T^{1/2}$ plot may give a graphical determination for the coefficients, γ and β. (2) From a purely experimental point of view, the higher temperature results could be fitted by a $T^{3/2}$ representation over a larger temperature range than by a T^3 representation. The solid line in Fig. 1 represents this $T^{3/2}$ contribution. (3) As mentioned previously, each domain has required either antiferromagnetic or ferrimagnetic interactions owing to the negative Curie temperature as determined from high-temperature susceptibility measurements.[2] Since each domain carries a net magnetic moment, one can speculate that ferrimagnetic interactions dominate and that spinwave excitations may be present. These excitations would give rise to a specific heat contribution proportional to $T^{3/2}$ whereas antiferromagnetic spinwave excitations would result in a T^3 contribution which is indistinguishable from the phonon contribution.

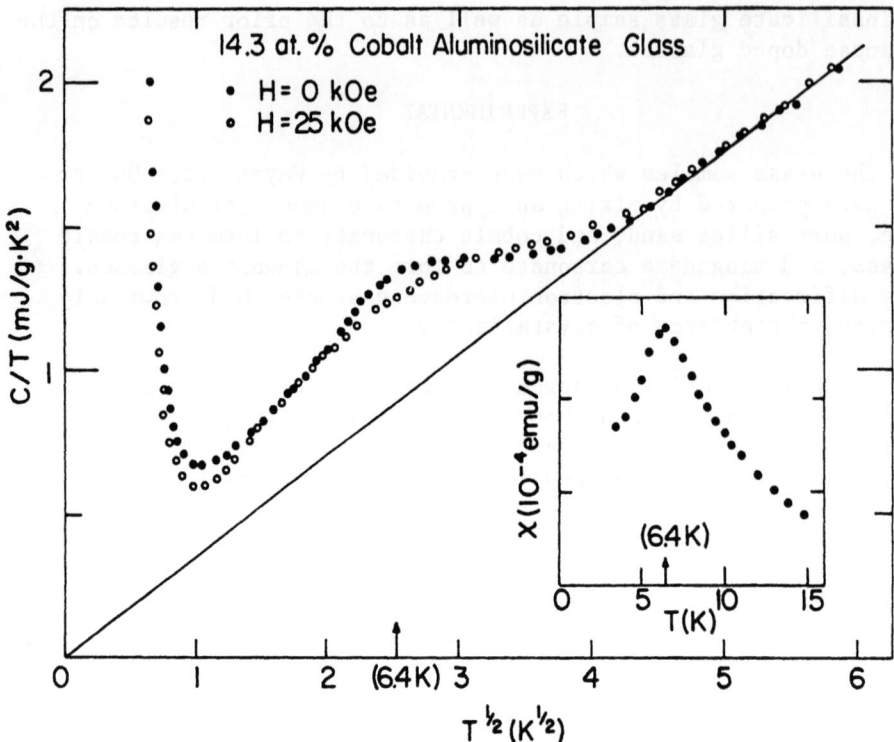

Fig. 1. Specific heat of 14.3 at.% cobalt aluminosilicate glass in 0 and 25 kOe fields between 0.4 and 36 K. The solid line represents the $T^{3/2}$ contribution. The inset shows the low-field susceptibility. The arrows indicate the peak temperature of 6.4 K.

At temperatures below 1 K, the specific heat begins to increase with decreasing temperatures. This lowest temperature specific heat has the appearance of the tail of a Schottky contribution, i.e., proportional to T^{-2}. This contribution is probably nuclear in origin as a result of a hyperfine field ($\sim$ 435 kOe) removing the degeneracy of the ^{59}Co nucleus (I = 7/2). However, the $\sim$ 3% decrease in this Schottky contribution due to the application of a 25 kOe field is puzzling unless the external field effectively reduces the strength of the hyperfine field acting on the nuclei.

After subtracting the T^{-2} and $T^{3/2}$ contributions from the specific heat of the glass sample, the remaining specific heat is shown in Fig. 2 as a plot of C_{excess} vs T. This excess specific heat in zero magnetic field shows a rather broad maximum centered at 8.2 K. In a 25 kOe field, its center has shifted to 10.3 K and its height has decreased by approximately 6%.

A CALORIMETRIC INVESTIGATION OF A CoO·Al$_2$O$_3$·SiO$_2$ GLASS

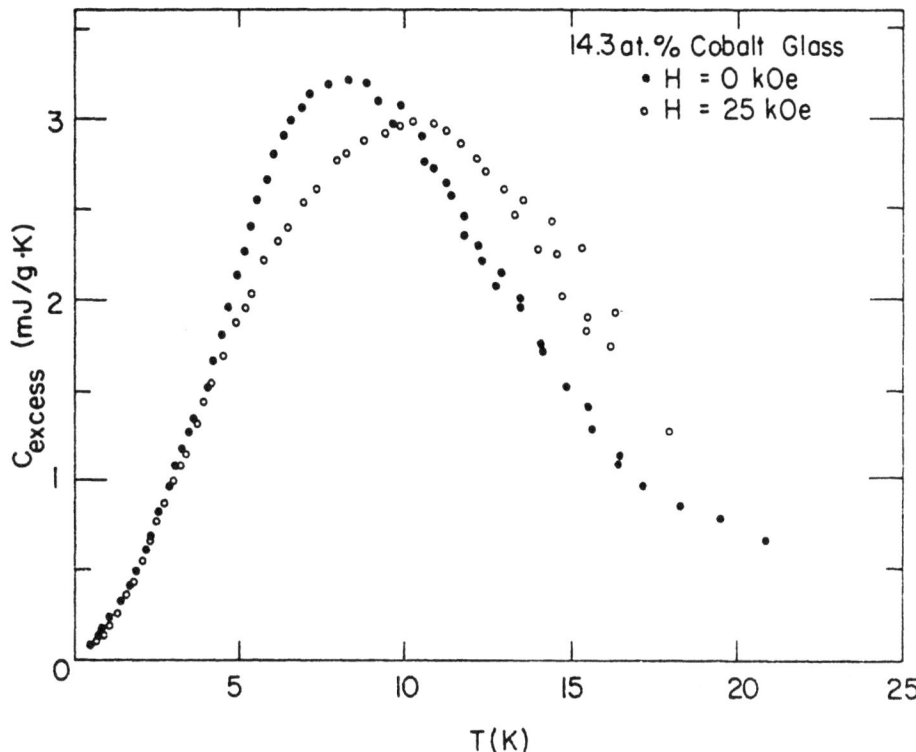

Fig. 2. The excess specific heat of the 14.3 at .% cobalt glass in 0 and 25 kOe fields.

This excess specific heat can be compared to the calculated specific heat from the domain model. Two possibilities for the value of the spin exist: (1) The first assumes an infinite spin and use Eqn. (2) to determine the parameters KV and ν (the number of domains per gram). This is similar to the method employed in the interpretation of the previous study on the manganese aluminosilicate glass. (2) The second uses a finite spin as determined from high-temperature susceptibility results. Deducing a spin of 5/2 from the effective magnetic moment for the cobalt glass, the specific heat can be calculated from statistical mechanics with the Hamiltonian defined in Eqn. (1). Figure 3 shows a comparison between the zero field results and the calculated specific heats for finite and infinite spin values. Values for the parameters KV and ν are listed in Table I for both possibilities. Note at the lowest temperatures, the experimental data is significantly larger than that calculated for S = 5/2 and follows a linear instead of an exponential temperature dependence. This might result from a spread in anisotropy values such that at low energies the distribution is a constant. This will give rise to linear T dependence in the specific heat as seen previously in amorphous materials.[10]

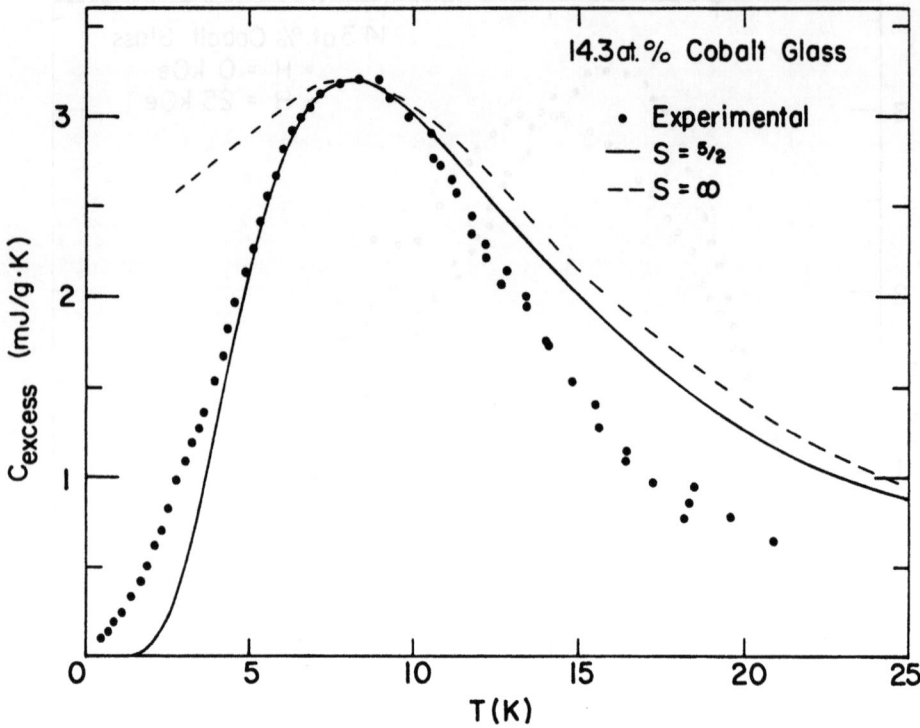

Fig. 3. The excess specific heat of the 14.3 at .% cobalt glass in zero field compared to the calculated specific heats for S equal to 5/2 and ∞.

TABLE I

Specific Heat Parameters

at.%	γ(mJ/g·K^2)	β(mJ/g·K$^{5/2}$)	S = ∞ ν(g^{-1})	KV(K)	S = finite ν(g^{-1})	KV(K)	S
14.3 Co	----	.347	1.8x10^{20}	51.7	3.2x10^{20}	26.4	5/2
14.3 Mn	.963	.263	1.2x10^{20}	32.8	1.8x10^{20}	19.1	3
13.3 Mn	.755	.263	2.4x10^{20}	27.8	4.3x10^{20}	14.5	5/2

Evidence for this type of distribution has been suggested from Mössbauer results.[7] Also note that the calculated specific heat for infinite spin approaches a finite value at absolute zero which is a consequence of this classical treatment.

Owing to the good agreement between experimental data and the specific heat calculation for finite spin, the specific heats of the manganese glasses were reanalyzed. The data were similarly plotted as C/T vs $T^{1/2}$ in order to determine the coefficients γ and β. After subtracting these contributions these excess specific heats were compared to the specific heat calculation for finite spin value. The resulting fitting parameters are also shown in Table I. Note that the anisotropy energies, KV, are approximately 50% smaller for the finite spin case while ν the number of domains is substantially increased.

Table II shows a comparison of the anisotropy energies for the glasses as determined from susceptibility (χ) and calorimetric (C) measurements. Although the anisotropy energies are of the same order of magnitude, there is a significant difference between those determined from the susceptibility and those from the specific heat, especially for infinite spin. Although this difference in anisotropy energies as well as the corresponding difference in magnetic moments of the domains are unexplainable, the domain model should not be discredited. The model in its present form follows directly from the behavior of the susceptibility. At high temperatures, the susceptibility follows a Curie-Weiss behavior, $1/(T + \theta)$, which indicates the presence of a system of independent, paramagnetic particles of finite spin S. Around 50 K, the susceptibility begins to increase more rapidly than $1/(T + \theta)$ and correspondingly signifies the formation of domains and large magnetic moments. Then at the lowest temperatures, the susceptibility peak requires the freezing of the magnetic moments and occurs at a temperature related to the anisotropy energy. Based on these observations, the model was formulated with a discrete value for the anisotropy energy and one for the effective magnetic moment of the domains. Thus, the present domain model is still very simplistic and will require further improvements in order to describe all magnetic properties.

In conclusion, the specific heat results have been compared to that calculated from the domain model for finite and infinite spin values. Even though these values represent two limits, the calculated specific heats for both show reasonable agreement with the experimental results. In addition, the number of domains ν required in the specific heat calculations is substantially lower than the number of magnetic ions present indicating formation of magnetic clusters. This formation of magnetic clusters is also essential to explain the ferrimagnetic spinwave specific heat which is observed as a $T^{3/2}$ contribution. Thus, these observations lead to the conclusion that clustering of the magnetic ions (possible formation of domains) occurs with the value of the spin lying between 5/2 and infinity.

TABLE II

Comparison of Anisotropy Energies

at.%	T^{χ}_{max}(K)	T^{C}_{max}(K)	χ(Ref.2)	KV(K) from $C(S = \infty)$	$C(S$ = finite)
14.3 Co	6.4	8.2	18.4	51.7	26.4
14.3 Mn	4.25	5.2	12.6	32.8	19.1
13.3 Mn	2.95	4.4	8.7	27.8	14.5

ACKNOWLEDGMENTS

We wish to thank Professor A. M. de Graaf for kindly providing the cobalt and manganese aluminosilicate glass samples for this investigation. We are most grateful to Professor de Graaf for his fruitful discussions concerning the domain model and its application to the calorimetric results.

REFERENCES

1. Samples kindly provided by A. M. de Graaf, Wayne State University.

2. R. A. Verhelst, R. W. Kline, A. M. de Graaf and H. O. Hooper, Phys. Rev. B11, 4427 (1975).

3. T. J. Moran, N. K. Batra, R. A. Verhelst and A. M. de Graaf, Phys. Rev. B11, 4436 (1975).

4. L. Néel, in Low Temperature Physics (Les Houches, 1961), ed. by C. de Witt, B. Dreyfus and P. G. de Gennes (Gordon and Breach, New York, 1962).

5. V. Cannella and J. A. Mydosh, Phys. Rev. B6, 4220 (1972).

6. R. W. Kline, A. M. de Graaf, L. E. Wenger and P. H. Keesom, AIP Conf. Proc. 29, 169 (1976).

7. L. H. Bieman, P. F. Kenealy and A. M. de Graaf, AIP Conf. Proc. 34, (1976) and following paper.

8. J. D. Livingston and C. P. Bean, J. Appl. Phys. 32, 1964 (1961).

9. L. E. Wenger, Ph.D. thesis (Purdue University, 1975).

10. R. C. Zeller and R. O. Pohl, Phys. Rev. $\underline{4}$, 2029 (1971);
    R. B. Stephens, Phys. Rev. B$\underline{8}$, 2896 (1973).

DISCUSSION

N. Y. Rivier: Is the specific heat expected to be linear at low temperatures?

L. E. Wenger: Yes, in the manganese bearing glasses the linear dependence is indeed characteristic of that system.

M. P. O'Horo: Did you make an estimate of the number of clusters from your calculation?

L. E. Wenger: That depends on the value of the spin, a typical estimate would be around 10^{20}.

L. N. Mulay: What is your estimate of the number of Co ions per cluster?

L. E. Wenger: I would say around twenty.

EXCHANGE FIELDS AND CRYSTAL FIELDS IN $CoO \cdot Al_2O_3 \cdot SiO_2$ GLASS:
A MÖSSBAUER STUDY[*]

L.H. Bieman, P.F. Kenealy, and A.M. de Graaf

Department of Physics, Wayne State University

Detroit, Michigan, 48202

ABSTRACT

In an effort to understand more fully the remarkable low temperature properties of cobalt aluminosilicate glasses, Co-57 Mössbauer source experiments on two samples of concentrated $CoO \cdot Al_2O_3 \cdot SiO_2$ glass (17 and 25 at.% Co, respectively) were performed. These glasses exhibit a sharply peaked susceptibility at ∼5°K and ∼7°K, respectively, seeming to indicate the onset of long range antiferromagnetic order. However, the specific heat and the sound velocity do not show cooperative-type anomalies. It has previously been shown that these results may be understood qualitatively by assuming that below ∼50°K these glasses are divided into small magnetically ordered domains whose net moments relax in a superparamagnetic fashion. At room temperature the Mössbauer spectra consist of three sharp lines, indicating the presence of a well defined electric field gradient at the cobalt sites. Below ∼20°K the spectra are broadened dramatically by magnetic hyperfine fields as large as 48 Teslas. It will be shown that the Mössbauer spectra can be completely understood within the framework of the superparamagnetic domain model.

INTRODUCTION

In a recent series of papers attention has been drawn to the remarkable low temperature magnetic, ultrasonic, and thermal properties of cobalt and manganese aluminosilicate glasses.[1-3] These glasses exhibit a relatively sharp peak in the low field susceptibility,[1] very similar to the peak observed in the susceptibility of

the so-called spin glass alloys.[4] There is also a broad dip in
the sound velocity[2] and a broad maximum (superimposed on an anomalously large nearly linear background) in the specific heat,[3]
which correlate with the susceptibility peak. It has been suggested[1-3] that these observations may be accounted for by assuming
that the glasses below about 50°K consist of small magnetically
ordered regions (domains) whose net magnetic moments, μ, relax in
a superparamagnetic fashion with a relaxation time

$$\tau = \tau_o \exp(KV/kT). \tag{1}$$

Here, K is the anisotropy energy density of the domains, and V
the volume. At a temperature sufficiently below KV/k the domain
moments freeze in random directions, resulting in a sharply
reduced susceptibility. This mechanism also leads to a broad dip
in the sound velocity, and a corresponding anomaly in the specific
heat, in agreement with experiment. The present Co-57 Mössbauer
study of two Co-aluminosilicate glasses was undertaken to further
test the superparamagnetic domain model. It will be shown that
the low temperature Mössbauer spectra of these two glasses can
indeed be explained within the framework of this model. The
broadening of the spectra which is observed below $\sim$20°K is then
due to the fact that τ becomes greater than the nuclear lifetime.
Computer fitting of the low temperature spectra has yielded values
of the anisotropy parameter KV which are in reasonable agreement
with values obtained from the other measurements. A preliminary
account of this work was reported at the Joint MMM-Intermag
Conference held in Pittsburgh in June 1976.[5]

EXPERIMENT

Two Mössbauer sources were prepared by firing a mixture of
cobalt carbonate doped with 5 millicuries of Co-57, aluminum oxide,
and pure silica sand on a quartz disk in an arc-image furnace.
The samples contained 17 at.% Co and 25 at.% Co, respectively.
Both samples had a mass of approximately 50 mg and an area of
approximately 0.5 cm^2. Similarly prepared glasses showed no crystallinity when examined by powder X-ray diffraction and electron
microscopy. A Fe-57, 90% enriched $Na_4Fe(CN)_6 \cdot 10H_2O$ absorber of
thickness 0.25 mg Fe-57/cm^2 was used for the measurements. Data
were taken on a standard Mössbauer spectrometer with the source
placed in a variable temperature cryostat, while the absorber was
kept at room temperature. A laser interferometer was used to calibrate the velocity of the spectrometer.

RESULTS AND DISCUSSION

A series of Co-57 Mössbauer spectra taken on the 17 at.% Co aluminosilicate glass sample are shown in Fig. 1. A similar series of spectra taken on the 25 at.% Co aluminosilicate glass are displayed in Fig. 2. The outstanding features are the three surprisingly sharp lines observed at room temperature and the dramatic broadening setting in below about 20°K. Both the room temperature and the low temperature observations can be explained by making the following assumptions:

1. Upon decaying the Co-57 ion stabilizes into either Fe^{2+} or Fe^{3+}. This assumption is consistent with the results obtained by Mullen and Ok[6] who showed that in a CoO crystal the Co-57 nucleus decays into Fe^{3+} if a vacancy is nearby. Otherwise, Fe^{2+} is formed. In a glass we expect the presence of many vacancies. Also, the close agreement between the Fe^{2+} and Fe^{3+} isomer shifts in the Co aluminosilicate glass and the corresponding isomer shifts in the CoO crystal supports this assumption. The values of the isomer shifts are listed in Table 1.

2. The electric field gradient at the Co sites is well defined. This assumption is based on the interpretation of the room temperature spectra in terms of a Fe^{2+} quadrupole doublet superimposed on a Fe^{3+} quadrupole doublet, as shown in Figs. 1a and 2a. The linewidths observed in the Co aluminosilicate glass are only about three times larger than the widths observed in crystalline CoO.[6] This fact implies that if the increase in linewidth is due to a distribution of electric field gradients in the glass, this distribution must be surprisingly narrow. Values of the linewidths and quadrupole splittings are collected in Table 1.

3. Below about 50°K the glass breaks up into many superparamagnetic domains. At this temperature the rapid fluctuation of the magnetic hyperfine fields associated with the superparamagnetic relaxation prevents the observation of the magnetic hyperfine splitting in the Mössbauer spectrum. The effects of the magnetic hyperfine fields on the Mössbauer spectrum become evident when the relaxation time τ, Eq. 1, is comparable to the nuclear life time. The maximum broadening is observed when this relaxation time becomes so large that the magnetic hyperfine fields may be considered static.

4. Below ~20°K the values of the magnetic hyperfine fields do not change appreciably with temperature. This assumption is based on the fact that the domains form at about 50°K, so that their net magnetization should be near saturation at ~20°K. Because the Fe^{3+} ion is in a S-state, only the electron spins contribute to the magnetic hyperfine fields and this contribution

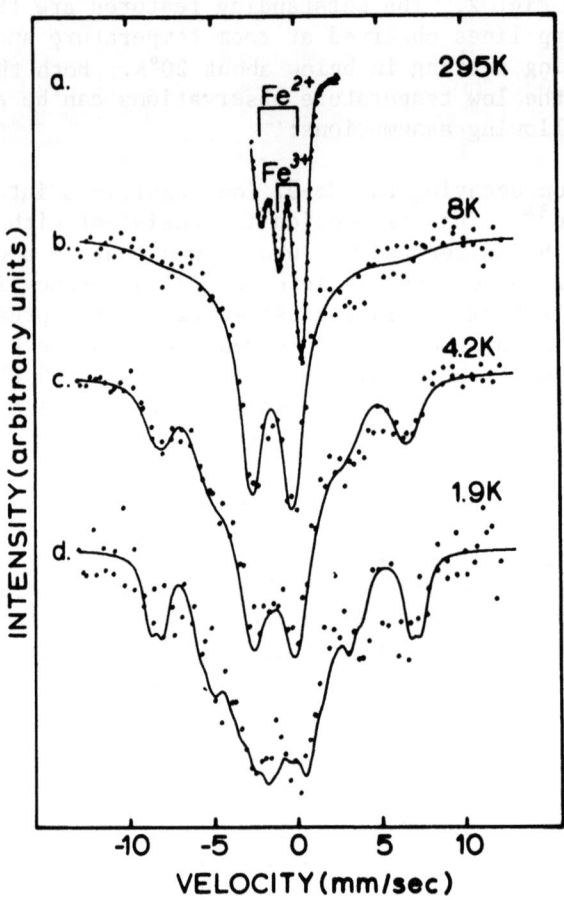

FIG. 1. Cobalt aluminosilicate glass (17 at.% Co) Mössbauer spectra at various temperatures (dots are experimental points, solid lines are fitted curves).

EXCHANGE FIELDS AND CRYSTAL FIELDS 591

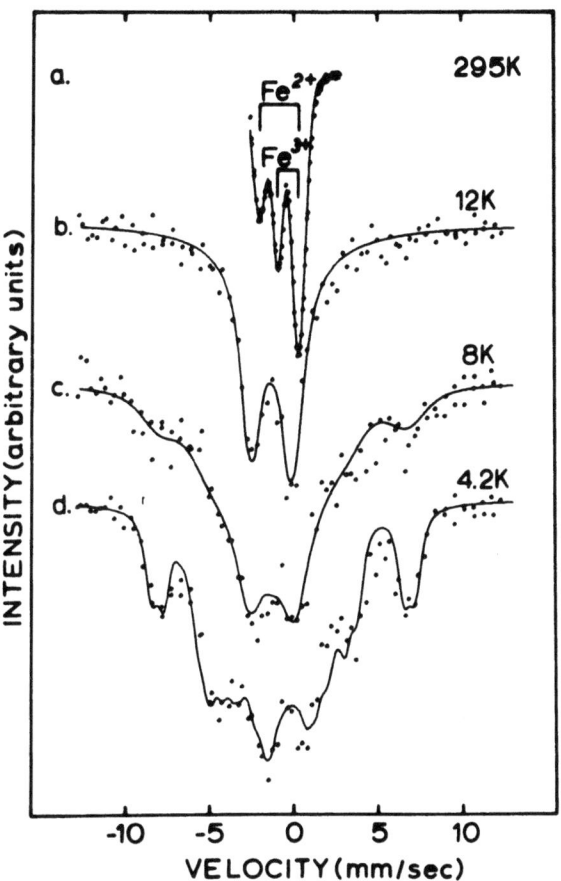

FIG. 2. Cobalt aluminosilicate glass (25 at.% Co) Mössbauer spectra at various temperatures (dots are experimental points, solid lines are fitted curves).

is not sensitive to the chemical environment. Hence, a sharply defined magnetic hyperfine field is expected at the Fe^{3+} sites.[7] On the other hand, for the Fe^{2+} ion the 3d electron angular momentum contribution is significant.[7] This contribution will vary from site to site in the glass, giving rise to a distribution of magnetic hyperfine fields at the Fe^{2+} sites.

5. The angle between the electric field gradient and the magnetic hyperfine fields is random.

The fitting of the observed Mössbauer spectra was carried out in two stages. First the spectra of the 17 at.% Co glass at 1.9°K and the 25 at.% Co glass at 4.2°K were fitted using the above assumptions. At these temperatures the relaxation time τ is sufficiently long so that the magnetic hyperfine fields may be considered static. The magnetic hyperfine field distributions for the two glasses obtained from this fitting procedure are shown in Fig. 3. The fitted spectra are compared with the experimental spectra in Figs. 1d and 2d. Considering the limited knowledge of the glass structure the agreement is excellent. Next, the spectra at the higher temperatures were fitted incorporating relaxation effects utilizing a method developed by Blume and Tjon.[8] In fitting these spectra only the relaxation time was allowed to vary. The fitted spectra are compared with the experimental spectra in

TABLE 1. Linewidths, isomer shifts, and quadrupole splittings of Fe^{2+} and Fe^{3+} in Co-aluminosilicate glass and crystalline CoO.

	17 at.% Co Glass		25 at.% Co Glass		CoO (Ref. 6)	
	Fe^{2+}	Fe^{3+}	Fe^{2+}	Fe^{3+}	Fe^{2+}	Fe^{3+}
Line Width	1.02	0.77	0.97	0.72	0.29	>0.4
Isomer Shift	−1.01	−0.48	−1.00	−0.47	−1.12	−0.45
Quadrupole Splitting	2.30	1.25	2.26	1.20	--	--

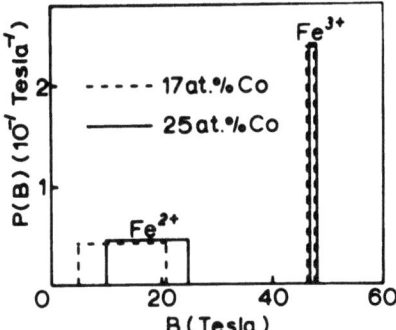

FIG. 3. Distribution of magnetic hyperfine fields in the 17 at.% Co and 25 at.% Co aluminosilicate glass at the Fe^{2+} and Fe^{3+} sites as obtained from fitting the Mössbauer spectra in Figs. 1d and 2d.

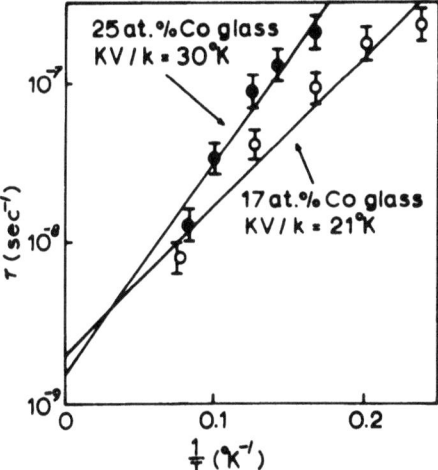

FIG. 4. Relaxation time τ of the superparamagnetic domains vs 1/T in the 17 at.% Co and 25 at.% Co aluminosilicate glass. The data (dots) were obtained from fitting the Mössbauer spectra. The solid lines represent the linear least square fits to the data.

Figs. 1b, 1c, 2b, and 2c. Again, the agreement is excellent. The values of τ obtained are shown in Fig. 4. In order to compare these results with Eq. 1, the values of $\ell n \tau$ were least square fitted to straight lines. From the slopes a value of $KV/k = 21°K$ for the 17 at.% Co glass and a value of $KV/k = 30°K$ for the 25 at.% Co glass were obtained. These values are within a factor of 2 of those obtained from susceptibility, sound velocity and specific heat measurements.[1-3] The values of τ_o (2.0×10^{-9} and 1.5×10^{-9} sec.) are comparable to values quoted in the literature.[9]

ACKNOWLEDGMENT

We would like to thank Dr. G.B. Beard for his assistance and constant interest, and Dr. R.B. Hahn for his help with the preparation of the sources.

REFERENCES

*Work supported in part by the National Science Foundation.

1. R.A. Verhelst, R.W. Kline, A.M. de Graaf and H.O. Hooper, Phys. Rev. B11, 4427 (1975).

2. T.J. Moran, N.K. Batra, R.A. Verhelst and A.M. de Graaf, Phys. Rev. B11, 4436 (1975).

3. R.W. Kline, A.M. de Graaf, L.E. Wenger and P.H. Keesom, AIP Conference Proceedings 29, 169 (1976).

4. V. Cannella and J.A. Mydosh, Phys. Rev. B6, 4220 (1972).

5. L.H. Bieman, P.F. Kenealy, and A.M. de Graaf, AIP Conference Proceedings 34 (1976).

6. J.G. Mullen and H.N. Ok, *Mössbauer Effect Methodology*, edited by I. Gruverman (Plenum, New York, 1968), Vol. 4, p. 103; H.N. Ok and J.G. Mullen, Phys. Rev. 168, 550 (1968).

7. N. Greenwood and T. Gibb, *Mössbauer Spectroscopy*, (Chapman and Hall, Ltd., London, 1971), p. 103.

8. M. Blume and H.A. Tjon, Phys. Rev. 165, 446 (1968).

9. L. Néel, in *Low Temperature Physics* (Les Houches, 1961), edited by C. de Witt, B. Dreyfus, and P.G. de Gennes (Gordon and Breach, New York, 1962); W.F. Brown, Jr., Phys. Rev. 130 1677 (1963); A. Aharoni, Phys. Rev. B7, 1103 (1973).

DISCUSSION

A. Bishay: I have difficulties understanding why the Co^{57} ion would preferentially choose to decay to Fe^{3+} ions.

A. M. de Graaf: It does not. Upon decaying the Co^{57} ion stabilizes into either Fe^{2+} or Fe^{3+} depending on the overall composition of the glass.

M. P. O'Horo: Could I ask why the clusters are formed in that system?

A. M. de Graaf: Because of the presence of super-exchange interactions.

M. P. O'Horo: From the magnetic results you have indicated that the clusters are very uniform in size. Could that be due to just a random interaction?

A. M. de Graaf: No, what I showed was the hyperfine field distribution which is not necessarily the same as the domain size distribution. I think they are related but we do not know what the relation is.

R. A. Levy: Did you attempt to examine how the Mössbauer parameters, that is, the isomer shift quadrupole splitting and linewidth vary with temperature?

A. M. de Graaf: We are in the process of doing that.

M. P. O'Horo: Why are the resonant peaks of the Fe^{2+} ions appear to be spread out relative to the peaks of the Fe^{3+} ions in these glasses?

A. M. de Graaf: The model we used in this study explains well that preferential spreading.

OXIDATION - REDUCTION EQUILIBRIUM IN GLASS FORMING MELTS

Dr. A. Paul

Department of Ceramics, Glasses and Polymers

University of Sheffield

ABSTRACT

The effect of temperature, oxygen fugacity, and melt composition on the oxidation-reduction equilibrium in glass-forming oxide melts has been discussed. The thermodynamics of mutual oxidation-reduction in glass, and the effect of heat treatment at lower temperatures has also been described.

Oxidation-reduction processes play a very important part in glass manufacture, particularly in the preparation of homogeneous glass free from bubbles and in making coloured glasses containing transition metal ions. Transition metals, due to their unique electronic structure, exhibit variable valence states and co-ordination geometries in glass forming oxide melts. When a transition metal oxide is introduced into an oxide melt, very often it distributes into different states of oxidation depending upon the time and temperature of melting, the melt composition, and the furnace atmosphere. In a given set of conditions, after sufficient time of melting, a melt comes to equilibrium with the partial pressure of oxygen in the ambient atmosphere and the relative concentrations of the different oxidation states reach equilibrium values; a typical example of iron (II)-iron (III) equilibrium is shown in Figure 1.

The oxidation-reduction reaction in an oxide melt may be represented in different ways[1]. Taking iron (II)-iron (III) as an example, the reaction may be written in

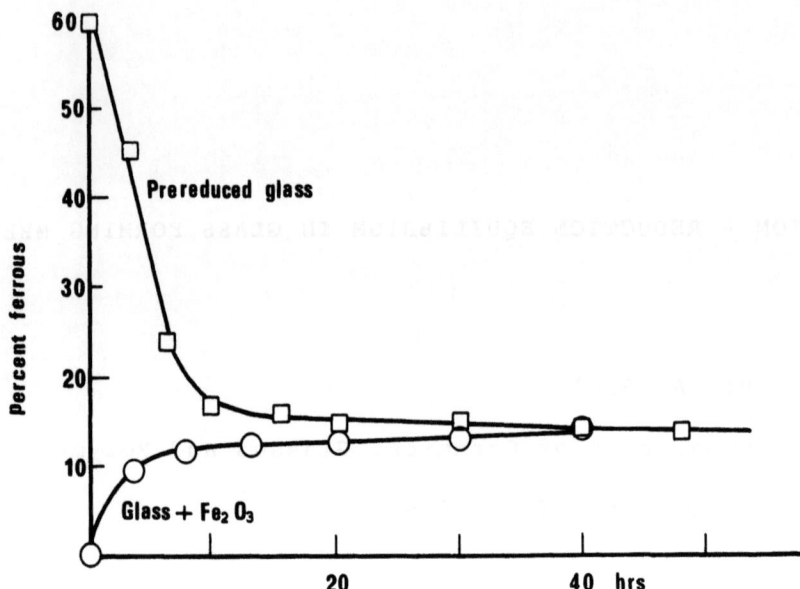

Figure 1: Attainment of iron(II)-iron(III) Equilibrium in $30Na_2O, 70SiO_2$ melt with time at $1400°C$ (total iron = 0.50 wt.% as Fe_2O_3)

terms of pure oxides:

$$4Fe_3O_4 \text{ (melt)} + O_2 \text{ (melt)} \rightleftharpoons 6Fe_2O_3 \text{ (melt)} \quad (1)$$

It should be pointed out that in the solution of the melt Fe_3O_4 and Fe_2O_3 may not be present as molecular species as written in the above equation; it is only a convenient way of expressing the chemical stoichiometry, and from there to calculate the Raoultian activity of the participating oxides. In principle the same equilibrium conditions in the melt can be represented by any one of the following relations:

$$4FeO \text{ (melt)} + O_2 \text{ (melt)} \rightleftharpoons 2Fe_2O_3 \text{ (melt)} \quad (2)$$

or in terms of the ionic species present in the system (which sometimes can be identified with spectroscopic and electrochemical techniques):

$$4Fe^{2+} \text{ (melt)} + O_2 \text{ (melt)} \rightleftharpoons 4Fe^{3+} \text{ (melt)} + 2O^{2-} \text{ (melt)} \quad (3)$$

Equation (3) is formally equivalent to equation (2) but suffers from the disadvantage of introducing species

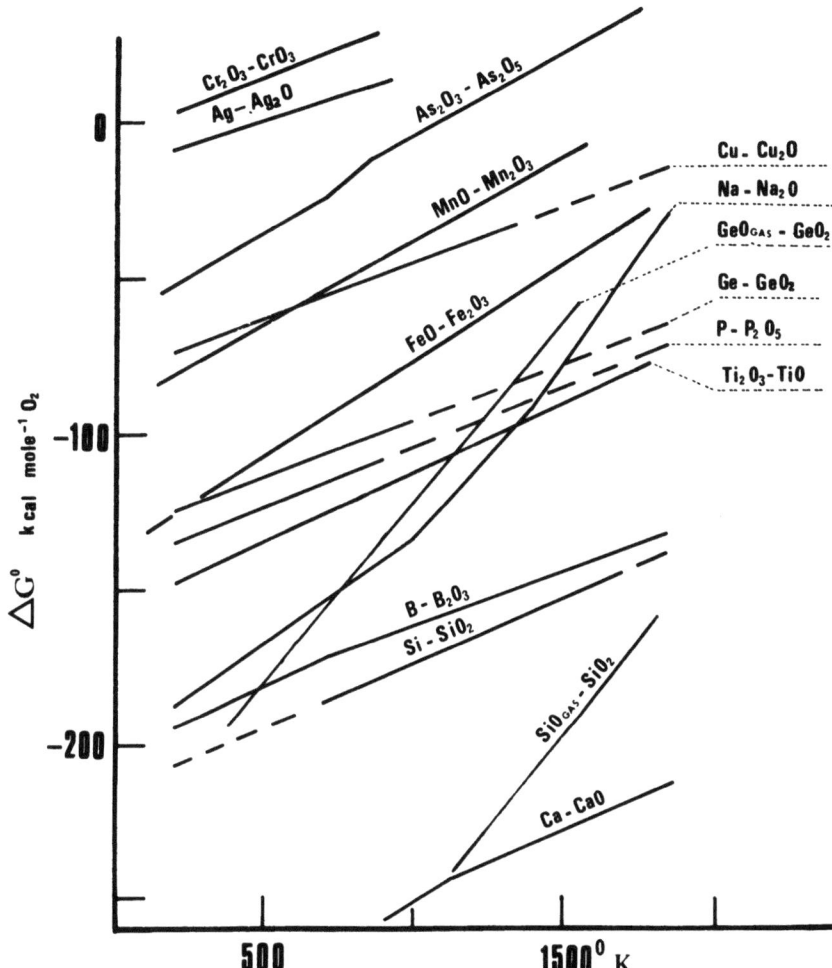

Figure 2: Ellingham diagram for some oxides

whose activity has not yet been satisfactorily measured. With the present state of knowledge choosing any one of (1), (2) or (3) is a matter of personal choice and convenience.

The standard free energy of formation of different pure oxides can be determined by various physico-chemical techniques (gas equilibration, emf measurements etc.,) and is the basis of the well-known Ellingham diagram. The Ellingham diagram for some oxides of interest to glass technology is shown in Figure 2. It can be seen that the easily reducible oxides are on the top of the diagram, the transition metal oxides flock together at the middle, and the conventional glass forming oxides like CaO, B_2O_3, SiO_2 lie at the bottom portion of the diagram. This explains why in a glass forming oxide melt containing say iron oxide, iron (III) can be reduced to iron (II) prior to any reduction of CaO or SiO_2. However, the Ellingham diagram is valid only for pure oxides. Transition metal oxides do not mix ideally with glass forming melts, and consequently activity corrections have to be made in Figure 2 before it can be used profitably in the case of glass forming melts. Such corrections for FeO and Na_2O in a sodium silicate melt is shown in Figure 3.

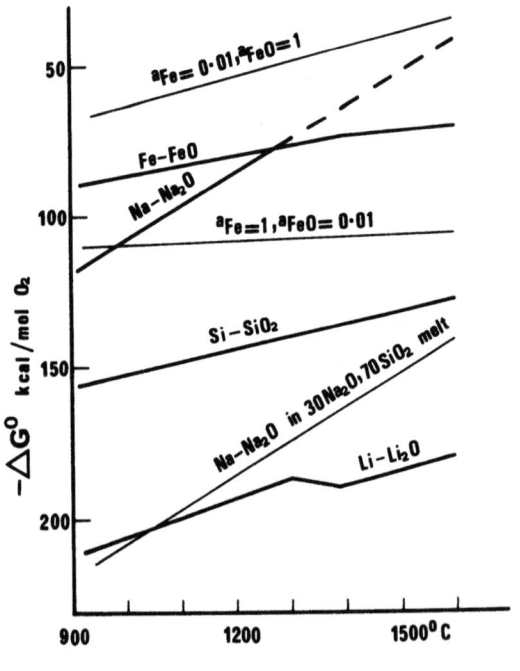

Figure 3:

Effect of changing iron and sodium oxide activity on the Free energy diagram.

OXIDATION

From previous discussions it is clear that the factors determining oxidation-reduction equilibrium in an oxide melt are:

a) fugacity of oxygen in the melt,
b) equilibrium constant which is a function of standard free energy of the reaction, and temperature, and
c) activity of the transition metal oxides in the melt.

a) Effect of pO_2 on the Oxidation-Reduction Equilibrium in Glass.

As expected, the oxidation-reduction equilibrium in an oxide melt moves towards the oxidised side with increasing oxygen fugacity. Under equilibrium, the fugacity of oxygen in the furnace atmosphere and that in the melt are same. For simplicity the fugacity of oxygen in the furnace atmosphere is generally replaced by the partial pressure of oxygen. This replacement of fugacity by partial pressure is permissible only at low pressures. At high pressures deviations are quite frequent especially in the case of chemical solubility of the gas.

A number of workers[2] have studied the effect of pO_2 on the oxidation-reduction equilibrium in glass forming melts, and the relationship expected from equation (3) has always been observed. Some typical examples are shown in Figure 4.

b) Effect of Temperature on the Oxidation-Reduction Equilibrium in Glass

The oxidation-reduction equilibrium in an oxide melt moves towards the reduced side with increasing temperature of melting. This is expected from the Ellingham diagram (Figure 2) where all the free energy lines have positive slopes. The slope of the free energy line against temperature is a measure of entropy change, for $\Delta G^° = \Delta H^° - T\Delta S^°$. In redox reactions in glass, as written in equations (1) or (2), one mole of gaseous oxygen is converted into the condensed liquid phase, thus the entropy of the system is lowered. This also explains the change of slope of the free energy line at a temperature of any phase transformation in the system. Over a limited temperature range $a_{O^{2-}}$ (melt) remains virtually unchanged for a melt of constant chemical composition[3]. Thus from equation (3)

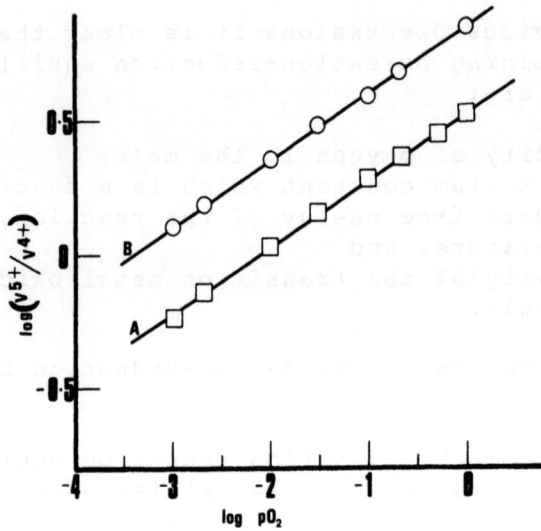

Figure 4: Variation of [vanadium (V)/vanadium(IV)] ratio with pO_2 in a $40BaO, 60B_2O_3 + V_2O_5$ melt at 1200°C. A - 23.8 wt.% V_2O_5, B - 2.38 wt.% V_2O_5

$$4 \log (Fe^{3+}/Fe^{2+}) = \log K + \text{constant} \qquad (4)$$

From Van't Hoff Isochore

$$\log K = -\Delta H/(4.575\ T) + \text{constant} \qquad (5)$$

Combining (4) and (5)

$$4 \log (Fe^{3+}/Fe^{2+}) = -\Delta H/(4.575\ T) + \text{constant} \qquad (6)$$

Thus a plot of $\log (Fe^{3+}/Fe^{2+})$ against $1/T$ is expected to be linear. Such plots for ferrous-ferric equilibrium in different glasses are shown in Figure 5.

c) Activity of Transition Metal Oxides in Glass Forming Melts

Figure 6 shows the variation of iron (III)/iron (II) concentration ratio in different simple silicate glasses when equilibrated at 1400°C with air as the furnace atmosphere. From equation (2)

OXIDATION 603

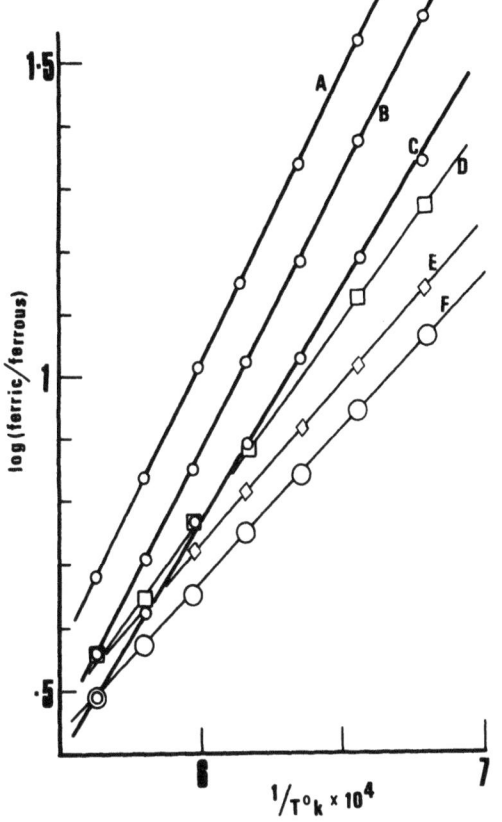

Figure 5:

Variation of [iron(III)/iron(II)] ratio with temperature in different silicate melts

A - $30K_2O, 70SiO_2$, B - $30Na_2O, 70SiO_2$, C - $30Li_2O, 70SiO_2$
D - $25Na_2O, 5CaO, 70SiO_2$,
E - $20Na_2O, 10CaO, 70SiO_2$,
F - $15Na_2O, 15CaO, 70SiO_2$.

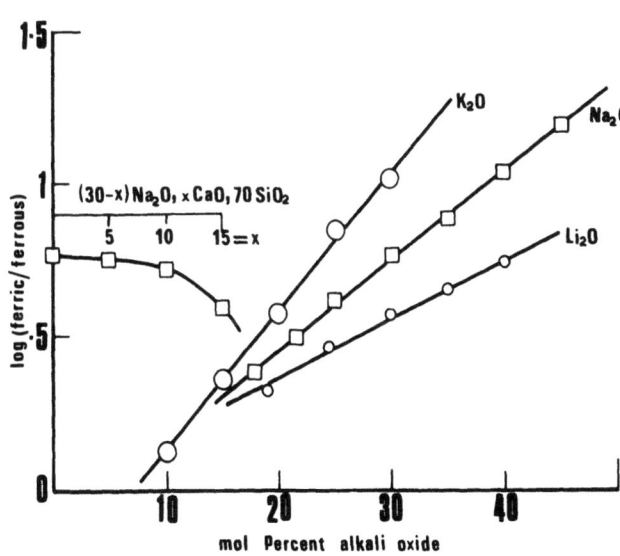

Figure 6:

[iron(III)/iron(II)] ratio in different silicate melts equilibrated at 1400°C with air.

$$K = \frac{a_{Fe_2O_3}^2}{a_{FeO}^4} \cdot \frac{1}{pO_2} \qquad (7)$$

All these experiments were made with a constant temperature and pO_2 (which fixes K and pO_2 in equation (7)). Since the concentration ratio of ferric and ferrous iron changes with melt composition the conclusion must be that the activity coefficients γFe_2O_3 and γFeO change with the melt composition.

The activity and consequently the activity coefficient of any component oxide in solution is related to the excess free energy of mixing. In the case of transition metal oxides this excess free energy of mixing originates mainly from two sources: the Madelung energy and the ligand field stabilisation energy. In the case of a solution of two ionic oxides with known structure, the excess free energy of mixing can be calculated[4]; some typical examples are shown in Figure 7.

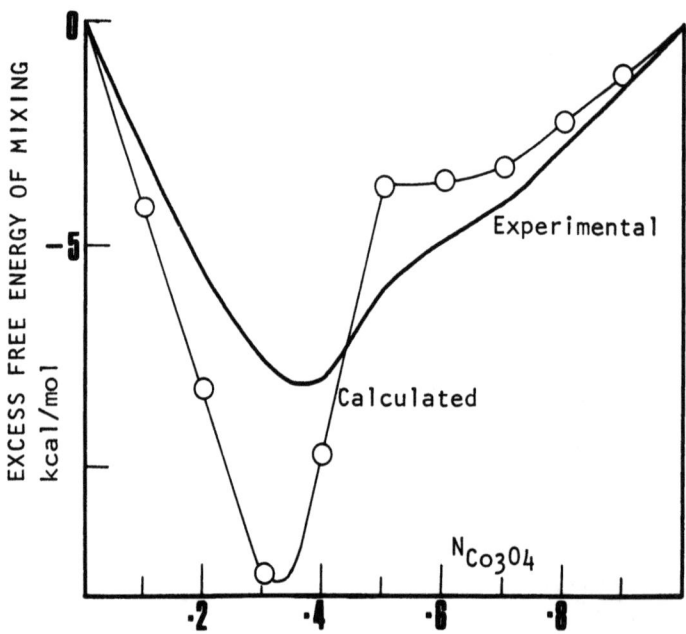

Figure 7: Calculated and experimentally determined Excess Free energy in the solid solution system: $(1-x)Fe_3O_4$, xCo_3O_4.

However, glass forming oxide melts are far from ionic, and no such simple approach to calculate the excess free energy of mixing has yet been developed. Some experimentally determined activity of a few transition metal oxides in different glass forming melts are shown in Figures 8 and 9. It may be seen in Figure 9(b) that at low concentrations of Cu_2O in the melt, $N_{Cu_2O} \propto \sqrt{a_{Cu_2O}}$ indicating that Cu_2O dissociates into isolated Cu^+ ions in this melt at low concentrations.

Transition metal ions in glass forming melts can occur in different coordination symmetries depending upon composition and temperature of the melt[5]. Very often these coordination geometries are classified into two broad groups: "octahedral" and "tetrahedral". In oxide systems the site preference energy of any particular cation can be calculated using the electrostatic approximation; some typical results of calculation are given in Table 1.

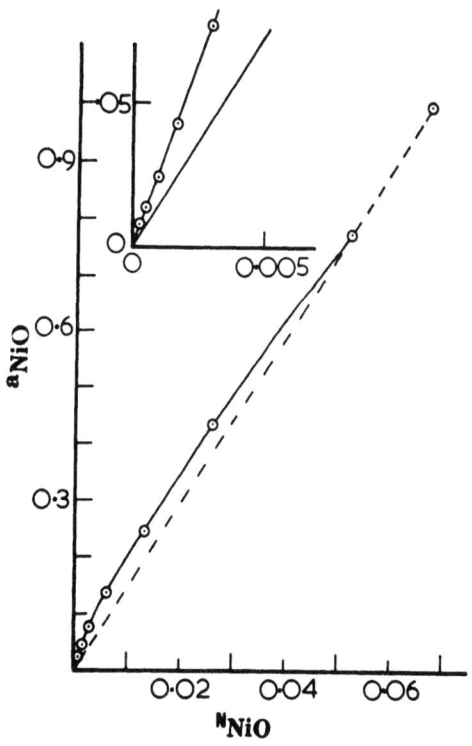

Figure 8:

Activity-concentration relationship of NiO in $Na_2O, 4B_2O_3$ melts at 1000°C.

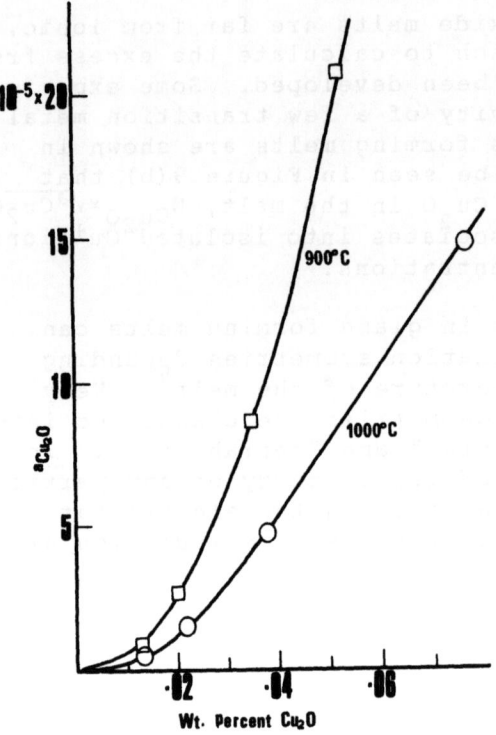

(a)

Figure 9:

Activity-concentration relationship of Cu_2O in $3Na_2O,7B_2O_3$ melts.

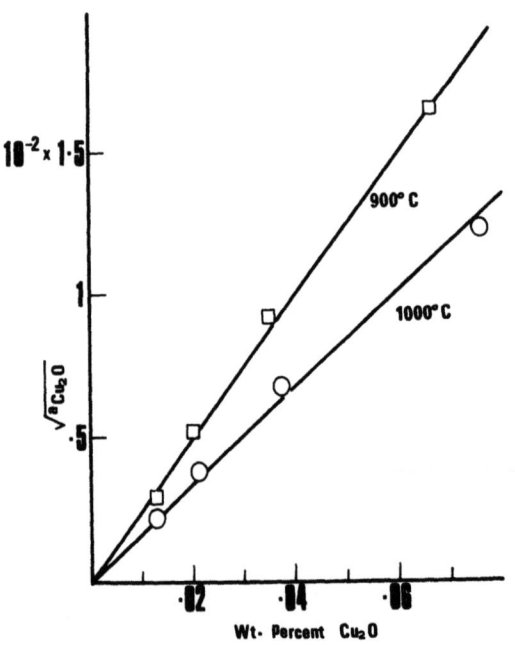

(b)

OXIDATION

INTERACTION BETWEEN TWO REDOX OXIDES IN GLASS

The oxidation state of commercial glasses is usually controlled not by altering the furnace atmosphere or temperature of melting but by adding an oxidising or reducing oxide to the melt. As for example, the use of arsenic or cerium oxides for oxidising FeO in glass forming melts is a common practice.

As discussed before when a melt is equilibrated at a temperature, T, with an oxygen pressure, $pO_2 = x$, this gives an invariant point on the Ellingham diagram. Such a point for air ($pO_2 = 0.21$ atmosphere) at $1400°C$ corresponds to $-5,188$ cals and is shown as A in Figure 10. Now if a redox oxide is added to this melt and brought to equilibrium, the activities of the redox oxides will be adjusted to satisfy the following condition:

$$-5,188 = \Delta G_T^o + RT \ln K \qquad (8)$$

where ΔG_T^o is the standard free energy at $T°k$ as shown on the Ellingham diagram. If two redox oxides are added to the same glass together, each oxidation-reduction equil-

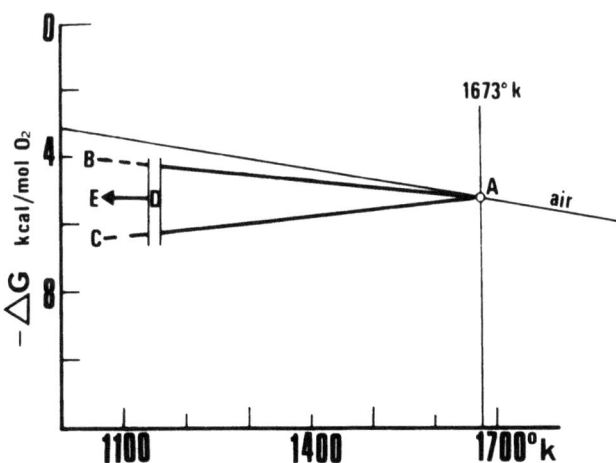

Figure 10: Schematic Cooling path of a mixed redox system in a glass-forming oxide melt.

ibrium will be separately adjusted to comply with the relation (8). When this melt is cooled, the melt becomes virtually cut off from atmospheric oxygen, for the diffusion of molecular oxygen inside the glass forming melt is a very slow process. Now since the activity coefficient of different redox oxides dissolved in a glass forming melt changes differently with temperature (activity coefficient of a redox oxide also changes with concentration and melt composition), the free energy of different redox systems will follow different paths on cooling, as shown schematically by lines AB and AC in Figure 10. It is clear from Figure 10 that the free energy difference between the two redox systems increases with increase of cooling, and the two redox systems start interacting (mutual oxidation-reduction). However, this interaction can not continue for a long time for eventually the temperature of the melt falls below a certain critical value and all long-range readjustment in structure becomes restricted due to the high viscosity of the melt. Thus below a "fictive temperature" the system cools down to a thermodynamically metastable state along some line DE shown in Figure 10.

Let us take an example to explain this point. In Figure 2 the $MnO-Mn_2O_3$ line lies well below the As_2O_3-As_2O_5 line. This means that in the case of pure oxides As_2O_5 will oxidise MnO to Mn_2O_3. However, in an oxide melt arsenic oxide is well known to reduce the pink colour due to Mn_2O_3. How is this possible?

The variation of activity coefficients of manganese and arsenic oxides dissolved in a silicate melt with temperature is shown in Figure 11. Now if we equilibrate this silicate melt at $T = 1673°k$ and $pO_2 = 0.21$ atmosphere containing a certain amount of manganese oxide dissolved in it, the equilibrium condition will be represented by

$$-5,188 = \Delta G^\circ_{1763}(MnO-Mn_2O_3) + 4.575 \times 1673 \log \frac{N^2_{Mn_2O_3}}{N^4_{MnO}} \cdot \frac{\gamma^2_{Mn_2O_3}}{\gamma^4_{MnO}} \cdot \frac{1}{0.21} \quad (9)$$

If we now isolate the melt from the atmosphere and cool, since no oxygen either enters or leaves the system, then $N_{Mn_2O_3}$ and N_{MnO} can not change and, since $\gamma_{Mn_2O_3}$

OXIDATION

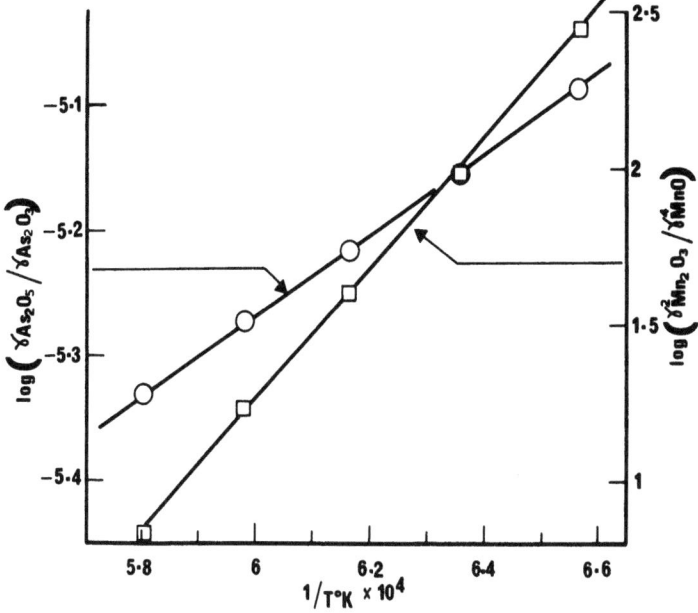

Figure 11: Variation of activity coefficients of manganese and arsenic oxides with temperature in $30Na_2O, 70SiO_2$ melt.

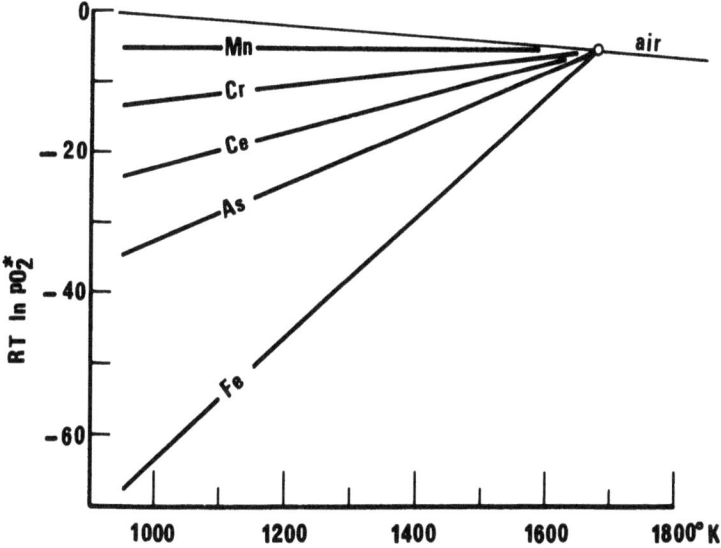

Figure 12: Pseudo oxygen potential of different redox oxides in $30Na_2O, 70SiO_2$ melt.

and γMnO will change with temperature, the system will acquire a pseudo equilibrium pO_2^* which can be calculated from a knowledge of $\Delta G°$ (MnO-Mn$_2$O$_3$) at the lower temperature, and the value of the activity coefficient ratio, $\dfrac{\gamma^2_{Mn_2O_3}}{\gamma^4_{MnO}}$ at that temperature. Such lines of psuedo pO_2^* in a silicate melt for various redox systems are shown in Figure 12. It is clear from this figure that although the standard free energy line of the MnO-Mn$_2$O$_3$ system is below that of As$_2$O$_3$-As$_2$O$_5$ line, the pseudo oxygen pressure produced by arsenic oxides in this glass on cooling is much lower than that due to manganese oxides dissolved in it. Thus on cooling As$_2$O$_3$ reduces Mn$_2$O$_3$ in this melt, and this process will continue throughout the interaction zone until the high viscosity at lower temperature intervenes and stops any further mutual oxidation-reduction.

Thus the important factors determining mutual oxidation-reduction in glass forming melts are:

1) The initial conditions of equilibrium, this fixes the point A in Figure 10,
2) The variation of activity coefficient ratio with temperature for the individual redox system; the larger the difference in variation between the two systems the greater is the mutual interaction,
3) The rate of cooling of the melt. If the rate of cooling is faster than the time necessary for electron transfer between two redox ions and the consequent structural rearrangements involved with it then the extent of interaction will be sensitive to further heat treatment around the "fictive temperature".

Table 1

Octahedral Site Preference Energies
(P in kcal/g atomic weight) for
Various Cations

Ion	P	Ion	P	Ion	P
Li^{1+}	− 3.6	Mg^{2+}	− 5.0	Al^{3+}	− 2.5
Cu^{1+}	− 8.6	Ca^{2+}	−30.7	Ti^{3+}	−21.9
Ag^{1+}	−19.6	Mn^{2+}	−14.7	V^{3+}	−11.6
		Fe^{2+}	− 9.9	Cr^{3+}	+16.6
		Co^{2+}	−10.5	Mn^{3+}	+ 3.1
		Ni^{2+}	+ 9.0	Fe^{3+}	−13.3
		Cu^{2+}	− 0.1	Ga^{3+}	−15.4
		Zn^{2+}	−31.6	In^{3+}	−40.0
		Cd^{2+}	−29.1		

REFERENCES

1. A. Paul and D. Lahiri, J. Amer. Ceram. Soc.,
   49 (1966) 565.
2(a). S. Banerjee and A. Paul, J. Amer. Ceram. Soc.,
   57 (1974) 286.
  (b). W.D. Johnston and A. Chelko, J. Amer. Ceram. Soc.,
   49 (1966) 562.
  (c). W.D. Johnston, J. Amer. Ceram. Soc.,
   47 (1964) 198
   48 (1965) 184
  (d). F. Irmann, J. Amer. Chem. Soc., 74 (1952) 4767.
3. R.W. Douglas, P. Nath and A. Paul, Phys. Chem. Glasses,
   6 (1965) 216.
4. A. Paul and S. Basu, Trans. Jour. Brit. Ceram. Soc.,
   73 (1974) 167.
5(a). A. Paul and R.W. Douglas, Phys. Chem. Glasses,
   8 (1967) 233
   9 (1968) 21.
  (b). A. Paul, Jour. Non-Cryst. Solids. 15 (1974) 517.

Table 1

Octahedral Site 'Preference' Energies
(in kcal/g-atom's weight) for
Various Cations

Ion	P	Ion	P	Ion	P
		Mg	5.0	Al	
		Fe		Ti	
		Mn		Mn	
		Zn			

EFFECT OF TEMPERATURE AND OXYGEN PARTIAL PRESSURE ON COORDINATION AND VALENCE STATES OF FE CATIONS IN CALCIUM SILICATE GLASSES: A MÖSSBAUER STUDY[*]

R. A. Levy

Rensselaer Polytechnic Institute, Troy, N. Y. 12181

Introduction

Current models of silicate liquids are generally discussed from a structural standpoint in terms of an irregular assemblage of O^{2-} anions in which the cations are distributed over the available cavities. In these models, it is generally considered that "network-formers", such as the Si^{4+} cations, occupy tetrahedral sites while "network-modifiers" such as the Ca^{2+}, Fe^{2+} or Na^+ cations occupy octahedral sites. Amphoteric cations such as Fe^{3+} of Al^{3+} can occupy both positions. In a study of a wide range of sodium and calcium silicate glasses Pargamin et al.[1] and Levy et al.[2] reported on the usefulness of the Mössbauer effect in determining simultaneously the valence states and coordination of Fe cations present as a major constituent in these amorphous systems. In a continuing effort to provide a more complete basis for understanding the structure of oxide glasses, we consider in this Mössbauer study the effect of equilibration temperature and oxygen partial pressure on the distribution of Fe cations between valences and coordination sites in two calcium silicate glasses. The nominal compositions of the samples (expressed in weight per cent) are as follows:

Group I $[(SiO_2)_{65}(CaO)_{35}]_{65}[Fe_2O_3]_{35}$

These samples were equilibrated in an atmosphere of air at 1350°C, 1391°C, 1436°, 1503°, 1550° and 1569°.

Group II $[(SiO_2)_{45}(CaO)_{55}]_{65}[Fe_2O_3]_{35}$

These samples were equilibrated at a constant temperature of 1550°C in an oxygen pressure of 1.0×10^0, 2.1×10^{-1}, 8.8×10^{-2}, 2.2×10^{-2} and 1.2×10^{-3} atm.

These two compositions were selected because of their relatively low melting points ($\sim 1300°C$) and their ability to remain vitreous upon quenching. Group I was earmarked for the present temperature investigation because of the difference in the fraction of tetrahedrally coordinated Fe^{3+} cations observed[2] in two samples equilibrated at 1390° and 1550°C. No such temperature changes were observed for the glass composition of Group II. The proximity of the composition of sample I to a two-phase region on the SiO_2-CaO-"Fe_2O_3" phase diagram[3] coupled with the observation that at 1390°C the fraction of tetrahedrally coordinated Fe^{3+} cations was close in value to that of the CaO-rich glass matrix of Sample II had led us to suspect microscopic precipitation of a metastable silica rich phase. The fact that no evidence for such precipitates was present in the case of Group II was the compelling reason for conducting the oxygen partial pressure measurements on that set of samples.

Experimental Methods

Starting materials used in preparation of all our samples* were comprised of 99.92% silica, 99.95% calcium carbonate and 99.5% ferric oxide. All constituents underwent standard drying procedures before being weighed, thoroughly mixed, sintered, and melted. All samples were equilibrated in air at temperatures varying from 1350°C to 1569°C in a vertical furnace and then rapidly quenched in water. All temperatures were measured and controlled to within an estimated accuracy of ± 5°C. Desired oxygen pressures for samples of Group II were attained by passing mixtures of O_2 and N_2 in a furnace adequately sealed from the atmosphere. The oxygen pressures were monitored at the entrance of the furnace with the aid of calibrated flowmeter tubes and at the exit with a calcia stabilized zirconia electrolyte cell described by G. Bernard[4]. The uncertainty in the calculation of the oxygen pressure was estimated to be within 0.15 $\log_{10}$ units. Parts of the quenched melts were ground into a 325-mesh powder, diluted with a boron nitride powder and pressed into a lucite cavity holder. An investigation of the optimum thickness of the glass samples revealed no discrepancies in the Mössbauer parameters up to a thickness corresponding to 0.30 mg/cm^2 of Fe^{57}. All samples discussed in this work were weighed in proportion to boron nitride to yield a thickness corresponding to

---

*Samples were prepared at Carnegie-Mellon University with support provided by the National Science Foundation.

0.20 mg/cm^2 of Fe^{57}. X-ray analyses conducted on the powdered specimens revealed all samples to be completely vitreous.

The Mössbauer spectra were obtained at room temperature from a constant acceleration transmission spectrometer. The radioactive source used consisted of 25 mCi Co^{57} diffused into a copper matrix. The Doppler shifted 14.4 keV photons traversing the samples were detected with a proportional counter filled with a gas mixture of Kr and CO$_2$. The data then was accumulated in a 256-channel analyser operating in the time-sequence-storage mode. The known spectrum of a sodium nitroprusside specimen was used for calibration purposes. The absorption spectra were plotted vs relative velocity scale equivalent to the energy of the incident γ-rays. The Mössbauer parameters quoted in this study are in units of mm/sec and all isomer shifts are reported relative to Co^{57} in Cu.

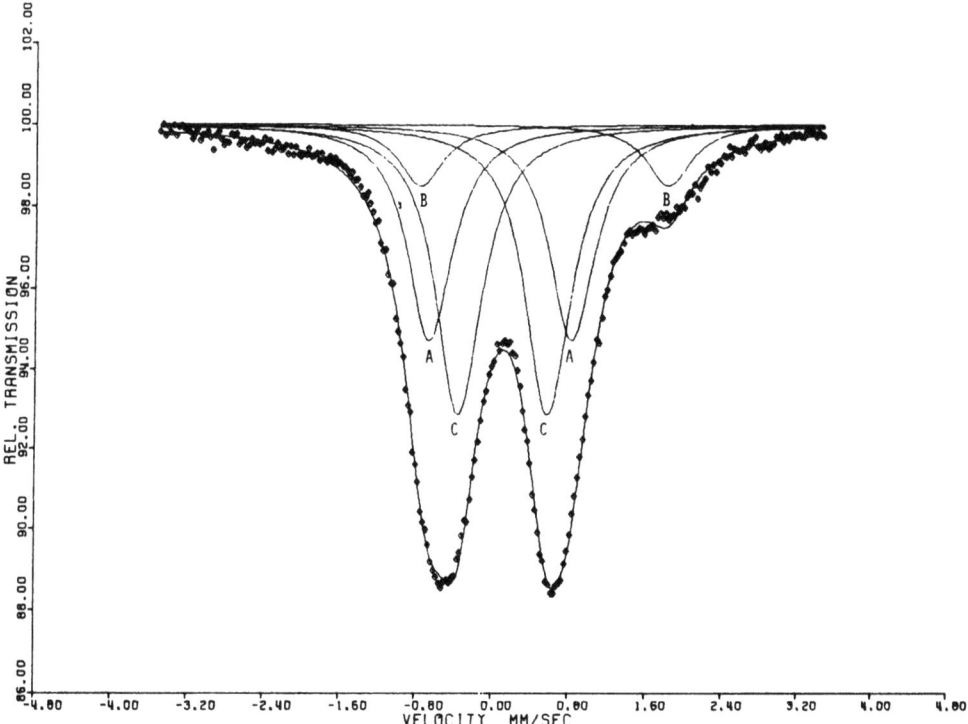

Fig. 1. Mössbauer spectrum of a glass from Group I equilibrated at 1350°C and rapidly quenched in water. The pairs A and B correspond to the contribution of the Fe$^{3+}_{tet}$ and Fe$^{3+}_{oct}$ cations respectively, the pair B corresponds to the contribution of the Fe^{2+} cations.

Results and Discussion

A typical Mössbauer spectrum of a calcium silicate glass containing Fe_2O_3 is shown in Fig. 1. The common feature to all observed spectra was the presence of three broadened asymmetric peaks which we fitted to a sum of six lorentzian-shaped components: a doublet arising from the contribution of the Fe^{2+} cations and two other pairs corresponding to the Fe^{3+} cations with tetrahedral and octahedral coordination. Details of the fitting procedure can be found in Refs. (1) and (2). For a given Mössbauer spectrum, the Fe^{2+} doublet is observed to emerge with the largest isomer shift and quadrupole splitting.

The isomer shift is the result of a difference in s-electron density at the nuclei of the source and absorber. The difference in density translates in a difference in the energy of emission and absorption of the γ-rays that results in a displacement of the absorption spectrum from relative zero velocity. In the case of Fe^{57}, the isomer shift depends strongly on the oxidation state of iron and less strongly on the nature of the coordinating anions and the degree of covalency. Quadrupole splitting results from the interaction of a nuclear quadrupole moment with a local electric field gradient (caused by electrons in the Mössbauer isotope or from surrounding ions) that splits the single transition line of Fe^{57} into a two-peak (doublet) spectrum. The quadrupole splitting is sensitive to local symmetry and coordinating ligands around the iron atoms.

The observed values of the isomer shift are explained by considering the fact that the Fe^{2+} cation ($3d^6$, $^5D_4$) has one d-electron more than the Fe^{3+} cation ($3d^5$, $^6S_{5/2}$) causing more screening of s-electrons from the nucleus resulting in a lower electron density at the nucleus and higher value of the isomer shift. The larger quadrupole splitting observed for Fe^{2+} cations is, as first suggested by Kurkjian[5], the result of axial or rhombic symmetry distortions generated by the d-electron outside the half-filled spherical shell. In the case of the Fe^{3+} cations, the electrons are spherically symmetric and would not contribute to the electric field gradient at the nucleus. For an ideal tetrahedral or octahedral coordination the symmetry is cubic and no quadrupole splitting should be observed. Thus the presence of quadrupole doublets are indicators in that case of marked deviations from cubic symmetry. Assignments of doublets to respective oxidation states and coordination sites was made in accordance with existing surveys for values of the isomer shift and quadrupole splitting of Fe^{2+} and Fe^{3+} cations in crystals and glasses.[5] In those surveys the isomer shift values of Fe^{3+} cations were observed to fall into two categories. For values of δ less than about 0.15 mm/sec (relative to Co^{57} in Cu) the Fe^{3+} cations were observed to occupy tetrahedral sites while for values of δ larger than about 0.15 mm/sec preference for octahedral sites was indicated. In all cases the quadrupole splitting of the Fe^{3+} cations in tetrahedral coordination was <u>always</u> roughly twice as large than that of the Fe^{3+} cations in octahedral coordination. In the present work if the best fitted spectrum showed two doublets with an isomer shift near 0.15 mm/sec, the doublet with the smaller isomer shift and larger quadrupole splitting was selected to correspond to Fe^{3+} cation in tetrahedral coordination. By integrating the areas under a given doublet we were able to determine the fraction of Fe^{2+} cations and the fraction of Fe^{3+} cations in octahedral coordination, referred to here as R.

Figure 2 shows a plot of the fraction of Fe^{2+} cations as a function of equilibration temperature for the glasses of Group I. The observed data indicates a rapid increase in the fraction of Fe^{2+} cations from an average value of 7.61% at $1315°C^2$ to 23.44%

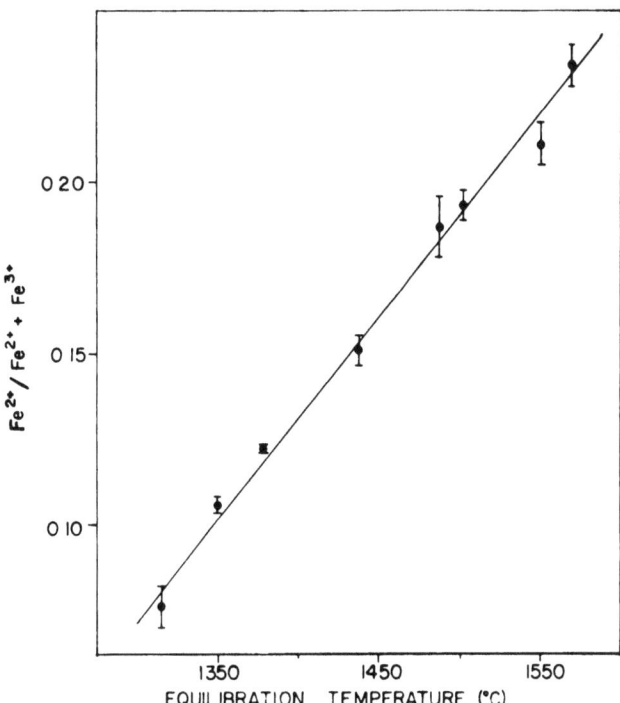

Fig. 2. Mössbauer fraction of Fe^{2+} cations as a function of equilibration temperature for glasses of Group I. Data points shown at 1315°C and 1483°C are taken from Ref. 2.

at 1569°C. This is explained by considering the equilibrium reaction

$$2O^{2-} + 4Fe^{3+} \rightleftarrows 4Fe^{2+} + O_2 \quad (1)$$

that yields a concentration equilibrium constant

$$K = (Fe^{2+}/Fe^{3+})^4 \; p_{O_2} \quad (2)$$

which displaces the above redox reaction toward the reduced side with increasing temperature of the melt. The equilibrium constant as written in Eq. 2 assumes that the reaction proceeds in a melt of constant chemical composition where the O^{2-} activity is assumed to remain unchanged and in a constant environment where the activity coefficients of the Fe^{2+} and Fe^{3+} cations may also be assumed to be constant. For simplicity the activity of oxygen has been replaced by the partial pressure of oxygen which is permissible only at low pressures.

Table 1

Fraction of ferrous ions, ferric ions in tetrahedral and octahedral sites and value of R_{Fe} for the $[(SiO_2)_{65}(CaO)_{35}]_{65}[Fe_2O_3]_{35}$ glass equilibrated in air at various temperatures.

Equilibration Temperature (°C)	Fe^{2+} (%)	S.D.	Fe^{3+}_{tetr} (%)	S.D.	Fe^{3+}_{oct} (%)	S.D.	R_{Fe} (%)	S.D.
1350	10.93	0.38	37.62	1.19	51.45	1.12	57.76	1.29
	10.22	0.37	40.05	1.06	49.66	1.01	55.37	1.13
1391	11.17	0.38	40.44	1.07	48.39	1.02	54.47	1.18
	12.76	0.41	36.27	1.21	50.97	1.13	58.42	1.33
	12.10	0.46	38.01	1.38	49.89	1.29	56.78	1.50
	11.88	0.39	38.71	1.09	49.41	1.03	56.09	1.19
	12.66	0.41	35.35	1.17	51.99	1.08	59.52	1.28
	12.69	0.44	39.23	1.26	48.08	1.18	55.06	1.40
	12.83	0.38	38.96	1.01	48.21	0.95	55.31	1.10
1436	15.11	0.45	36.01	1.22	48.88	1.13	57.58	1.36
1503	19.28	0.63	32.00	1.53	48.72	1.41	60.33	1.82
	19.32	0.62	30.73	1.56	49.95	1.39	61.89	1.81
1550	21.08	0.64	29.04	1.55	49.87	1.36	63.19	1.81
1569	23.33	0.86	22.78	2.09	53.89	1.77	70.29	2.56
	23.55	0.93	23.42	2.22	53.02	1.89	69.36	2.72

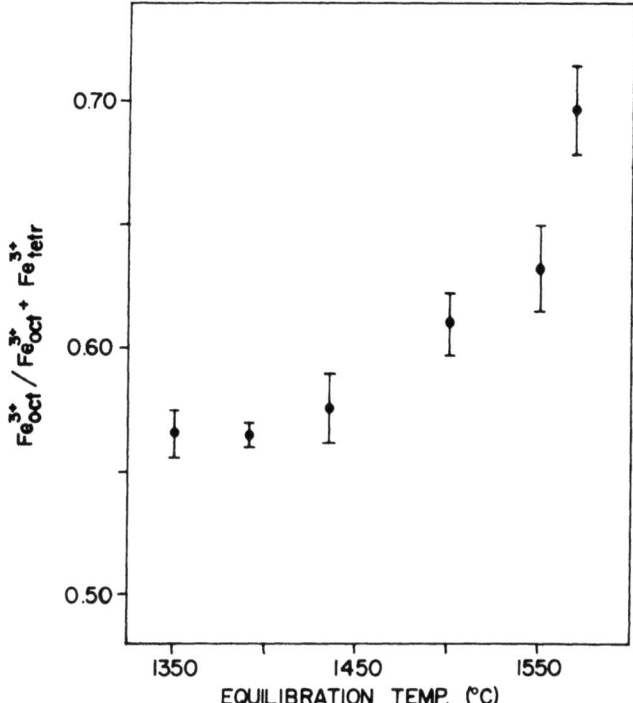

Fig. 3. Mössbauer fraction of ferric cations in octahedral sites as a function of equilibration temperature for glasses of Group I.

A linear least square fit through the data yields in the considered temperature range, and at this fixed oxygen partial pressure of 2.1×10^{-1} atm, a rate of increase for the fraction of Fe^{2+} cations equal to $5.95 \times 10^{-4}/°C$. It is interesting to note that the increase in the Fe^{2+} fraction with increasing equilibration temperature is compensated by a preferential decrease in the tetrahedrally coordinated Fe^{3+} cations as seen in Table 1 and Fig. 3. This behavior which was not observed in the case of samples from Group II equilibrated at different temperatures is believed to be caused by the presence of a metastable silica-rich phase that becomes more abundant in the samples of Group I equilibrated at lower temperatures. Evidence of this is provided by x-ray diffraction data taken on a sample melted at 1391°C and heat treated for 12 hours at 800°, 900°, 950°, 1000°, 1050°, and 1100°C. Results of these measurements[6] revealed the presence of a unidentified metastable phase that disappears during the heat treatment period at 950° to give way to stable structures of cristobalite, hematite and andratite. Further evidence for the presence of this metastable phase can be obtained from a comparison of the R value for samples from Group I and Group II. In the case for instance of the samples equilibrated at 1391°C, one finds an average R = 0.56 for the glass composition with the high SiO_2/CaO ratio quite close in value to the average R = 0.55 observed[2] for the glass composition with the low SiO_2/CaO ratio. The lack of dependence of R on basicity of the melt can only be explained by considering precipitation of an SiO_2-rich phase containing little or no iron. These observations coupled with the fact that the composition of samples in Group I lies near a two phase region on the ternary SiO_2-CaO-"Fe_2O_3" phase diagram[3] lend strong support to our conclusion.

For glasses of Group II the fraction of Fe^{2+} cations is plotted as function of log p_{O_2} in Fig. 4. The observed increase in this fraction with lower oxygen partial pressures is a direct

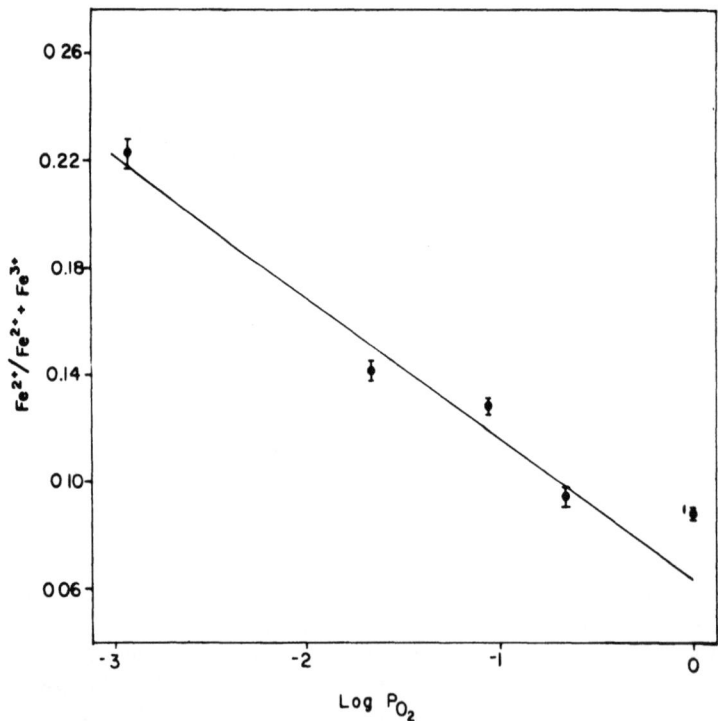

Fig. 4. Mössbauer fraction of ferrous cations as a function of log p_{O_2} for glasses of Group I.

consequence of the shift in the oxidation-reduction equilibrium (Eq. 1) in these glasses toward the right. This behavior is expected since the equilibrium constant k (in Eq. 2) is temperature dependent so that any decrease in oxygen partial pressure must be compensated by a corresponding increase in the Fe^{2+}/Fe^{3+} concentration ratio in the melt.

At constant equilibration temperature Eq. 2 can be written

$$\log K = 4 \log \frac{Fe^{2+}}{Fe^{3+}} + \log p_{O_2}$$

which yields a slope of 4 in a linear plot of $-\log p_{O_2}$ vs log Fe^{2+}/Fe^{3+}. A linear least square fit of the data, shown for the glasses of Group II in Table 2, yields a slope of 4.8 which is in good agreement with the expected result considering the fact that K is assumed here not to vary with compositional changes. A comparison between the fraction of Fe^{2+} cations observed in the glasses from Group I and Group II equilibrated at 1550°C in air reveal a decrease in this fraction with increased basicity of the melt (i.e., lower SiO_2/CaO ratios). From the general equation of equilibrium of a redox pair (Eq. 1) the thermodynamic equilibrium constant is defined as

Table 2

Fractions of ferrous ions, ferric ions in tetrahedral and octahedral sites, and values of R_{Fe} for the $[(SiO_2)_{45}(CaO)_{55}]_{65}[Fe_2O_3]_{35}$ glass equilibrated in a varying oxygen partial pressure at 1550°C.

p_{O_2}	Fe^{2+} (%)	S.D.	Fe^{3+}_{tet} (%)	S.D.	Fe^{3+}_{oct} (%)	S.D.	R_{Fe} Ratio	S.D.
1.0×10^0	7.16	0.34	42.40	0.89	50.44	0.86	54.34	0.92
	7.58	0.31	41.35	0.85	51.07	0.82	55.26	0.89
	6.24	0.45	43.76	1.07	50.00	1.05	53.32	1.10
	6.44	0.30	42.25	0.71	51.31	0.75	54.84	0.80
	6.40	0.34	41.73	0.93	51.86	0.90	55.41	0.96
2.1×10^{-1}	9.78	0.33	40.44	0.78	49.78	0.77	55.18	0.84
	9.06	0.36	39.86	1.02	51.08	0.97	56.17	1.08
8.8×10^{-2}	13.13	0.45	39.34	1.25	47.53	1.17	54.73	1.36
	12.65	0.41	36.58	1.21	50.77	1.12	58.12	1.32
2.2×10^{-2}	13.79	0.46	36.36	1.35	49.85	1.25	57.85	1.47
	14.50	0.51	35.68	1.42	49.82	1.31	58.27	1.58
1.2×10^{-3}	21.15	0.77	32.72	1.80	46.14	1.63	58.51	2.15
	21.83	0.92	32.53	2.17	45.64	1.97	58.38	2.61
	23.95	1.05	26.19	2.40	49.85	2.07	65.56	2.94

$$K \equiv \frac{[Fe^{2+}]^4}{[Fe^{3+}]^4} \frac{[O_2]}{[O^{2-}]^2} \qquad (3)$$

where the brackets denote activities in the melt. Since in this comparison between samples from the two groups the oxygen partial pressure and the equilibration temperature are kept the same one may be tempted to conclude that the observed decrease in the Fe^{2+}/Fe^{3+} ratio with increased basicity must be due to a decrease in the oxygen ion activity with increased basicity. However, the results of all experiments conducted to measure the oxygen ion activity in oxide melts[7] have proved this parameter to increase with increased basicity. This apparent paradox can be resolved by recognizing that K must vary with composition as well as temperature in these glass systems.

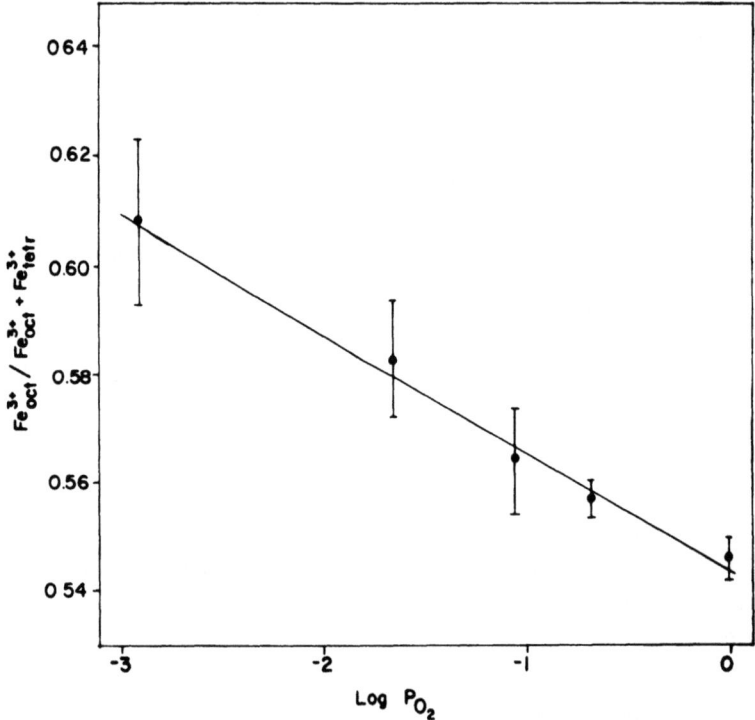

Fig. 5. Mössbauer fraction of ferric cations in octahedral sites as a function of log p_{O_2} for glasses of Group II.

The fraction of Fe^{2+} cations in glasses of Group II is observed (from Table 2 and Fig. 5) to increase at the expense of the fraction of Fe^{3+} cations tetrahedrally coordinated. The preferential decrease in the Fe^{3+} cations tetrahedrally coordinated with lower oxygen partial pressures can be explained using simple geometrical considerations. A complete glass network can be formed in an oxide if the ratio of the number of oxygen anions over the number of cations is equal to 2, as in the case of SiO_2. In a mixture of oxides the oxygen anions from the modifiers (Ca^{2+} or Fe^{2+}) attach themselves to Si atoms interrupting the Si-O network, while the cations remain in the interstials in order to provide electrical neutrality. In the glass systems considered in the present investigation, the oxygen contributed by CaO is used by part of the Fe^{3+} cations to build up their tetrahedra. With decreasing oxygen partial pressures in the furnace atmosphere, the availability of oxygen in the melt must decrease accordingly causing the observed displacement of the $Fe^{3+}_{oct} \rightleftarrows Fe^{3+}_{tet}$ reaction toward the left and the preferential increase in the fraction of Fe^{2+} cations.

The average values of the Mössbauer parameters for the glasses of Group I and II are listed respectively in Tables 3 and 4. An examination of the isomer shifts associated with the Fe^{2+} doublet and each of the Fe^{3+} doublets for glasses of Group I reveal a systematic increase in the average value of this parameter with higher equilibration temperatures. This behavior is consistent with the gradual depletion of the silica-rich phase as

Table 3

Average values of Mössbauer parameters for the $[(SiO_2)_{65}(CaO)_{35}]_{65}$-$[Fe_2O_3]_{35}$ glass equilibrated in air at various temperatures.

Equilibration Temperature (°C)	Fe^{2+} Isomer Shift (mm/sec)	Fe^{2+} Quadrupole Splitting (mm/sec)	Fe^{3+}_{tetr} Isomer Shift (mm/sec)	Fe^{3+}_{tetr} Quadrupole Splitting (mm/sec)	Fe^{3+}_{Oct} Isomer Shift (mm/sec)	Fe^{3+}_{Oct} Quadrupole Splitting (mm/sec)	Half Width ± 0.01 (mm/sec)
1350	0.520 ±.012	2.52 ±.04	0.086 ±.003	1.44 ±.01	0.113 ±.001	0.90 ±.01	0.277 ±.002
1391	0.517 ±.006	2.52 ±.01	0.102 ±.002	1.43 ±.01	0.123 ±.001	0.88 ±.01	0.281 ±.001
1436	0.527 ±.010	2.50 ±.01	0.106 ±.005	1.41 ±.02	0.132 ±.002	0.87 ±.01	0.282 ±.003
1503	0.557 ±.009	2.44 ±.02	0.125 ±.006	1.38 ±.02	0.155 ±.002	0.85 ±.01	0.289 ±.002
1550	0.552 ±.009	2.45 ±.02	0.135 ±.006	1.38 ±.02	0.155 ±.002	0.85 ±.01	0.291 ±.003
1569	0.573 ±.010	2.40 ±.02	0.135 ±.011	1.38 ±.04	0.162 ±.011	0.86 ±.01	0.299 ±.003

Table 4

Average values of Mössbauer parameters for the [SiO$_2$
Average values of Mössbauer parameters for the [(SiO$_2$)$_{45}$(CaO)$_{55}$]$_{65}$-[Fe$_2$O$_3$]$_{35}$ glass equilibrated in a varying oxygen partial pressure at 1550°C.

P_{O_2}	Fe^{2+}		Fe$^{3+}_{tetr}$		Fe$^{3+}_{Oct}$		
	Isomer Shift (mm/sec)	Quadrupole Splitting (mm/sec)	Isomer Shift (mm/sec)	Quadrupole Splitting (mm/sec)	Isomer Shift (mm/sec)	Quadrupole Splitting (mm/sec)	Half Width (mm/sec)
1.0 x 10^0	0.436 ±.008	2.57 ±.01	0.077 ±.001	1.48 ±.01	0.098 ±.001	0.915 ±.004	0.274 ±.001
2.1 x 10^{-1}	0.466 ±.007	2.53 ±.01	0.088 ±.002	1.46 ±.01	0.111 ±.002	0.899 ±.006	0.283 ±.002
8.8 x 10^{-2}	0.496 ±.011	2.44 ±.02	0.099 ±.004	1.41 ±.01	0.133 ±.001	0.856 ±.008	0.297 ±.002
2.2 x 10^{-2}	0.499 ±.012	2.41 ±.02	0.100 ±.005	1.42 ±.02	0.139 ±.002	0.851 ±.010	0.306 ±.002
1.2 x 10^{-3}	0.538 ±.006	2.34 ±.01	0.152 ±.006	1.28 ±.02	0.204 ±.002	0.748 ±.011	0.352 ±.003

reflected here by the relative decrease in the basicity of the iron-bearing calcium silicate matrix at higher equilibration temperatures. A decrease in the basicity of the melt has been previously reported[2] to result in higher average values for the isomer shift. This is clearly seen for instance in the case of a sample from Group I equilibrated at 1350°C which yields shift values ($\delta_{Fe^{2+}}$ = 0.520 mm/sec, $\delta_{Fe^{3+}_{tet}}$ = 0.086 mm/sec, $\delta_{Fe^{3+}_{oct}}$ = 0.113 mm/sec, that are characteristic of values observed in a CaO-rich sample of Group II equilibrated in air at 1550°C ($\delta_{Fe^{2+}}$ = 0.466 mm/sec, $\delta_{Fe^{3+}_{tet}}$ = 0.088 mm/sec, $\delta_{Fe^{3+}_{oct}}$ = 0.111 mm/sec). In the case of the quadrupole splitting, the apparent lack of any marked dependence of this parameter on equilibration temperature reflects in these glasses the absence of significant changes in the structural distortions around the Fe sites. For samples of Group II, the average values of the isomer shift associated with the Fe^{2+} doublet and the two Fe^{3+} doublets (Table 4) are seen to increase with decreasing oxygen partial pressures. Since an increase in the average value of the isomer shift reflects a decrease in the average s-electron density at the Fe site, the observed change in attributed here to an expansion of the orbital volume of the s-electrons caused by the systematic removal of oxygen anions. During this gradual process one expects the single-bonded oxygen anions to be removed first causing less and less of an interruption in the Si-O network. This is evidenced here in the average values of the quadrupole splitting and half width (Table 4) which are shown to reflect a tendency toward smaller distortions in symmetry of the coordination polyhedra and a wider distribution of site parameters with lower oxygen partial pressure.

Acknowledgments

The author wishes to acknowledge W. L. Swanton for experimental assistance as well as both C. H. P. Lupis and P. A. Flinn for many helpful discussions during the various stages of this project.

References

*Work partially supported by the National Aeronautics and Space Administration.

1. L. Pargamin, C. H. P. Lupis and P. A. Flinn, Metall. Trans. 3, 2093 (1972).
2. R. A. Levy, C. H. P. Lupis and P. A. Flinn, Physics Chem. Glasses, 17, 94 (1976).
3. B. Phillips and A. Muan, J. Am. Ceram. Soc. 42, 413 (1959).
4. G. Bernard, "Influence of Oxygen on the Surface Tension of Liquid Silver and its Alloys", Ph.D. thesis, Carnegie-Mellon University (1970).
5. C. R. Kurkjian and E. A. Sigety, Physics Chem. Glasses 9, 73 (1968).
6. R. A. Levy (to be published).
7. R. W. Douglas, P. Nath and A. Paul, Physics Chem. Glasses 6, 216 (1965).

DISCUSSION

R. Hasegawa: How do you explain the fact that only the isomer shift varies with temperature and not quadrupole splitting?

R. A. Levy: Our results are indicating that on the average the s-electron density of a given Fe^{3+} ion is decreasing with increasing equilibration temperature due to the precipitation of a silica-rich phase while the average distortion around that ion remains, roughly speaking, temperature independent and insensitive to this precipitated phase.

A. Bishay: At what temperature were the Mössbauer spectra taken?

R. A. Levy: All the spectra discussed here were taken at room temperature on glasses quenched from various equilibration temperatures. Our next object would be to take those glasses to low temperatures and study relaxation phenomena.

A MÖSSBAUER STUDY ON THE CLUSTERING AND CRYSTALLIZATION PHENOMENA

IN BaO-4B_2O_3 GLASSES CONTAINING DILUTE CONCENTRATIONS OF Fe_2O_3

K. J. Kim,* M. P. Maley,** and R. K. MacCrone

Department of Physics and Division of Materials

Engineering, Rensselaer Polytechnic Institute

Troy, New York 12181

ABSTRACT

The Mössbauer measurements of the clustering and crystallization phenomena in $(1-x)$ $(0.2BaO-0.8B_2O_3)-xFe_2O_3$, $x \leq 10$ mole %, glasses are made in the quenched and heat treated samples. The results indicate that small clusters of two and three iron ions are responsible for the magnetic inhomogeneities observed in the quenched samples. Two superparamagnetic crystalline particles coexist in the heat treated samples whose relative intensities are dependent on the iron concentration and are rather insensitive to heat treating temperatures above 700°C. An epitaxial mode of catalyzed nucleation and crystallization is discussed with reference to the clustering phenomenon and their consecutive developments into crystalline ferrite oxides.

INTRODUCTION

The effect of transition metal ions on catalyzed nucleation and crystallization has been observed in a glass matrix in several studies (1-3), but the mechanism of their influence, such as the type and size of nuclei and their consecutive evolution to crystalline form, has not been well understood. The first task of this problem is to identify the early stage of the nucleation process, and the second requirement includes a close examination of successive stages leading to a crystalline formation in an effort to clarify the proper mechanism of nucleation and crystallization if it exists. For this purpose, structure sensitive techniques such as electron paramagnetic resonance (EPR), magnetic susceptibility measurement, and Mössbauer effect spectroscopy are employed to probe the magnetic properties and microstructures shown by the transition metal ions in glasses.

The glass system, $BaO-B_2O_3$, containing iron-oxide has been studied by Fahmy, et al (4) in the intermediate concentration regime (15 mole % of Fe_2O_3) and Tanigawa, et al (5) in the higher concentrations (from 20 to 55 mole % of Fe_2O_3). Both works utilized magnetic susceptibility and small angle X-ray scattering, and the former identified mixed iron-oxide precipitates by prolonged heat treatment, while the latter observed the formation of barium-ferrite precipitates by suitable heat treatments. Recently, in the dilute concentration regime (less than 10 mole % of Fe_2O_3), the magnetic inhomogeneities and microstructural distributions in the same glass system have been investigated by using EPR, magnetic susceptibility, and Mössbauer effect measurements (6-7). The present study includes a brief discussion of these observations and a close examination of crystallized glasses. By heat treating the quenched metastable oxide glasses, an effort is made to elucidate the mechanism involved in the early stage of nucleation and the consecutive crystallization process.

EXPERIMENTAL

The base glass samples are prepared by mixing reagent grade B_2O_3, BaO, and α-Fe_2O_3 powders in a 20 ml platinum crucible at 1250°C for 1.5 hours and by following the procedure described by Moon, et al (6). The nucleated and crystallized specimens are prepared by heat treating the quenched specimen at the temperature, T_n, for two hours. The prepared glasses are crushed into a powder in a ball mill, mixed with liquid polyester resin, and then molded into a solid disc, 0.6 cm in diameter and 1.5 mm in thickness for the Mössbauer measurements. The resulting absorber has 0.2 mg/cm^2 of Fe^{57} for the 5 mole % specimen which is thin enough to avoid any broadening of the resonance lines due to the blackness effect. The chemical compositions for the quenched specimens are given in Table I.

A conventional electro-mechanical constant-acceleration Mössbauer spectrometer equipped with a standard 1024 multi-channel analyzer is used to collect the data. A 60 mCi source of Co^{57} in copper is used for all measurements. This gives the line-width of 0.23 mm/sec for the centroid of α-iron metal foil. The details of the measurements at low temperatures and in the applied magnetic fields can be found in the literature (8).

TABLE I

Chemical Composition of $(1-x)[.2BaO- .8B_2O_3] -xFe_2O_3$ Glasses

Concentration x mole %	BaO g	B_2O_3 g	Fe_2O_3, g	
			natural	enriched
10	2.760	5.013	1.407	68% enriched 0.190
7	2.852	5.180	0.698	68% enriched 0.413
5	2.913	5.291	0.703	68% enriched 0.091
3	2.975	5.403	0.429	97% enriched 0.05

RESULTS AND ANALYSIS

A. Preliminay - Clustering Phenomina in the Quenched Specimen

Mössbauer measurements are made at temperatures ranging from 1.7 to 300 K and in magnetic fields of up to 55 KOe for the quenched specimens. A typical room temperature spectrum is reproduced in Fig. 1(a) for the 5 mole % specimen in which two doublets are resolved using a curve-fitting computer routine assuming a superposition of Lorentzian lines. The dominant doublet (A) has a quadrupole splitting (QS) of 1.11 mm/sec and an isomer shift (IS) of 0.32 mm/sec which may be identified as the Mössbauer parameters for Fe^{3+} ions in the glass. On the other hand, the minor doublet (B) has a QS of 2.94 mm/sec and an IS of 1.01 mm/sec which may be attributed to from Fe^{2+} ions. From the temperature dependence of the spectral behavior, the superparamagnetic nature of iron ions is inferred in the glass matrix. Furthermore, the resultant Mössbauer spectra taken in the applied magnetic field are analyzed giving three paramagnetic hyperfine contributions: (i) a contribution from isolated Fe^{3+} ions having a ground state spin $\frac{5}{2}$ with H_{sat} (saturation internal hyperfine field) of 550 KOe, (ii) a contribution from pairs of Fe^{3+} ions formed through a strong antiferromagnetic exchange coupling having a ground state spin 0 with a magnetically perturbed doublet, and (iii) a contribution from three of Fe^{3+} ions coupled by an antiferromagnetic exchange interaction forming an isosceles triangle having a ground state spin $\frac{1}{2}$ with two saturation magnetic fields of 185 KOe and -160 KOe (negative sign denotes the direction of the magnetic field parallel to the spin-up direction)(7).

This observation is supplemented with magnetic susceptibility and EPR measurements. Fig. 1(b) shows the reciprocal susceptibility, $1/\chi$, vs. T behavior for the quenched specimens containing 3, 5, and 7 mole % Fe_2O_3. The general trend of the curves indicates a predominantly antiferromagnetic interaction being the basic character of the magnetic property of the quenched specimen. The measured Curie constant and Curie temperature depend on the iron concentration as well as on the measuring temperatures, indicating that a varying degree of interactions amongst iron ions is present in the glass. A typical dX''/dH vs. H curve is reproduced for the 5 mole % quenched specimen in Fig. 1(c). As seen in this figure, three resonances are observed at the g=2.0, 4.2, and 6.0 which are attributed to originating from clusters (pairs or triads), isolated Fe^{3+} free ions, and clusters (pairs or triads), respectively. EPR data are characterized by the appearance of a relatively sharp resonance at the g=4.2 for the low concentration which is replaced by an increasingly intense broad resonance at the g=2.0 for the higher concentration. These measurements for the quenched specimens are satisfactorily interpreted with an isolated cluster model as outlined in the previous paragraph, a non-random distribution of iron ions in a group of two (diad) and three (triad).

This preliminary result can be summarized as: (1) the iron ions exist as clusters of two or three ions in addition to a few isolated single ions, (2) the average cluster sizes are less than 15Å, and (3) the relative number of iron ions contributing to singles, pairs, and triads are approximately 33%, 41% and 27%, respectively for the 5 mole % specimen. Interested readers are referred to a detailed analysis reported elsewhere (6,7).

The main effort is made in analyzing the consecutive development of these clusters by heat treating the quenched specimen at various temperatures and the result of which follows in the next section.

B. Crystalline Formation in the Heat Treated Specimen

Mössbauer measurements are made at room temperature for the specimens heat treated at 550°C, 600, 650, 700 and 800°C. Typical spectra for the 7 mole % samples are presented in Fig. 2 and Fig. 4. A feature that can be seen immediately in Fig. 2 is that the spectrum does not change its shape, but remains as a quadrupole doublet up to the heat treating temperature of 550°C. The concentration dependence of IS and QS is presented in Fig. 3 for the specimens heat treated at 550°C. It is seen in this figure that

CLUSTERING AND CRYSTALLIZATION PHENOMENA IN $BaO-4B_2O_3$ GLASSES 631

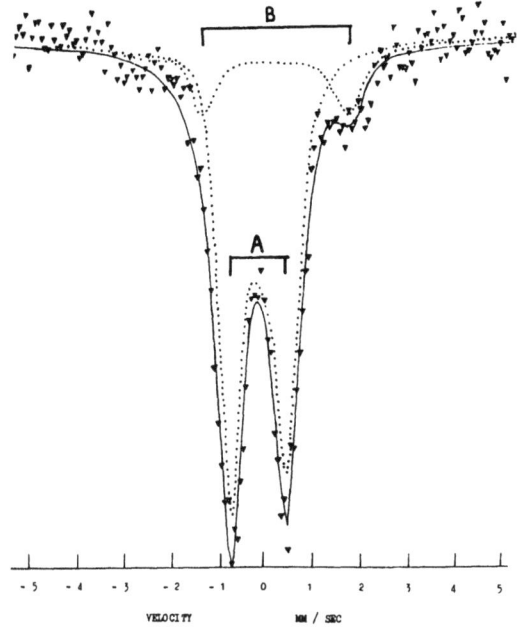

Fig. 1(a)

Mössbauer spectrum taken at room temperature for the 5 mole % specimen fitted to two Lorentzian doublets. The velocity scale is with respect to iron metal foil.

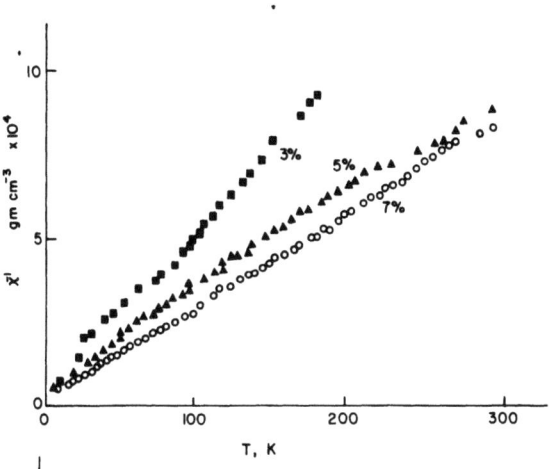

Fig. 1(b)

The inverse magnetic susceptibility plotted as a function of temperature for the 3, 5 and 7 mole % specimens.

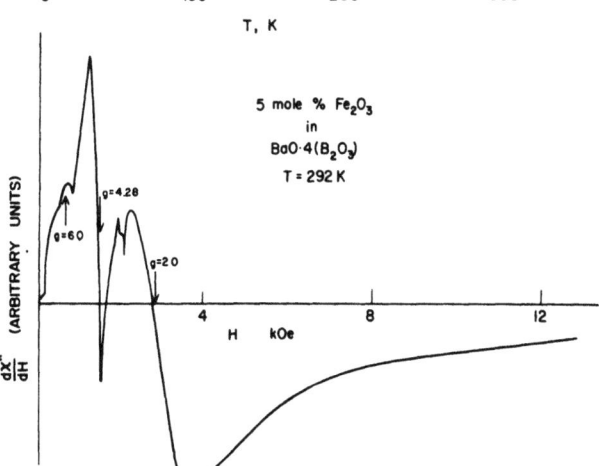

Fig. 1(c)

Typical dX''/dH vs. H curve for the 5 mole % specimen.

the IS value increases over all concentrations on heat treatment indicating that the coordination of oxygen may change from a tetrahedral (ferric ions in the quenched specimen) to octahedral environment which is prerequisite to forming spinel structures, and that the QS value decreases demonstrating a smaller electric field gradient, an indication of more ordered symmetry. There are apparent differences in the structural units from the quenched stage

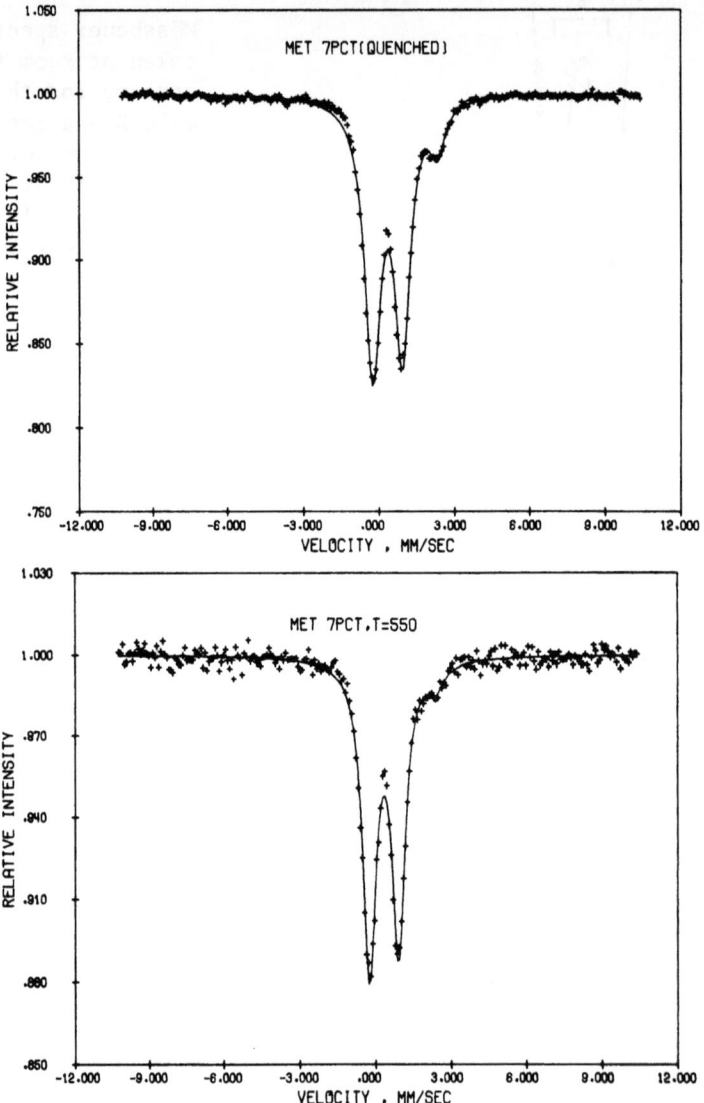

Fig. 2. Mössbauer spectra for the 7 mole % specimens heat treated at different temperatures. The quenched specimen is compared to the one heat treated at 550C. The velocity scale is with respect to iron metal.

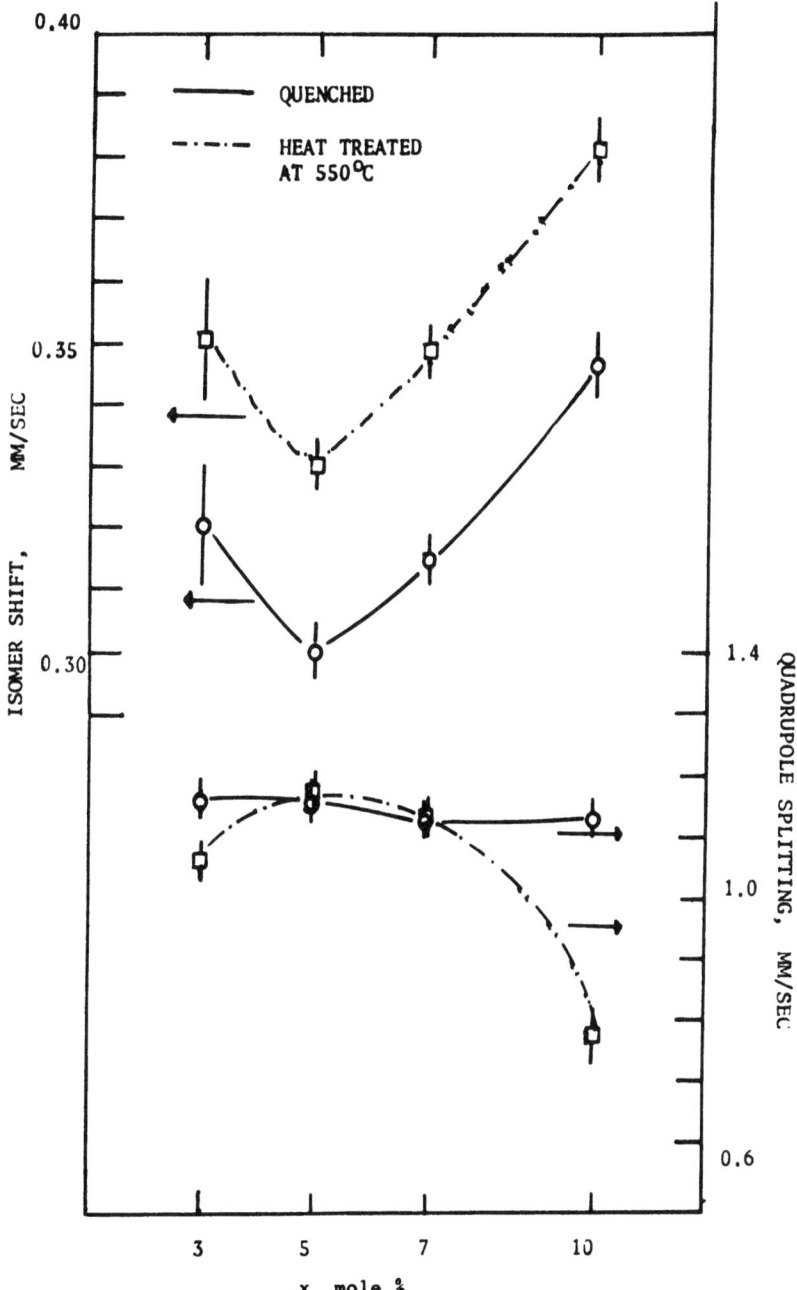

Fig. 3. Mössbauer parameters for the specimens heat treated at 550°C as a function of concentration. Those for the quenched specimens are shown for comparison. The isomer shifts are with respect to the centeroid of α-iron metal foil.

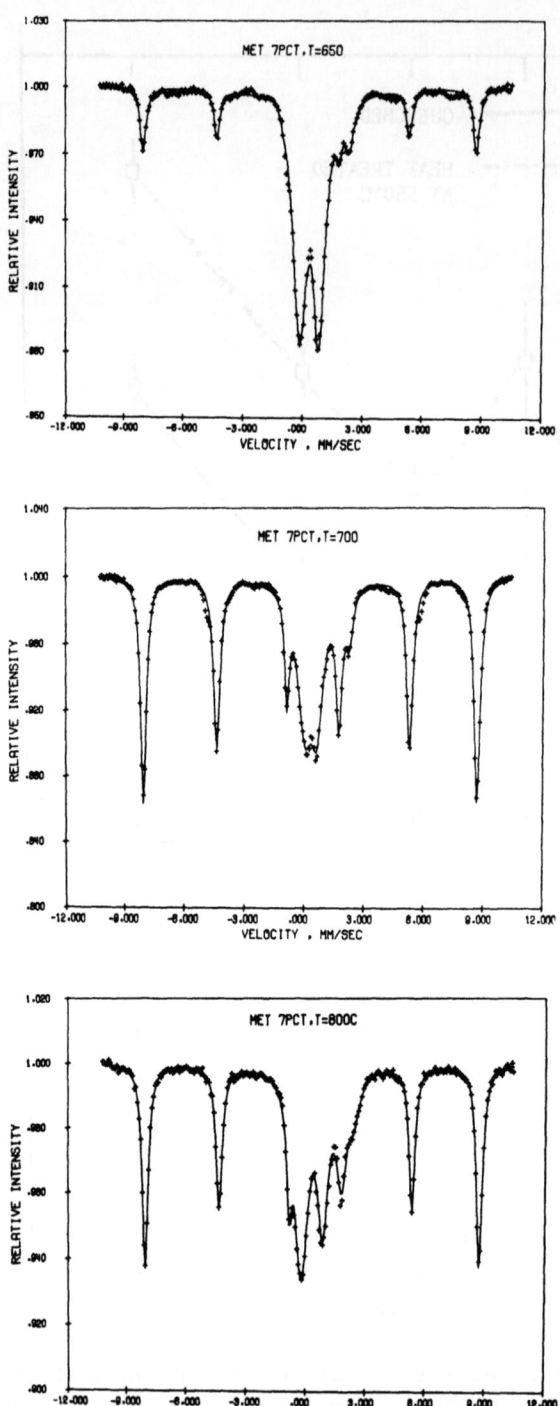

Fig. 4. Mössbauer spectra for the 7 mole % specimens heat treated at different temperatures as indicated in the figure. The velocity scale is with respect to the centeroid of α-iron metal foil.

which give rise to the assumption that the simple doublet shown by the heat treated specimen may ariginate from superparamagnetic fine particles centered on diads, or triads since we established the structure of the quenched specimen as a small cluster. This doublet remains consistent up to the heat treating temperature of 630°C.

When the heat treating temperature reaches 650°C, an apparent magnetic hyperfine splitting developes which grows into a more intense hyperfine pattern with further heat treatment leading to a saturation intensity above 700°C as shown in Fig. 4. The results obtained for the 5 and 10 mole % samples give conclusions similar to this feature with a noticeable exception: the temperature at which the hyperfine splitting starts to appear increases with decreasing concentration.

Mössbauer parameters determined from the spectra for three specimens are summarized in Table II, in which isomer shifts, quadrupole splittings, line-widths and internal magnetic fields are shown as a function of iron concentrations and heat treating temperature. The internal magnetic fields are the same for all spectra regardless of ironconcentrations, and the value is (517 ± 5) KOe

TABLE II

Mössbauer Parameters for Heat-Treated Specimens

Concentration x mole %	Nucleating Temp. T_n, °C	Hyperfine Structure				Center Peaks			Portion of HFS Tot.Area %
		IS mm/sec	QS mm/sec	Γ* mm/sec	H_{int} Koe	IS mm/sec	QS mm/sec	Γ mm/sec	
5	600					0.46	1.28	0.78	0
	700	0.37	0.07	0.37	518	0.50	0.64	0.82	62
	800	0.36	0.06	0.41	518	0.41	0.82	0.73	69
7	600	-	-	-	-	0.45	1.18	0.72	0
	650	0.32	0.07	0.38	514	0.45	0.99	0.81	24
	700	0.36	0.07	0.45	516	0.50	0.55	0.72	59
	800	0.32	0.07	0.41	514	0.41	1.08	0.72	58
10	650	0.32	0.09	0.45	509	0.41	0.99	0.82	11
	750	0.32	0.09	0.45	508	0.41	0.99	0.86	28
	800	0.32	0.11	0.41	511	0.41	0.99	0.86	35
Error		0.03	0.03	0.05	5	0.03	0.03	0.05	2

*Full width at half-maximum of the outer peaks of six-line.

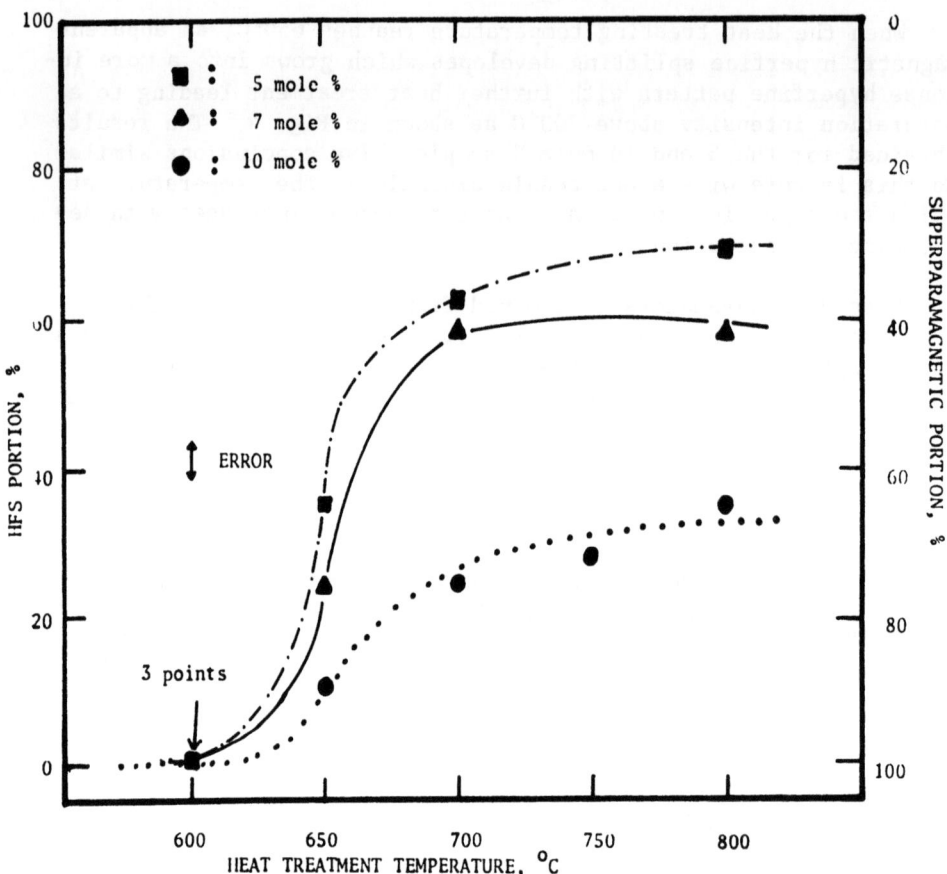

Fig. 5. The ratio of the magnetically split portion to the whole spectrum for the 5, 7 and 10 mole % sample as a function of the heat treatment temperature. The Mössbauer fraction of iron ions in two phases is assumed to be the same.

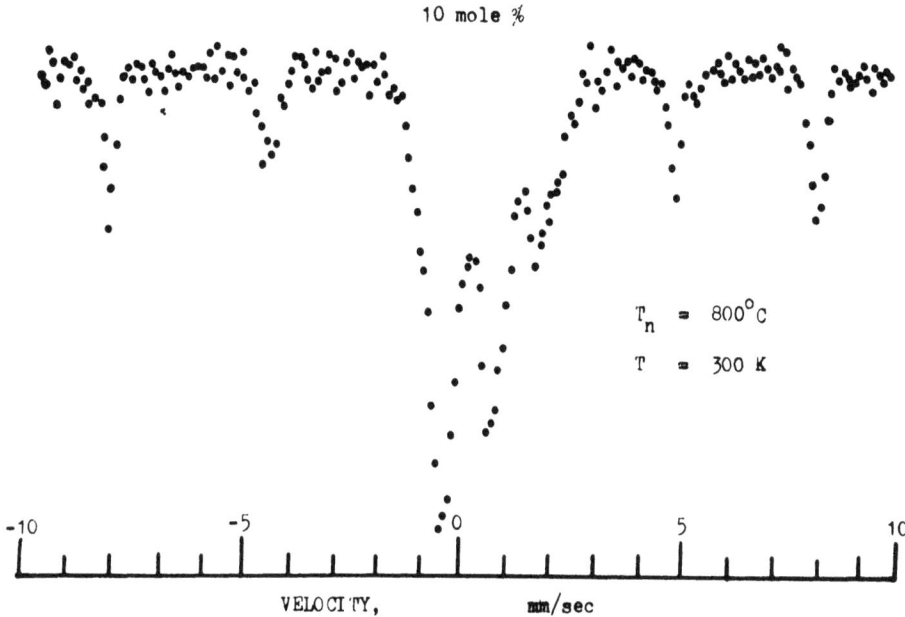

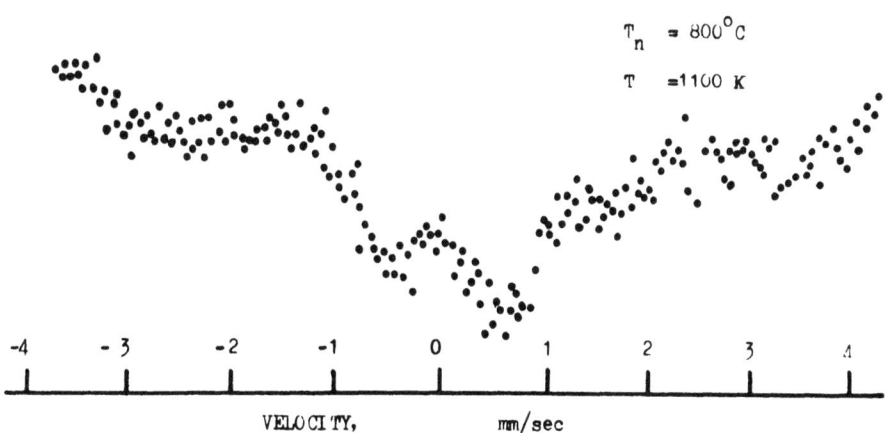

Fig. 6. High temperature spectrum taken at 800°C for the 10 mole % sample. The room temperature spectrum is shown for comparison.

which is equivalent to the internal field of the Fe^{3+} in the octahedral site of α-Fe_2O_3 (9). The values of IS and QS for the hyperfine lines confirm the identification of the newly developed structural unit as α-Fe_2O_3 form. Center peaks which are always present in the spectra may be understood to have risen from the remaining clusters which had not joined the already formed structural units. They may be presumed to be fine particles of subcritical size behaving like superparamagnets (10).

The intensity of the hyperfine lines approximated from the area under the resonance peaks shows a significant concentration dependence. By assuming the same Mössbauer fraction for the iron ions in the hyperfine split portion (HFS) and in the superparamagnetic portion, the ratio of the HFS to the whole iron ions can be estimated from the corresponding areal ratio. Fig. 5 contains the hyperfine portion (%HFS) or the superparamagnetic portion plotted as a function of the heat treating temperature. Why this implication is significant will be discussed in the next section.

An attempt is made to elucidate the kinetics during the structural evolution by taking Mössbauer measurements at high temperatures. A typical spectrum is shown in Fig. 6 for the 10 mole % sample in which a spectrum taken at 1100K is compared with one taken at room temperature for the sample heat treated at 800°C. It becomes evident that the high temperature spectrum does not show any HFS while the room temperature spectrum contains a distinct HFS ascribed to Fe^{3+} ions in α-Fe_2O_3. Two aspects of the result are noteworthy: (1) the absence of HFS in the high temperature is to be interprested as evidence for the superparamagnetic behavior of α-Fe_2O_3 particles since it would be expected if the structural unit present had such a size that its magnetic moment rotates around its easy axis under the influence of the thermal agitation and (2) the resultant spectrum does not give any kinetic information since the high temperature spectra become identical for the quenched and heat treated specimens. An alternate interpretation for the absence of the HFS at high temperature may be found to be the reduction of the Mössbauer fraction. This possibility is eliminated based on the observation that a sudden reduction of the Mössbauer fraction does not occur until the measuring temperature reaches 1150K (11). The paramagnetic Neel temperature of α-Fe_2O_3 is about 970K (12), and the HFS disappears when measured at 800K. This confirms the superparamagnetic nature of the magnetically split particles.

Since the superparamagnetic nature of the crystalline particles is established in the heat treated specimens, the particle sizes can be estimated from their superparamagnetic relaxation scheme, i.e.

$$D = \left[(\ln(\tau/2\tau_0)) \times (3kT/\pi K) \right]^{1/3}$$

where D is the average particle diameter blocked at temperature T giving characteristic time τ, K is the anisotropy constant at T, k is the Boltzman constant, and τ_o the prefrequency factor (13). Here a spherical particle shape is assumed. This relation implies that the superparamagnetic relaxation can be detected in two ways: by varying the temperature with a constant volume and by varying the volume of the particle keeping the temperature constant. For α-Fe_2O_3 particles, K=4.4 x 10^4 erg/cm^3, τ_o^{-1} = 3.39 x 10^8 sec^{-1} (13), and the characteristic time becomes comparable to the nuclear Larmor precession time (2.5 x 10^{-8} sec) when the transition from the six-line to the doublet (or vice versa) occurs. For the 7 mole % specimen as shown in Fig. 4, T=300K, and τ=2.5 x 10^{-8} sec which by substituting into the above equation, D=160Å is estimated which is the average particle size formed by heat treating at 650°C for the 7 mole % sample. For the spectrum shown in Fig. 6, the transition to doublet takes place at about 800K, and the corresponding τ is 2.5 x 10^{-8} sec which gives D=216Å, the particle size formed at T_n=800°C for the 10% sample. This estimation assumes the K value to be constant over the temperature range 300-800K, however the anisotropy energy usually shows some temperature dependence.

DISCUSSION

The experimental results suggests that upon heat treating the quenched specimen a distinct crystalline structure starts to appear in the glass. When the crystalline precipitates observed in the higher concentration is considered (4,5), this is not a surprising result, although there are several interesting features. It is to be noted that crystalline particles develop rather abruptly at about 650°C and the amount of iron ions which form the crystalline increase rapidly with heat treating temperatures reaching some maximum value that shows a significant concentration dependence as analyzed in Fig. 5. It is also interesting to note that the on-set temperature of about 650°C is 50° higher than the one reported by MacCrone (14) for the intermediate concentration regime.

According to the classical nucleation theory, the nucleation rate should reach some maximum as the temperature increases, and the crystallization should continue to its completion. The observed result cannot be interpreted by this argument and a model of oxydation controlled exitaxial mode of catalyzed nucleation is proposed. In this model, it is believed that the clusters already present in the quenched state act by precipitating from the glass as fine particles during the early stage of the nucleation, i.e. the formation of superparamagnetic fine particles of less than 50Å as seen in Fig. 2. They then may serve as sites for crystallization of the main crystalline phase at a higher heat treating temperature, i.e. the crystalline phase of α-Fe_2O_3 as seen in Fig. 4. In the course of crystallization, the sudden development and saturation of its magnitude suggests that there exist two particle sizes which follow a step-function type growth and nucleation with temperature.

The superparamagnetic nature of crystalline development suggests that the cluster phase gradually moves into the superparamagnetic fine particles of growing sizes by simply clumping the near neighbor clusters. This process may be controlled by the oxygen diffusion and activity at a given heat treating temperature. It is expected that the inter-cluster exchange interaction becomes important as particles are formed. It is believed that glass in glass phase separation is not a precurser to the observed processes.

One may expect that the triads serve as sites for a better epitaxy than single ions in the glass. The higher on-set temperature of crystalline appearance may be attributable to the dilution of iron ions which gives rise to more isolated ions. This may result in increasing difficulties in forming particles and elevating the on-set temperature as the concentration decreases.

The saturation of crystalline ferrites above 700°C suggests a step-function type nucleation and growth mechanism different from the classical theory. The concentration dependence of this maximum indicates that the 10 mole % sample forms relatively large particles limiting the nucleation of new particles which result in low saturation value. Relatively high Fe^{2+} concentration (about 15%) may be partially responsible for this. However, further work is needed for the understanding of concentration dependence.

CONCLUSION

It is known that crystallization does not occur easily in barium-borate glasses. By adding a dilute concentration of iron oxide, catalyzed crystalline ferrite particles are observed, and the resultant processes can be summarized as:
(1) a non-random distribution of iron ions in the form of small clusters is responsible for the magnetic inhomogeneities present in the quenched stage.
(2) superparamagnetic fine particles of less than 50Å develop through heat treating below 650°C,
(3) large crystalline ferrite, α-Fe_2O_3, are formed by heat treating the quenched specimen above 650°C which shows superparamagnetic behavior, i.e. particle size of 160Å for the 7% sample heated at 650°C, and

(4) an epitaxial mode of catalyzed nucleation and crystallization with a step-function type nucleation is proposed.

ACKNOWLEDGEMENTS

We would like to express thanks to Drs. D. W. Moon and J. Aiken for the magnetic susceptibility and EPR measurements, and the experimental assistants of Mr. D. Cifeli and Mr. C. Chiou. Useful discussion with Dr. L. H. Schwartz and his permission to use computer facilities at Northwestern University is greatfully acknowledged.

REFERENCES

* Present address: Materials Science and Engineering, Northwestern University, Evanston, Illinois 60201.
** Present address: Q26, Los Alamos Scientific Laboratory, Los Alamos, New Mexico 87544

1. R. R. Shaw and J. H. Heasley, J. Amer. Ceramic Soc. 50, 297 (1967).
2. P. S. Rogers and J. Williamson, Glass Technol. 2, 128 (1969).
3. B. T. Shirk and W. R. Buessam, J. Amer. Ceramic Soc. 53, 292 (1969).
4. M. Fahmy, M. J. Park, M. Tomozawa, and R. K. MacCrone, J.Phys. Chem. Glasses, 13, 21 (1972).
5. M. Tanigawa and H. Tanako, Osaka Kogyo Gigutsu Shikenjo, 15, 285 (1964).
6. D. W. Moon, J. Aitken, G. Cieloszyk, and R. K. MacCrone, Annual Report, ONR Contract NR-032-538 (1974).
7. K. J. Kim, M. P. Maley, and R. K. MacCrone, to be published.
8. P. P. Craig, O. E. Nagle, W. A. Steuert, and R. D. Taylor, Phys. Rev. Letters 9, 12 (1962).
9. R. S. Hargrove and W. Kundig, Solid State Commun. 8, 303 (1970)
10. T. Nakamura, T. Shinjo, Y. Enkoh, N. Yamamoto, M. Shiga, and Y. Nakamura, Phys. Letters 12, 178 (1964).
11. C. F. Chiou, Master's Thesis, Rensselaer Polytechnic Institute (1973).
12. D. E. Cox, G. Shirane, and S. L. Ruby, Phys. Rev. 125, 1158 (1962).
13. W. Kundig, H. Bommel, G. Constabaris, and R. H. Lindquist, Phys. Rev. 142 (2), 327 (1966).
14. R. K. MacCrone, *Amorphous Magnetism*, ed. H. O. Hooper and A. M. de Graaf (Plenum, New York, 1972), p. 77.

DISCUSSION

M. P. O'Horo: Are all spectra shown here including those exhibiting hyperfine splitting taken at room temperature?

K. J. Kim: Yes, those exhibiting hyperfine splitting represent the bunch of samples that have been heat treated.

M. P. O'Horo: Can you tell anything about the size of the crystallites formed in your samples?

K. J. Kim: The crystallites are estimated to be around 200 Å for the 10 mole % specimen and around 260 Å for the 7 mole % specimen.

R. A. Levy: How do you account for the observed change in the isomer shift as a function of heat treatment?

K. J. Kim: It is a structural effect. The overall increase is due to an apparent decrease in the s-electron density.

D. Stauffer: Is it possible to measure the growth of these precipitates as a function of time?

K. J. Kim: Yes it is. We are planning such a study in the near future.

MAGNETIC PROPERTIES OF $(Fe_2O_3\text{-}TiO_2)$ IN $BaO\text{-}B_2O_3\text{-}SiO_2$ OXIDE GLASSES

L. Trombetta, J. Williams and R. K. MacCrone

Materials Division
Rensselaer Polytechnic Institute
Troy, New York 12181

INTRODUCTION

Glasses containing dilute amounts of transition metal ions may be studied using structure sensitive magnetic measurements. In many glass systems, transition metal additives can promote and control the devitrification of the glass (1). In some cases the use of two different transition metals is more effective in producing crystallization than is the use of either of the individual transition metals (2). The present study is one in which a $BaO\text{-}B_2O_3\text{-}SiO_2$ base glass doped with Fe_3O_4 or Fe_3O_4 plus Ti_2O_3 oxides is studied using static magnetization and electron paramagnetic resonance (EPR). Past studies (3,4,8) of single transition metal in glasses enable meaningful comparison with this mixed transition metal glass.

When the Fe ions are in the minority, an enhanced clustering of Fe^+ ions is observed. When the fraction of Ti ions is lower than the fraction of Fe ions, there is little effect on the behavior of the Fe ions.

An equimolar barium borate silicate glass was selected because it does not devitrify under any conditions encountered in this work and its electrical properties are also being studied (5). The total amount of transition metal ion was held constant at 6 mol % and the ratio

$$\frac{Fe^{+3} + Ti^{+4}}{Fe^{+2} + Ti^{+3}}$$

was fixed at 2:1 for each sample. The concern was not to disrupt the glass network by adding different total amounts of transition metal ion. The effect of the Ti on the Fe ions was being sought, not the ions effect on the glass structure.

SAMPLE PREPARATION AND EXPERIMENTAL METHODS

The equimolar $BaO-B_2O_3-SiO_2$ base glass was prepared by mixing reagent grade barium carbonate, boron oxide and silica powders. This mixture was melted in a platinum crucible in a globar furnace in air for one hour at 1240°C and then poured onto a steel plate. To insure homogeneity, the glass pieces were crushed to a 200 mesh size and remixed before samples were removed for production of the iron and titanium containing samples. Iron and titanium were added as reagent grade Fe_3O_4 and TiO_2 or Ti_2O_3 powders, respectively, and the samples remelted in a globar furnace in argon for one hour at 1240°C. During the melt the mixture was stirred twice with a pure silica rod to further insure homogeneity. The glass was quenched by being pressed between steel plates at room temperature and then annealed at 580°C in air for one hour before being allowed to cool in the annealing furnace overnight. Samples for both EPR and static magnetization were ground to a 100 mesh powder before being weighed out and packed into sample holders.

Static magnetization measurements were done with a vibrating sample magnetometer (Princeton Applied Research Corp. model FM-1) in applied fields of up to H = 10 kOe. The magnetic field was swept from 0 to 10 kOe and back to 0 kOe at about 1 kOe min^{-1}, the field was then reversed and the sweep repeated. Plots of the magnetization, M, and the applied field, H, were directly recorded on an X-Y plotter. Calibration of the magnetometer was done by measuring the saturation magnetization of a pure nickel sphere of known mass. Electron paramagnetic resonance experiments were done using a Varian spectrometer operated at a microwave power level of one milliwatt and a frequency of 9.5 GHz.

RESULTS AND DISCUSSION

Figure 1 shows the magnetization vs. applied field results for a sample at both 300K and 77K. Samples containing less than 3 mol % of iron were diamagnetic at room temperature. The diamagnetism of the base glass was determined and subsequently subtracted from the sample measurements. The paramagnetic component remaining after this subtraction is due to the iron and titanium ions and is the concern of this paper. No hysteresis was observed for any of the samples.

If the magnetic ions exist in the glass as noninteracting

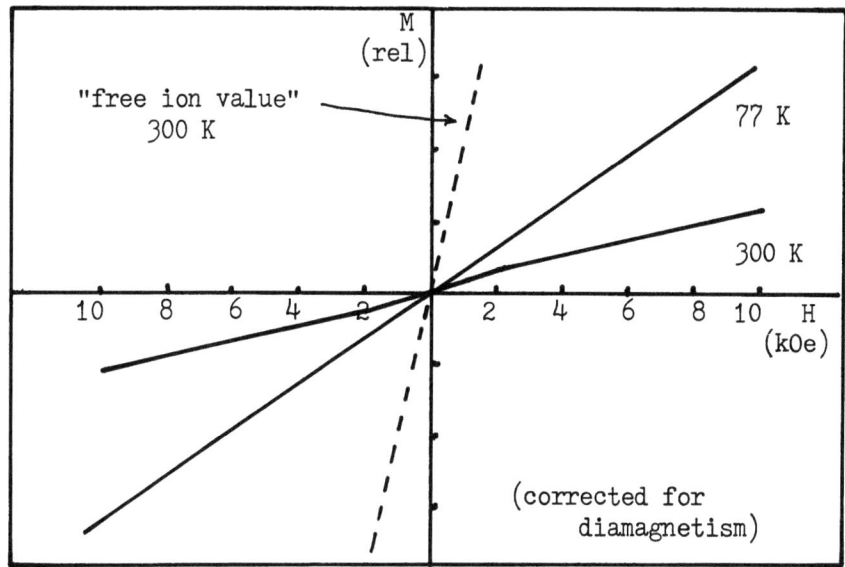

Figure 1. Magnetization against magnetic field for barium-borate-silicate glass containing 1 mol% Fe and 5 mol% Ti.

individuals, the simple paramagnetic system resulting would have a magnetization, $M = \chi H$ with $\chi = C/T$ where C, the Curie constant, is given by a suitable sum of the form

$$C = \sum_i \frac{N_i g^2 \beta^2 J(J+1)}{3K} \qquad (1)$$

The experimental values, however, of the magnetization of the BaO-B_2O_3-SiO_2 glass containing dilute quantities of iron are more than an order of magnitude below "free ion" prediction. This result would be surprising if previous studies of an iron doped BaO-$4B_2O_3$ glass (3) and an iron doped calcium-magnesium-alumino silicate glass (8) were not available. The fundamental result of these studies was that the iron ions exist as free ions, and also as pairs and also as triads; ions within the diads and triads couple antiferromagnetically via the superexchange interaction. The static magnetism of these glasses and their EPR behavior can be interpreted using the progressive ordering of these diads and triads as the temperature is reduced. Figure 1 shows magnetic behavior of the iron doped BaO-B_2O_3-SiO_2 glass, which is much less magnetic than the previously studied glasses: the spins in this mixed glass must be more strongly coupled. There are apparently two causes for this strong coupling: one is ionic affinity within the melt, the other is the effect of the titanium ions on the iron ions.

The effect of the Ti is illustrated in Figure 2a and 2b. These figures plot x/X vs. x where x is the total Fe ion concentration and X is the magnetic susceptibility. For a simple paramagnetic system these plots would be horizontal lines; deviations from the horizontal show magnetic coupling. In Figure 2a, in which only iron is being added to the $BaO-4B_2O_3$ system, the antiferromagnetic coupling is made more effective as x increases as one would expect. Figure 2b, in which iron is being added to the $BaO-B_2O_3-SiO_2$ system, there is a critical value of x at x = 2.5 mol % beyond which an increase in x causes a decrease in x/X. At low values of x, for which the Ti concentration is relatively high, the Fe ions are more effectively antiferromagnetically coupled than for higher values of x, for which the Ti concentration is lower.

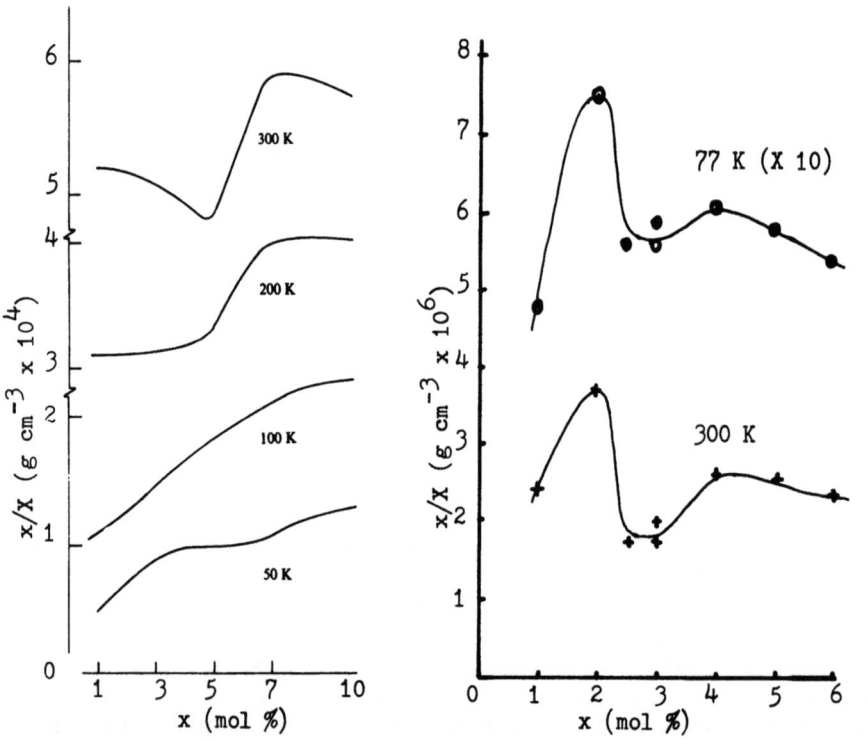

Figure 2a and 2b. Reduced reciprocal susceptibility, x/X, against molar concentration of Fe, x, at various temperatures for
a. $BaO-4B_2O_3$ base glass with Fe only.
b. $BaO-B_2O_3-SiO_2$ base glass with Fe and Ti. Here the Fe + Ti concentration is constant @ 6 mol%.

The electron paramagnetic resonance (EPR) absorption derivative data for several of these glasses at room temperature are shown in Figures 3a and 3b. Analysis of the EPR spectra (6,7) identify the g = 4.28 resonance with the isolated Fe^{+3} ions. The source of the g = 2 resonance is ions coupled by magnetic interactions and the source of the g = 6 resonance is weak crystal field terms in the Hamiltonian. The behavior of the g = 2 resonance in Figure 3a shows the drastic effect on the coupling of the iron moments produced by the addition of titanium. Samples which have been doped with Fe_2O_3 or Fe_3O_4 have a broad g = 2 resonance indicating that the ions are strongly interacting via dipole-dipole interactions. The sample containing Ti ions in addition to the Fe ions has a much reduced and narrowed g = 2 line. The

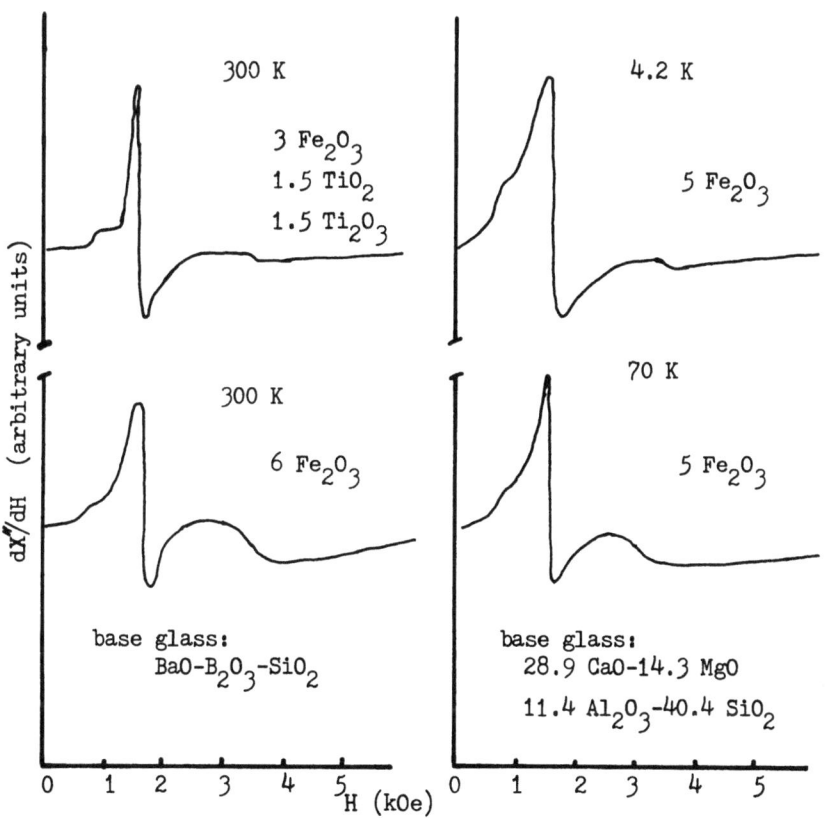

Figure 3a. and 3b. Electron paramagnetic resonance spectra for indicated glasses. Compositions in mol %.

g = 4.28 line is also much narrower in the Ti containing glass. We propose that the Ti has forced the iron ions to strongly couple, forming an amorphous clump of magnetically ordered glass. What resonance is seen is produced by ions at the edge of the magnetically blocked clump. Further evidence for this mechanism is seen in EPR results for a calcium-magnesium-alumino silicate glass doped with Fe_2O_3 (8). The temperature dependence of the EPR spectrum for one of these samples is illustrated in Figure 3b. At liquid helium temperatures, the g = 2 line is small and narrow as the exchange coupled Fe ions are blocked. At 30K the line has greatly broadened and at 70K it has broadened still further as more and more magnetically ordered groups of ions unblock. The low and high temperature EPR spectra for this glass agree strikingly with the with Ti and without Ti EPR spectra for the iron-doped BaO-B_2O_3-SiO_2 glasses.

CONCLUSION

This rather cursory study indicates that in a mixed Fe-Ti glass of constant TMI concentration Ti ions enhance the clustering of Fe ions when the former are in a majority. However, when the Ti ions are in a minority, the usual clustering of Fe is apparently unchanged.

The EPR behavior suggests that in the enhanced clustering regime, the antiferromagnetic coupling is much stronger than that occurring when Fe only is present in the glass.

ACKNOWLEDGEMENTS

We would like to thank Dr. J. T. Warden for taking the EPR spectra and the National Science Foundation for financial support under Grant GH-34548.

REFERENCES

1. P. S. Rogers and J. Williamson, Glass Tech. 10, No. 5, 128 (1969).
2. P. Topping, 77th Annual Meeting of the American Ceramic Society, Washington, D. C. (1975).
3. D. W. Moon, J. M. Aitken, R. K. MacCrone and G. S. Cieloszyk, Physics and Chemistry of Glass 16, No. 5, 91 (1975).
4. D. W. Moon, J. C. Williams, R. K. MacCrone and K. Kim, Technical Report #2, Office of Naval Research # N00014-67-0117-9917 (1975).

5. B. S. Rawal and R. K. MacCrone, to be published in the Proceedings of the Fourth International Conference on the Physics of Non-Crystalline Solids, Clausthal-Zellerfeld, Germany (1976).
6. H. O. Hopper, et al., Amorphous Magnetism, edited by H. O. Hooper and A. M. de Graaf, Plenum Press, New York (1973).
7. H. H. Wickman, M. P. Klien and D. A. Shirly, J. Chem. Physics 42, 2113 (1965).
8. D. W. Moon, J. C. Williams and R. K. MacCrone, 78th Annual Meeting of the American Ceramic Society, Cincinnati, Ohio (1976).

DISCUSSION

M. P. O'Horo: Are the clusters you talk about in this study the same as the diade and triade clusters discussed in the case of the barium borate glasses?

J. C. Williams: No, they are not the same. The diade and triade clusters discussed in the case of the barium borate glasses block at much lower temperatures.

K. J. Kim: Has any body done Mössbauer spectroscopy on these glasses?

J. C. Williams: No, the only other piece of work we have done on these glasses was to examine whether or not there was a phase separation effect and the evidence is that there was not any.

M. P. O'Horo: Have you examined the products of crystallization in these glasses?

J. C. Williams: We are in the process of doing that.

MAGNETIC PROPERTIES OF AN IRON BOROSILICATE GLASS

Michael P. O'Horo, James F. O'Neill

Xerox Corporation

Webster Research Center, Webster, New York 14580

INTRODUCTION

Transition Metal Ion (TMI) oxide glasses are known to show unusual magnetic properties near the solubility limit of the TMI oxide, associated with the presence of antiferromagnetic TMI clusters[1-4]. Heat treatment of the quenched glasses in this region can produce magnetic crystalline precipitates[4,5] with superparamagnetic properties. This paper discusses an investigation of the magnetic properties of both these states in an Fe borosilicate glass system. The purpose of this investigation was to elucidate the relationship between the precipitates and magnetic clusters, and determine the magnetic parameters of the initial precipitated phase.

PREPARATION

The specific base composition of the glass system investigated was 40 m% SiO_2, 30 m% B_2O_3, 20 m% CaO and 10 m% Al_2O_3. Iron cations were added in the form of Fe_2O_3 from 0.5 to 20 m%, such that the relative molar proportions of the base glass were maintained. The glasses were melted at 1350°C for three hours in an air atmosphere and then quenched onto a graphite plate at room temperature. The Fe^{3+}/Fe^{2+} ratio in the quenched glasses ranged from 2.0 to 2.5. The quenched samples were amorphous and macroscopically homogeneous up to 15 m% Fe_2O_3. Above this concentration a number of crystalline phases, i.e., α-Fe_2O_3, Fe_3O_4 and $Ca_3Fe_2Si_3O_{12}$ appeared on quenching. In none of the samples termed amorphous could any trace of crystallinity be detected by X-ray or electron diffraction methods. Density and DTA measurements showed the pres-

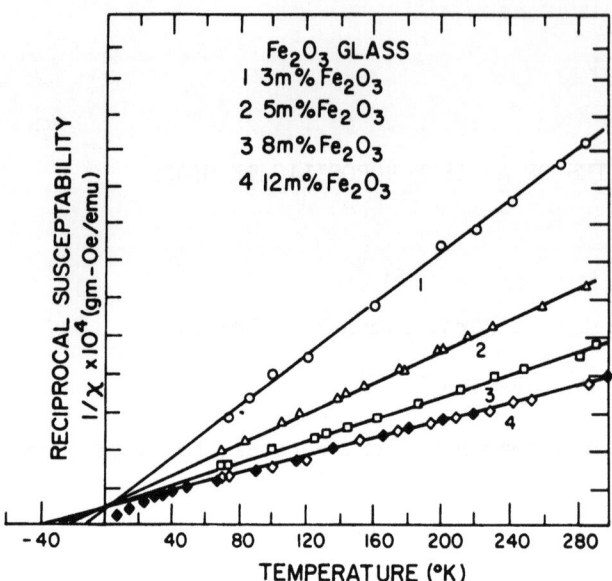

Figure 1. Reciprocal susceptability of Fe_2O_3 glasses as a function of temperature.

TABLE I

Magnetic Parameters of Fe_2O_3 Glasses

m% Fe_2O_3	N_{Fe} $\times 10^{21}$ (cm^{-3})	R_o $\times 10^{-8}$ (cm)	$\chi_{RT} \times 10^6$ (emu/gm)	θ (K°)	$\mu_{eff}(\mu_B)$ Experimental
12	5.5	5.68	26.7	-42	4.7
8	3.6	6.52	21.6	-34	5.0
5	2.3	7.59	16.0	-24	5.1
3	1.4	2.60	9.0	-16	5.1

ence of amorphous phase separation in the quenched glasses containing more than 5 m% Fe_2O_3. TEM micrographs of HF fracture surfaces in a glass containing 12 m% Fe_2O_3 displayed a very fine scaled (>100Å), uniformly distributed heterogeneity with a morphology similar to that of phase separation due to spinodial decomposition[6].

Precipitation of a ferrite phase was produced in the high concentration Fe_2O_3 glasses (>5m%) by heat treatments above the glass transition temperature Tg. Below 5 m% Fe_2O_3 no crystallization of any sort could be produced. TEM micrographs of the 12 m% Fe_2O_3 glass heat treated at Tg (=600°C) showed the appearance after twenty minutes of a uniform distribution of small (~30Å), equiaxed particles displaying a narrow size distribution. With further heat treatment the particles grew in volume but remained essentially constant in number. The initial precipitates had a complex diffraction pattern containing a number of unindexed lines as well as the major spinel lines. With heat treatments at higher temperatures (T>700°C) the unknown lines disappeared leaving a definite spinel pattern which has been identified as Fe_3O_4.

RESULTS AND DISCUSSION

A. Quenched Glasses

In Figure 1 all of the glasses containing Fe_2O_3 display Curie-Weiss behavior down to approximately 50°K. In the temperature range above 50°K plots of σ versus H/T show good superposition, indicative of paramagnetic behavior. The room temperature susceptibility χ_{RT}, paramagnetic Curie temperature θ, and the effective magnetic moment μ_{eff} for the various glasses are presented in Table I. The magnitude of θ is seen to increase with Fe_2O_3 concentration and argues against interpreting the magnetic properties of the glasses as being due to the inclusion of small antiferromagnetic crystallites[1]. At higher Fe_2O_3 concentration the value of χ no longer scales with Fe content, showing an increasing negative deviation from the expected linear behavior. Also the values of μ_{eff} in this range are significantly lower than the expected values based on the known concentration of Fe ions and the magnetic moments of Fe^{3+} and Fe^{2+} ions (5.9 and 4.9μ_B). An obvious source of this usual magnetic behavior would be the presence within the glass of a large number of antiferromagnetically coupled Fe ions. There is no evidence of antiferromagnetic crystallites in any of the glasses investigated; however, the presence of fine (>100Å) clusters of antiferromagnetically coupled Fe ions is very possible at high Fe_2O_3 concentrations[7]. Clusters of this type have been proposed by several investigators[1,2,3,4] to explain the magnetic properties of specific TMI oxide glass systems. If nearest neighbor proximity of the TMI's in a random distribution is sufficient for antiferromagnetic coupling, then it can be shown that in the high Fe_2O_3 glasses, the formation of cluster of 2,3 and higher order Fe ions becomes highly probable[7]. Moon, et al.,[2] has interpreted the magnetic properties of a Fe_2O_3-BaO-B_2O_3 glass system in terms of isolated paramagnetic Fe ions and antiferromagnetic diad and triad Fe ion clusters. A specific feature of this model is the temperature dependence of μ_{eff} due to small binding energy of the triad and higher order clusters. Temperature effects of this type are also seen in the glass system being discussed. A qualitative explanation of the concentration and temperature dependence of the magnetic properties can be given in terms of the small cluster model. However, a quantitative fit was not possible, particularly at high Fe_2O_3 concentration. The analysis is complicated by the presence of amorphous phase separation in these regions, since the assumption of random distribution of Fe ion is no longer valid. If one of the amorphous regions is Fe rich, the formation of larger clusters is greatly enhanced.

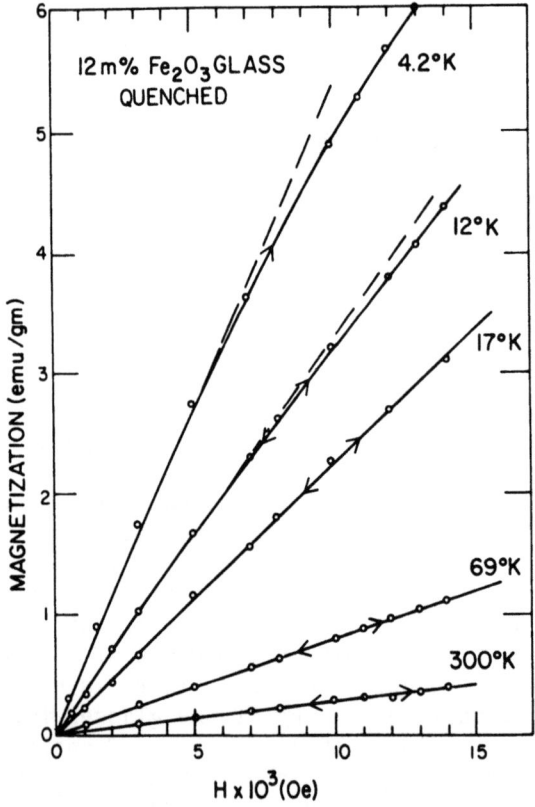

Figure 2. Magnetization versus applied field for the 12 m% Fe_2O_3 glass at various temperatures.

Evidence of larger clusters is seen in the low temperature magnetization. Figure 2 shows the field dependence of the 12 m% Fe_2O_3 glass at various temperatures. Down to 12°K curves are linear as expected, for paramagnetic behavior. At 4.2°K the curve shows a definite deviation from linearity beginning at 50kOe. The theoretical Langevin (or Brillouin) curve for a paramagnetic ion with a moment of 5-6μ_B has only slight curvature above 10kOe at this temperature and would not account for the curvature in Figure 2. A good fit of the Langevin function to the curvature shown in the magnetic data is possible with a moment of 10μ_B. Further evidence of a magnetic moment of this magnitude is provided by the presence of a small (~0.05 emu/gm) remanence in the 12 m% Fe_2O_3 sample at 4.2°K. The remanence is found to be time dependent, decaying to zero in approximately five minutes. This behavior is consistent with the behavior of a SPM particle near its blocking temperature T_B[7]. At these temperatures the relaxation time τ is of the order of the measurement time. The relaxation time is given by the expression

$$\tau = \tau_0 \exp(KV/kT) \qquad (1)$$

τ_0 is 10^{-9} sec and K is the magnetic anistropy constant of the particle. The clusters responsible for this behavior have a net moment sufficient for superparamagnetic behavior yet small enough so that $T_B \leq 4.2°K$. As will be seen in the next section, a superparamagnetic Fe cluster, approximately 30Å in diameter has a T_B of 12°K.

The above results indicate the presence in the quenched glasses of a assembly of superparamagnetic moments with a magnitude greater than the Fe paramagnetic moment (5-6μ_B). The only likely magnetic interactions in these oxide materials is antiferromagnetic superexchange, so that a clustering of four or more Fe ions would be required to produce the larger moments. Clusters of this size may arise due to random interactions in the melt, particularly in the higher Fe_2O_3 glasses near the solubility limit. However, amorphous phase separation into Fe rich and depleted regions upon quenching is an ubiquitous phenomena in such complex glasses and must also be considered. This type of process is known to precede the precipitation of a crystalline phase in a number of glass systems[9]. A process in which Fe clusters formed in the Fe-rich phase serve as heterogeneous nuclei has been proposed for this system based on precipitation studies[4]. At present the model for cluster formation based on incipient phase separation seems more likely, however, a more exact determination of the cluster distribution must be made to resolve the question.

B. Precipitated Glasses

In order to investigate the initial stages of the precipitation process, glasses containing 12 m% Fe_2O_3 were heated at Tg (600°C) for various times. The room temperature magnetization curves for these samples are shown in Figure 3. After a twenty minute heat treatment the magnetization displays a definite departure from linear paramagnetic behavior, and saturation effects are evident after one hour. None of the samples shows hysteresis and the curves have a Langevin-like shape characteristic of superparamagnetism (SP)[7]. The amount of precipitated magnetic phase has been monitored by magnetization measurements, and determined to have a sigmodial behavior as a function of heat treatment time. An analysis of the results in terms of the Avrami relation[9] shows precipitation to be in the growth stage with no new nucleation. TEM micrographs are supportive of this interpretation. The magnetization increase with heat treatment has been related to the growth of the ferrite particle's volume.

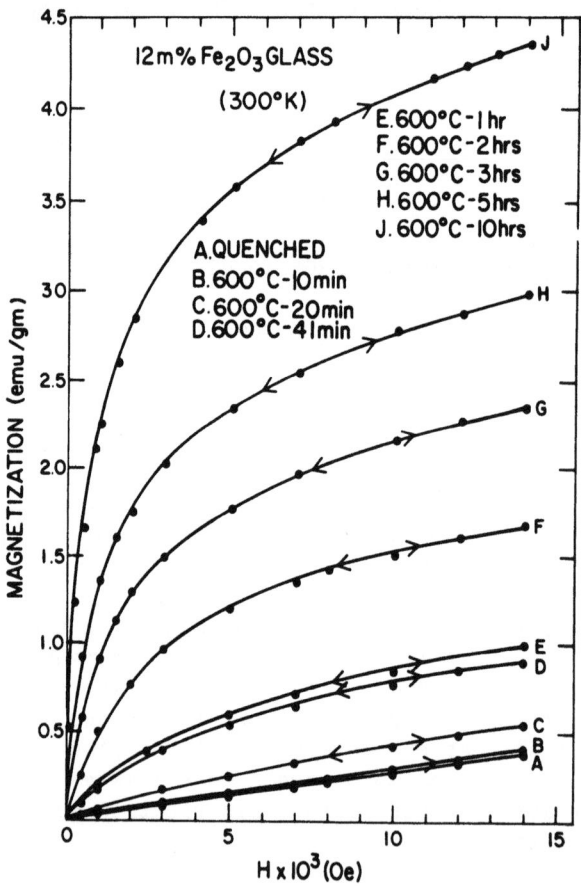

Figure 3. Magnetization versus applied field at 300°K for various heat treated glasses.

MAGNETIC PROPERTIES OF AN IRON BOROSILICATE GLASS

The initial magnetization of the earliest heat treated glasses at low temperature is shown in Figure 4. The precipitated glasses display a maximum in the initial magnetization after having been cooled to 4.2°K in zero field. The maximum is a consequence of the "unfreezing" of the SP cluster from their random orientations frozen in on cooling. When cooled to 4.2°K in a high field (10kOe) the hysteresis curves of these glasses at 4.2°K including the quenched glass show a unaxial displacement relative to the hysteresis curves of the 0 field cooled runs. Both of these effects are a consequence of the superparamagnetic nature of the glasses and indicate the similarity of the quenched and initially precipitated glasses. Verhelst, et al.,[3] has found similar behavior in a CoO-SiO$_2$-Al$_2$O$_3$ system in which amorphous clusters (~50Å) of SP Co ion has been postulated.

The magnetic behavior of the glasses discussed above can be described by the relation

$$M = N_p \frac{\mu^2 H}{3kT} + \int N_{sp} I_s \, VL(x) dV \qquad (2)$$

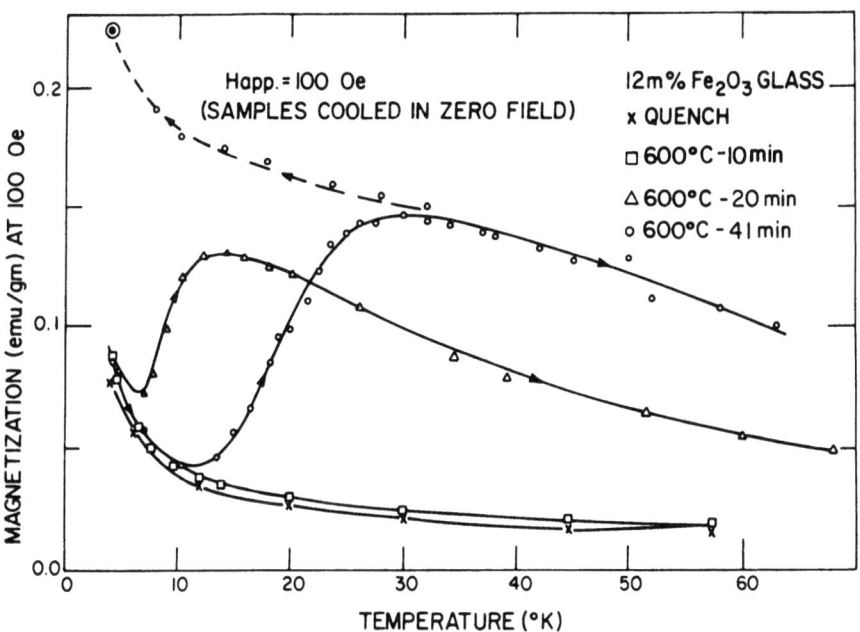

Figure 4. Initial magnetization of glasses containing initial precipitates as a function of temperature.

where $x = I_s VH/kT$ and the first term is due to the isolated paramagnetic ions, while the second is a consequence of the SP clusters and precipitates. In Figure 5 the magnetization of a 12 m% Fe_2O_3 glass heat treated at five hours is shown at various temperatures. The magnetization below T_B (~70°K) has ferrimagnetic behavior, while that above corresponds to the behavior given by equation 2. The magnetic parameters of the various precipitated glasses were measured and listed in Table II. The volumes were obtained from TEM micrographs. Also listed are magnetic parameters determined by analyzing the data in terms of equation 2, i.e., N, Is, and K. It was assumed in this analysis that only one superparamagnetic moment due to a homogeneous equisized precipitate was present in the glass. The values of N are consistent with the TEM results, however, Is and K show very unusual behavior with a maximum value in both cases after three hours heat treatment. The magnetization behavior for the precipitated glasses as a function of temperature is shown in Figure 6. The glasses precipitated in the 600°C range appear to show two Curie temperatures. The lower Curie temperature T_{c1} (listed in parenthesis in Table II over T_{c2}) is identified with the poorly defined but definite inflection in the curve at lower temperatures. For comparisons a curve of a precipitated glass containing only one magnetic phase (Fe_3O_4) is included in Figure 6. As can be seen there is no trace of an inflection in this curve. Both Curie temperatures increase with heat treatment time, indicative of a magnetic phase with changing composition. The T_{c1} temperature is assumed to be associated with the unknown metastable phase.

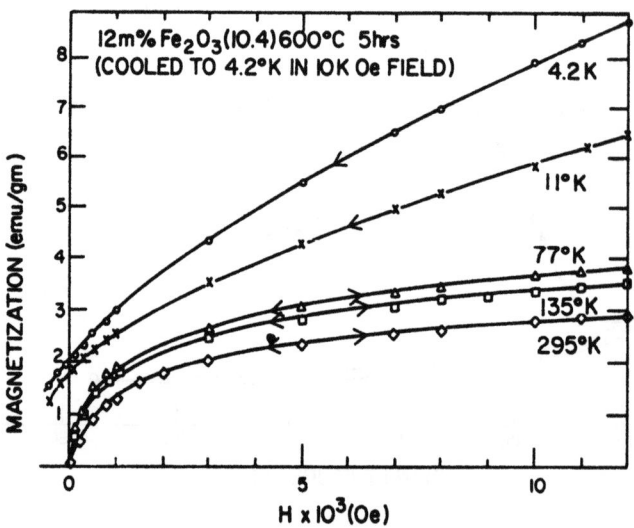

Figure 5. Magnetization versus applied field for 12 m% Fe_2O_3 glass heat treated at 600°C for five hours.

MAGNETIC PROPERTIES OF AN IRON BOROSILICATE GLASS

TABLE II

Magnetic Parameters of the Precipitated Glasses

Heat Treatment Time	N (cm^{-3} x 10^{17})	V (cm^3 x 10^{20})	Is (emu/cm^3)	T_B (°K)	K x 10^6 (ergs/cm^3)	T_c (°K)
20 min.	1.4	2.2	62	12	1.88	453
1 hour	1.9	4.8	238	32	2.30	573
5 hours	1.9	11.3	336	69	2.10	(433) 738
10 hours	2.3	38	252	102	0.92	(470) 762
24 hours	2.6	66	168	116	0.70	(485) 783

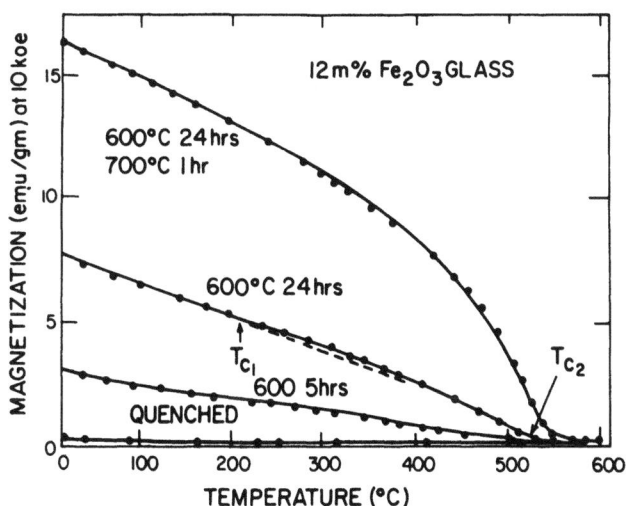

Figure 6. Magnetization of the various heat treated glasses as a function of temperature.

The unusual behavior of magnetic parameters of the heat treated glasses can be discussed in terms of inhomogeneities in the initial precipitated phase. The precipitates originate with the amorphous Fe clusters present in the quenched glass. The initial precipitates most likely contain Ca, Al, Si as well as Fe ions and have a metastable structure distinct from the equilibrium cubic spinel structure. During growth the particle gains Fe ions from the glass matrix loosing the other cations. The stable Fe_3O_4 phase begins to appear, most probably, in the core of the particle. The interfacial shell region can then be considered as a second magnetic phase with a metastable concentration and structure. After long heat treatment only the stable ferrite core remains. This morphology is similar to that proposed by M. Fahmy, et al.[5] for the magnetic precipitates in a B_2O_3-BaO-Fe_2O_3 glass. In this system the shell phase was α-Fe_2O_3.

If the precipitates are magnetically heterogenous in the manner described above, than the SP analysis used is obviously inadequate, which may account for the unusual behavior of I_S and K. An analysis of the results has also been made using a bimodal distribution of magnetic moments in the superparamagnetic Langevin relation. The magnetization curves of the samples heat treated for more than one hour were able to be fit by this method, however, the analysis is still incomplete and will be published at a later date.

In conclusion, it has been shown that the quenched glasses containing high Fe_2O_3 concentrations although amorphous and paramagnetic contain antiferromagnetic Fe clusters. In addition to the anticipated diad and triad clusters, the presence of a significant number of larger superparamagnetic clusters ($\mu \sim 10\mu_B$) has been found. The proposed origin of these clusters is the phase separation of the glass into amorphous Fe-rich and Fe-depleted regions on quenching.

The heat treatment of the 12 m% Fe_2O_3 glass at T_g precipitated a uniform distribution of equisized superparamagnetic ferrite crystallites. The morphology of the precipitate growth, strongly indicates that the SPM Fe cluster in the Fe-rich phase serve as heterogeneous nucleation sites for the precipitated crystallites. The magnetic properties also indicates that the main difference between the amorphous clusters and the initial precipitates is their size and the consequent superparamagnetic blocking temperatures T_B. An analysis of the magnetic parameters of the initial precipitates show the presence of an inhomogeneous composition which changes with particle growth. It is proposed that the initial precipitate is a metastable phase and can be viewed as a particle with a stable ferrite spinel core $(1-x)Fe_3O_4 \cdot xFe_{8/3}O_4)$ and a interfacial region of an unknown magnetic phase. The interfacial region decreases relative to the ferrite core and eventually disappears as the particle grows.

REFERENCES

1. H.O. Hooper, et al., *Amorphous Magnetism*, Plenum Press New York, 47, (1973).
2. D.W. Moon, J.M. Atkins, R.K. MacCrone, Phys. Chem. Glasses, 16, 158, (1975).
3. R.A. Verhelst, R.W. Kline, A.M. de Graaf, Phys. Rev. B., 11, 4427, (1975).
4. M.P. O'Horo, J.F. O'Neill, Bull. Am. Phys. Soc., 20, 292, (1975).
5. M. Fahmy, M.J. Park, M. Tamazawa, R.K. MacCrone, Phys. Chem. Glasses, 13, 21, (1972).
6. J.W. Cahn, R.J. Charles, Phys. Chem. Glasses, 6, 181, 1965.
7. D.K. Duff, V. Cannella, *Amorphous Magnetism*, Plenum Press, New York, 207, (1973).
8. C.P. Bean, J.D. Livingston, J. Appl. Phys. 30, 120S, (1959).
9. S.T. Suleimenov, et al., *The Structure of Glass: Phase Separation Phenomena*, Ed. EA. Porai-Koshits, Consultants Bureau, New York, Vol. 8, 162, (1973).
10. Christian, *Transformation of Metals and Alloys*, Pergamon Press, New York, 186, (1965).

DISCUSSION

A. Bishay: Can you relate the various Curie temperatures you observe to specific phases that are present in your system?

M. P. O'Horo: I think what is happening in the system is a gradual change in the composition of the precipitate. We start with a metastable phase which I am unable to index from the diffraction lines but which most likely is a calcium ferrite that becomes upon further heat treatment the cubic spinel Fe_3O_4.

A. Bishay: How much platinum dissolved because of the type of crucible you used in your glass?

M. P. O'Horo: I am not sure, I have never analyzed for platinum and did not worry about it too much.

A. Bishay: I would, because melting a glass containing such high concentration of iron in platinum would certainly modify your nominal composition, and might possibly affect your results.

L. N. Mulay: How do you know that whatever the specie you had in there is indeed antiferromagnetic?

M. P. O'Horo: The only magnetic species I have are the iron ions, that is the Fe^{3+} or Fe^{2+} ions. The magnetic moment of either one of these is $5\mu_B$ or less so that the only conceivable way I can imagine of getting $10\mu_B$, aside from impurity which I do not think I have, would be an antiferromagnetic cluster of these ions in sufficient number so that the net moment of the cluster would be $10\mu_B$.

D. J. Sellmyer: It is then a ferrimagnetic cluster, right?

M. P. O'Horo: It is a ferrimagnetic cluster, but when one considers the assembly of these clusters, they exhibit a superparamagnetic behavior.

K. J. Kim: How did you analyze for the Fe^{2+}/Fe^{3+} ratio in these glasses?

M. P. O'Horo: I have analyzed the glasses chemically before the heat treatment and that is where I got the ratios for the quenched glasses. I have analyzed some of the glasses after heat treatment and it did not appear to be a significant change in the ratios although the Mössbauer spectra indicate that the contribution of the Fe^{2+} ions is somehow smaller than expected. At higher heat treatment temperatures there is a more obvious increase in the concentration of the Fe^{3+} ions.

Paramagnetic Impurity Concentrations in Amorphous Polymers[†]

M. Centanni[*] and P. A. Casabella

Rensselaer Polytechnic Institute

Troy, New York 12181

Abstract

Paramagnetic oxygen molecules are present as impurities in most polymers and are easily detected by nuclear magnetic resonance and magnetic susceptibility measurements at liquid helium temperature. The spin-lattice relaxation time (T_1) of the protons in the polymers is found to increase from about one second to hundreds of seconds when the oxygen is removed by gentle heating in a vacuum. When the polymers are subsequently exposed to air at room temperature the oxygen again diffuses into them gradually lowering T_1. Polyethylene and polystyrene have been studied by this technique, and the oxygen diffusion is found to proceed quite differently in the two materials. The magnetic susceptibility measurements have been used to determine the amount of oxygen in the polymers. Both polyethylene and polystyrene were found to have diamagnetic susceptibilities of the order of 10^{-6} at liquid helium temperature. When the oxygen was removed the susceptibility increased by an amount comparable to the original susceptibility. This increase in susceptibility is attributed to the removal of approximately 10^{18} oxygen molecules per cubic centimeter of sample. Polyethylene samples of differing crystallinities were studied by this technique and the data suggests that more oxygen is present in the more amorphous samples.

[†]Work partially supported by the National Aeronautics and Space Administration.

In an earlier study, the spin-lattice relaxation time of protons in several polymers was found to be anomalously short at liquid helium temperatures due to the presence of paramagnetic impurities.[1] Paramagnetic oxygen molecules that diffuse into the material from the air are believed to be the impurities responsible for this effect. This paper describes experiments designed to test that hypothesis by measuring the amount of oxygen present in the polymers and the time required for it to diffuse into the polymers from the atmosphere.

The experiments were performed on samples of three polymers: polystyrene, low density polyethylene and high density polyethylene. The quantities measured were: proton spin-lattice relaxation time at 4.2 K, magnetic susceptibility at 4.2 K, and mass. In each case, the results of measurements made on a specimen immediately after it was removed from a vacuum oven at 70°C were compared with measurements made on a specimen that was exposed to the atmosphere for some time. Unfortunately, because of differing geometrical requirements it was not possible to measure all three quantities on a single sample.

A pulsed nuclear magnetic resonance spectrometer was used to determine the proton spin-lattice relaxation time (T_1). For short relaxation times the conventional 180° - 90° pulse sequence was used. However, for relaxation times longer than a few seconds, it was more convenient to saturate the spin system and then study the recovery of the resonance signal by applying a single 90° pulse after a time t. In that case, the signal strength after the 90° pulse is proportional to the nuclear magnetization which is described by the equation

$$M(t) = M(\infty)(1 - e^{-t/T_1}) . \qquad (1)$$

In this equation, $M(t)$ is the nuclear magnetization after a time t, and $M(\infty)$ is the equilibrium magnetization toward which the system relaxes. T_1 can be obtained from a semi-log plot of $\Delta M/M$ versus t where

$$\frac{\Delta M}{M} = \frac{M(\infty) - M(t)}{M(\infty)} = e^{-t/T_1} \qquad (2)$$

This plot will be a straight line with a slope of $- 1/T_1$.

Figure 1 shows plots of this sort for a single polystyrene sample that was baked out in the vacuum oven for one week. There are five different sets of data plotted corresponding to different lengths of time that the sample was exposed to air at room temperature after removal from the vacuum oven. The times range from zero to 19.5 hours. The sample was cooled to liquid helium

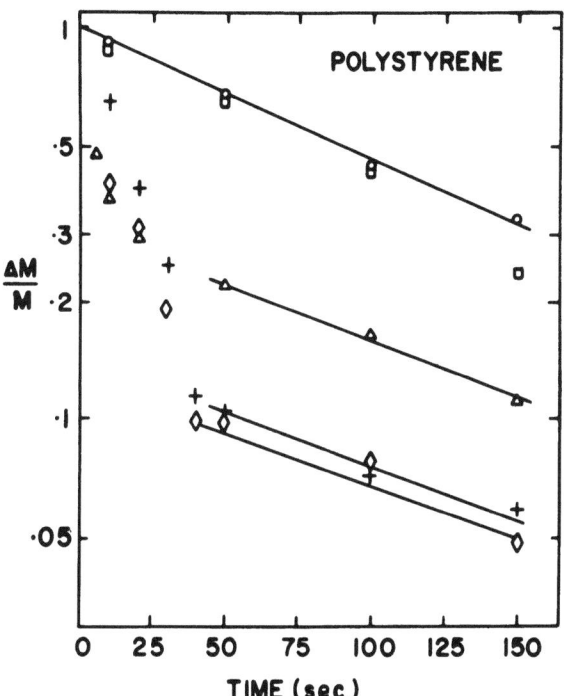

Fig. 1. Evolution of proton magnetization at 4.2 K in polystyrene for five samples exposed to air for different times after baking in vacuum. Circles were measured immediately after baking, squares were measured after one hour, triangles after 5.5 hours, crosses after 17.5 hours, and diamonds after 19.5 hours.

temperature to make the measurements, but otherwise was kept in the laboratory at room temperature.

Note that most of the data cannot be fitted with Eq. (2). Instead it is necessary to postulate that there are two regions of the sample each having its own relaxtion time. Then $\Delta M/M$ is described by the equation

$$\frac{\Delta M}{M} = \frac{M_A}{M} e^{-t/T_{1A}} + \frac{M_B}{M} e^{-t/T_{1B}} \qquad (3)$$

where $M = M_A + M_B$.

Here M_A and T_{1A} are, respectively, the equilibrium nuclear magnetization and proton spin-lattice relaxation time in region A, and M_B and T_{1B} are the same quantities in region B. M is the total nuclear magnetization of the sample at equilibrium.

Equation 3 has been fitted to the data of Fig. 1, and the resulting values of the parameters are given in Table 1. In this model, region B is assumed to be that portion of the sample into which oxygen has diffused from the atmosphere, and region A is the interior portion that is relatively free of oxygen. Since the protons that produce the resonance are distributed uniformly throughout the sample, M_B is proportional to the volume of the sample into which oxygen has diffused. M_B/M is then the fraction of the sample containing oxygen and M_A/M is the fraction that is still relatively free of oxygen.

Table 1

Proton Spin-lattice Relaxation Times in Polystyrene at 4.2 K

Time in Air	T_{1A}(sec)	T_{1B}(sec)	M_A/M	M_B/M
0	132	-	1.00	0
1 hr	145	-	1.00	0
5.5 hr	146	?	0.31	0.69
17.5 hr	145	14.5	0.16	0.84
19.5 hr	145	12.6	0.14	0.86

The relaxation time for the oxygen free region (T_{1A}) is seen to remain relatively constant as would be expected. No detectable oxygen has diffused into the sample after a one hour exposure to air but after 5.5 hours oxygen it has penetrated into 31% of the sample. The data were not sufficiently linear to yield a value of T_{1B} after 5.5 hours, but the values for 17.5 hours and 19.5 hours indicate that T_{1B} is decreasing as more oxygen diffuses into the material. However, even after 19.5 hours the oxygen concentration is far from its equilibrium value. Before putting the sample in the vacuum oven a single value of T_1 of 0.8 seconds was measured. It will probably require several days before T_{1B} is reduced to that value, and M_A goes to zero.

The data for the two polyethylene samples are shown in Fig. 2 and Fig. 3. In these cases each graph shows the results before the sample is baked out, immediately after removal from the vacuum oven, and after a two hour exposure to air at room temperature. The results for the two samples are quite similar, differing only in that the T_1 values for the high density sample are higher than the corresponding values for the low density sample.

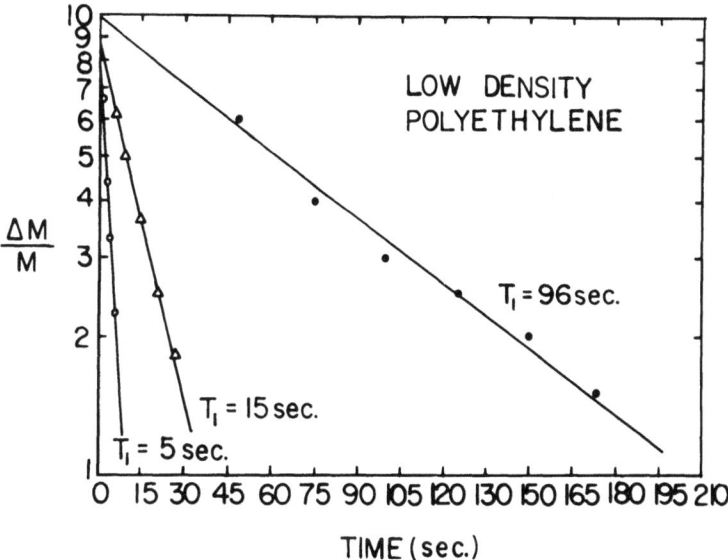

Fig. 2. Evolution of proton magnetization at 4.2 K in low density polyethylene. Open circles are for sample exposed to air since manufacture. Closed circles are for sample just removed from vacuum oven, and triangles were measured two hours later.

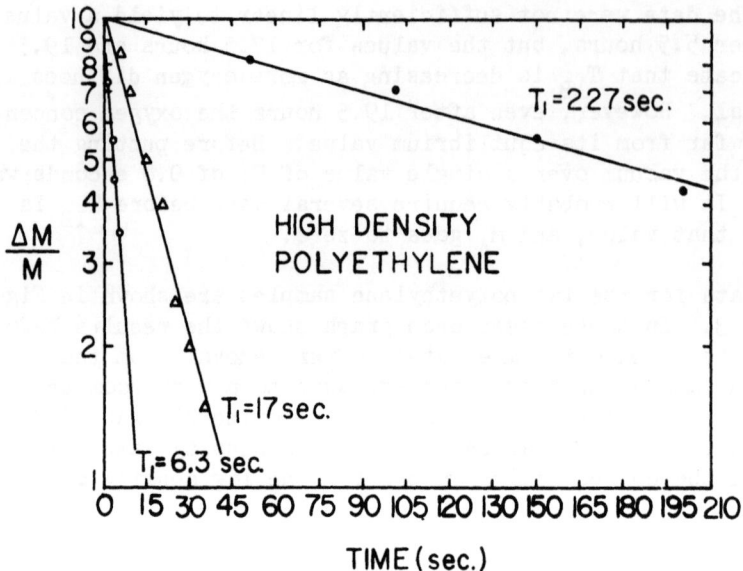

Fig. 3. Evolution of proton magnetization at 4.2 K in high density polyethylene. Open circles are for sample exposed to air since manufacture. Closed circles are for sample just removed from vacuum oven, and triangles were measured two hours later.

In both cases, Eq. 2 fits the data quite well yielding a single relaxation time for each sample and indicating that the oxygen is uniformly distributed through the material. The values of T_1 can be seen to approach the equilibrium values much faster than in the case of polystyrene. These results indicate that oxygen diffuses into polyethylene much faster than into polystyrene.

Although these experiments provide a very sensitive method for detecting molecular oxygen in these polymers, they are not very helpful in estimating how many oxygen molecules are present in each sample. Therefore the magnetic susceptibility at 4.2 K was measured on samples exposed to the atmosphere since manufacture and on samples which had been baked out at 70°C in a vacuum oven for one week. The removal of oxygen should make the samples less paramagnetic or more diamagnetic. If the oxygen is treated as a paramagnetic gas, then the removal of N oxygen molecules per cubic centimeter of sample is related to the change in susceptibility $\Delta\chi$ by the Curie relation

PARAMAGNETIC IMPURITY CONCENTRATIONS IN AMORPHOUS POLYMERS

$$\Delta\chi = \frac{N\mu^2}{3kT}.$$

Here μ is the magnetic moment of the oxygen molecule, k is Boltzmann's constant and T is the absolute temperature.

The measurements were made on a vibrating sample magnetometer. The high density polyethylene sample was found to be weakly paramagnetic, while the other two samples were diamagnetic. Figure 4 shows the results of the susceptibility measurements on low density polyethylene and illustrates the magnitude of the change observed. Table 2 lists the measured susceptibilities and the number of oxygen molecules that have been removed per cubic centimeter. In all three cases the density of oxygen removed is on the order of 10^{18} molecules per cubic centimeter or more.

Table 2

Change of Magnetic Susceptibility of Polymers due to Removal of Oxygen

	Magnetic Susceptibility		Number of O_2 molecules removed per cubic centimeter
	Exposed to air since manufacture	Baked in vacuum	
Low density polyethylene	-9.1×10^{-7}	-4.9×10^{-6}	10×10^{18}
High density polyethylene	1.1×10^{-6}	7.2×10^{-7}	2.6×10^{18}
Polystyrene	-3.0×10^{-6}	-4.9×10^{-6}	13.7×10^{18}

A quick calculation indicates that this change in density should produce a change in mass that can be detected with a sensitive balance. 10^{18} oxygen molecules have a mass of about 5×10^{-5} gm. However, if we assume that the oxygen is accompanied by about four times as many nitrogen molecules then the mass change would be over 10^{-4} gm per cubic centimeter of sample.

Some preliminary mass measurements have been made with a microbalance. Three samples were again put in the vacuum oven at 70°C for one week and immediately weighed upon removal. The samples were then stored in the laboratory and weighed every twenty four hours until the mass stabilized. The approximate mass

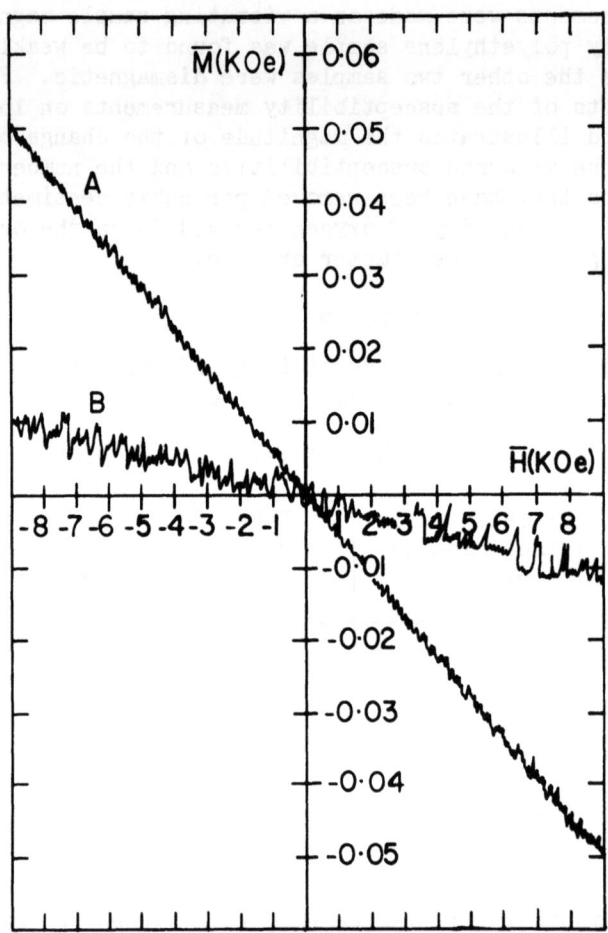

Fig. 4. Plot of M vs H for low density polyethylene. Curve A is for sample just removed from vacuum oven and curve B is for sample exposed to air since manufacture. The slopes give the magnetic susceptibility of these samples.

increases observed per cubic centimeter of sample were: 10^{-4} gm for low density polyethylene, 7×10^{-5} gm for high density polyethylene and 1.6×10^{-4} gm for polystyrene. Since different samples were used for the susceptibility measurements and the mass measurements, the agreement between the two is satisfactory. The greater mass increase observed for low density polyethylene compared to high density polyethylene is to be expected because of the more porous structure of the low density sample.

Furthermore, the two polyethylene samples required only two days to reach their maximum mass, while the polystyrene sample took about one week. These results agree with the observed differences in diffusion rates noted in the T_1 measurements earlier.

In conclusion, it appears that the paramagnetic impurities in polyethylene and polystyrene are oxygen molecules that have diffused into the samples from the atmosphere. These materials normally contain of the order of 10^{18} oxygen molecules per cubic centimeter of material, and these molecules diffuse into polyethylene considerably faster than into polystyrene.

References

*Based upon the dissertation submitted by M. Centanni in partial fulfillment of the requirements for the Ph.D. degree at Rensselaer Polytechnic Institute.

1. R. A. Oliva, Ph.D. Thesis, Rensselaer Polytechnic Institute (1973).

LIST OF PARTICIPANTS

The following list includes the names and affiliations of those participants who attended the Second International Symposium on Amorphous Magnetism, August 25-27, 1976, Rensselaer Polytechnic Institute. The asterisk indicates the names of the speakers.

T. W. Adair, III. Texas A & M U., College Station, Texas 77843
*A. Amamou, California Institute of Technology, Pasadena, CA 91125
J. A. Ambrose, GTE Labs, Waltham, Mass. 02154
P. Anderson, Drexel University, Philadelphia, PA 19104
*P. W. Anderson, Bell Laboratories, Murray Hill, N. J. 07974
*S. Arajs, Clarkson College of Technology, Potsdam, N. Y. 13676
A. Bishay, The American University in Cairo, Cairo, Egypt
J. J. Becker, GE R & D Center, Schenectady, N. Y. 12301
C. Blackway, Rensselaer Polytechnic Institute, Troy, N.Y. 12181
*D. S. Boudreaux, Allied Chemical Corp., Morristown, N.J. 07960
E. Brown, Rensselaer Polytechnic Institute, Troy, N. Y. 12181
*J. I. Budnick, University of Connecticut, Storrs, Conn. 06268
E. Callen, American University, Washington, D. C. 20016
P. C. Campbell, Rensselaer Polytechnic Institute, Troy, N.Y.12181
*G. S. Cargill, IBM Watson Research Center, Yorktown Heights,N.Y.10598
*P. A. Casabella, Rensselaer Polytechnic Institute, Troy, N.Y.12181
M. A. Centanni, Rensselaer Polytechnic Institute, Troy, N.Y. 12181
D-Y. Chen, GE R & D Center, Schenectady, N. Y. 12301
*C. L. Chien, Johns Hopkins University, Baltimore, MD 21218
*W-Y. Ching, University of Wisconsin, Madison, Wisc. 53706
*R. W. Cochrane, McGill University, Montreal, Quebec, Canada
*P. J. Cote, Watervliet Arsenal, Watervliet, N. Y. 12189
J. R. Cullen, National Science Foundation, Washington, D.C. 20550
*A. M. de Graaf, Wayne State University, Detroit, Mich. 48202
L. Didier, Stanford University, Stanford, CA 94305
*J. Durand, California Institute of Technology, Pasadena, CA 91125
P. E. Duwez, California Institute of Technology, Pasadena, CA 91125
*A. R. Ferchmin, Polish Academy of Sciences, Poznan, Poland
*A. T. Fiory, Bell Laboratories, Murray Hill, N. J. 07974
*D. W. Forester, Naval Research Laboratory, Washington, D.C. 20375
*H. Fujimori, Tohoku University, Sendai, Japan
F. S. Gardner, Office of Naval Research, Boston, Mass. 02210
*B. C. Giessen, Northeastern University, Boston, Mass. 02215

D. I. Gordon, Naval Surface Weapons Ctr., Silver Springs, MD 20910
S. Gregory, Cornell University, Ithaca, N. Y. 14853
*H. J. Güntherodt, University of Basel, Basel, Switzerland
J. C. Gustafson, GTE Labs, Waltham, Mass. 02154
*R. Hasegawa, Allied Chemical Corp., Morristown, N. J. 07960
*G. F. Hawkins, Wayne State University, Detroit, Mich. 48203
N. D. Heiman, IBM, Morgan Hill, CA 95037
*W. A. Hines, University of Connecticut, Storrs, Conn. 06268
*D. L. Huber, University of Wisconsin, Madison, Wisc. 53706
H. B. Huntington, Rensselaer Polytechnic Institute, Troy,N.Y. 12181
*C. M. Hurd, National Research Council, Ottawa, Ontario, Canada
*T. Ichikawa, University of Pennsylvania, Philadelphia, PA 19174
*L. T. Kabacoff, University of Connecticut, Storrs, Conn. 06268
*T. Kaneyoshi, Nagoya University, Nagoya, Japan
E. S. Kirkpatrick, IBM Watson Research Ctr., Yorktown Heights, N.Y. 10598
*K. J. Kim, Northwestern University, Evanston, Ill. 60201
*M. W. Klein, Bar Ilan University, Ramat-Gan, Israel
*R. J. Kobliska, IBM Watson Research Ctr., Yorktown Heights, N.Y. 10598
W. J. Kossler, College of William and Mary, Williamsburg, VA 23185
T. Kubaska, Rensselaer Polytechnic Institute, Troy, N. Y. 12181
W. F. Lankford, George Mason University, Fairfax, VA 22030
*U. Larsen, University of Copenhagen, Copenhagen Ø, Denmark
*R. A. Levy, Rensselaer Polytechnic Institute, Troy, N. Y. 12181
*M. J. Lin, Wayne State University, Detroit, Mich. 48201
*F. E. Luborsky, GE R & D Center, Schenectady, N. Y. 12301
S. H. Macomber, Rensselaer Polytechnic Institute, Troy, N.Y. 12181
S. P. McAlister, National Research Council, Ottawa, Ont., Canada
L. I. Mendelsohn, Allied Chemical Corp., Morristown, N. J. 07960
D. L. Mitchell, National Science Foundation, Washington, DC 20550
*T. Mizoguchi, IBM Watson Research Ctr., Yorktown Heights, N.Y. 10598
*L. N. Mulay, Penn State University, University Park, Pa. 16802
*J. A. Mydosh, Kamerlingh Onnes Laboratorium der Rijksuniversiteit, Leiden, The Netherlands
*K. Nagamine, TRIUMF, U. of British Columbia, Vancouver, Ont., Can.
*R. C. O'Handley, Allied Chemical Corp., Morristown, N. J. 07960
*M. P. O'Horo, Xerox Corporation, Webster, N. Y. 14580
*W. P. Pala, Jr., American University, McLean, VA 22101
*A. Paul, Sheffield University, Sheffield S10 2TZ, England
*D. I. Paul, Columbia University, New York, N. Y. 10027
*C. H. Perry, IBM Corporation, Kingston, N. Y. 12401
*G. E. Peterson, Bell Labs, Murray Hill, N. J. 07974
*S. J. Pickart, University of Rhode Island, Kingston, R.I. 02881
*R. O. Pohl, Cornell University, Ithaca, N. Y. 14850
*S. J. Poon, California Institute of Technology, Pasadena, CA 91125
J. Prater, University of Pennsylvania, Philadelphia, PA 19174
*K. Raj, University of Connecticut, Storrs, Conn. 06268
R. Raman, Northeastern University, Boston, Mass. 02115

PARTICIPANTS

B. Rawal, Rensselaer Polytechnic Institute, Troy, N. Y. 12181
*J. A. Rayne, Carnegie-Mellon University, Pittsburgh, PA. 15213
R. Resnick, Rensselaer Polytechnic Institute, Troy, N.Y. 12181
*N. Y. Rivier, Imperial College, London SW7, England
*L. M. Roth, State University of New York, Albany, N. Y. 12222
*R. R. Saxena, University of California, Berkeley, CA 94720
M. P. Sarachik, City College of New York, New York, N. Y.10010
*A. S. Schaafsma, University of Groningen, Groningen, The Netherlands
J. S. Schilling, Universität Bochum, Bochum, Germany
P. Schlottman, University of California, Berkeley, CA 94720
*D. J. Sellmyer, University of Nebraska, Lincoln, Neb.68588
M. Shalmon, McGill University, Montreal, Quebec, Canada
I. W. Shepherd, University of Manchester, Manchester, England
*E. Siegel, Public Service Electric & Gas Co., New York, N.Y. 10011
*R. F. Soohoo, University of California, Davis, CA 95616
*D. Stauffer, Saar State University, Saarbrücken, W. Germany
P. Stoler, Rensselaer Polytechnic Institute, Troy, N. Y. 12181
*M. Takahashi, Tohoku University, Aramaki, Sendai, Japan
*R. L. Thomas, Wayne State University, Detroit, Mich. 48202
R. E. Tompkins, GE R & D Center, Schenectady, N. Y. 12301
*R. Tournier, CNRS - CRTBT, Grenoble, Cedex, France
*C. C. Tsuei, IBM Watson Research Center, Yorktown Heights, N.Y. 10598
*D. Turnbull, Harvard University, Cambridge, Mass. 02138
*F. van der Woude, University of Groningen, Groningen, The Netherlands
R. A. Verhelst, University of Maine, Orono, ME 04473
*L. E. Wenger, Purdue University, West Lafayette, Ind. 47907
*J. C. Williams, Rensselaer Polytechnic Institute, Troy, N. Y. 12181
J. Wong, GE R & D Center, Schenectady, N. Y. 12301
R. Y. H. Wong, Rutgers University, New Brunswick, N. J. 08903
D. Yarkony, Massachusetts Institute of Technology, Cambridge, MA 02139
*A. P. Young, Institut Laue Langevin, Grenoble, Cedex, France
H. Zaiss, Stanford University, Stanford, CA 94305

SUBJECT INDEX

A

Activity coefficients
  in oxide glasses 604
Alumino-silicate glasses
  571, 577
Amorphous polymers 663
Anisotropy
  in amorphous alloys 380, 393, 493
  in amorphous thin films 425
  in spin glasses 56, 161
Antiferromagnetism
  in amorphous alloys 243, 247, 265
  in oxide glasses 550
Arrott plots 253, 503
$\underline{Au}$Cr 117
$\underline{Au}$Fe 74, 86, 90, 95
$\underline{Au}$Mn 47, 90, 95

B

Bernal Model 217
Borate glasses 627, 643

C

$CaO \cdot SiO_2 Fe_2O_3$ 613
Chemisorption activities 415
$CoO \cdot Al_2O_3 \cdot SiO_2$ glass 577, 587, 657
Coordination numbers 465, 613
Crystallization
  in amorphous alloys 278, 294, 319, 327, 341, 371, 398, 519
  in oxide glasses 571, 627

$\underline{Cu}$Mn 74, 90, 95
Curie-Weiss behavior 236, 267, 551, 653

D

Dense random packing model 207, 459, 463, 469, 521, 529
Deltamax 349
Dipolar coupling
  in polymers 155
Domain wall pinning theory 403

E

Edwards-Anderson theory 5, 63, 88
Electron microscopy 497, 653
Ellingham diagram 599
EPR-ESR
  in oxide glasses 535, 549, 630, 648

F

$Fe_{80}B_{20}$ 289
$Fe\underline{C}$ alloys 489
$Fe_5Co_{70}Si_{15}B_{10}$ 370
FeP alloys 495
$Fe_{80}P_{13}C_7$ 406, 485
$FePB$ alloys 275
Ferromagnetic resonance
  in oxide glasses 561

G

GdCo alloys 459
$GdCo_4$ 459

H

Hall effect measurements
    in amorphous alloys 259
    in amorphous thin films 447
    in spin glasses 47
Harris-Plischke-Zuchermann
    model 425
Heisenberg model 39, 123
$HoFe_2$ 484
Hyperfine field distributions
    in amorphous alloys 222
    in oxide glasses 593

I

Interference function 526
Iron-borosilicate glass 651
Iron silicate glasses 561
Ising model 40, 123, 529
Isomer shift of Fe
    in oxide glasses 592, 616, 629
Isothermal aging
    in amorphous alloys 335

K

Knight shift in amorphous alloys
    B^{10} 224
    B^{11} 224
    Fe^{57} 224
    P^{31} 209, 224
Kondo effect 80, 86, 186, 241

L

LaCe 90
$La_{100-x}Gd_x)_{80}Au_{20}$ 245
Lennard-Jones potential 471

M

Magnetic viscosity 60
Magnetite 566
Magnetization measurements
    in amorphous alloys 235, 265, 275, 305, 345, 369, 485, 667
    in oxide glasses 643, 651
Magnetostriction
    in metallic glasses 379
Mattis model 105
Mean field theory 13, 123
Metastable states 22
Metglas type alloys 289, 319, 327, 369, 381, 408
Methanation activities 415
Mictomagnetism 2, 14, 73, 250, 273
$MnO \cdot A_2O_3 \cdot SiO_2$ glass 571
MoFe 90, 95
Monte Carlo calculations 42, 80, 108
Mo-Permalloy 348
Morse potential 477
Mössbauer effect
    in amorphous alloys 135, 221, 289, 335
    in oxide glasses 587, 613, 627
    in spin glasses 78
Muon knight shift 30
Muon-spin resonance 29

N

$Nb_{50}Ni_{50-x}Fe_x$ 235
$(Ni_{100-x}Fe_x)_{79}P_{13}B_8$ 305
$(Ni_{100-x}Mn_x)_{78}P_{14}B_8$ 265
NiP alloys 499
NMR
    in amorphous polymers 207, 221, 663
    in oxide glasses 535

O

Oxidation-reduction equilibrium 597, 618
Oxygen fugacity 597

P

Pair-correlation functions 463
-distribution functions 523
$Pd_{77.5}Cu_6Si_{16.5}$ 257
$\underline{Pd}Fe$ 29
$\underline{Pd}Mn$ 74, 169
$\overline{Pd_{81}Si_{19}}$ 257
$Pd_{80}Si_{20}$ 522
Percolation theory 17
Phase separation
  in oxide glasses 561, 620, 640
Positron annihilation 516
Pressure effects in spin glasses 95
Pseudo-dipolar model 385
Pseudo-inverse operator 545

Q

Quadrupole splitting of Fe
  in oxide glasses 592, 616, 629
Quantum nucleation 21

R

Radial distribution function 469, 521, 531
Remanence
  in amorphous alloys 369
  in spin glasses 55, 62, 155, 169, 175
Resistivity measurements
  in amorphous alloys 181, 254, 258, 286, 297, 305, 319, 327, 503, 518
  in spin glasses 78, 95
Resistivity minimum 181, 286, 297, 316, 319

S

Silectron 349
Silicate glasses 598, 613
Skew scattering 47
Small angle scattering 479
Specific heat
  in amorphous alloys 500, 515
  in oxide glasses 571, 578
  in spin glasses 78
Speromagnetism 139
Spinel structure 564, 653
Spin glass behavior 2, 23, 40, 55, 66, 73, 85, 105, 271
Spin nodals 21
Spin waves resonance
  in amorphous bubble films 435
Square Permalloy 348
$\underline{Sr}SEu$ 161
Supermalloy 348
Supermendur 349
Superparamagnetism
  in nickel on alumina dispersions 416
  in oxide glasses 578, 589, 638, 656
  in spin glasses 78
Susceptibility
  in amorphous alloys 197, 235, 260, 305, 492, 502
  in oxide glasses 549, 630, 646, 652
  in spin glasses 78, 164, 172

T

Taggart, Tahir-Kheli and Shiles model 425
$TbFe_2$ 479
Thouless-Palmer-Anderson model 8
Titanium phosphate glasses 549

U

Ultrasonic velocity in $\underline{Au}Cr$ 117

X

X-ray measurements
  in amorphous alloys 262, 336,
  459, 479, 486, 500, 514

Y

YFe_2 135, 484

Z

$Zr_{40}Cu_{60-x}Fe_x$ 235
$Zr_4Cu_{54}Gd_6$ 235

MIX
Papier aus verantwortungsvollen Quellen
Paper from responsible sources
FSC® C105338

If you have any concerns about our products,
you can contact us on
ProductSafety@springernature.com

In case Publisher is established outside the EU,
the EU authorized representative is:
**Springer Nature Customer Service Center GmbH
Europaplatz 3, 69115 Heidelberg, Germany**

Printed by Libri Plureos GmbH
in Hamburg, Germany